Algebraic Groups: Structure and Actions

Proceedings of Symposia in
PURE MATHEMATICS

Volume 94

Algebraic Groups: Structure and Actions

2015 Clifford Lectures
Algebraic Groups: Structure and Actions
March 2–5, 2015
Tulane University, New Orleans, Louisiana

Mahir Bilen Can
Editor

American Mathematical Society
Providence, Rhode Island

2010 *Mathematics Subject Classification.* Primary 12F10, 14C15, 14C35, 14C40, 14E07, 14E08, 14J70, 14L15, 14L30, 14M17, 14M25, 20G15; Secondary 11G05, 11R34, 14G05, 14K05, 14K30, 14M27, 20G15.

Library of Congress Cataloging-in-Publication Data

Names: Can, Mahir, editor.
Title: Algebraic groups : structures and actions : 2015 Clifford lectures on algebraic groups, structures and actions, March 2-5, 2015, Tulane University, New Orleans, LA / Mahir Bilen Can, editor.
Description: Providence, Rhode Island : American Mathematical Society, [2017] | Series: Proceedings of symposia in pure mathematics ; volume 94 | Includes bibliographical references.
Identifiers: LCCN 2016021970 | ISBN 9781470426019 (alk. paper)
Subjects: LCSH: Differential algebraic groups–Congresses. | Linear algebraic groups–Congresses. | Group theory–Congresses. | Geometry, Algebraic–Congresses. | AMS: Field theory and polynomials – Field extensions – Separable extensions, Galois theory. msc | Algebraic geometry – Cycles and subschemes – (Equivariant) Chow groups and rings; motives. msc | Algebraic geometry – Cycles and subschemes – Applications of methods of algebraic K-theory. msc | Algebraic geometry – Birational geometry – Birational automorphisms, Cremona group and generalizations. msc | Algebraic geometry – Surfaces and higher-dimensional varieties – Hypersurfaces. msc | Algebraic geometry – Algebraic groups – Group schemes. msc | Algebraic geometry – Special varieties – Homogeneous spaces and generalizations. msc | Algebraic geometry – Special varieties – Toric varieties, Newton polyhedra. msc | Algebraic geometry – Birational geometry – Birational automorphisms, Cremona group and generalizations. msc | Group theory and generalizations – Linear algebraic groups and related topics – Linear algebraic groups over arbitrary fields. msc
Classification: LCC QA247.4 .A47 2017 | DDC 512/.2–dc23 LC record available at https://lccn.loc.gov/2016021970

DOI: `http://dx.doi.org/10.1090/pspum/94`

∞ The paper used in this book is acid-free and falls within the guidelines
established to ensure permanence and durability.
Visit the AMS home page at `http://www.ams.org/`

10 9 8 7 6 5 4 3 2 1   22 21 20 19 18 17

Contents

Preface

The prominent (semi)group theorist Alfred Hoblitzelle Clifford (1908–1992) joined Tulane University in 1955. Since 1984, honoring his contributions, the Mathematics Department at Tulane has hosted the annual Clifford Lectures, a weeklong series of talks by a distinguished mathematician. A mini-conference is held in conjunction with each of the Clifford Lecture series. The theme of the 2015 Clifford Lecture series was Algebraic Groups: Structure and Actions, and the main speaker was Michel Brion. This volume presents the proceedings of the associated mini-conference.

The theory of algebraic groups forms a very active research area in contemporary mathematics. It has rich relations to many other areas, including algebraic geometry, number theory, and representation theory. The topics that were covered in the Clifford Lectures contributed widely to this spectrum. They included pseudo-reductive groups, structure theory for algebraic groups, groups of birational transformations, the Tschirnhaus transformations and applications, algebraic theory of quadratic forms, geometry of classifying spaces, and G-torsors, as well as operational K-theory and its applications.

The papers in this volume not only present new results on the aforementioned themes, but also provide much awaited exposés of the foundational results of algebraic group theory recast in the language of schemes at the desired generality.

We gratefully acknowledge National Science Foundation Workshop grant DMS–1522969.

Mahir Bilen Can

Proceedings of Symposia in Pure Mathematics
Volume **94**, 2017
http://dx.doi.org/10.1090/pspum/094/01619

Computing torus-equivariant K-theory of singular varieties

Dave Anderson

1. Introduction

Vector bundles on algebraic varieties are basic objects of study. Among the many questions one can ask, some fundamental ones are these: What are the global sections of a vector bundle E on a variety X? How can they be computed? Does X carry any nontrivial vector bundles at all?

Somewhat more tractable than the space of global sections is the Euler characteristic $\chi(X,E) := \sum(-1)^i \dim(H^i(X,E))$, which makes sense whenever these dimensions are finite—e.g., when X is complete. This function is additive on short exact sequences, so one is led to consider the *Grothendieck group of vector bundles*,

$$K^\circ_{\mathrm{vb}}(X) := \left\langle [E] \,\middle|\, [E] = [E'] + [E''] \text{ whenever } 0 \to E' \to E \to E'' \to 0 \right\rangle,$$

i.e., the free abelian group on isomorphism classes of vector bundles, modulo the given relation for each short exact sequence. The Euler characteristic thus defines a function $K^\circ_{\mathrm{vb}}(X) \to \mathbb{Z}$, when X is complete.

If X is nonsingular and comes with an action of an algebraic group—say, a torus $T = (\mathbb{G}_m)^n$—then one can often take advantage of the group action to simplify many calculations, including Euler characteristics. Indeed, there is a version of the Atiyah-Bott localization formula in this context: assuming for simplicity that X has finitely many T-fixed points,

$$\chi_T(X,E) = \sum_{p\in X^T} \frac{[E_p]}{(1-[L_1(p)^*])\cdots(1-[L_d(p)^*])}. \tag{1}$$

Here, for any finite-dimensional T-representation V, $[V]$ denotes its graded character, or equivalently, its class in the representation ring $R(T)$. In the numerator on the right-hand side, E_p is the fiber of the equivariant vector bundle E at the fixed point p; in the denominator, the L_i's form a decomposition of the tangent space T_pX into one-dimensional weight spaces for the T-action. On the left-hand side, the equivariant Euler characteristic is defined as $\chi_T(X,E) := \sum(-1)^i[H^i(X,E)]$ in $R(T)$. By forgetting the T-action and remembering only dimension, one gets a homomorphism $R(T) \to \mathbb{Z}$ which takes χ_T to χ.

The localization formula (1) thus reduces the computation of Euler characteristics to a finite calculation, and one which is often quite easy. As a toy example,

2010 *Mathematics Subject Classification.* Primary 14C35, 14C40, 14C15, 14M25, 14L30.
This work was partially supported by NSF DMS-1502201.

consider $T = \mathbb{G}_m$ acting on $\mathbb{P}^1$ by $z \cdot [a, b] = [a, zb]$. For $n \in \mathbb{Z}$, let us write $\mathrm{e}^{nt} \in R(T)$ for the one-dimensional representation given by $z \cdot v = z^n v$, for $z \in T$. The action on $\mathbb{P}^1$ has two fixed points, $0 = [1, 0]$ and $\infty = [0, 1]$, and it induces a natural action on the line bundle $E = \mathcal{O}(1)$, so that $[E_0] = \mathrm{e}^t$ and $E_\infty = \mathrm{e}^0 = 1$. The tangent spaces are $[T_0\mathbb{P}^1] = \mathrm{e}^t$ and $[T_\infty\mathbb{P}^1] = \mathrm{e}^{-t}$, so their duals are $[L(0)^*] = \mathrm{e}^{-t}$ and $[L(\infty)^*] = \mathrm{e}^t$. Putting all this into the formula, we have

$$\begin{aligned}\chi_T(\mathbb{P}^1, \mathcal{O}(1)) &= \frac{\mathrm{e}^t}{1 - \mathrm{e}^{-t}} + \frac{1}{1 - \mathrm{e}^t} \\ &= 1 + \mathrm{e}^t,\end{aligned}$$

which, taking $t \mapsto 0$, recovers the familiar fact that $\mathcal{O}(1)$ has two sections (once one knows it has vanishing H^1).

When X is singular, $K^\circ_{\mathrm{vb}}(X)$ may be rather complicated, even for relatively simple varieties, such as complete toric varieties. The combinatorial structure of such varieties makes many of their invariants finite and computable, yet there are examples of (projective, three-dimensional) toric varieties X such that $K^\circ_{\mathrm{vb}}(X)$ contains a copy of the ground field (and, in particular, may be uncountably generated) [**Gu**]. On the other hand, on a general (singular, non-projective) toric variety, it is not known if there are any non-trivial vector bundles at all.

The purpose of this note is to survey some basic properties and applications of *operational K-theory*, especially as applied to varieties with the action of a (split) torus. The groups $\mathrm{op}K^\circ_T(X)$ (and the non-equivariant version, $\mathrm{op}K^\circ(X)$) are defined rather abstractly, but they turn out to be more computable than the "geometric" theories $K^\circ_{\mathrm{perf}}(X)$ and $K^\circ_{\mathrm{vb}}(X)$. Furthermore, any Euler characteristic computation, such as the one exhibited above, can be carried out operationally: there will be canonical maps $K^\circ_{\mathrm{vb}}(X) \to K^\circ_{\mathrm{perf}}(X) \to \mathrm{op}K^\circ(X)$, and when X is complete, the Euler characteristic $\chi\colon K^\circ_{\mathrm{vb}}(X) \to \mathbb{Z}$ factors though these homomorphisms.

Operational cohomology was introduced by Fulton and MacPherson [**FM**] to serve as a contravariant counterpart to Chow homology groups, since no other option was available. For K-theory, there are already several contravariant counterparts to the "homology" K-theory of coherent sheaves. They all map to $\mathrm{op}K^\circ$, so it is natural to study this theory as well. We will also see that properties of operational K-theory yield applications to ordinary K-theory: Corollary 4.5 implies that the usual K-theory of a toric threefold is always nontrivial, and Proposition 5.3 gives an example of a projective toric variety which has K-theory classes not lifting to equivariant classes.

Acknowledgements. The author is grateful to Michel Brion and Mahir Can for organizing the Clifford Lectures at Tulane University. The author also thanks his collaborators on this project, Richard Gonzales and Sam Payne, and the referee for helpful comments on the manuscript.

2. Some background

All schemes will be separated and of finite type over an algebraically closed field k. A torus T has character group $M = \mathrm{Hom}_{\mathrm{alg.gp.}}(T, \mathbb{G}_m) \cong \mathbb{Z}^n$.

2.1. Equivariant coherent sheaves. When T acts on a scheme X, a T-action on a coherent sheaf $\mathscr{F}$ is defined as follows: writing $a\colon T \times X \to X$ for the action map, and $p\colon T \times X \to X$ for the projection, one must specify isomorphisms

$a^*\mathscr{F} \cong p^*\mathscr{F}$, satisfying some natural compatibilities. (See, e.g., [**CG**, §5].) A coherent sheaf equipped with a T-action in this way is called an *equivariant coherent sheaf.* A homomorphism of equivariant coherent sheaves is one that respects the actions. We will write $(T\text{-}\mathbf{Coh}_X)$ for the resulting abelian category of T-equivariant coherent sheaves on X.

The K-theory of equivariant coherent sheaves, $K_\circ^T(X)$, is defined to be the Grothendieck group of $(T\text{-}\mathbf{Coh}_X)$.

2.2. Equivariant Chow groups. The *Chow group* $A_i(X)$ is defined as i-dimensional cycles modulo rational equivalence: $Z_i(X)$ is the free abelian group on i-dimensional subvarieties of X, and $R_i(X)$ is the subgroup generated by cycles of the form $[\mathrm{div}_W(f)]$, for f a rational function on some $(i+1)$-dimensional subvariety W. For schemes equipped with the action of an algebraic group, Edidin and Graham defined *equivariant Chow groups* $A_*^G(X)$ to be Chow groups of quotients constructed from Totaro's approximations to the classifying space [**EG1, To1**]. If the scheme X is smooth, these groups fit together to form a graded ring under intersection product; one often uses cohomological grading and writes $A_G^*(X)$ in this case.

When the group is a torus T, as it will be here, Brion gave a concrete characterization of Edidin-Graham's equivariant Chow groups. Let

$$\Lambda = \Lambda_T = \mathrm{Sym}_{\mathbb{Z}}^* M \cong \mathbb{Z}[t_1, \ldots, t_n];$$

it is a basic fact that $A_*^T(\mathrm{pt}) = A_T^*(\mathrm{pt}) = \Lambda$.

THEOREM 2.1 ([**Br**, Theorem 2.1]). *The equivariant Chow group $A_*^T(X)$ is identified with the Λ-module generated by $[Y]$ for $Y \subseteq X$ a T-invariant subvariety, subject to relations $[\mathrm{div}_W(f)] - \lambda \cdot [W]$, for W an invariant subvariety and f a rational function on W, which is an eigenfunction of weight $\lambda \in M$.*

2.3. Vector bundles and perfect complexes. In order to have a K-theory with good local-to-global properties, like Mayer-Vietoris and localization sequences, one has to work not with the Grothendieck groups of vector bundles, but of perfect complexes. A *perfect complex* of $\mathcal{O}_X$-modules is one which is locally quasi-isomorphic to a bounded complex of vector bundles. This notion was introduced in [**SGA6**], and the basic properties of their K-theory were established in the remarkable paper [**TT**]. (The equivariant analogues of these foundational results have been recently developed in [**KrRa**].) Because it possesses the expected properties in general, while K_{vb}° does not, one usually writes $K^\circ(X) := K_{\mathrm{perf}}^\circ(X)$. Here we will generally preserve the subscripts for clarity—the exception being when we have restricted to quasi-projective schemes, for in this case it is known that $K_{\mathrm{vb}} = K_{\mathrm{perf}}$.

The question of the relationship between vector bundles and perfect complexes remains open for more general schemes. Certainly any vector bundle is a perfect complex, and there is a natural homomorphism $K_{\mathrm{vb}}^\circ(X) \to K_{\mathrm{perf}}^\circ(X)$. However, local quasi-isomorphisms need not glue to a global isomorphism, so the nature of this homomorphism is generally ill understood. Implicit in [**TT**, §3] is the fact that $K_{\mathrm{vb}}^\circ(X) \to K_{\mathrm{perf}}^\circ(X)$ is an isomorphism whenever X has the *resolution property*: every coherent sheaf on X is the quotient of a vector bundle. This is made explicit in [**To3**], where the resolution property is discussed in more detail; see also [**Pa2**], which deals with the case of toric varieties.

The reader who is content to restrict to quasi-projective schemes may safely ignore this technical point about perfect complexes. On the other hand, there are singular, non-projective, complete toric threefolds which, until very recently, were not known to carry any nontrivial vector bundles at all; but it follows from results we shall discuss later that K°_{perf} is nontrivial for every such variety. (Perling and Schröer recently proved that every complete toric threefold carries vector bundles with arbitrarily large third Chern class, so in fact, they have many nontrivial vector bundles [**PS**].)

The takeaway of this remark is the following: the main results to be described here do not directly imply anything about vector bundles on a scheme, except in cases where vector bundles are known to exist for independent reasons.

3. Bivariant theories

To provide a framework for analyzing and proving Riemann-Roch type theorems, Fulton and MacPherson introduced the notion of a *bivariant theory*. Fix a class of fiber squares of schemes, called "independent squares", which includes all squares where one side is the identity. To each morphism of schemes $X \to Y$, the theory assigns a graded abelian group $B(X \to Y)$, along with three basic operations:

(1) Given a composition of morphisms $X \to Y \to Z$, there is a *product* homomorphism

$$B(X \to Y) \otimes B(Y \to Z) \xrightarrow{\cdot} B(X \to Z).$$

(2) If $f\colon X \to Y$ is proper, and $g\colon Y \to Z$ is any morphism, there is a *pushforward* homomorphism

$$B(X \to Z) \xrightarrow{f_*} B(Y \to Z).$$

(3) For any morphism $g\colon Y' \to Y$ such that the resulting fiber square

$$\begin{array}{ccc} X' & \xrightarrow{f'} & Y' \\ \downarrow & & \downarrow g \\ X & \xrightarrow{f} & Y \end{array}$$

is independent, there is a *pullback* homomorphism

$$g^*\colon B(X \to Y) \to B(X' \to Y').$$

Particular cases are the associated "homology" theory $B_*(X) = B(X \to \text{pt})$, which is covariant (via pushforward) for proper morphisms; and the associated "cohomology" theory, $B^*(X) = B(\text{id}\colon X \to X)$, which is contravariant (via pullback) for arbitrary morphisms, and is a ring under the product operation. The three operations are required to satisfy a number of axioms, which we will not list here; for complete descriptions, see the original [**FM**] or the summaries in [**AP**, **GK**].

Elements of a bivariant group $B(X \to Y)$ can be understood as generalized Gysin homomorphisms, and from this point of view, the axioms model the usual behavior of Gysin maps. Indeed, $\alpha \in B(f\colon X \to Y)$ defines a "wrong-way" Gysin pullback $f^\alpha\colon B_*(Y) \to B_*(X)$, by

$$f^\alpha(y) = \alpha \cdot y,$$

and if f is proper, also a Gysin pushforward $f_\alpha\colon B^*(X) \to B^*(Y)$, by

$$f_\alpha(x) = f_*(x \cdot \alpha).$$

There are three main examples of bivariant theories that play a role in the present story: operational Chow theory, operational K-theory, and bivariant K-theory of perfect complexes.

Relatively perfect complexes. Suppose X and Y are quasi-projective. Then any morphism $f\colon X \to Y$ factors as a closed embedding $\iota\colon X \hookrightarrow P$, followed by a smooth projection $p\colon P \to Y$. An *f-perfect complex* of sheaves on X is $\mathscr{F}^\bullet$ such that $\iota_*\mathscr{F}^\bullet$ is quasi-isomorphic to a bounded complex of vector bundles on P. Defining $K^\circ_{\mathrm{perf}}(X \to Y)$ to be the Grothendieck group of f-perfect complexes on X yields a bivariant theory, whose independent squares are *Tor-independent* [**FM**]. (A fiber square

$$\begin{array}{ccc} X' & \longrightarrow & Y' \\ \downarrow & & \downarrow \\ X & \longrightarrow & Y \end{array}$$

is Tor-independent if $\mathrm{Tor}_i^Y(\mathcal{O}_X, \mathcal{O}_{Y'}) = 0$ for all $i > 0$.)

Operational theories. Looking for a cohomological counterpart to Chow homology, Fulton and MacPherson defined an *operational bivariant theory* by imitating the relationship between singular (Borel-Moore) homology and cohomology in topology. In topology, if $g\colon Y \to X$ is any continuous map, then an element $c \in H^i(X)$ acts as a homomorphism $H_j(Y) \to H_{j-i}(Y)$, sending $\alpha \in H_j(Y)$ to $g^*(c) \cap \alpha$. Proceeding backward from this property, for a scheme X, a class in operational Chow cohomology $c \in A^i(X)$ is defined to be a collection of homomorphisms $c_g\colon A_j(Y) \to A_{j-i}(Y)$, one for each morphism $g\colon Y \to X$; these are required to satisfy some basic compatibilities (projection formula, etc.), modelled on the topological situation.

For the rest of this section, we will work with the category of schemes with T-action ("T-schemes") and equivariant morphisms. The definitions and properties of operational Chow and K-theory are quite similar, so we will focus on the latter. The purpose here is to give a general impression of these bivariant theories; complete definitions can be found in [**AP**].

The independent squares for operational K-theory are all fiber squares. A class $c \in \mathrm{op}K^\circ_T(X \to Y)$ is a collection of homomorphisms $c_g\colon K^T_\circ(Y') \to K^T_\circ(X')$, one for each morphism $g\colon Y' \to Y$ (with $X' = X \times_Y Y'$). Product is given by composition of homomorphisms. In addition to compatibility with pushforward and pullback, we require that classes in $\mathrm{op}K^\circ_T(X \to Y)$ *commute* with Gysin pullbacks for flat maps and regular embeddings.

To see what this means, suppose $h\colon W \to Z$ is a regular embedding, and $Y' \to Z$ is any morphism, with $Y'' = Y' \times_Z W$. One can define a homomorphism $h^!\colon K^T_\circ(Y') \to K^T_\circ(Y'')$ by sending $[\mathscr{F}]$ to $\sum(-1)^i[\mathrm{Tor}_i^{\mathcal{O}_Z}(\mathscr{F}, \mathcal{O}_W)]$, a finite sum since for regular embeddings Tor^i vanishes for large enough i. When h is flat, it is even easier to define such a homomorphism: all higher Tor vanishes, so $h^![\mathscr{F}] = [\mathscr{F} \otimes_{\mathcal{O}_Z} \mathcal{O}_W]$. Commuting with Gysin pullbacks means that in Figure 1, the diagram of fiber squares on the left produces a commutative diagram on the right.

If X is an n-dimensional variety, there are always homomorphisms $A^i_T(X) \to A^T_{n-i}(X)$ and $\mathrm{op}K^\circ_T(X) \to K^T_\circ(X)$, sending an operator to its value on $[X]$ or

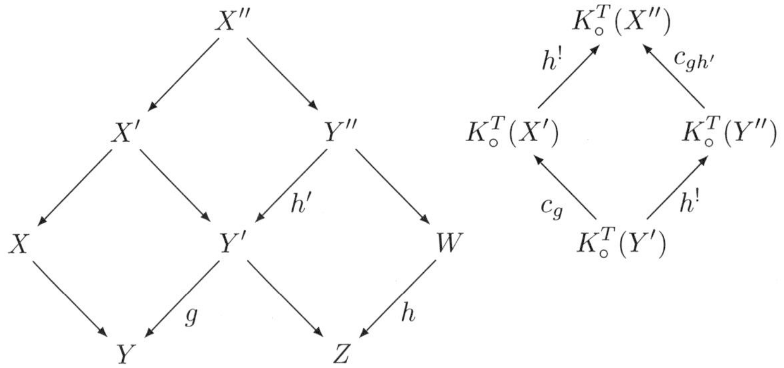

FIGURE 1. Commuting with Gysin pullback, for h flat or regular embedding.

$[\mathcal{O}_X]$, respectively. An important property of operational groups is that these are "Poincaré isomorphisms" when X is smooth:

PROPOSITION 3.1 ([**Fu3**, Proposition 17.4.2],[**EG1**, Proposition 4],[**AP**, Proposition 4.3]). *Suppose X is smooth. The natural homomorphisms $A_T^i(X) \to A_{n-i}^T(X)$ and $\mathrm{op}K_T^\circ(X) \to K_\circ^T(X)$ are isomorphisms.*

More generally, if $f\colon X \to Y$ is any morphism, and $g\colon Y \to Z$ is smooth, then there is a distinguished "orientation" class $[g] \in \mathrm{op}K_T^\circ(Y \to Z)$ giving rise to a natural isomorphism

$$\mathrm{op}K_T^\circ(X \to Y) \xrightarrow{\cdot[g]} \mathrm{op}K_T^\circ(X \to Z).$$

(Taking $X = Y$ and $Z = \mathrm{pt}$, one recovers Proposition 3.1.) A similar statement holds for bivariant Chow theory.

When X is complete, composing with the equivariant Euler characteristic defines a homomorphism

$$\mathrm{op}K_T^\circ(X) \to \mathrm{Hom}_{R(T)}(K_\circ^T(X), R(T)),$$
$$c \mapsto (\alpha \mapsto \chi_T(c_{\mathrm{id}}(\alpha))).$$

In general, this is neither injective nor surjective. However, when X is a T*-linear variety*—a class which includes toric varieties, spherical varieties, and Schubert varieties—it is an isomorphism. This is an echo of similar statements for Chow cohomology ([**FMSS**, **To2**]).

THEOREM 3.2 ([**AP**, Theorem 6.1]). *For a complete T-linear variety X, we have $\mathrm{op}K_T^\circ(X) \cong \mathrm{Hom}_{R(T)}(K_\circ^T(X), R(T))$.*

In particular, since $K_\circ^T(X)$ is a finitely generated $R(T)$-module for such varieties, so is $\mathrm{op}K_T^\circ(X)$. This stands in contrast to $K^\circ(X)$, which may contain a copy of the base field, and thus be uncountably generated [**Gu**].

4. Kimura's exact sequences and the Kan extension property

The definitions of operational Chow and K-theory may appear unwieldy: a class in $\mathrm{op}K_T^\circ(X)$ is a collection of compatible endomorphisms of $K_\circ^T(Y)$, as Y ranges over all schemes mapping equivariantly to X. However, the main tool for computing—introduced for (non-equivariant) Chow theory by Kimura and developed for equivariant K-theory in [**AP**]—turns out to be quite effective, especially when one has access to resolution of singularities.

An *equivariant envelope* is a T-equivariant proper map $X' \to X$ such that every invariant subvariety of X is the birational image of an invariant subvariety of X'.

PROPOSITION 4.1 ([**Ki**, Theorem 2.3],[**AP**, Proposition 5.3]). *Let $X' \to X$ be an equivariant envelope. The* first Kimura sequences

$$0 \to A_T^*(X) \to A_T^*(X') \to A^*(X' \times_X X') \qquad \textit{and} \tag{2}$$

$$0 \to \mathrm{op}K_T^\circ(X) \to \mathrm{op}K_T^\circ(X') \to \mathrm{op}K_T^\circ(X' \times_X X') \tag{3}$$

are exact.

The key ingredients in the proof are the corresponding exact sequences for homology theories,

$$A_*^T(X' \times_X X') \to A_*^T(X') \to A_*^T(X) \to 0 \qquad \text{and}$$
$$K_\circ^T(X' \times_X X') \to K_\circ^T(X') \to K_\circ^T(X) \to 0.$$

These were established (in the nonequivariant case) by Kimura [**Ki**] and Fulton-Gillet [**FG**, **Gi**], respectively, and extended to the equivariant setting in [**AP**, Appendix].

When the envelope is birational, a more precise statement is available. An *abstract blowup diagram* is a fiber square

$$\begin{array}{ccc} E & \hookrightarrow & X' \\ \downarrow & & \downarrow \\ S & \hookrightarrow & X, \end{array}$$

with $X' \to X$ proper, $S \subseteq X$ closed, and $X' \smallsetminus E \to X \smallsetminus S$ an isomorphism.

PROPOSITION 4.2 ([**Ki**, Theorem 3.1],[**AP**, Proposition 5.4]). *Given an abstract blowup square as above, suppose additionally that $X' \to X$ is an equivariant envelope. Then the* second Kimura sequences

$$0 \to A_T^*(X) \to A_T^*(X') \oplus A_T^*(S) \to A_T^*(E) \qquad \textit{and} \tag{4}$$

$$0 \to \mathrm{op}K_T^\circ(X) \to \mathrm{op}K_T^\circ(X') \oplus \mathrm{op}K_T^\circ(S) \to \mathrm{op}K_T^\circ(E) \tag{5}$$

are exact.

Combined with the "Poincaré isomorphisms" of Proposition 3.1, the second Kimura sequence allows computations of operational theories to be carried out by reduction to the smooth case. A useful consequence is the *Kan extension* property. In a precise sense, $\mathrm{op}K_T^\circ(X)$ is the universal target for contravariant "cohomology" theories on schemes, which when restricted to smooth schemes, agree with $K_T^\circ(X)$. Regarding K_T° as a functor

$$K_T^\circ \colon (T\text{-}\mathbf{Sm})^{\mathrm{op}} \to (R(T)\text{-}\mathbf{Mod}),$$

this is the statement that equivariant operational K-theory is the (right) Kan extension of K_T° along the inclusion $(T\text{-}\mathbf{Sm})^{\mathrm{op}} \hookrightarrow (T\text{-}\mathbf{Sch})^{\mathrm{op}}$ of smooth schemes in all schemes. (This is a basic notion in category theory; in fact, most familiar constructions—limits, colimits, etc.—can be realized as certain Kan extensions. See [**Mac**].) Explicitly:

PROPOSITION 4.3 ([**AP**, Theorem 5.8]). *Assume the base field has characteristic zero. Let L°_T be any contravariant functor from T-schemes to $R(T)$-modules, whose restriction to smooth schemes admits a natural transformation to K°_T. Then there is a unique extension of this transformation to a transformation $\eta\colon L^\circ_T \to \mathrm{op}K^\circ_T$.*

The same proof works, *mutatis mutandis*, to show that in characteristic zero (or whenever there exist suitable resolutions of singularities), equivariant operational Chow cohomology $A^*_T(X)$ is the right Kan extension of the equivariant intersection ring on smooth T-varieties. This means that there is a natural homomorphism

$$\text{“}A_T\text{”}(X) := \lim_{X\to Y} A^*_T(Y) \to A^*_T(X),$$

where the limit in the source is taken over all maps of X to smooth schemes Y. (This construction was offered as a substitute for the intersection ring in [**Fu1**], prior to the introduction of the operational Chow ring. In fact, “A”(X) is the *left* Kan extension of the intersection ring, essentially by definition.)

More interestingly, defining “K_T”(X) as the analogous limit, there are natural homomorphisms

$$\text{“}K_T\text{”}(X) \to K^\circ_{T,\mathrm{vb}}(X) \to K^\circ_{T,\mathrm{perf}}(X) \to KH^\circ_T(X) \to \mathrm{op}K^\circ_T(X), \tag{6}$$

all of which are isomorphisms when X is smooth. Here $KH^\circ_T(X)$ is (the degree-zero part of) Weibel's homotopy K-theory (see [**We**])—or more precisely, its equivariant version, constructed very recently in [**KrRa**].

Non-equivariant homotopy K-theory possesses a descent property for abstract blowup squares (*cdh-descent*, proved in [**Ha**]), which in particular implies that there is a natural exact sequence

$$KH^1(E) \to KH^0(X) \to KH^0(X') \oplus KH^0(S) \to KH^0(E).$$

Combining this with the second Kimura sequence (5) and some basic facts about toric varieties, one can prove something about the rightmost map of (6):

THEOREM 4.4 ([**AP**, Theorem 7.1]). *If X is a three-dimensional toric variety, then the natural homomorphism $KH^\circ(X) \to \mathrm{op}K^\circ(X)$ is surjective.*

For general toric varieties, it is proved in [**CHWW**] that the map $K^\circ_{\mathrm{perf}}(X) \to KH^\circ(X)$ is a split surjection. This, together with Theorem 3.2, lets us deduce that $K^\circ_{\mathrm{perf}}(X)$ is nontrivial for complete toric threefolds; more specifically:

COROLLARY 4.5 ([**AP**, Theorem 1.4]). *For any complete three-dimensional toric variety, the homomorphism $K^\circ_{\mathrm{perf}}(X) \to \mathrm{Hom}(K_\circ(X), \mathbb{Z})$ is surjective.*

At the time this was proved, we did not know whether all complete toric varieties admit nontrivial vector bundles, even in dimension three (see [**Pa2**] for a discussion of this question). Results of Gharib and Karu [**GhKa**] gave conditions on the fan of a toric threefold guaranteeing nontrivial $K^\circ_{\mathrm{perf}}(X)$ for many such varieties; the corollary above extends this fact to all complete toric threefolds. Almost concurrently, Perling and Schröer [**PS**] proved that in fact complete toric threefolds do carry nontrivial vector bundles, but their methods, like ours, seem to run into difficulties in higher dimension.

It would be very interesting to know cdh-descent for equivariant homotopy K-theory. As pointed out in [**KrRa**], this would have many applications; one might be the equivariant analogue of Corollary 4.5.

5. Riemann-Roch theorems

As mentioned before, the original motivation for introducing bivariant theories was to unify several Riemann-Roch type theorems. For this section, we restrict to the category of quasi-projective schemes. In this context, Fulton [**Fu3**, §18] showed how to construct natural homomorphisms

$$K^{\circ}(X \to Y) \xrightarrow{\tau} A^*(X \to Y)$$

which induce commuting squares

$$\begin{array}{ccc} K_{\circ}(X) & \xrightarrow{\tau} & A_*(X)_{\mathbb{Q}} \\ {\scriptstyle f^!}\Big\uparrow & & \Big\uparrow{\scriptstyle \mathrm{td}(T_f)\cdot f^!} \\ K_{\circ}(Y) & \xrightarrow{\tau} & A_*(Y)_{\mathbb{Q}} \end{array} \tag{7}$$

when f is smooth, and

$$\begin{array}{ccc} K^{\circ}(X) & \xrightarrow{\tau} & A^*(X)_{\mathbb{Q}} \\ {\scriptstyle f_!}\Big\downarrow & & \Big\downarrow{\scriptstyle f_!(\,\cdot\mathrm{td}(T_f))} \\ K^{\circ}(Y) & \xrightarrow{\tau} & A^*(Y)_{\mathbb{Q}} \end{array} \tag{8}$$

when f is smooth and proper. (We often write $A_*(X)_{\mathbb{Q}}$ for $A_*(X)\otimes\mathbb{Q}$.) Here T_f is the relative tangent bundle of f, and td is the *Todd class*. It can be characterized formally by setting

$$\mathrm{td}(L) = \frac{x}{1-\mathrm{e}^{-x}}$$

for a line bundle L with first Chern class x, and requiring the property $\mathrm{td}(E) = \mathrm{td}(E')\cdot\mathrm{td}(E'')$ whenever $0 \to E' \to E \to E'' \to 0$ is an exact sequence of vector bundles.

When Y is a point, the homomorphism $f_!$ is identified with the Euler characteristic, and the second diagram expresses the classical Hirzebruch-Riemann-Roch formula.

The Riemann-Roch transformation τ was constructed in the equivariant setting by Edidin and Graham [**EG2**]. Here one must also take certain completions: let $\widehat{A}^T_*(X) = \prod_{i\in\mathbb{Z}} A^T_i(X)$, and let $\widehat{K}^T_{\circ}(X)$ be the completion with respect to the kernel of the homomorphism $R(T)\to\mathbb{Z}$ sending each e^t to 1.

THEOREM 5.1 ([**EG2**]). *There is a natural homomorphism* $\tau\colon K^T_{\circ}(X) \to \widehat{A}^T_*(X)_{\mathbb{Q}}$, *which induces an isomorphism* $\widehat{K}^T_{\circ}(X)_{\mathbb{Q}} \xrightarrow{\sim} \widehat{A}^T_*(X)_{\mathbb{Q}}$.

This can be extended to the bivariant setting:

THEOREM 5.2 ([**AGP**]). *There are natural homomorphisms* $\mathrm{op}K^{\circ}_T(X\to Y) \xrightarrow{t} \widehat{A}^*_T(X\to Y)_{\mathbb{Q}}$, *inducing isomorphisms* $\widehat{\mathrm{op}K^{\circ}_T}(X\to Y)_{\mathbb{Q}} \xrightarrow{t} \widehat{A}^*_T(X\to Y)_{\mathbb{Q}}$. *These commute with the forgetful homomorphisms* $\mathrm{op}K^{\circ}_T \to \mathrm{op}K^{\circ}$ *and* $A^*_T \to A^*$.

The proof is mostly formal, building on Theorem 5.1. In the non-equivariant case, this factors the perfect complex transformation: for quasi-projective schemes, Fulton's argument in fact constructs a transformation $K^\circ \to \mathrm{op}K^\circ$, such that the composition

$$K^\circ(X \to Y) \to \mathrm{op}K^\circ(X \to Y) \xrightarrow{t} A^*(X \to Y)_{\mathbb{Q}}$$

is the transformation τ. It would be interesting to know if this can be extended to general schemes.

The Riemann-Roch theorems can be used to exhibit an example of a 3-dimensional projective toric variety X such that $K_T^\circ(X) \to K^\circ(X)$ is not surjective. This should be viewed as contrasting with $K_\circ^T(X) \to K_\circ(X)$, which is surjective for any variety [**Me**].

PROPOSITION 5.3 ([**AGP**]). *Let X be the toric mirror dual to $(\mathbb{P}^1)^3$, i.e., corresponding to the fan over the faces of a cube. Then $K_T^\circ(X) \to K^\circ(X)$ is not surjective.*

PROOF. It is shown in [**KP**] that $A_T^*(X)_{\mathbb{Q}} \to A^*(X)_{\mathbb{Q}}$ is not surjective, so neither is $\alpha\colon \widehat{A}_T^*(X)_{\mathbb{Q}} \to A^*(X)_{\mathbb{Q}}$. Now consider the diagram

$$\begin{array}{ccccccc}
K_T^\circ(X) & \longrightarrow & \mathrm{op}K_T^\circ(X) & \hookrightarrow & \widehat{\mathrm{op}K_T^\circ}(X)_{\mathbb{Q}} & \xrightarrow{\sim} & \widehat{A}_T^*(X)_{\mathbb{Q}} \\
\gamma\downarrow & & \downarrow & & \downarrow & & \downarrow\alpha \\
K^\circ(X) & \xrightarrow{\beta} & \mathrm{op}K^\circ(X) & \hookrightarrow & \mathrm{op}K^\circ(X)_{\mathbb{Q}} & \xrightarrow{\sim} & A^*(X)_{\mathbb{Q}}.
\end{array}$$

The homomorphism $\beta\colon K^\circ(X) \to \mathrm{op}K^\circ(X)$ is surjective by Theorem 4.4, so a diagram chase shows that γ cannot be. □

6. Localization theorems

One of the most useful features of equivariant K- and Chow theory is the possibility of computing by localizing at T-fixed points. At the foundation of this technique are isomorphisms

$$\overline{S}^{-1} A_*^T(X^T) \xrightarrow{\sim} \overline{S}^{-1} A_*^T(X)$$

and

$$S^{-1} K_\circ^T(X^T) \xrightarrow{\sim} S^{-1} K_\circ^T(X),$$

where $\overline{S} \subseteq \Lambda_T$ is the multiplicative set generated by all nonzero $\lambda \in M$, and $S \subseteq R(T)$ is generated by elements of the form $1 - \mathrm{e}^{-\lambda}$, for nonzero $\lambda \in M$. (These isomorphisms were established in [**Br**, §2.3, Corollary 2] and [**Th2**, Théorème 2.1], respectively.) These can be extended to the bivariant setting:

THEOREM 6.1 ([**AGP**]). *There are natural homomorphisms*

$$\overline{S}^{-1} A_T^*(X \to Y) \xrightarrow{\mathrm{loc}^A} \overline{S}^{-1} A_T^*(X^T \to Y^T) \quad \textit{and}$$

$$S^{-1}\mathrm{op}K_T^\circ(X \to Y) \xrightarrow{\mathrm{loc}^K} S^{-1}\mathrm{op}K_T^\circ(X^T \to Y^T),$$

inducing isomorphisms of $\overline{S}^{-1}\Lambda_T$-modules and $S^{-1}R(T)$-modules, respectively, and commuting with the basic bivariant operations (product, pushforward, and pullback).

This bivariant version of the localization theorem is formally similar to the Riemann-Roch theorem (Theorem 5.2); in fact, both are deduced from a general statement about transformations of operational bivariant theories. Origins of this formal similarity can be found in the Lefschetz-Riemann-Roch theorem of Baum, Fulton, and Quart [**BFQ**].

The Riemann-Roch formulas expressed by diagrams (7) and (8) have localization analogues. Suppose $f\colon X \to Y$ is a proper flat equivariant map, with the induced map $\overline{f}\colon X^T \to Y^T$ also flat. In K-theory, one asks for classes $\varepsilon^K(f) \in \mathrm{op}K_T^\circ(X^T)$ making the diagrams

$$\begin{array}{ccc} S^{-1}K_\circ^T(X) & \xrightarrow{\mathrm{loc}^K} & S^{-1}K_\circ^T(X^T) \\ f^! \uparrow & & \uparrow \varepsilon^K(f)\cdot \overline{f}^! \\ S^{-1}K_\circ^T(Y) & \xrightarrow{\mathrm{loc}^K} & S^{-1}K_\circ^T(Y^T) \end{array} \tag{9}$$

and

$$\begin{array}{ccc} S^{-1}\mathrm{op}K_T^\circ(X) & \xrightarrow{\mathrm{loc}^K} & S^{-1}\mathrm{op}K_T^\circ(X^T) \\ f_! \downarrow & & \downarrow \overline{f}_!(\,\cdot\varepsilon^K(f)) \\ S^{-1}\mathrm{op}K_T^\circ(Y) & \xrightarrow{\mathrm{loc}^K} & S^{-1}\mathrm{op}K_T^\circ(Y^T), \end{array} \tag{10}$$

commute. (In Chow theory, one has the corresponding problem of finding classes $\varepsilon^A(f) \in A_T^*(X^T)$.)

THEOREM 6.2 ([**AGP**]). *In the above setting, if $\overline{f}\colon X^T \to Y^T$ is smooth, there exist unique classes $\varepsilon^K(f)$ and $\varepsilon^A(f)$ fitting into commutative diagrams as depicted.*

The classes $\varepsilon^K(f)$ and $\varepsilon^A(f)$ are called (K- or Chow-theoretic) *total equivariant multiplicities* of f. To justify the name, let us consider the case where $Y = \mathrm{pt}$ and X^T consists of finitely many *nondegenerate* fixed points, meaning that for each $p \in X^T$, the zero character does not occur among the weights of the T-action on T_pX. (In this situation, the scheme-theoretic fixed locus X^T is reduced, so $\overline{f}$ is smooth; see, e.g., [**CGP**, Proposition A.8.10(2)].) Let us write $\varepsilon_p^K(X)$ and $\varepsilon_p^A(X)$ for the restrictions of the total multiplicities to $S^{-1}R(T) = S^{-1}\mathrm{op}K_T^\circ(p)$ and $\overline{S}^{-1}\Lambda_T = \overline{S}^{-1}A_T^*(p)$, respectively.

PROPOSITION 6.3 ([**AGP**]). *Let $\lambda_1, \dots, \lambda_n$ be the weights of T acting on the Zariski tangent space T_pX, and let $C = C_pX \subseteq T_pX$ be the tangent cone. Then*

$$\varepsilon_p^K(X) = \frac{[\mathcal{O}_C]}{(1-\mathrm{e}^{-\lambda_1})\cdots(1-\mathrm{e}^{-\lambda_n})} \quad \textit{and} \quad \varepsilon_p^A(X) = \frac{[C]}{\lambda_1\cdots\lambda_n},$$

as elements of $S^{-1}R(T) = S^{-1}K_T^\circ(T_pX)$ and $\overline{S}^{-1}\Lambda_T = \overline{S}^{-1}A_T^*(T_pX)$, respectively.

In particular, $\varepsilon_p^A(X)$ is the Brion-Rossmann equivariant multiplicity of X at p [**Br**].

Continuing this basic setup, so $Y = \mathrm{pt}$ and all fixed points of X are nondegenerate, diagram (10) expresses an Atiyah-Bott-type localization formula:

COROLLARY 6.4. *Given a T-equivariant vector bundle E of rank r on X, its equivariant Euler characteristic may be computed as*

$$\chi_T(E) = \sum_{p \in X^T} (\mathrm{e}^{\chi_1(p)} + \cdots + \mathrm{e}^{\chi_r(p)}) \cdot \varepsilon_p^K(X),$$

where the fiber of E is a T-representation with weights $\chi_1(p), \ldots, \chi_r(p)$.

The computation done in the introduction is a simple special case.

The total equivariant multiplicities play a role in localization analogous to that of the Todd class in Riemann-Roch. In fact, they are directly related, at least in the case where X and Y are smooth, with finitely many fixed points: for each $p \in X^T$, one has

$$\mathrm{td}(T_f)|_p = \frac{\varepsilon_p^K(f)}{\varepsilon_p^A(f)}$$

in an appropriate localization of $\widehat{R(T)} = \widehat{\Lambda}_T$.

7. Other directions

7.1. Chang-Skjelbred and GKM theorems. Part of the initial motivation for introducing operational K-theory was to find a geometric interpretation for the ring of *piecewise exponential functions* on the fan Δ associated to a toric variety $X = X(\Delta)$. An example of such a function is shown in Figure 2; the fan has three maximal (two-dimensional) cones, with the southwest-pointing ray passing through $(-1, -2)$. The corresponding toric variety is the weighted projective space $\mathbb{P}(1, 1, 2)$.

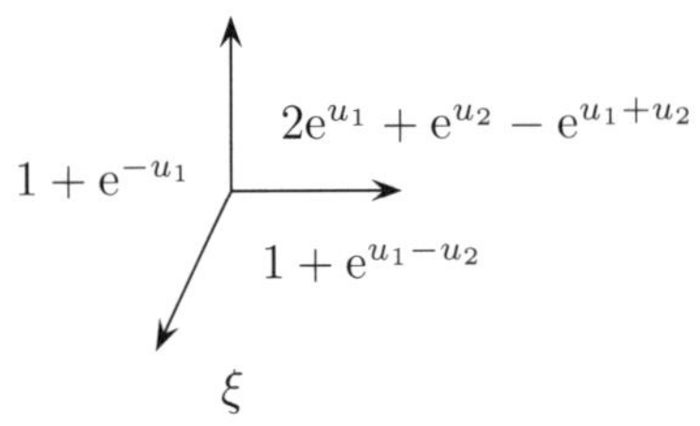

FIGURE 2. A piecewise exponential function for $\mathbb{P}(1, 1, 2)$

In [**VV**], it was shown that when $X = X(\Delta)$ is nonsingular, $K_T^\circ(X)$ is isomorphic to the ring PExp(Δ). (A similar result was proved in [**BV**] for more general simplicial toric varieties, taking rational coefficients. For equivariant Chow cohomology, analogous results were proved by Brion in the smooth case [**Br**], and Payne in the general case [**Pa1**].) For general $X(\Delta)$, one should take operational K-theory on the geometric side: there is always an isomorphism $\mathrm{op}K_T^\circ(X) \cong \mathrm{PExp}(\Delta)$ [**AP**, Theorem 1.6]. A similar result was proved by Harada-Holm-Ray-Williams for the topological equivariant K-theory and cobordism rings of weighted projective spaces, under some conditions on the weights [**HHRW**].

This can be regarded as an instance of a Chang-Skjelbred or GKM-type theorem. Over fields of characteristic zero, R. Gonzales has shown that this phenomenon is quite general in operational K-theory. A *T-skeletal variety* is one such that both X^T and the set of one-dimensional T-orbits are finite.

THEOREM 7.1 ([**Go**, Theorem 5.4]). *For a complete T-skeletal variety X, $\mathrm{op}K_T^\circ(X) \to \mathrm{op}K_T^\circ(X^T)$ induces an isomorphism onto the subring*

$$\mathrm{PExp}(X) := \{(f_p)_{p\in X^T} \mid f_p - f_q \textit{ is divisible by } (1 - \mathrm{e}^{\chi_{p,q}})\};$$

here $\chi_{p,q}$ is the character of the one-dimensional orbit connecting fixed points p and q.

Furthermore, if X is any complete T-variety, the restriction homomorphism $\mathrm{op}K_T^\circ(X) \to \mathrm{op}K_T^\circ(X^T)$ is injective [**Go**, Proposition 3.7].

7.2. Bivariant algebraic cobordism. The algebraic cobordism theory $\Omega_*(X)$ of Levine and Morel acts as a covariant "homology" theory with respect to proper maps. J. González and K. Karu have defined a corresponding equivariant operational bivariant theory, Ω_T^*, developed its properties in order to compute for toric varieties: $\Omega_T^*(X(\Delta))$ is isomorphic to a ring of piecewise graded power series on Δ [**GK**, Theorem 7.3].

It would be interesting to know more about the relation of Ω_T^* with $\mathrm{op}K_T^\circ$ and A_T^*; for instance, one might look for a Riemann-Roch type transformation from Ω_T^* to all other such bivariant theories.

References

[AP] Dave Anderson and Sam Payne, *Operational K-theory*, Doc. Math. **20** (2015), 357–399. MR3398716

[AGP] D. Anderson, R. Gonzales, and S. Payne, "Operational Riemann-Roch and localization theorems," in preparation.

[BFQ] Paul Baum, William Fulton, and George Quart, *Lefschetz-Riemann-Roch for singular varieties*, Acta Math. **143** (1979), no. 3-4, 193–211, DOI 10.1007/BF02392092. MR549774

[SGA6] P. Berthelot, A. Grothendieck, and L. Illusie, *Seminaire de Géométrie Algébrique 6: Théorie des intersections et théorème de Riemann-Roch*, 1966-1967.

[Br] M. Brion, *Equivariant Chow groups for torus actions*, Transform. Groups **2** (1997), no. 3, 225–267, DOI 10.1007/BF01234659. MR1466694

[BV] Michel Brion and Michèle Vergne, *An equivariant Riemann-Roch theorem for complete, simplicial toric varieties*, J. Reine Angew. Math. **482** (1997), 67–92. MR1427657

[CG] Neil Chriss and Victor Ginzburg, *Representation theory and complex geometry*, Birkhäuser Boston, Inc., Boston, MA, 1997. MR1433132

[CGP] Brian Conrad, Ofer Gabber, and Gopal Prasad, *Pseudo-reductive groups*, New Mathematical Monographs, vol. 17, Cambridge University Press, Cambridge, 2010. MR2723571

[CHWW] G. Cortiñas, C. Haesemeyer, Mark E. Walker, and C. Weibel, *The K-theory of toric varieties*, Trans. Amer. Math. Soc. **361** (2009), no. 6, 3325–3341, DOI 10.1090/S0002-9947-08-04750-8. MR2485429

[EG1] Dan Edidin and William Graham, *Equivariant intersection theory*, Invent. Math. **131** (1998), no. 3, 595–634, DOI 10.1007/s002220050214. MR1614555

[EG2] Dan Edidin and William Graham, *Riemann-Roch for equivariant Chow groups*, Duke Math. J. **102** (2000), no. 3, 567–594, DOI 10.1215/S0012-7094-00-10239-6. MR1756110

[Fu1] William Fulton, *Rational equivalence on singular varieties*, Inst. Hautes Études Sci. Publ. Math. **45** (1975), 147–167. MR0404257

[Fu2] William Fulton, *Introduction to toric varieties*, Annals of Mathematics Studies, vol. 131, Princeton University Press, Princeton, NJ, 1993. The William H. Roever Lectures in Geometry. MR1234037

[Fu3] William Fulton, *Intersection theory*, 2nd ed., Ergebnisse der Mathematik und ihrer Grenzgebiete. 3. Folge. A Series of Modern Surveys in Mathematics [Results in Mathematics and Related Areas. 3rd Series. A Series of Modern Surveys in Mathematics], vol. 2, Springer-Verlag, Berlin, 1998. MR1644323

[FG] William Fulton and Henri Gillet, *Riemann-Roch for general algebraic varieties* (English, with French summary), Bull. Soc. Math. France **111** (1983), no. 3, 287–300. MR735307

[FM] William Fulton and Robert MacPherson, *Categorical framework for the study of singular spaces*, Mem. Amer. Math. Soc. **31** (1981), no. 243, vi+165, DOI 10.1090/memo/0243. MR609831

[FMSS] W. Fulton, R. MacPherson, F. Sottile, and B. Sturmfels, *Intersection theory on spherical varieties*, J. Algebraic Geom. **4** (1995), no. 1, 181–193. MR1299008

[GhKa] Saman Gharib and Kalle Karu, *Vector bundles on toric varieties* (English, with English and French summaries), C. R. Math. Acad. Sci. Paris **350** (2012), no. 3-4, 209–212, DOI 10.1016/j.crma.2011.12.013. MR2891113

[Gi] Henri Gillet, *Homological descent for the K-theory of coherent sheaves*, Algebraic K-theory, number theory, geometry and analysis (Bielefeld, 1982), Lecture Notes in Math., vol. 1046, Springer, Berlin, 1984, pp. 80–103, DOI 10.1007/BFb0072019. MR750678

[Go] R. Gonzales, "Localization in equivariant operational K-theory and the Chang-Skjelbred property," `arXiv:1403.4412v1`.

[GK] José Luis González and Kalle Karu, *Bivariant algebraic cobordism*, Algebra Number Theory **9** (2015), no. 6, 1293–1336, DOI 10.2140/ant.2015.9.1293. MR3397403

[GKM] Mark Goresky, Robert Kottwitz, and Robert MacPherson, *Equivariant cohomology, Koszul duality, and the localization theorem*, Invent. Math. **131** (1998), no. 1, 25–83, DOI 10.1007/s002220050197. MR1489894

[Gu] Joseph Gubeladze, *Toric varieties with huge Grothendieck group*, Adv. Math. **186** (2004), no. 1, 117–124, DOI 10.1016/j.aim.2003.07.009. MR2065508

[Ha] Christian Haesemeyer, *Descent properties of homotopy K-theory*, Duke Math. J. **125** (2004), no. 3, 589–620, DOI 10.1215/S0012-7094-04-12534-5. MR2166754

[HHRW] M. Harada, T. Holm, N. Ray, and G. Williams, "The equivariant K-theory and cobordism rings of divisive weighted projective spaces," preprint, 2013, arXiv:1306.1641v1.

[KP] Eric Katz and Sam Payne, *Piecewise polynomials, Minkowski weights, and localization on toric varieties*, Algebra Number Theory **2** (2008), no. 2, 135–155, DOI 10.2140/ant.2008.2.135. MR2377366

[Ki] Shun-ichi Kimura, *Fractional intersection and bivariant theory*, Comm. Algebra **20** (1992), no. 1, 285–302, DOI 10.1080/00927879208824340. MR1145334

[KR] Ioanid Rosu, *Equivariant K-theory and equivariant cohomology*, Math. Z. **243** (2003), no. 3, 423–448, DOI 10.1007/s00209-002-0447-1. With an appendix by Allen Knutson and Rosu. MR1970011

[KrRa] A. Krishna and C. Ravi, "On the K-theory of schemes with group scheme actions," `arXiv:1509.05147v1`.

[Mac] Saunders Mac Lane, *Categories for the working mathematician*, 2nd ed., Graduate Texts in Mathematics, vol. 5, Springer-Verlag, New York, 1998. MR1712872

[Me] Alexander S. Merkurjev, *Equivariant K-theory*, Handbook of K-theory. Vol. 1, 2, Springer, Berlin, 2005, pp. 925–954, DOI 10.1007/978-3-540-27855-9_18. MR2181836

[Pa1] Sam Payne, *Equivariant Chow cohomology of toric varieties*, Math. Res. Lett. **13** (2006), no. 1, 29–41, DOI 10.4310/MRL.2006.v13.n1.a3. MR2199564

[Pa2] Sam Payne, *Toric vector bundles, branched covers of fans, and the resolution property*, J. Algebraic Geom. **18** (2009), no. 1, 1–36, DOI 10.1090/S1056-3911-08-00485-2. MR2448277

[PS] M. Perling and S. Schröer, "Vector bundles on proper toric 3-folds and certain other schemes," `arXiv:1407.5443v2`.

[Th1] R. W. Thomason, *Algebraic K-theory of group scheme actions*, Algebraic topology and algebraic K-theory (Princeton, N.J., 1983), Ann. of Math. Stud., vol. 113, Princeton Univ. Press, Princeton, NJ, 1987, pp. 539–563. MR921490

[TT] R. W. Thomason and Thomas Trobaugh, *Higher algebraic K-theory of schemes and of derived categories*, The Grothendieck Festschrift, Vol. III, Progr. Math., vol. 88,

Birkhäuser Boston, Boston, MA, 1990, pp. 247–435, DOI 10.1007/978-0-8176-4576-2_10. MR1106918

[Th2] R. W. Thomason, *Une formule de Lefschetz en K-théorie équivariante algébrique* (French), Duke Math. J. **68** (1992), no. 3, 447–462, DOI 10.1215/S0012-7094-92-06817-7. MR1194949

[To1] Burt Totaro, *The Chow ring of a classifying space*, Algebraic K-theory (Seattle, WA, 1997), Proc. Sympos. Pure Math., vol. 67, Amer. Math. Soc., Providence, RI, 1999, pp. 249–281, DOI 10.1090/pspum/067/1743244. MR1743244

[To2] Burt Totaro, *Chow groups, Chow cohomology, and linear varieties*, Forum Math. Sigma **2** (2014), e17, 25, DOI 10.1017/fms.2014.15. MR3264256

[To3] Burt Totaro, *The resolution property for schemes and stacks*, J. Reine Angew. Math. **577** (2004), 1–22, DOI 10.1515/crll.2004.2004.577.1. MR2108211

[VV] Gabriele Vezzosi and Angelo Vistoli, *Higher algebraic K-theory for actions of diagonalizable groups*, Invent. Math. **153** (2003), no. 1, 1–44, DOI 10.1007/s00222-002-0275-2. MR1990666

[We] Charles A. Weibel, *Homotopy algebraic K-theory*, Algebraic K-theory and algebraic number theory (Honolulu, HI, 1987), Contemp. Math., vol. 83, Amer. Math. Soc., Providence, RI, 1989, pp. 461–488, DOI 10.1090/conm/083/991991. MR991991

DEPARTMENT OF MATHEMATICS, THE OHIO STATE UNIVERSITY, COLUMBUS, OHIO 43210
E-mail address: anderson.2804@math.osu.edu

Proceedings of Symposia in Pure Mathematics
Volume **94**, 2017
http://dx.doi.org/10.1090/pspum/094/01620

Algebraic structures of groups of birational transformations

Jérémy Blanc

Abstract. A priori, the set of birational transformations of an algebraic variety is just a group. We survey the possible algebraic structures that we may add to it, using in particular parametrised family of birational transformations.

1. Introduction

Let X be an algebraic variety defined over an algebraically closed field k. We denote by $\mathrm{Bir}(X)$ the group of birational transformations of X, and by $\mathrm{Aut}(X)$ its subgroup of automorphisms (biregular morphisms).

If X is projective, it is known that $\mathrm{Aut}(X)$ has a natural structure of group scheme, maybe with infinitely many components ([**Mat1958**], see also [**MO1967**, **Han1987**]). In particular, it is a scheme of finite dimension.

This is false in general for $\mathrm{Bir}(X)$, which can be much larger. In this note, we give a survey on the following question:

What kind of algebraic structure can we put on $\mathrm{Bir}(X)$*?*

As usual in algebraic geometry, even if one does not know the structure of $\mathrm{Bir}(X)$, one can define what is a morphism $A \to \mathrm{Bir}(X)$, where A is an algebraic variety, or more generally a locally noetherian scheme (see §2.1). This corresponds to a functor

$$(\text{locally noetherian schemes}) \to (\text{Sets}),$$

introduced by M. Demazure [**Dem1970**], which is unfortunately not representable by a scheme, or more generally an ind-scheme, as we explain in §2.2. The functor is representable for $\mathrm{Aut}(X)$, if X is projective, and gives the classical group scheme structure explained before. It is also representable if X is affine, but by an ind-algebraic group (see §2.2).

Even if we do not know what kind of structure one can put on $\mathrm{Bir}(X)$, the morphisms introduced define a Zariski topology on $\mathrm{Bir}(X)$, as explained by J.-P. Serre in [**Ser2010**]. We recall this topology in §2.3, and describe some of its properties. We then finish Section 2 by recalling what is usually called algebraic

2010 *Mathematics Subject Classification.* Primary 14E07, 14L30.

Key words and phrases. Birational transformations, representability of functors, Zariski topology, algebraic groups.

The author gratefully acknowledges support by the Swiss National Science Foundation Grants "Birational Geometry" PP00P2_153026 and "Algebraic subgroups of the Cremona groups" 200021_159921.

subgroup of $\mathrm{Bir}(X)$, and by explaining the relation with the topology and the functors/morphisms defined.

Section 3 consists in looking at a sub-functor of the above one, introduced by M. Hanamura in [**Han1987**]. It corresponds to flat families of birational transformations, and has the advantage of being representable by a scheme (§3.1). The structure is compatible with the composition and behaves quite well if the variety X is not uniruled (§3.2). This is however not the case if X is an arbitrary algebraic variety. We briefly describe the case where $X = \mathbb{P}^n$ in §3.3.

The author thanks Michel Brion and Jean-Philippe Furter for interesting discussions during the preparation of the article and also thanks the anonymous referee for his precious remarks that improved the exposition of the text.

2. Structures given by families of transformations

2.1. Functors Bir_X and Aut_X. In [**Dem1970**], M. Demazure introduced the following functor (that he called Psaut, for pseudo-automorphisms, the name he gave to birational transformations):

DEFINITION 2.1. Let X be an irreducible algebraic variety and A be a locally noetherian scheme. We define

$$\begin{array}{rcl} \mathrm{Bir}_X(A) & = & \left\{ \begin{array}{c} A\text{-birational transformations of } A \times X \text{ inducing an} \\ \text{isomorphism } U \to V, \text{ where } U, V \text{ are open subsets} \\ \text{of } A \times X, \text{ whose projections on } A \text{ are surjective} \end{array} \right\}, \\ \mathrm{Aut}_X(A) & = & \left\{ A\text{-automorphisms of } A \times X \right\} = \mathrm{Bir}_X(A) \cap \mathrm{Aut}(A \times X). \end{array}$$

The above families were also introduced and studied before in [**Ram1964**], at least for automorphisms. Definition 2.1 implicitly gives rise to the following notion of families, or morphisms $A \to \mathrm{Bir}(X)$ (as in [**Ser2010, Bla2010, BF2013, PR2013**]):

DEFINITION 2.2. Taking A, X as above, an element $f \in \mathrm{Bir}_X(A)$ and a k-point $a \in A(\mathrm{k})$, we obtain an element $f_a \in \mathrm{Bir}(X)$ given by $x \dashrightarrow p_2(f(a,x))$, where $p_2 \colon A \times X \to X$ is the second projection.

The map $a \mapsto f_a$ represents a map from A (more precisely from the $A(k)$-points of A) to $\mathrm{Bir}(X)$, and will be called a *morphism* from A to $\mathrm{Bir}(X)$.

REMARK 2.3. We can similarly define morphisms $A \to \mathrm{Aut}(X)$, and observe that these are exactly the morphisms $A \to \mathrm{Bir}(X)$ having image in $\mathrm{Aut}(X)$.

REMARK 2.4. If X, Y are two irreducible algebraic varieties and $\psi \colon X \dashrightarrow Y$ is a birational map, the two functors Bir_X and Bir_Y are isomorphic, via ψ. In other words, morphisms $A \to \mathrm{Bir}(X)$ corresponds, via ψ, to morphisms $A \to \mathrm{Bir}(Y)$.

If ψ is moreover an isomorphism, then it also induces an isomorphism between the two functions Aut_X and Aut_Y. Equivalently, morphisms $A \to \mathrm{Aut}(X)$ corresponds, via ψ, to morphisms $A \to \mathrm{Aut}(Y)$.

As we will see, the functor $A \to \mathrm{Bir}_X(A)$ is not representable by a scheme, if X is an arbitrary algebraic variety (for example if $X = \mathbb{P}^2$).

Firstly, taking $X = \mathbb{P}^2$, one can construct very large families:

EXAMPLE 2.5. For each $m \geq 1$, the following $\mathbb{A}^m$-birational map of $\mathbb{A}^m \times \mathbb{P}^2$

$$\mathbb{A}^m \times \mathbb{P}^2 \dashrightarrow \mathbb{A}^m \times \mathbb{P}^2$$

$$((a_1,\dots,a_m),[x:y:z]) \mapsto \left((a_1,\dots,a_m),\left[xz^{m-1} : yz^{m-1} + \sum_{i=1}^{m} a_i x^i z^{m-i} : z^m\right]\right)$$

which restricts, on the open subset where $z = 1$, to the automorphism

$$\begin{array}{rcl} \mathbb{A}^m \times \mathbb{A}^2 & \to & \mathbb{A}^m \times \mathbb{A}^2 \\ ((a_1,\dots,a_m),(x,y)) & \mapsto & \left((a_1,\dots,a_m),\left(x, y + \sum_{i=1}^{m} a_i x^i\right)\right) \end{array}$$

yields injective morphisms $\mathbb{A}^m \to \mathrm{Bir}(\mathbb{P}^2)$ and $\mathbb{A}^m \to \mathrm{Aut}(\mathbb{A}^2)$, whose images contain the identity.

Of course, the same kind of example generalises to $\mathbb{P}^n$ and $\mathbb{A}^n$ for any $n \geq 2$. It shows that neither $\mathrm{Bir}(\mathbb{P}^2)$ nor $\mathrm{Aut}(\mathbb{A}^2)$ can be endowed with the structure of a locally noetherian scheme, compatible with the above families (morphisms), or equivalently says that the functor $\mathrm{Bir}_{\mathbb{P}^2}$ and $\mathrm{Aut}_{\mathbb{A}^2}$ are not representable by a locally noetherian scheme.

2.2. Ind-varieties and ind-groups. One way to avoid the problem of noetherianity consists of studying ind-schemes, which are inductive limits of locally noetherian schemes. One of the first articles in this direction is [**Sha1966**], which introduces the notion of "infinite dimensional algebraic varieties", or simply ind-variety, as given by a formal inductive limit of closed embeddings of algebraic varieties $X_i \hookrightarrow X_{i+1}$.

DEFINITION 2.6. An *ind-scheme* (resp. *ind-variety*) is given by a countable union $(X_i)_{i\in\mathbb{N}}$ of schemes (resp. algebraic varieties) together with closed embeddings $X_i \hookrightarrow X_{i+1}$.

A *morphism* between two ind-schemes $(X_i)_{i\in\mathbb{N}}$ and $(Y_i)_{i\in\mathbb{N}}$ is given by a collection of morphisms $\rho_i \colon X_i \to Y_{j_i}$, where $\{j_i\}_{i\in\mathbb{N}}$ is a sequence of indices, which is compatible with the inclusions.

The aim of this construction was to study the groups $\mathrm{Aut}(\mathbb{A}^n)$, which are ind-algebraic varieties, as shown in [**Sha1982**]. The group structure being compatible, the groups $\mathrm{Aut}(\mathbb{A}^n)$ are then shown to be ind-algebraic groups (see again [**Sha1982**]), even if the X_i are not subgroups. One can moreover observe that this structure gives the representability of the functor $\mathrm{Aut}_{\mathbb{A}^n}$ by an ind-algebraic group [**Bla2015**, Lemma 2.7].

More generally, for any affine algebraic variety X, the group $\mathrm{Aut}(X)$ can be seen as an ind-group [**KM2005**]. This again gives the representability of the functor Aut_X by an ind-algebraic group (see [**KM2005**, Theorem 3.3.3] or [**FK2014**]).

After having introduced this new category, the natural question to ask is wether the functor Bir_X can always be represented by an ind-scheme. This what I.R. Shafarevich asked in [**Sha1966**, §3]: "Can one introduce a universal structure of an infinite-dimensional group in the group of all automorphisms (resp. all birational automorphisms) of an arbitrary algebraic variety?"

The answer, given in [**BF2013**], is negative, and can again been shown explicitly for the case of $\mathbb{P}^2$. The problem does not come from the infinite dimension but from the degenerations of birational maps of high degree to maps of smaller degree. Let us give the following example ([**BF2013**, Example 3.1]):

EXAMPLE 2.7. Let $\hat{V} = \mathbb{P}^2 \backslash \{[0:1:0],[0:0:1]\}$, let $\rho\colon \hat{V} \to \mathrm{Bir}(\mathbb{P}^2)$ be the morphism given by

$$\begin{array}{ccc} \hat{V} \times \mathbb{P}^2 & \dashrightarrow & \hat{V} \times \mathbb{P}^2 \\ ([a:b:c],[x:y:z]) & \mapsto & ([a:b:c],[x(ay+bz):y(ay+cz):z(ay+cz)]) \end{array}$$

and define $V \subseteq \mathrm{Bir}(\mathbb{P}^2)$ to be the image of ρ. The map $\rho\colon \hat{V} \to V$ sends the line $L \subseteq \hat{V}$ corresponding to $b = c$ to the identity, and induces a bijection $\hat{V} \backslash L \to V \backslash \{\mathrm{id}\}$.

REMARK 2.8. The above map corresponds, on the affine plane where $z = 1$, to

$$\begin{array}{ccc} \hat{V} \times \mathbb{A}^2 & \dashrightarrow & \hat{V} \times \mathbb{A}^2, \\ ([a:b:c],(x,y)) & \mapsto & \left([a:b:c],(x \cdot \frac{ay+b}{ay+c}, y)\right). \end{array}$$

With this example, one can see that the structure of $V \subset \mathrm{Bir}(\mathbb{P}^2)$ should be the quotient $\hat{V} \to V$, i.e. the quotient of $\hat{V}$ modulo the equivalence relation that identifies all points of L [**BF2013**, Lemma 3.3]. As this line is equivalent to any other general line, the structure obtained is not the one of an algebraic variety, or even of an algebraic space. It shows that $\mathrm{Bir}_{\mathbb{P}^2}$ is not representable by an ind-variety, or even an ind-algebraic space or ind-algebraic stack [**BF2013**, Proposition 3.4]. We summarise it here:

THEOREM 2.9. ([**BF2013**, Theorem 1]) *For each $n \geq 2$, there is no structure of ind-algebraic variety* (*or algebraic variety*) *on* $\mathrm{Bir}(\mathbb{P}^n)$, *such that morphisms* $A \to \mathrm{Bir}(\mathbb{P}^n)$ *correspond to morphisms of ind-algebraic varieties* $A \to \mathrm{Bir}(\mathbb{P}^n)$.

Despite of this, it could be interesting to study equivalence classes on algebraic varieties. If the relation is closed and étale, one obtains an algebraic space [**Art1971**, Definition 2.3]. One could then seek for generalisations of this, by admitting non-étale equivalence relations, like the one induced by the above example. It can however introduce some pathologies: the local ring at the special point of $\mathrm{id} \in V$ corresponds to functions defined on a open set of $\hat{V}$ containing L and would then be the ring of constant functions, since rational functions on $\mathbb{P}^2$ defined on an open subset containing a line are constant. Can we anyway work with this category and obtain results not too far from algebraic schemes? Working then with inductive limit of the corresponding categories, we could then maybe be able to describe the groups of birational transformations in a functorial way. This gives rise to the following question:

QUESTION 2.10. Can we enlarge the category of ind-scheme to a "not too nasty" category in order to be able to represent the functor $\mathrm{Bir}_{\mathbb{P}^2}$? (or Bir_X in general)?

Another question would be to determine the varieties X for which Bir_X can be represented by an ind-scheme. Until now, we did not see any example, except the "trivial ones" where the structure comes from a group scheme. In particular, the following question arises:

QUESTION 2.11. Is there an algebraic variety X such that Bir_X can be represented by an ind-scheme, but not by a group scheme?

2.3. Group structure and Zariski topology on $\mathrm{Bir}(X)$**.** Note that the inverse map yields an isomorphism of functors from Bir_X to itself. Similarly, we can define a functor $\mathrm{Bir}_X \times \mathrm{Bir}_X$, in the same way as for Bir_X, and then observe that the composition is a morphism of functors. The notion of families given by Bir_X is then compatible with the group structure.

Even if Bir_X is not representable, we can define a topology on the group $\mathrm{Bir}(X)$, given by this functor. This topology was called *Zariski topology* by J.-P. Serre in [**Ser2010**]:

DEFINITION 2.12. Let X be an algebraic variety. A subset $F \subseteq \mathrm{Bir}(X)$ is *closed in the Zariski topology* if for any algebraic variety A (or more generally any locally noetherian algebraic scheme) and any morphism $A \to \mathrm{Bir}(X)$ the preimage of F is closed.

In the case where Bir_X is represented by an algebraic group, then the above topology is compatible with the Zariski topology of the algebraic group. Moreover, even if $\mathrm{Bir}(X)$ is not an algebraic group, then its topology and group structure behave not so far from algebraic groups. For instance, we can define the Zariski topology on $\mathrm{Bir}(X) \times \mathrm{Bir}(X)$, using morphisms as above, and check that the composition law yields a continuous map $\mathrm{Bir}(X) \times \mathrm{Bir}(X) \to \mathrm{Bir}(X)$. Moreover, the map sending an element on its inverse is a homeomorphism $\mathrm{Bir}(X) \to \mathrm{Bir}(X)$. Similarly, taking powers, left and right-multiplications and conjugation are homeomorphisms (see for example [**Bla2014**, Lemma 2.3]). Using such properties, one can see for instance that the closure of a subgroup is again a subgroup, and that the closure of an abelian subgroup (for example a cyclic group) is abelian.

For $n \geq 2$, the Zariski topology of $\mathrm{Bir}(\mathbb{P}^n)$ is not the one of any algebraic variety, or even ind-variety [**BF2013**, Theorem 2]. The obstruction follows from the bad topology of the set V constructed in Example 2.7: it contains a point where all closed subsets of positive dimension pass through.

However, we can describe the topology of $\mathrm{Bir}(\mathbb{P}^n)$, using maps of low degree.

DEFINITION 2.13. For each $\varphi \in \mathrm{Bir}(\mathbb{P}^n)$, the *degree* of φ is the degree $\deg(\varphi)$ of the pull-back of a general hyperplane. Equivalently, it is the degree of the polynomial that define φ, when these are taken without common factor.

We define by $\mathrm{Bir}(\mathbb{P}^n)_d$ (respectively by $\mathrm{Bir}(\mathbb{P}^n)_{\leq d}$) the set of elements of $\mathrm{Bir}(\mathbb{P}^n)$ of degree exactly d (respectively of degree $\leq d$).

REMARK 2.14. We have $\mathrm{Bir}(\mathbb{P}^n)_1 = \mathrm{Bir}(\mathbb{P}^n)_{\leq 1} = \mathrm{Aut}(\mathbb{P}^n)$.

We can first remark that $\mathrm{Bir}(\mathbb{P}^n)_{\leq d}$ is closed in $\mathrm{Bir}(\mathbb{P}^n)$ for each d [**BF2013**, Corollary 2.8]. This is the semi-continuity of the degree, which was also proved in [**Xie2015**, Lemma 4.1] for arbitrary surfaces. Then, the topology of $\mathrm{Bir}(\mathbb{P}^n)$ can be deduced from its subsets of bounded degree:

LEMMA 2.15. [**BF2013**, Proposition 2.10] *The topology of* $\mathrm{Bir}(\mathbb{P}^n)$ *is the inductive limit topology given by the Zariski topologies of* $\mathrm{Bir}(\mathbb{P}^n)_{\leq d}$, $d \in \mathbb{N}$, *which* *are the quotient topology of* $\pi_d\colon H_d \to \mathrm{Bir}(\mathbb{P}^n)_{\le d}$*, where* H_d *is an algebraic variety, endowed with its Zariski topology.*

The algebraic varieties H_d are given by $(n+1)$-uples of homogeneous polynomials of degree d inducing birational maps. The map $\pi_d\colon H_d \to \mathrm{Bir}(\mathbb{P}^n)_{\le d}$ restricts then to a bijection on $(\pi_d)^{-1}(\mathrm{Bir}(\mathbb{P}^n)_d)$, but not on maps of smaller degree, that can be represented in many different ways in H_d, by multiplying each coordinate by the same factor. These distinct possible factors are responsible of the fact that the Zariski topology of $\mathrm{Bir}(\mathbb{P}^n)_{\le d}$ is not the one of an algebraic variety.

Note that $\mathrm{Bir}(\mathbb{P}^n)$ is connected for each n [**Bla2010**], and that $\mathrm{Bir}(\mathbb{P}^2)_d$ is connected for $d \le 6$ [**BCM2015**]. Moreover, $\mathrm{Bir}(\mathbb{P}^2)$ does not contain any closed normal subgroup [**Bla2010**], even if it is not simple, viewed as an abstract group [**CL2013**].

The Zariski topology of $\mathrm{Bir}(X)$, for an arbitrary algebraic variety X, it still not well understood.

2.4. Algebraic subgroups. Studying biregular actions of algebraic groups on algebraic varieties is a very classical subject of algebraic geometry. More generally, one can study rational actions of algebraic groups. This was done for example in [**Wei1955, Ros1956**]. Using the notion of morphism $A \to \mathrm{Bir}(X)$ of Definition 2.2, the algebraic actions and algebraic subgroups of $\mathrm{Bir}(X)$ can be naturally defined:

DEFINITION 2.16. Let X be an irreducible algebraic variety and G be an algebraic group. A *birational group action* (respectively *biregular group action*) of G on X is a morphism $G \to \mathrm{Bir}(X)$ (respectively $G \to \mathrm{Aut}(X)$) which is also a group homomorphism. The image of this morphism is a subgroup of $\mathrm{Bir}(X)$ (respectively of $\mathrm{Aut}(X)$) which is called *algebraic subgroup*.

Note that any birational map $X \dashrightarrow Y$ conjugate birational group actions on X to birational group actions on Y. This allows sometimes to obtain biregular group actions:

THEOREM 2.17. ([**Wei1955**, Theorem page 375], [**Ros1956**, Theorem 1]) *Let* X *be an irreducible algebraic variety,* G *be an algebraic group and* $G \to \mathrm{Bir}(X)$ *a birational group action. Then, there exists a birational map* $X \dashrightarrow Y$*, where* Y *is another algebraic variety, that conjugates this action to a biregular group action.*

In this theorem, we can moreover assume Y to be projective, using equivariant completions (see [**Sum74**]). In particular, studying connected rational algebraic actions on a variety X amounts to study the connected components of the group scheme $\mathrm{Aut}(Y)$, where Y is a projective algebraic variety Y birational to X. This allows for instance to show that maximal connected subgroups of $\mathrm{Bir}(\mathbb{P}^2)$ are $\mathrm{Aut}(\mathbb{P}^2)$, $\mathrm{Aut}(\mathbb{P}^1 \times \mathbb{P}^1)^\circ$, $\mathrm{Aut}(\mathbb{F}_n)$, $n \ge 2$.

One can characterise the algebraic subgroups of $\mathrm{Bir}(\mathbb{P}^n)$ only using the Zariski topology defined in §2.3. These are the closed subgroups of bounded degree:

THEOREM 2.18. ([**BF2013**, Corollary 2.18, Proposition 2.19])

(1) *Every algebraic subgroup of* $\mathrm{Bir}(\mathbb{P}^n)$ *is closed (for the Zariski topology) and of bounded degree.*

(2) *For each closed algebraic subgroup $G \subset \mathrm{Bir}(\mathbb{P}^n)$ of bounded degree, there is a unique algebraic group structure on G, compatible with the group structure of* $\mathrm{Bir}(\mathbb{P}^n)$*, and such that morphisms $A \to \mathrm{Bir}(\mathbb{P}^n)$ having image in G correspond to morphisms of algebraic varieties $A \to G$.*

There is also a characterisation of connected algebraic subgroups of $\mathrm{Bir}(X)$, for any irreducible algebraic variety X:

THEOREM 2.19. ([**Ram1964**]) *Let X be an irreducible algebraic variety and $G \subset \mathrm{Aut}(X)$ be a subgroup having the following properties:*

(1) *(connectedness) For any $f \in G$, there is a morphism $A \to \mathrm{Aut}(X)$, where A is an irreducible algebraic variety, whose image contains f and the identity.*
(2) *(bounded dimension) There is an integer d such that for any injective morphisms $A \to \mathrm{Aut}(X)$ having image contained in G, we have $\dim A \leq d$.*

Then, there is a unique structure of algebraic group on G, compatible with the group structure of $\mathrm{Aut}(X)$*, such that morphisms $A \to \mathrm{Aut}(X)$ having image into G correspond to morphism of algebraic varieties $A \to G$.*

This nice result gives in particular the following corollary:

COROLLARY 2.20. *For each algebraic subgroup $G \subset \mathrm{Bir}(X)$, there is a unique structure of algebraic group on G, compatible with the group structure of* $\mathrm{Bir}(X)$*, such that morphisms $A \to \mathrm{Bir}(X)$ having image into G correspond to morphism of algebraic varieties $A \to G$.*

Moreover, the restriction of the Zariski topology of $\mathrm{Bir}(X)$ *on G is the Zariski topology of the algebraic group obtained.*

PROOF. By definition of an algebraic subgroup of $\mathrm{Bir}(X)$, there is an algebraic group H and a a morphism $H \to \mathrm{Bir}(X)$, which is also a group homomorphism, whose image is G; this corresponds to a birational group action of H on X (by definition). Using Theorem 2.17, there exists a birational map $\varphi\colon X \dashrightarrow Y$, where Y is another algebraic variety, that conjugates the action of H to a biregular group action. We can thus replace X with Y and assume that we have a biregular action $H \to \mathrm{Aut}(X)$ whose image is G.

We then denote by H° the connected component of H, and denote by $G' \subset G$ the normal subgroup of G corresponding to the image of H°. Then, the subgroup $G' \subset \mathrm{Aut}(Y)$ has the properties needed by Theorem 2.19 (the connectedness is given by H° and the bounded dimension is given by the fact that the image of an algebraic morphism is of bounded degree). This gives to G' a unique structure of algebraic group such that compatible with the group structure of $\mathrm{Aut}(X)$, such that morphisms $A \to \mathrm{Aut}(X)$ (or equivalently $A \to \mathrm{Bir}(X)$) having image into G correspond to morphism of algebraic varieties $A \to G'$. This implies that the Zariski topology of $\mathrm{Bir}(X)$ on G is the Zariski topology of the algebraic group obtained.

Since G/G' is finite, we can put a natural structure of algebraic variety on the finitely many translated of G' in G. □

It also seems that every algebraic subgroup of $\mathrm{Bir}(X)$ is closed, as stated in [**Pop2013a, Pop2013b**]. The case of $\mathbb{P}^n$ is given by Theorem 2.18 above but we did not find a proof of this statement for a general algebraic variety X.

3. Flat families and scheme structure

3.1. The functor $\mathrm{Bir}_X^{\mathrm{flat}}$. Another way of studying (bi)-rational maps between projective algebraic varieties consists of studying graphs. This was the viewpoint of [**Han1987**]. Let us recall the following basic notions:

DEFINITION 3.1. Let X, Y be irreducible algebraic varieties and $f\colon X \dashrightarrow Y$ a rational map. The *graph of f* is denoted Γ_f and is the closure of

$$\{(x, f(x)) \mid x \in \mathrm{dom}(f)\}$$

in $X \times Y$.

LEMMA 3.2. *Let X be a locally noetherian scheme, and denote by $\pi_i\colon X\times X \to X$ the i-th projection, for $i = 1, 2$. Then, the following maps are bijective:*

$$\begin{array}{ccc} \mathrm{Bir}(X) & \to & \left\{ \begin{array}{l} \textit{irreducible closed subsets } Y \subset X \times X \\ \textit{such that } \pi_i\colon Y \to X, \textit{ is a birational} \\ \textit{morphism, for } i = 1, 2. \end{array} \right\}, \\ f & \mapsto & \Gamma_f. \\ \mathrm{Aut}(X) & \to & \left\{ \begin{array}{l} \textit{irreducible closed subsets } Y \subset X \times X \\ \textit{such that } \pi_i\colon Y \to X, \textit{ is an} \\ \textit{isomorphism, for } i = 1, 2. \end{array} \right\}, \\ f & \mapsto & \Gamma_f. \end{array}$$

Applying this to $\mathrm{Bir}_X(A)$ (see Definition 2.1), we obtain the following:

LEMMA 3.3. *Let X be an irreducible algebraic variety and A be a locally noetherian scheme. We have a bijection*

$$\begin{array}{ccc} \mathrm{Bir}_X(A) & \to & \left\{ \begin{array}{l} \textit{irreducible closed subsets } Y \subset A \times X \times X \\ \textit{admitting a dense open subset } W \subset Y \\ \textit{such that the projection } W \to A \textit{ is surjective, and} \\ \textit{such that the two projections } A \times X \times X \to A \times X \\ \textit{restrict to open immersions } W \to A \times X. \end{array} \right\}, \\ f & \mapsto & \textit{closure of } \{(a, x, \pi_2(f(a, x))) \mid (a, x) \in \mathrm{dom}(f)\}. \end{array}$$

PROOF. The set $\mathrm{Bir}_X(A)$ corresponds to a subset of $\mathrm{Bir}(A\times X)$. By Lemma 3.2, this latter is in bijection with irreducible closed subsets $Y \subset (A\times X)\times(A\times X)$ such that $\pi_i\colon Y \to A\times X$ is a birational morphism, for $i = 1, 2$. Moreover, $f \in \mathrm{Bir}(A\times X)$ is sent onto the closure of $\{((a, x), f(a, x)) \mid (a, x) \in \mathrm{dom}(f)\}$.

As $\mathrm{Bir}_X(A)$ only consists of A-birational maps, we can forget one copy of A and obtain the closure of $\{((a, x), \pi_2(f(a, x))) \mid (a, x) \in \mathrm{dom}(f)\}$ in $A \times X \times X$, which is an irreducible closed subset $Y \subset A \times X \times X$ such that the two projections to $A\times X$ are birational. As before, every such subset provides in turn an A-birational map of $A \times X$.

A A-birational map f yields an element of $\mathrm{Bir}_X(A)$ if and only if there exist two open subsets $U, V \subset A\times X$, whose projections on A are surjective and such that the map f induces an isomorphism $U \to V$. Denoting by $\mu_1, \mu_2\colon A\times X\times X \to A\times X$ the two projections, the set $W = (\mu_1)^{-1}(U) = (\mu_2)^{-2}(V)$ is an open subset of Y, and the two projections give isomorphisms $\mu_1\colon W \to U$ and $\mu_2\colon W \to V$. Conversely, the existence of W and of two open embeddings to $A \times X$ yields U and V, and thus an element of $\mathrm{Bir}_X(A)$. □

Using these bijections, one can define the subfunctor $\mathrm{Bir}_X^{\mathrm{flat}}$ of Bir_X, corresponding to flat families:

DEFINITION 3.4. ([**Han1987**, Definition 2.1]) Let X be a projective algebraic variety and A a locally noetherian scheme. A *flat family of birational transformations* (resp. *of automorphisms*) of X over A is a closed subscheme $Y \subset A \times X \times X$, flat over A, such that for each $a \in A$, the fibre Y_a is the graph of an element of $\mathrm{Bir}_X(a)$ (respectively of $\mathrm{Aut}_X(a)$).

DEFINITION 3.5. Let X be an algebraic variety and A a locally noetherian scheme. We define $\mathrm{Bir}_X^{\mathrm{flat}}(A) \subset \mathrm{Bir}_X(A)$ as the set of elements $f \in \mathrm{Bir}_X(A)$ such that the corresponding graph in $A \times X \times X$ (see Lemma 3.3) is a flat family of birational transformations.

We similarly define $\mathrm{Aut}_X^{\mathrm{flat}}(A) = \mathrm{Bir}_X^{\mathrm{flat}}(A) \cap \mathrm{Aut}_X(A)$.

3.2. Representability of $\mathrm{Bir}_{\mathbb{P}^n}^{\mathrm{flat}}$. As M. Hanamura explains in [**Han1987**, Remark 2.10], the advantage of $\mathrm{Bir}_X^{\mathrm{flat}}$ over Bir_X is that it is representable.

REMARK 3.6. Recall that $\mathrm{Hilb}(X \times X)$ is an algebraic scheme (locally noetherian but with infinitely many components) that represents the functor $A \to \mathrm{Hilb}_{X\times X}(A)$, where

$$\mathrm{Hilb}_{X\times X}(A) \quad = \quad \left\{ \text{ closed subsets } Y \subset A \times X \times X \text{ that are flat over } A \right\}.$$

Hence, $\mathrm{Aut}_X^{\mathrm{flat}}$ and $\mathrm{Bir}_X^{\mathrm{flat}}$ are subfunctors of $\mathrm{Hilb}_{X\times X}$.

PROPOSITION 3.7 ([**Han1987**]). *Let X be an irreducible algebraic variety. For each locally noetherian scheme A, $\mathrm{Aut}_X^{\mathrm{flat}}(A)$ and $\mathrm{Bir}_X^{\mathrm{flat}}(A)$ are open subsets of $\mathrm{Hilb}_{X\times X}(A)$. Hence, both $\mathrm{Aut}_X^{\mathrm{flat}}$ and $\mathrm{Bir}_X^{\mathrm{flat}}$ are representable by the schemes $\mathrm{Aut}(X)$ and $\mathrm{Bir}(X)$, viewed as open subschemes of $\mathrm{Hilb}(X \times X)$.*

However, $\mathrm{Bir}_X^{\mathrm{flat}}$ has some "nasty properties", as M. Hanumara explains: "It turns out, however, that the scheme $\mathrm{Bir}(X)$ has some nasty properties; it is not a group scheme in general; even when X and X' are birationally equivalent, $\mathrm{Bir}(X)$ and $\mathrm{Bir}(X')$ may not be isomorphic." Another problem is that the composition law $\mathrm{Bir}(X) \times \mathrm{Bir}(X) \to \mathrm{Bir}(X)$ is not a morphism in general (see Corollary 3.15). The essential reason for these "nasty properties" is that the flatness of the graphs is not invariant under birational maps $X \dashrightarrow Y$ and even under birational transformations of X.

One example is given in [**Han1987**, (2.9)], comparing an abelian variety A of dimension $n \geq 2$ and the blow-up $\tilde{A} \to A$ at one point. Then, $\dim \mathrm{Bir}^\circ(A) = n$ but $\dim \mathrm{Bir}^\circ(\tilde{A}) = 0$, hence $\mathrm{Bir}(A)$ and $\mathrm{Bir}(\tilde{A})$ are not isomorphic. Moreover, $\mathrm{Bir}(\tilde{A})$ is not even equi-dimensional. In §3.3, we will describe more precisely the case of $\mathbb{P}^n$.

In the case where $\mathrm{char}(\mathrm{k}) = 0$ and where X is a terminal model, it is however proved in [**Han1987**] that the scheme obtained has a group scheme structure, compatible with the group structure of $\mathrm{Bir}(X)$. This has been generalised in [**Han1988**], in the case of non-uniruled varieties.

THEOREM 3.8. ([**Han1988**, Theorem 2.1]) *Let X be a non-uniruled projective variety over an algebraically closed field* k *of characteristic* 0, *and let us put on* $\mathrm{Bir}(X)$ *the scheme structure that represents* $\mathrm{Bir}_X^{\mathrm{flat}}$ *(see Proposition* 3.7). *Then, the following hold:*

(1) $\dim \mathrm{Bir}(X) \leq \min\{\dim X, q(X)\}$, *where $q(X)$ denotes the irregularity of a non-singular model of X.*
(2) *There exists a projective variety Y (which may be taken non-singular), birational to X, such that $\mathrm{Bir}(Y)_{\mathrm{red}}$ has a natural structure of a group scheme, locally of finite type over* k.
(3) $\mathrm{Bir}(Y)_{\mathrm{red}}$ *contains* $\mathrm{Aut}(Y)$ *as an open and closed group subscheme;* $\mathrm{Bir}^\circ(Y)$ *coincides with* $\mathrm{Aut}^\circ(Y)$ *and is an abelian variety.*

THEOREM 3.9. ([**Han1988**, Theorem 2.2]) *Let X and Y be as in Theorem* 3.8. *Then, the following hold:*

(1) *Let G be a group scheme, locally of finite type over* k. *Then to give a birational action of G on X (Definition* 2.16) *is equivalent to a homomorphism of group schemes* $G \to \mathrm{Bir}(Y)_{\mathrm{red}}$.
(2) *Let Y' be another projective variety birational to X with the property that* $\mathrm{Bir}(Y')_{\mathrm{red}}$ *is also a group scheme. Then,* $\mathrm{Bir}(Y)_{\mathrm{red}}$ *and* $\mathrm{Bir}(Y')_{\mathrm{red}}$ *are isomorphic as group schemes.*

3.3. The functors $\mathrm{Bir}_{\mathbb{P}^n}^{\mathrm{flat}}$. As explained before, the functor $\mathrm{Bir}_X^{\mathrm{flat}}$ is representable by a scheme, for any algebraic variety X (Proposition 3.7). Let us illustrate the structure that we obtain, in the case where $X = \mathbb{P}^n$. Using the notion of degree of a birational map of $\mathbb{P}^n$ (Definition 2.13), one can define a subfunctor $\mathrm{Bir}_{\mathbb{P}^n}^{\deg}$ of $\mathrm{Bir}_{\mathbb{P}^n}$. For each A we define $\mathrm{Bir}_{\mathbb{P}^n}^{\deg}(A) \subset \mathrm{Bir}_{\mathbb{P}^n}(A)$ as the elements $f \in \mathrm{Bir}_X(A)$ such that the corresponding morphism $A \to \mathrm{Bir}_{\mathbb{P}^n}(A)$ has constant degree on connected components of A. Similarly, we can define $\mathrm{Bir}_{\mathbb{P}^n}^d$, for each integer d, by taking only maps of degree d.

LEMMA 3.10. *Let $n \geq 2$ be an integer.*

(1) *For each $d \geq 1$, the functor $\mathrm{Bir}_{\mathbb{P}^n}^d$ is representable by an algebraic variety. This gives to the set $\mathrm{Bir}(\mathbb{P}^n)_d$ a natural structure of algebraic variety.*
(2) *The functor $\mathrm{Bir}_{\mathbb{P}^n}^{\deg}$ is representable by an algebraic scheme.*

PROOF. The first part is the statement of [**BF2013**, Proposition 2.15(b)]. The second part follows from the first one, by taking the disjoint union of the $\mathrm{Bir}(\mathbb{P}^n)_d$. □

REMARK 3.11. Note that the structure of algebraic variety of $\mathrm{Bir}(\mathbb{P}^n)_d$ is obtained by associating to an element

$$\begin{array}{rccc} f\colon & \mathbb{P}^n & \dashrightarrow & \mathbb{P}^n \\ & [x_0 : \cdots : x_n] & \mapsto & [f_0(x_0,\ldots,x_n) : \cdots : f_n(x_0,\ldots,x_n)] \end{array}$$

its coordinates $[f_0 : \cdots : f_n]$, that lives in the projective space parametrising the $(n+1)$-uples of polynomials of degree d, up to scalar multiplication (see [**BF2013**] or [**BCM2015**, §1] for more details).

The notion of degree can be generalised: we can associate to any element $f \in \mathrm{Bir}(\mathbb{P}^n)$ a sequence of integers $(d_1,\ldots,d_{n-1})$ called *multidegree* of f in [**Pan2000**], [**Dolg2012**, §7.1.3] (or characters in [**ST1968**]). By definition, d_i is equal to the degree of $f^{-1}(H_i)$, where $H_i \subset \mathbb{P}^n$ is a general linear subspace of codimension i. In particular, $d_1 = \deg(f)$ and $d_{n-1} = \deg(f^{-1})$. Another way to see the multidegree is to observe that the graph $\Gamma_f \subset \mathbb{P}^n \times \mathbb{P}^n$ is equal, in the chow ring of $\mathbb{P}^n \times \mathbb{P}^n$, to $\sum_{i=0}^n d_i h_{n-i,i}$, where $h_{i,j}$ denotes the class of a linear subspace $\mathbb{P}^i \times \mathbb{P}^j$, where $d_0 = d_n = 1$ and where $(d_1, \ldots, d_{n-1})$ is the multidegree of f. See [**Pan2000**] or [**Dolg2012**, §7.1.3] for more details.

LEMMA 3.12. *Let $n \geq 2$ be an integer.*

(1) *For each locally noetherian scheme A and each $f \in \mathrm{Bir}^{\mathrm{flat}}_{\mathbb{P}^n}(A)$, the induced morphism $A \to \mathrm{Bir}(\mathbb{P}^n)$ has constant multidegree on connected components of A.*
(2) *The functor $\mathrm{Bir}^{\mathrm{flat}}_{\mathbb{P}^n}$ is a subfunctor of $\mathrm{Bir}^{\mathrm{deg}}_{\mathbb{P}^n}$.*

PROOF. Let A be a locally noetherian scheme, and let $f \in \mathrm{Bir}_{\mathbb{P}^n}(A)$, which corresponds to a morphism $\rho_f \colon A \to \mathrm{Bir}(\mathbb{P}^n)$, and to an irreducible subset Y of $A \times \mathbb{P}^n \times \mathbb{P}^n$, given by the closure of $\{(a, x, \pi_2(f(a,x))) \mid (a,x) \in \mathrm{dom}(f)\}$ (see Lemma 3.3). Moreover, $\pi_2(f(a,x)) = \rho_f(a)(x)$ for each $(a,x) \in \mathrm{dom}(f)$.

By definition, $f \in \mathrm{Bir}^{\mathrm{flat}}_{\mathbb{P}^n}(A)$ if and only if Y is flat over A and the fibre Y_a of each $a \in A$ is the graph of an element of $\mathrm{Bir}_X(a)$.

Since Y is flat over A, the class of Y_a in the chow ring of $\mathbb{P}^n \times \mathbb{P}^n$ is locally constant. In particular, the multidegree of ρ_f is constant on connected components of A. This implies that $\mathrm{Bir}^{\mathrm{flat}}_{\mathbb{P}^n}(A) \subset \mathrm{Bir}^{\mathrm{deg}}_{\mathbb{P}^n}(A)$. □

EXAMPLE 3.13. We choose $A = \mathbb{A}^1$ and consider the morphism $\kappa \colon A \to \mathrm{Bir}(\mathbb{P}^2)$ given by

$$\kappa(t)\colon [x:y:z] \mapsto [-xz + ty^2 : yz : z^2].$$

For $t \neq 0$, $\kappa(t)$ is a quadratic birational involution of $\mathbb{P}^2$, but $\kappa(0)$ is equal to the linear automorphism $[x:y:z] \mapsto [-x:y:z]$. As the degree drops, the corresponding family is not flat over $A = \mathbb{A}^1$.

We can observe this by looking at the corresponding graph:

$$Y = \left\{ ([x:y:z],[X:Y:Z],t) \in A \times \mathbb{P}^2 \times \mathbb{P}^2 \;\middle|\; \begin{array}{lcl} Yz & = & Zy \\ Xz^2 & = & Z(-xz+ty^2) \\ Xyz & = & Y(-xz+ty^2) \end{array} \right\}.$$

When $t \neq 0$, the fibre Y_t is the graph of $\kappa(t)$, which is an irreducible surface in $\mathbb{P}^2 \times \mathbb{P}^2$. When $t = 0$, the fibre Y_0 is the union of the graph of $\kappa(0)$ and of the surface given by $z = 0, Z = 0$.

EXAMPLE 3.14. We choose again $A = \mathbb{A}^1$ and consider the morphism $\nu \colon A \to \mathrm{Bir}(\mathbb{P}^n)$, given by

$$\nu(t)\colon [x_0 : \cdots : x_n] \mapsto \left[\frac{1}{x_0 + tx_n} : \frac{1}{x_2} : \cdots : \frac{1}{x_n}\right].$$

This morphism corresponds to the composition of the standard transformation $\nu(0)$ with a family of automorphism and is thus flat by [**Han1987**, Proposition 2.5]. We can also verify this by looking at the corresponding graph and observing that the fibre of $t \in \mathbb{A}^1$ is the graph of $\nu(t)$.

COROLLARY 3.15. *Putting on* $\mathrm{Bir}(\mathbb{P}^n)$ *the scheme structure provided by the representability of* $\mathrm{Bir}^{\mathrm{flat}}_{\mathbb{P}^n}$, *the following hold:*

(1) *The set* $\mathrm{Bir}(\mathbb{P}^n)_d$ *is open in* $\mathrm{Bir}(\mathbb{P}^n)$, *for each* d.
(2) *For* $n \geq 2$, *the multiplication map* $\mathrm{Bir}(\mathbb{P}^n) \times \mathrm{Bir}(\mathbb{P}^n) \to \mathrm{Bir}(\mathbb{P}^n)$ *is not a morphism: it is not even continuous.*

PROOF. The part (1) follows from Lemma 3.12(1).

To see (2), we consider the morphism $\nu\colon \mathbb{A}^1 \to \mathrm{Bir}(\mathbb{P}^n)$ given in Example 3.14. Since the family is flat over $\mathbb{A}^1$, it corresponds to a morphism of schemes. We then define

$$\begin{array}{rccl} \nu'\colon & \mathbb{A}^1 & \to & \mathrm{Bir}(\mathbb{P}^n) \\ & t & \mapsto & \nu(t) \circ \nu(0) \end{array}$$

which is a morphism in the sense of Definition 2.2, but not a morphism of schemes as it corresponds to an element of $\mathrm{Bir}_{\mathbb{P}^n}(\mathbb{A}^1) \setminus \mathrm{Bir}^{\mathrm{flat}}_{\mathbb{P}^n}(\mathbb{A}^1)$. Indeed, $\nu'(0)$ is the identity, which is of degree 1, but for $t \neq 0$, the element $\nu'(t) \in \mathrm{Bir}(\mathbb{P}^n)$ is the quadratic transformation

$$\begin{array}{rcl} [x_0 : \cdots : x_n] & \mapsto & \left[\frac{1}{1/x_0 + t/x_n} : x_1 : \cdots : x_n\right] \\ & = & [x_0 x_n : x_1(x_n + t x_0) : \cdots : x_n(x_n + t x_0)]\,. \end{array}$$

In particular, ν' is not continuous, as $(\nu')^{-1}(\mathrm{Bir}(\mathbb{P}^n)_1) = \{0\}$ is not open, so the composition map mult: $\mathrm{Bir}(\mathbb{P}^n) \times \mathrm{Bir}(\mathbb{P}^n) \to \mathrm{Bir}(\mathbb{P}^n)$ is not continuous. □

We finish this text by comparing the two scheme structures on $\mathrm{Bir}(\mathbb{P}^n)$ given by the functors $\mathrm{Bir}^{\mathrm{flat}}_{\mathbb{P}^n}$ and $\mathrm{Bir}^{\mathrm{deg}}_{\mathbb{P}^n}$.

LEMMA 3.16. *The functors* $\mathrm{Bir}^{\mathrm{flat}}_{\mathbb{P}^n}$ *and* $\mathrm{Bir}^{\mathrm{deg}}_{\mathbb{P}^n}$ *are not equal if* $n \geq 3$.

PROOF. If $n \geq 3$, we can easily find some families of constant degree but having inverse of non-constant degree. This shows that the functors $\mathrm{Bir}^{\mathrm{flat}}_{\mathbb{P}^n}$ and $\mathrm{Bir}^{\mathrm{deg}}_{\mathbb{P}^n}$ are not equal. Take for example the family of automorphisms of $\mathbb{A}^n$ given by

$$\begin{array}{lrcl} \xi(t)\colon & (x_1, \ldots, x_n) & \mapsto & (x_1 + (x_2)^2, x_2 + t(x_3)^2, x_3, \ldots, x_n), \\ \xi(t)^{-1}\colon & (x_1, \ldots, x_n) & \mapsto & (x_1 - (x_2 - t(x_3)^2)^2, x_2 - t(x_3)^2, x_3, \ldots, x_n), \end{array}$$

and extend it to a family of birational transformations of $\mathbb{P}^n$. We then find $\deg(\xi(t)) = 2$, $\deg(\xi(t)^{-1}) = 4$ for each $t \neq 0$, but $\deg(\xi(0)) = \deg(\xi(0)^{-1}) = 2$. □

REMARK 3.17. It seems to us that $\mathrm{Bir}^{\mathrm{flat}}_{\mathbb{P}^2} = \mathrm{Bir}^{\mathrm{deg}}_{\mathbb{P}^2}$. One reason for this is that the Hilbert polynomial of the graph of an element $f \in \mathrm{Bir}(\mathbb{P}^2)_d$ is $P(x) = x^2(d+1) + 3x + 1$ (when we view this graph in $\mathbb{P}^8$ via the Segre embedding $\mathbb{P}^2 \times \mathbb{P}^2 \to \mathbb{P}^8$), and is then only dependent of degree d.

QUESTION 3.18. Is $\mathrm{Bir}^{\mathrm{flat}}_{\mathbb{P}^n}$ corresponding to algebraic families with a fixed multidegree (on connected components)?

References

[Art1971] Michael Artin, *Algebraic spaces*, Yale University Press, New Haven, Conn.-London, 1971. A James K. Whittemore Lecture in Mathematics given at Yale University, 1969; Yale Mathematical Monographs, 3. MR0407012

[BCM2015] Cinzia Bisi, Alberto Calabri, and Massimiliano Mella, *On plane Cremona transformations of fixed degree*, J. Geom. Anal. **25** (2015), no. 2, 1108–1131, DOI 10.1007/s12220-013-9459-9. MR3319964

[Bla2010] Jérémy Blanc, *Groupes de Cremona, connexité et simplicité* (French, with English and French summaries), Ann. Sci. Éc. Norm. Supér. (4) **43** (2010), no. 2, 357–364. MR2662668

[Bla2014] Jérémy Blanc, Algebraic elements of the Cremona groups in *From Classical to Modern Algebraic Geometry, Corrado Segre's Mastership and Legacy*, Series: Trends in the History of Science (2016) Birkauser.

[Bla2015] Jérémy Blanc, *Conjugacy classes of special automorphisms of the affine spaces.* Algebra Number Theory **10** (2016), no. 5, 939–967.

[BF2013] Jérémy Blanc and Jean-Philippe Furter, *Topologies and structures of the Cremona groups*, Ann. of Math. (2) **178** (2013), no. 3, 1173–1198, DOI 10.4007/annals.2013.178.3.8. MR3092478

[CL2013] Serge Cantat and Stéphane Lamy, *Normal subgroups in the Cremona group*, Acta Math. **210** (2013), no. 1, 31–94, DOI 10.1007/s11511-013-0090-1. With an appendix by Yves de Cornulier. MR3037611

[Dem1970] Michel Demazure, *Sous-groupes algébriques de rang maximum du groupe de Cremona* (French), Ann. Sci. École Norm. Sup. (4) **3** (1970), 507–588. MR0284446

[Dolg2012] Igor V. Dolgachev, *Classical algebraic geometry*, Cambridge University Press, Cambridge, 2012. A modern view. MR2964027

[FK2014] Jean-Philippe Furter and Hanspeter Kraft, *On the geometry of the automorphism group of affine n-space.* manuscript in preparation

[Han1987] Masaki Hanamura, *On the birational automorphism groups of algebraic varieties*, Compositio Math. **63** (1987), no. 1, 123–142. MR906382

[Han1988] Masaki Hanamura, *Structure of birational automorphism groups. I. Nonuniruled varieties*, Invent. Math. **93** (1988), no. 2, 383–403, DOI 10.1007/BF01394338. MR948106

[Har1977] Robin Hartshorne, *Algebraic geometry*, Springer-Verlag, New York-Heidelberg, 1977. Graduate Texts in Mathematics, No. 52. MR0463157

[KM2005] Tatsuji Kambayashi and Masayoshi Miyanishi, *On two recent views of the Jacobian conjecture*, Affine algebraic geometry, Contemp. Math., vol. 369, Amer. Math. Soc., Providence, RI, 2005, pp. 113–138, DOI 10.1090/conm/369/06808. MR2126658

[Mat1958] T. Matsusaka, *Polarized varieties, fields of moduli and generalized Kummer varieties of polarized abelian varieties*, Amer. J. Math. **80** (1958), 45–82. MR0094360

[MO1967] Hideyuki Matsumura and Frans Oort, *Representability of group functors, and automorphisms of algebraic schemes*, Invent. Math. **4** (1967), 1–25. MR0217090

[Pan2000] Ivan Pan, *Sur le multidegré des transformations de Cremona* (French, with English and French summaries), C. R. Acad. Sci. Paris Sér. I Math. **330** (2000), no. 4, 297–300, DOI 10.1016/S0764-4442(00)00142-7. MR1753297

[PR2013] Ivan Pan and Alvaro Rittatore, *Some remarks about the Zariski topology of the Cremona group.* http://arxiv.org/abs/1212.5698

[Pop2013a] Vladimir L. Popov, *Some subgroups of the Cremona groups*, Affine algebraic geometry, World Sci. Publ., Hackensack, NJ, 2013, pp. 213–242, DOI 10.1142/9789814436700_0010. MR3089039

[Pop2013b] V. L. Popov, *Tori in the Cremona groups* (Russian, with Russian summary), Izv. Ross. Akad. Nauk Ser. Mat. **77** (2013), no. 4, 103–134; English transl., Izv. Math. **77** (2013), no. 4, 742–771. MR3135700

[Ram1964] C. P. Ramanujam, *A note on automorphism groups of algebraic varieties*, Math. Ann. **156** (1964), 25–33. MR0166198

[Ros1956] Maxwell Rosenlicht, *Some basic theorems on algebraic groups*, Amer. J. Math. **78** (1956), 401–443. MR0082183

[ST1968] J. G. Semple and J. A. Tyrrell, *Specialization of Cremona transformations*, Mathematika **15** (1968), 171–177. MR0237490

[Ser2010] Jean-Pierre Serre, *Le groupe de Cremona et ses sous-groupes finis* (French, with French summary), Astérisque **332** (2010), Exp. No. 1000, vii, 75–100. Séminaire Bourbaki. Volume 2008/2009. Exposés 997–1011. MR2648675

[Sha1966] I. R. Shafarevich, *On some infinite-dimensional groups*, Rend. Mat. e Appl. (5) **25** (1966), no. 1-2, 208–212. MR0485898

[Sha1982] I. R. Shafarevich, *On some infinite-dimensional groups. II* (Russian), Izv. Akad. Nauk SSSR Ser. Mat. **45** (1981), no. 1, 214–226, 240. MR607583

[Sum74] Hideyasu Sumihiro, *Equivariant completion*, J. Math. Kyoto Univ. **14** (1974), 1–28. MR0337963

[Wei1955] André Weil, *On algebraic groups of transformations*, Amer. J. Math. **77** (1955), 355–391. MR0074083

[Xie2015] Junyi Xie, *Periodic points of birational transformations on projective surfaces*, Duke Math. J. **164** (2015), no. 5, 903–932, DOI 10.1215/00127094-2877402. MR3332894

Departement Mathematik und Informatik, Universität Basel, Spiegelgasse 1, CH-4051 Basel, Switzerland

E-mail address: jeremy.blanc@unibas.ch

Proceedings of Symposia in Pure Mathematics
Volume **94**, 2017
http://dx.doi.org/10.1090/pspum/094/01621

The Hermite-Joubert problem over p-closed fields

Matthew Brassil and Zinovy Reichstein

ABSTRACT. An 1861 theorem of Ch. Hermite asserts that every field extension (and more generally, every étale algebra) E/F of degree 5 can be generated by an element $a \in E$ whose minimal polynomial is of the form

$$f(x) = x^5 + b_2x^3 + b_4x + b_5.$$

Equivalently, $\mathrm{tr}_{E/F}(a) = \mathrm{tr}_{E/F}(a^3) = 0$. A similar result for étale algebras of degree 6 was proved by P. Joubert in 1867. It is natural to ask whether or not these classical theorems extend to étale algebras of degree $n \geqslant 7$. Prior work of the second author shows that the answer is "no" if $n = 3^a$ or $n = 3^a + 3^b$, where $a > b \geqslant 0$.

In this paper we consider a variant of this question where F is required to be a p-closed field. More generally, we give a necessary and sufficient condition for an integer n, a field F_0 and a prime p to have the following property: Every étale algebra E/F of degree n, where F is a p-closed field containing F_0, has an element $0 \neq a \in E$ such that $F[a] = E$ and $\mathrm{tr}(a) = \mathrm{tr}(a^p) = 0$.

As a corollary (for $p = 3$), we produce infinitely many new values of n, such that the classical theorems of Hermite and Joubert do not extend to étale algebras of degree n. The smallest of these new values are $n = 13$, 31, 37, and 39.

1. Introduction

An 1861 theorem of Ch. Hermite [**He**] asserts that for every étale algebra E/F of degree 5 there exists an element $0 \neq a \in E$ whose characteristic polynomial is of the form

$$f(x) = x^5 + b_2x^3 + b_4x + b_5\,.$$

An easy application of Newton's formulas shows that this is equivalent to $\mathrm{tr}_{E/F}(a) = \mathrm{tr}_{E/F}(a^3) = 0$; see, e.g., [**Co**$_2$, section 1]. A similar result for étale algebras of degree 6 was proved by P. Joubert in 1867; see [**Jo**]. For modern proofs of these results, see [**Co**$_2$, **Kr**]. (Here we are assuming that F is an infinite field of characteristic $\neq 2$ or 3. As usual, by an étale algebra E/F of degree n we mean a direct product $E := E_1 \times \ldots \times E_r$, where each E_i is a separable field extension of F and $[E_1 : F] + \ldots + [E_r : F] = n$.)

2010 *Mathematics Subject Classification.* Primary 12F10, 14J70; Secondary 14G05, 11G05.

Key words and phrases. Hermite-Joubert problem, étale algebra, hypersurface, rational point, p-closed field, elliptic curve.

The second author has been partially supported by a Discovery Grant from the National Science and Engineering Board of Canada.

It is natural to ask if the above-mentioned theorems of Hermite and Joubert can be extended to $n \geqslant 7$; cf., e.g. [**Co**$_2$, Section 4]. The answer is "no" if n is of the form 3^k or $3^{k_1} + 3^{k_2}$, where $k_1 > k_2 \geqslant 0$; see [**Re**$_1$, Theorem 1.3] or [**RY**$_2$, Corollary 1.7(a) and Theorem 1.8]. For other values of n (in particular, for $n = 7$), this question remains open. One can also ask a similar (even more difficult) question for an arbitrary prime p.

HERMITE-JOUBERT PROBLEM 1.1. Let $n \geqslant 2$ be an integer, p be a prime, and F_0 be a base field. Which triples (F_0, p, n) have the following property: for every field extension F/F_0 and every étale algebra E/F of degree n, there exists an element $0 \neq a \in E$ such that $\operatorname{tr}(a) = \operatorname{tr}(a^p) = 0$?

We will usually want to choose the element $a \in E$ above so that $F[a] = E$, i.e., a generates E as an F-algebra. We will also consider a variant of this problem, where a is only required to satisfy $\operatorname{tr}(a^p) = 0$, rather than $\operatorname{tr}(a) = \operatorname{tr}(a^p) = 0$.

In this paper we will show that these questions become tractable if we restrict our attention to the case, where F is a p-closed field. Recall that a field F is called p-closed if the degree of every finite field extension of F is a power of p. Some authors use the term p-field in place of p-closed field; see, e.g., [**Pf**, Definition 4.1.11].

LOCAL HERMITE-JOUBERT PROBLEM 1.2. Let $n \geqslant 2$ be an integer, p be a prime, and F_0 be a base field. Which triples (F_0, p, n) have the following property: for every p-closed field F containing F_0 and every étale algebra E/F of degree n, there exists an element $0 \neq a \in E$ such that $\operatorname{tr}(a) = \operatorname{tr}(a^p) = 0$?

Equivalently, for an arbitrary field F and an étale algebra E/F of degree n, we are asking if there is a finite field extension F'/F of degree prime to p and an element $0 \neq a \in E' := E \otimes_{F_n} F'$ such that $\operatorname{tr}_{E'/F'}(a) = \operatorname{tr}_{E'/F'}(a^p) = 0$; see Lemma 3.1.

Before stating our main results, we recall the definition of the "general field extension" E_n/F_n of degree n. Let F_0 be a base field and $x_1, \ldots, x_n$ be independent variables. Set $L_n := F_0(x_1, \ldots, x_n)$, $F_n := L_n^{\mathrm{S}_n}$ and $E_n := L_n^{\mathrm{S}_{n-1}} = F_n(x_1)$, where S_n acts on L_n by permuting $x_1, \ldots, x_n$ and S_{n-1} by permuting $x_2, \ldots, x_n$.

Let p be a prime. We will say that

$$n = p^{k_1} + \ldots + p^{k_m}$$

is a *base p presentation* (or *base p expansion*) of n if each power of p appears in the sum at most $p-1$ times. It is well known that the base p expansion of n is unique.

THEOREM 1.3. *Let p be a prime, F_0 be a field of characteristic $\neq p$ containing a primitive pth root of unity ζ_p, and $n = p^{k_1} + \ldots + p^{k_m}$ be the base p expansion of an integer $n \geqslant 3$. Then the following conditions are equivalent.*

(1) For every p-closed field F containing F_0 and every n-dimensional étale algebra E/F, there exists an element $0 \neq a \in E$ such that $\operatorname{tr}_{E/F}(a^p) = 0$.

(2) There exists a finite field extension F'/F_n of degree prime to p and an element $0 \neq a \in E' := E_n \otimes_{F_n} F'$ such that $\operatorname{tr}_{E'/F'}(a^p) = 0$. Here E_n/F_n is the general field extension of degree n defined above.

(3) The equation

$$p^{k_1} y_1^p + \ldots + p^{k_m} y_m^p = 0 \tag{1.1}$$

has a solution $y = (y_1 : \ldots : y_m) \in \mathbb{P}^{n-1}(F_0)$.

($*$) *Moreover, if (3) holds, then the element* a *in parts (1) and (2) can be chosen so that* $E = F[a]$ *and* $E' = F'[a]$, *respectively.*

The implication (1) $\Longrightarrow$ (2) readily follows from the definition of a p-closed field. The proof of the implication (2) $\Longrightarrow$ (3), based on the fixed point method, is implicit in [**RY**$_2$, Sections 6] (where m is assumed to be 2). We will present a self-contained argument in Section 4. The implication (3) $\Longrightarrow$ (1) was initially motivated by [**DR**, Section 8]; the proof we present in Section 5 is entirely elementary.

THEOREM 1.4. *Let* p *be a prime,* F_0 *be a field of characteristic* $\neq p$ *containing a primitive pth root of unity* ζ_p, *and* $n = p^{k_1} + \ldots + p^{k_m}$ *be the base* p *expansion of an integer* $n \geqslant 3$. *Then the following conditions are equivalent.*

(1) For every p-closed field F *containing* F_0 *and every n-dimensional étale algebra* E/F, *there exists an element* $0 \neq a \in E$ *such that* $\mathrm{tr}_{E/F}(a) = \mathrm{tr}_{E/F}(a^p) = 0$.

(2) There exists a finite field extension F'/F_n *of degree prime to* p *and an element* $0 \neq a \in E' := E_n \otimes_{F_n} F'$ *such that* $\mathrm{tr}_{E'/F'}(a) = \mathrm{tr}_{E'/F'}(a^p) = 0$.

(3) The system of equations

$$\begin{cases} p^{k_1} y_1 + \ldots + p^{k_m} y_m = 0 \\ p^{k_1} y_1^p + \ldots + p^{k_m} y_m^p = 0 \end{cases} \tag{1.2}$$

has a solution $y = (y_1 : \ldots : y_m) \in \mathbb{P}^{m-1}(F_0)$.

($**$) *Moreover, assume that the equivalent conditions (1), (2) and (3) hold,* $(\mathrm{char}(F_0), p, n) \neq (2, 3, 3)$, $(2, 3, 4)$, *or* $(3, 2, 3)$, *and one of the following additional conditions is met:*

(i) there is a solution $y = (y_1 : \ldots : y_m) \in \mathbb{P}^{m-1}(F_0)$ *to* (1.2), *such that* $y \neq (1 : \ldots : 1)$, *or (ii)* $p > 2$ *and* $\mathrm{char}(F_0) \neq 2$.

Then the element a *in parts (1) and (2) can be chosen so that* $E = F[a]$ *and* $E' = F'[a]$, *respectively.*

Note that conditions (i) and (ii) are very mild. In particular, condition (i) is automatic if $\mathrm{char}(F_0)$ does not divide n, since in this case $(1 : \ldots : 1)$ is not a solution to (1.2).

We will prove Theorem 1.4 in Section 6 by modifying our proof of Theorem 1.3. Assertion ($*$) of Theorem 1.3 follows from a result of J.-L. Colliot-Thélène [**CT**]; see Section 8. Assertion ($**$) of Theorem 1.4 requires a variant of Colliot-Thélène's result, which is proved in Section 7.

For $p = 2$, Springer's theorem about rational points of quadric hypersurfaces allows us to remove the requirement that F is a 2-closed field in part (1) of Theorems 1.3 and 1.4. This leads to a nearly complete solution of the Hermite-Joubert Problem 1.1 for $p = 2$; see Corollary 10.1. The same arguments go through for $p = 3$, if we assume a long-standing conjecture of J. W. S. Cassels and P. Swinnerton-Dyer about rational points on cubic hypersurfaces; see Section 11.

It is natural to ask for which n, p and F_0 there exist non-trivial solutions to equation (1.1) and the system (1.2). Some partial answers to this question are given in Section 12. In particular, we show that for $p \geqslant 3$ the system (1.2) has a solution over any field F_0 if, when we write $n = [n_d n_{d-1} \ldots n_0]_p$ in base p, one of the digits n_k is $\geqslant 2$ or the number of non-zero digits is $\geqslant p + 3$; see Lemma 12.3.

(Here n_k is the number of times p^k occurs in the presentation $n = p^{k_1} + \ldots + p^{k_m}$ in Theorems 1.3 and 1.4. If each n_k is 0 or 1, then the number of non-zero digits is m.) This implies, in particular, that for "most" n the system (1.2) has a non-trivial solution over any base field F_0. That is, if we fix $p \geqslant 3$ and let S_N be the set of integers $n \in \{1, \ldots, N\}$ such that the system (1.2) has a non-trivial solution over every base field F_0, then $|S_N|/N$ will rapidly converge to 1, as $N \to \infty$.

On the other hand, it is easy to see that the system (1.2) has no non-trivial solutions if $n = p^k$ for any $k \geqslant 1$ or $n = p^{k_1} + p^{k_2}$, where $k_1 > k_2 \geqslant 0$ and $\mathrm{char}(F_0)$ does not divide $p^{(k_1-k_2)(p-1)} + (-1)^p$. This way we recover most of [**Re**$_1$, Theorem 1.3]. In Section 13 we will extend this result as follows (for $p = 3$ only).

THEOREM 1.5. *Let E_n/F_n be the general field extension of degree n, over the base field $F_0 = \mathbb{Q}$. Suppose $n = 3^{k_1} + 3^{k_2} + 3^{k_3}$, where $k_1, k_2, k_3 \geqslant 0$ are distinct integers such that $k_1 + k_2 + k_3 \equiv 0$ or $1 \pmod 3$. Then for any finite field extension F'/F_n of degree prime to 3 there does not exist an element $0 \neq a \in E'_n := E_n \otimes_{F_n} F'$ such that $\mathrm{tr}_{E'/F'}(a) = \mathrm{tr}_{E'/F'}(a^3) = 0$.*

This yields new examples, where the Hermite-Joubert Problem 1.1 has a negative answer in the classical setting (i.e., for $p = 3$ and $F_0 = \mathbb{Q}$). The smallest of these are $n = 13 = 3^2 + 3^1 + 3^0$, $n = 31 = 3^3 + 3^1 + 3^0$, and $n = 39 = 3^3 + 3^2 + 3^1$. We conjecture that Theorem 1.5 remains valid for all triples k_1, k_2, k_3 of distinct non-negative integers; see Conjecture 14.1. Some evidence in support of this conjecture is presented in Section 14. In particular, we show that the Hermite-Joubert Problem 1.1 (again, for $p = 3$ and $F_0 = \mathbb{Q}$) has a negative answer in the case, where $n = 37 = 3^3 + 3^2 + 3^0$, which is not covered by Theorem 1.5.

REMARK 1.6. Our approach to the Hermite-Joubert Problem 1.1 in this paper is to subdivide it into two parts: the Local Hermite-Joubert Problem, and the rest. In the language of [**Re**$_2$, Section 5], the Local Hermite-Joubert Problem is a Type 1 problem. The present paper is devoted to solving this Type 1 problem. In those cases, where the Local Hermite-Joubert Problem has a negative solution, so does the original Hermite-Joubert Problem 1.1 (e.g., as in Theorem 1.5).

In those cases, where the Local Hermite-Joubert Problem has a positive solution, the original Hermite-Joubert Problem becomes a "Type 2 question". This question remains open, except in a few special cases, such as the case considered in Section 10, where $p = 2$, or the cases studied by Hermite and Joubert, where $p = 3$ and $n = 5$ or 6.

Many questions concerning algebraic objects over fields F, can be subdivided into two parts in a similar manner: a Type 1 problem, where F is assumed to be a p-closed field for some prime p, and a Type 2 problem (the rest, in those cases, where the Type 1 problem has a positive solution). Existing techniques are often effective in addressing Type 1 problems but Type 2 problems tend to be out of reach, except in a few special cases. For a discussion of this phenomenon and numerous examples, see [**Re**$_2$, Section 5]. For further comments and remarks on Theorems 1.3 and 1.4, see Section 9.

2. Geometry of the hypersurfaces $X_{n,p}$ and $Y_{n,p}$

In this section we will prove some simple geometric properties of the hypersurfaces

$$X_{n,p} := \{(x_1 : \ldots : x_n) \mid x_1^p + \ldots + x_n^p = 0\} \subset \mathbb{P}^{n-1}$$

and

$$Y_{n,p} := \{(x_1 : \ldots : x_n) \,|\, x_1 + \ldots + x_n = x_1^p + \ldots + x_n^p = 0\} \subset \mathbb{P}^{n-2},$$

defined over the base field F_0.

Recall that a closed subvariety V of projective space is called *a cone* over a point $c \in V$ if V contains the line through c and c' for every $c \neq c' \in V$. We will say that V is a cone if it is a cone over one of its points.

Let Δ_n be the union of the "diagonal" hyperplanes $x_i = x_j$, over all $1 \leqslant i < j \leqslant n$.

LEMMA 2.1. *Assume* $\mathrm{char}(F_0) \neq p$. *Then*

(a) $X_{n,p}$ *is smooth.*

(b) The singular locus of $Y_{n,p}$ *is* $Y_{n,p} \cap \{(x_1 : \ldots : x_n) \,|\, x_1^{p-1} = \ldots = x_n^{p-1}\}$.

(c) $X_{n,p}$ *is absolutely irreducible if* $n \geqslant 3$.

(d) $Y_{n,p}$ *is absolutely irreducible if* $n \geqslant 5$.

(e) $X_{n,p}$ *is not contained in* Δ_n *for any* $n \geqslant 3$.

(f) $Y_{n,p}$ *is not contained in* Δ_n, *if* $n \geqslant 3$ *and* $(\mathrm{char}(F_0), p, n) \neq (2,3,3)$, $(2,3,4)$ *or* $(3,2,3)$.

(g) Let $(1 : \ldots : 1) \neq c \in Y_{n,p}$. *Then* $Y_{n,p}$ *is not a cone over* c.

(h) $Y_{n,p}$ *is not a cone if* $p > 2$ *and* $\mathrm{char}(F_0) \neq 2$.

PROOF. In order to prove the lemma we may, without loss of generality, pass to the algebraic closure of F_0, i.e., assume that F_0 is algebraically closed.

(a) and (b) readily follow from the Jacobian criterion.

(c) Assume the contrary. Then $X_{n,p}$ has at least two irreducible components, X_1 and X_2. Since $X_{n,p}$ is a hypersurface in $\mathbb{P}^{n-1}$, $\dim(X_1) = \dim(X_2) = n-2$, and $\dim(X_1 \cap X_2) = n-3$. Since we are assuming that $n \geqslant 3$, this implies that $X_1 \cap X_2 \neq \emptyset$. On the other hand, every point of $X_1 \cap X_2$ is singular in X, contradicting (a).

(d) Assume the contrary: $Y_{n,p}$ has at least two irreducible components, Y_1 and Y_2. Arguing as in (c), we see that $Y_1 \cap Y_2$ is a closed subvariety of the singular locus of Y, and $\dim(Y_1 \cap Y_2) = n-4$. On the other hand, by part (b), the singular locus of Y is 0-dimensional. Thus $n - 4 \leqslant 0$, as desired.

(e) Assume the contrary: $X_{n,p} \subset \Delta_n$. Recall that Δ_n is the union of the hyperplanes $H_{ij} \subset \mathbb{P}^{n-1}$ given by $x_i = x_j$, where $i \neq j$ and $1 \leqslant i, j \leqslant n$. Since $X_{n,p}$ is irreducible by part (c), it is contained in one of these hyperplanes. Since $X_{n,p}$ is invariant under the action of S_n, it is contained in every hyperplane H_{ij}. That is,

$$X_{n,p} \subset \bigcap_{1 \leqslant i < j \leqslant n} H_{ij} = \{(1 : \ldots : 1)\},$$

which is impossible, since $\dim(X_{n,p}) = n - 2 \geqslant 1$.

(f) First assume $n \geqslant 5$. Here $Y_{n,p}$ is irreducible by part (d), and the same argument as in part (e) shows that

$$Y_{n,p} \subset \bigcap_{1 \leqslant i < j \leqslant n} H_{ij} = \{(1 : \ldots : 1)\},$$

contradicting $\dim(Y_{n,p}) = n - 3 \geqslant 2$.

In the remaining cases, where $n = 3$ or 4, but $(\mathrm{char}(F_0), p, n) \neq (2,3,3)$, $(2,3,4)$ or $(3,2,3)$, we will exhibit a point $y \in Y_{n,p}$ which does not lie in Δ_n.

If $n = 4$ and $\mathrm{char}(F_0) \neq 2$, we can take $y := (1 : \zeta_4 : \zeta_4^2 : \zeta_4^3)$. Here $\zeta_4 \in F_0$ is a primitive 4th root of unity. (Recall that we are assuming F_0 to be algebraically closed.) If $\mathrm{char}(F_0) = 2$ but $p \neq 3$, set $y := (1 : \zeta_3 : \zeta_3^2 : 0)$. This covers all pairs $(\mathrm{char}(F_0), p)$, except for $(2, 3)$. (Recall that we are assuming that $\mathrm{char}(F_0) \neq p$ throughout.)

Now suppose $n = 3$. If $\mathrm{char}(F_0) \neq 2$ and $p \neq 2$, then we can take $y := (1 : -1 : 0)$. If $\mathrm{char}(F_0) \neq 3$ and $p \neq 3$, set $y := (1 : \zeta_3 : \zeta_3^2)$. This covers all pairs $(\mathrm{char}(F_0), p)$, except for $(2, 3)$ and $(3, 2)$. (Once again, we are assuming that $\mathrm{char}(F_0) \neq p$.)

(g) Assume the contrary. Since S_n acts on $Y_{n,p}$ by permuting coordinates, $Y_{n,p}$ is a cone over $g \cdot c$ for every $g \in \mathrm{S}_n$. Now it is easy to see that $Y_{n,p}$ contains the linear span of $\{g \cdot c \,|\, g \in \mathrm{S}_n\}$. Denote this linear span by L. Then L is an S_n-invariant linear subspace of $\mathbb{P}^{n-2}$. If $\mathrm{char}(F_0)$ does not divide n, then the S_n-representation on F_0^{n-1} is irreducible. Hence, the only S_n-invariant subspace of $\mathbb{P}^{n-2}$ is $\mathbb{P}^{n-2}$ itself. Thus $\mathbb{P}^{n-2} = L \subset Y_{n,p}$, a contradiction. If $\mathrm{char}(F_0)$ divides n, then the only other possibility is $L = \{(1 : 1 \ldots : 1)\}$. This is ruled out by our assumption that $c \neq (1 : \ldots : 1)$.

(h) Assume $Y_{n,p}$ is a cone over some point $c \in Y_{n,p}$. By part (d), $c = (1 : \ldots : 1)$[1]. We want to show that if $p > 2$ and $\mathrm{char}(F_0) \neq 2$ then $Y_{n,p}$ is not a cone over $c = (1 : \ldots : 1)$. Assume the contrary: whenever $Y_{n,p}$ contains a point $y = (y_1 : \ldots : y_n)$, it contains the entire line through c and y. That is,

$$(1 + ty_1)^p + \ldots + (1 + ty_n)^p = 0$$

as a polynomial in t. In particular, $p(y_1^{p-1} + \ldots + y_n^{p-1})$, which is the coefficient of t^{p-1} in this polynomial, should vanish for every $y = (y_1 : \ldots : y_n) \in Y_{n,p}$. Set $y := (-1 : 1 : 0 : \ldots : 0)$. Note that $y \in Y_{n,p}$, because $p > 2$. For this y, $p(y_1^{p-1} + \ldots + y_n^{p-1}) = 0$, reduces to $2p = 0$, contradicting our assumptions that $\mathrm{char}(F_0) \neq 2$ or p. □

3. Proof of Theorem 1.3: (1) $\Longrightarrow$ (2)

Recall that a field F is called p-closed if the degree of every finite field extension of L is a power of p. For every field F there exists an algebraic extension $F \subset F^{(p)}$, such that $F^{(p)}$ is p-closed field and, for every finite subextension $F \subset F' \subset F^{(p)}$, the degree $[F' : F]$ is prime to p. The field $F^{(p)}$ satisfying these conditions is unique up to F-isomorphism. We will refer to it as a *p-closure* of F. For details, see [**EKM**, Proposition 101.16].

LEMMA 3.1. *Let E/F be an étale algebra of degree n. Then*

(a) every element $a \in E \otimes_F F^{(p)}$ lies in the image of the natural map

$$\phi \colon E \otimes_F F' \hookrightarrow E \otimes_F F^{(p)}$$

for some intermediate field $F \subset F' \subset F^{(p)}$ (depending on a), such that $[F' : F]$ is finite (and thus automatically prime to p).

(b) $x \in E' := E \otimes_F F'$ generates E' over F' (i.e., $E' := F'[x]$) if and only if $\phi(x)$ generates $E \otimes_F F^{(p)}$ over $F^{(p)}$.

---

[1]The reader should keep in mind that the subsequent argument is only needed in the case where $\mathrm{char}(F_0)$ divides n. Otherwise the point $(1 : \ldots : 1)$ does not lie on $Y_{n,p}$, and part (h) follows from part (g).

PROOF. (a) Let $b_1, \dots, b_n$ be a basis of E, viewed as an F-vector space. Then

$$a = f_1(b_1 \otimes 1) + \dots + f_n(b_n \otimes 1)$$

for some $f_1, \dots, f_n \in F^{(p)}$, and thus lies in $E \otimes_F F'$, where $F' = F(f_1, \dots, f_n)$.

(b) Working in the basis $b_1 \otimes 1, \dots, b_n \otimes 1$, one readily checks that $1, x, \dots, x^{n-1}$ are linearly dependent over F' if and only if $1, \phi(x), \dots, \phi(x)^{n-1}$ are linearly dependent over $F^{(p)}$. $\square$

We are now ready to prove the implication (1) $\Longrightarrow$ (2) of Theorem 1.3. Applying (1) to the étale algebra $E_n \otimes_{F_n} F_n^{(p)}/F_n^{(p)}$ we see that there exists an element $a \in E_n \otimes_{F_n} F_n^{(p)}$ such that $\mathrm{tr}(a^p) = 0$. By Lemma 3.1(a), this element descends to $E_n \otimes_{F_n} F'$ for some intermediate field $F \subset F' \subset F^{(p)}$ such that $[F' : F]$ is finite (and hence, prime to p). $\square$

REMARK 3.2. Suppose $\phi(a') = a$. By Lemma 3.1(b), if a generates $E_n \otimes_{F_n} F_n^{(p)}$ as an algebra over $F^{(p)}$, then a' generates $E_n \otimes_{F_n} F'$ over F'.

4. Proof of Theorem 1.3: (2) $\Longrightarrow$ (3)

Choose F' and a as in (2), and let $d := [F' : F]$. Then $L' := L_n \otimes_{F_n} F'$ is an S_n-Galois algebra over F' and $E' := (L')^{\mathrm{S}_{n-1}}$ is an étale algebra of degree n over F'.

Let Z be birational model for the S_n-Galois algebra L', i.e., an F_0-variety with a S_n-action, whose F_0-algebra of rational functions $F_0(Z)$ is S_n-equivariantly isomorphic to L'. (Note that Z is not necessarily irreducible. If L' is the direct product of r field extensions of F', then Z has r irreducible components.) The S_n-equivariant inclusion

$$L_n \hookrightarrow L' = L_n \otimes_{F_n} F'$$

gives rise to a dominant S_n-equivariant rational map $Z \dashrightarrow \mathbb{A}^n$ of degree $d = [F' : F]$.

Now the element a gives rise to a S_n-equivariant rational map $f_a \colon Z \dashrightarrow \mathbb{P}^{n-1}$ defined as follows. Choose representatives $h_1, \dots, h_n$ of the left cosets of S_{n-1} in S_n, so that $h_i(1) = i$, and set

$$\begin{array}{cccc} f_a\colon & Z & \dashrightarrow & \mathbb{P}^{n-1} \\ & z & \mapsto & (h_1(a)(z), \dots, h_n(a)(z)) \,. \end{array}$$

Note that $h_1(a) = a, h_2(a), \dots, h_n(a)$ are the conjugates of a in L'. Since $a \in E' := (L')^{\mathrm{S}_{n-1}}$, $h_i(a) \in L'$ depends only on the coset $h_i \, \mathrm{S}_{n-1}$ (i.e., only on i) and not on the particular choice of h_i in this coset. Moreover, $h_1(a)^p + \dots + h_n(a)^p = \mathrm{tr}_{L'/F'}(a^p) = 0$, so the image of f_a lies in the hypersurface $X_{n,p} \subset \mathbb{P}^{n-1}$, given by $x_1^p + \dots + x_n^p = 0$, as in Section 2. In summary, we have the following diagram of S_n-equivariant rational maps:

(4.1)
$$\begin{array}{ccccc} Z & & & & \\ \Big\downarrow{\scriptstyle \text{generically } d:1} & \searrow^{f_a} & & & \\ \mathbb{A}^n & & X_{n,p} & \hookrightarrow & \mathbb{P}^{n-1} \end{array}$$

Note that since Z is only defined up to an S_n-equivariant birational isomorphism, we may assume without loss of generality that Z is projective.

Now consider the abelian subgroup

$$A := (\mathbb{Z}/p\mathbb{Z})^{k_1} \times \ldots \times (\mathbb{Z}/p\mathbb{Z})^{k_m}$$

of S_n. Recall from the statement of Theorem 1.3 that $n := p^{k_1} + \ldots + p^{k_m}$ is the base p presentation of n. We view A as a subgroup of S_n by embedding each factor $(\mathbb{Z}/p\mathbb{Z})^{k_i}$ into $\mathrm{S}_{p^{k_i}}$ via the regular representation. We now observe that that the origin is a smooth A-fixed F_0-point in $\mathbb{A}^n$. (In fact, this point is fixed by all of S_n.) Hence, by the "going up" theorem of J. Kollár and E. Szabó [**RY**$_1$, Proposition A.2], $X_{n,p}$ also has an A-fixed F_0-point[2]. In order to complete the proof of the implication (2) $\Longrightarrow$ (3) of Theorem 1.3, it remains to establish the following lemma.

LEMMA 4.1. *$X_{n,p}$ has an A-fixed point defined over F_0 if and only if equation* (1.1) *has a non-trivial solution in $\mathbb{P}^{m-1}(F_0)$.*

PROOF. An A-fixed point of $\mathbb{P}^{n-1}$ is the same thing as a 1-dimensional A-invariant linear subspace of $\mathbb{A}^n$. To find all 1-dimensional A-invariant linear subspaces, we will decompose the natural representation of A on F_0^n as a direct sum of irreducibles. First decompose F_0^n as a direct sum of A-invariant subspaces

$$F_0^n = F_0^{p^{k_1}} \oplus \ldots \oplus F_0^{p^{k_m}} ,$$

where A acts on $F_0^{p^{k_i}}$ by composing the natural projection $A \to (\mathbb{Z}/p\mathbb{Z})^{k_i}$ with the regular representation of $(\mathbb{Z}/p\mathbb{Z})^{k_i}$. It is natural to label the coordinates of $F_0^{p^{k_i}}$ by the elements $g_1, \ldots, g_{p^{k_i}}$ of $(\mathbb{Z}/p\mathbb{Z})^{k_i}$, rather than by the numbers $1, 2, \ldots, p^{k_i}$. In this notation, $F_0^{p^{k_i}}$ decomposes as a direct product of 1-dimensional invariant subspaces $\mathrm{Span}_{F_0}(R_\chi)$, one for each character $\chi\colon (\mathbb{Z}/p\mathbb{Z})^{k_i} \to F_0^*$, where

$$R_\chi = (\chi(g_1), \ldots, \chi(g_{p^{k_i}})).$$

Note that since we are assuming that $\zeta_p \in F_0$, every character χ and every vector R_χ are defined over F_0. Thus the irreducible decomposition of the natural representation of $A \subset \mathrm{S}_n$ on F_0^n is as follows:

$$F_0^n = V_0 \oplus \left(\bigoplus_{\chi_1} \mathrm{Span}_{F_0}(R_{\chi_1}, 0, \ldots, 0) \right) \oplus \ldots \oplus \left(\bigoplus_{\chi_m} \mathrm{Span}_{F_0}(0, \ldots, 0, R_{\chi_m}) \right) \tag{4.2}$$

where χ_i ranges over the non-trivial characters of $(\mathbb{Z}/p\mathbb{Z})^{k_i} \to F_0^*$. Here

$$V_0 := \{(x_1, \ldots, x_1, x_2, \ldots, x_2, \ldots, x_m, \ldots, x_m) \mid \text{ each } x_i \in F_0 \text{ repeats } p^{k_i} \text{ times}\}$$

is an m-dimensional subspace of F_0^n, where A acts trivially. On the other hand, A acts on the 1-dimensional subspace $\mathrm{Span}_{F_0}(0, \ldots, 0, R_{\chi_i}, 0, \ldots, 0)$ by the character $A \to (\mathbb{Z}/p\mathbb{Z})^{k_i} \xrightarrow{\chi_i} F_0^*$, so the 1-dimensional summands in the sum (4.2) are pairwise non-isomorphic. We conclude that the A-fixed points of $\mathbb{P}^{n-1}$ are either of the form

$$(y_1 : \ldots : y_1 : y_2 : \ldots : y_2 : \ldots : y_m : \ldots : y_m) \in \mathbb{P}(V_0) \tag{4.3}$$

for some $(y_1 : \ldots : y_m) \in \mathbb{P}^{m-1}$ or of the form

$$(0 : \ldots : 0 : R_{\chi_i} : 0 \ldots : 0) \tag{4.4}$$

for some non-trivial character $\chi_i \colon (\mathbb{Z}/p\mathbb{Z})^{k_i} \to F_0^*$.

---

[2]Note that [**RY**$_1$, Proposition A.2] assumes F_0 is algebraically closed. However, in the case where A is a finite abelian group of exponent p, the proof only requires that $\zeta_p \in F_0$; see [**RY**$_1$, Remark A.7].

A point of $\mathbb{P}^{n-1}$ of the form (4.4) has exactly p^{k_i} non-zero coordinates, and each of these non-zero coordinates is a pth root of unity. Hence the sum of the pth powers of the coordinates of this point is p^{k_i}, which is non-zero in F_0. Thus no A-fixed point of $\mathbb{P}^{n-1}$ of the form (4.4) lies on $X_{n,p}$. We conclude that every A-fixed point of $X_{n,p}$ is necessarily of the form (4.3). That is, $X_{n,p}$ has an A-fixed point defined over F_0 if and only if $X_{n,p}$ has an F_0-point of the form (4.3) or, equivalently, if and only if equation (1.1) has a solution in $\mathbb{P}^{m-1}(F_0)$. This completes the proof of Lemma 4.1 and thus of the implication (2) $\Longrightarrow$ (3) of Theorem 1.3. □

5. Proof of Theorem 1.3: (3) $\Longrightarrow$ (1)

In this section we will prove the following

PROPOSITION 5.1. *Let p be a prime, $n \geq 1$ be a positive integer, and $n = p^{k_1} + \cdots + p^{k_m}$ be the base p presentation of n.*

Suppose that F is a p-closed field. Then every n-dimensional étale algebra E/F can be written as $E \simeq E_1 \times \cdots \times E_m$, where each E_i is an étale algebra of degree p^{k_i} over F.

The implication (3) $\Longrightarrow$ (1) is an immediate consequence of this proposition. Indeed, assume that (3) holds. That is, there exists $y = (y_1 : \ldots : y_m) \in \mathbb{P}^{m-1}(F_0)$ such that $p^{k_1}y_1^p + \ldots + p^{k_m}y_m^p = 0$. Let F be a p-closed field and E/F be an étale algebra of degree n. By Proposition 5.1 we may assume that $E = E_1 \times \cdots \times E_m$, where E_i is an étale algebra of degree p^{k_i} over F. Now set $a := (y_1 1_{E_1}, \ldots, y_m 1_{E_m}) \in E$. Then

$$\operatorname{tr}_{E/F}(a^p) = \operatorname{tr}_{E_1/F}(a^p) + \cdots + \operatorname{tr}_{E_m/F}(a^p) = p^{m_1}y_1^p + \cdots + p^{m_k}y_m^p = 0\,,$$

as desired.

It thus remains to prove Proposition 5.1. Let $E = F_1 \times \cdots \times F_r$, where each F_i/F is a field extension of finite degree. Since F is a p-closed field, $[F_i : F] = p^{\mu_i}$ for some integer $\mu_i \geqslant 0$. We want to arrange the fields $F_1, \ldots, F_r$ into non-overlapping groups, $G_1, \ldots, G_m$ so that $E_i := \prod_{F_j \in G_i} F_j$ is an étale algebra of degree p^{k_i} over F. Clearly, the specific fields $F_1, \ldots, F_r$ are not important in this context. We are simply rearranging the integers $p^{\mu_1}, \ldots, p^{\mu_r}$ into m non-overlapping groups, so that the sum of the integers in group i is p^{k_i}. This allows us to restate Proposition 5.1 in a more elementary way, as Lemma 5.2 below. We begin by recalling some definitions.

A partition $\lambda = [\lambda_1, \ldots, \lambda_s]$ of n is an unordered collection of positive integers $\lambda_1, \ldots, \lambda_r$ such that $\lambda_1 + \cdots + \lambda_s = n$. There is a natural partial order on the set of partitions: $\mu = [\mu_1, \ldots, \mu_t] \preceq \lambda = [\lambda_1, \ldots, \lambda_s]$ if μ is obtaned from λ by partitioning each of the numbers $\lambda_1, \ldots, \lambda_n$. Equivalently, $\mu \preceq \lambda$ if $\mu_1, \ldots, \mu_t$ can be arranged into s disjoint groups, so that the sum of the numbers in group i is λ_i. For example, $[3, 2, 1] \preceq [3, 3]$ and $[3, 2, 1] \preceq [4, 2]$ but $[3, 3]$ and $[4, 2]$ are not compatible in this partial order.

Suppose p is a prime. We say that a partition $\lambda = [\lambda_1, \ldots, \lambda_s]$ of n is a p-partition if each λ_i is a power of p. Every integer $n \geqslant 1$ has a unique p-partition, $[p^{k_1}, \ldots, p^{k_m}]$, where each p^k occurs at most $p-1$ times. We shall refer to this partition as *the base p-partition of n* and will denote it by $[n]_p$. Using this terminology, Proposition 5.1 reduces to the following lemma.

LEMMA 5.2. *Let $\mu = [\mu_1, \ldots, \mu_r]$ be a p-partition of n. Then $\mu \preceq [n]_p$.*

PROOF. Let $\mathcal{P}$ be the set of all p-partitions of n that are $\succeq \mu$, and let $\lambda = [\lambda_1, \dots, \lambda_s]$ be a maximal element of $\mathcal{P}$. Note that a maximal element exists because $\mathcal{P}$ is non-empty ($\mu \in \mathcal{P}$) and has finitely many elements. We claim that no prime power p^i occurs in λ more than $p-1$ times. Since $[n]_p$ is the unique p-partition of n with this property, the claim implies that $\lambda = [n]_p$, and the lemma follows.

To prove the claim, assume the contrary, say $\lambda_1 = \dots = \lambda_p = p^i$. Then

$$\lambda = [\underbrace{p^i, \dots, p^i}_{p \text{ times}}, \lambda_{p+1}, \dots, \lambda_r] \prec [p^{i+1}, \lambda_{p+1}, \dots, \lambda_r],$$

contradicting the maximality of λ. This proves the claim, and thus Lemma 5.2, Proposition 5.1 and the implication (3) $\Longrightarrow$ (1) of Theorem 1.3. □

6. Proof of Theorem 1.4

The proof of Theorem 1.4 is largely similar to the proof of Theorem 1.3. We will outline the necessary modifications below.

Once again, the implication (1) $\Longrightarrow$ (2) readily follows from Lemma 3.1(b).

(2) $\Longrightarrow$ (3). Assumption (2) gives rise to a diagram

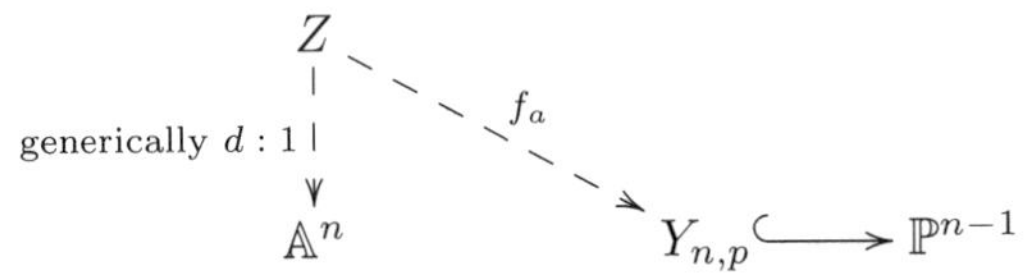

of S_n-equivariant dominant rational maps. Here $d = [F' : F_n]$ is prime to p. The only difference, compared to (4.1), is that we have replaced $X_{n,p}$ by $Y_{n,p}$. The "going up theorem" of Kollár and Szabó tells us that $Y_{n,p}$ has an A-fixed point defined over F_0. Here

$$A := (\mathbb{Z}/p\mathbb{Z})^{k_1} \times \dots \times (\mathbb{Z}/p\mathbb{Z})^{k_m} \subset \mathrm{S}_n,$$

as in Section 4. As we saw in the proof of Lemma 4.1, every A-fixed point of $X_{n,p}$ (and hence, of $Y_{n,p}$) defined over F_0 is of the form

$$(y_1 : \dots : y_1 : y_2 : \dots : y_2 : \dots : y_m : \dots : y_m)$$

for some $(y_1 : \dots : y_m) \in \mathbb{P}^{m-1}(F_0)$. Thus a point of this form has to lie on $Y_{n,p}$. Equivalently, the system (1.2) has a non-trivial solution over F_0, as desired.

(3) $\Longrightarrow$ (1): Let E/F be an étale algebra of degree n, such that $F_0 \subset F$ and F is a p-closed field. By Proposition 5.1, $E \simeq E_1 \times \dots \times E_m$, where each E_i is an étale algebra of degree p^{k_i} over F. Here $n = p^{k_1} + \dots + p^{k_m}$ is the base p expansion of n, as in the statement of Theorem 1.4. By (3), there exists a point $y = (y_1 : \dots : y_m) \in \mathbb{P}^{m-1}(F_0)$ such that

$$\begin{cases} p^{k_1} y_1 + \dots + p^{k_m} y_m = 0 \\ p^{k_1} y_1^p + \dots + p^{k_m} y_m^p = 0 \end{cases}$$

Set

$$a := (y_1 1_{E_1}, \dots y_m 1_{E_m}) \in E. \tag{6.1}$$

Then

$$\mathrm{tr}_{E/F}(a) = \mathrm{tr}_{E_1/F}(a) + \dots + \mathrm{tr}_{E_m/F}(a) = p^{m_1} y_1 + \dots + p^{m_k} y_m = 0$$

and

$$\mathrm{tr}_{E/F}(a^p) = \mathrm{tr}_{E_1/F}(a^p) + \cdots + \mathrm{tr}_{E_m/F}(a^p) = p^{m_1} y_1^p + \cdots + p^{m_k} y_m^p = 0\,,$$

as desired. □

7. Density of rational points on hypersurfaces

Let F be a p-closed field of charcateristic $\neq p$, and X be a smooth irreducible variety over F. J.-L. Colliot-Thélène [**CT**, p. 360] showed that if X has an F-point, then F-points are dense in X. The following variant of this results will play a key role in the next section.

PROPOSITION 7.1. *Let F be a p-closed field of characteristic $\neq p$. Suppose $X \subset \mathbb{P}^l$ is a closed hypersurface of degree $\leq p$ defined over a p-closed field F of characteristic $\neq p$. Assume that X has an F-point c such that X is not a cone over c. Then F-points are dense in X.*

Note that here we do not assume that X is either smooth or irreducible.

PROOF. Case 1: X is a hypersurface of degree $d < p$. Note that effective zero cycles of degree d are dense in X (these can be obtained by intersecting X with lines defined over F in $\mathbb{P}^l$). Since F is a p-closed field, every effective zero cycle of degree $d < p$ splits over F, i.e., is a sum of d F-points. Consequently, F points are dense in X.

Case 2: X is reducible over F, i.e. its irreducible components, $X_1, \ldots, X_r$ are defined over F and $r \geqslant 2$. Here each X_i is a hypersurface of degree $< p$. By Case 1, F-points are dense in each X_i; hence, they are dense in X.

Case 3: X is irreducible over F but reducible over $\overline{F}$. Note that since F is a p-closed field, and $\mathrm{char}(F) \neq p$, F is perfect. Hence, the irreducible components $X_1, \ldots, X_r$ of X are transitively permuted by the Galois group $\mathrm{Gal}(\overline{F}/F)$, which is a pro-$p$ group. Thus $r \geqslant 2$ is a power of p. Moreover, since $\deg(X) = \deg(X_1) + \ldots + \deg(X_r) \leqslant p$, we conclude that $r = p$ and $\deg(X_1) = \ldots = \deg(X_p) = 1$. In other words, X is a union of the hyperplanes $X_1, \ldots, X_p$. Now observe that $c \in X(F)$ is fixed by $\mathrm{Gal}(\overline{F}/F)$. After relabeling the components, we may assume that $c \in X_1$. Translating X_1 by $\mathrm{Gal}(\overline{F}/F)$, we see that c lies on every translate of X_1, i.e., on every X_i for $i = 1, \ldots, r$. Since each X_i is a hyperplane, we conclude that X is a cone over c, contradicting our assumption.

Case 4: X is absolutely irreducible. Choose a hyperplane $H \simeq \mathbb{P}^{l-1}$ in $\mathbb{P}^l$ such that H is defined over F and $c \notin H$. Let $\pi\colon X - \{c\} \to H$ be projection from c. Since X is not a cone over c, this map is dominant. In particular, there is a dense open subset $U \subset H$ such that π is finite over U. The preimage $\pi^{-1}(u)$ of any F-point $u \in U(F)$ is then an effective 0-cycle of degree $\leq p - 1$. Once again, every such 0-cycle splits over F, i.e., $\pi^{-1}(u)$ is a union of F-points. Taking the union of $\pi^{-1}(u)$, as u varies over $U(F)$, we obtain a dense set of F-points in X. □

If $p = 2$ or 3, then Proposition 7.1 remains true for all infinite fields F (not necessarily p-closed), under mild additional assumptions.

LEMMA 7.2. *(a) Let $X \subset \mathbb{P}^l$ be a hypersurface of degree 2 defined over a field F of characteristic $\neq 2$. Assume X has an F-point and X is not a cone. Then X is rational over F; in particular, F-points are dense in X.*

(b) (J. Kollár) Let $X \subset \mathbb{P}^l$ be an absolutely irreducible cubic hypersurface of dimension $\geqslant 2$ defined over a field F. Assume X has an F-point, and X is not a cone. If X is singular, assume also that $\mathrm{char}(F) \neq 2$ *or* 3. *Then X is unirational over F. In particular, if F is infinite, then F-points are dense in X.*

PROOF. (a) Suppose X is given by $q = 0$, where q is a quadratic form on F^{l+1}. Since X is not a cone, q is non-degenerate. Hence, X is irreduicible (and smooth). The stereographic projection from a point $c \in X(F)$ to a hyperplane $H \subset \mathbb{P}^l$ defined over F and not passing through c, gives rise to a birational isomorphism between X and H.

(b) If X is smooth, see [**Ko**, Theorem 1.1]. If X is singular and F is perfect, see [**Ko**, Theorem 1.2]. Finally, if X is singular and F is an imperfect field of characteristic $\neq 2, 3$, see the remark after the statement of Theorem 1.2 in [**Ko**, p. 468]. □

8. Proof of Assertions (∗) and (∗∗)

In this section we complete the proofs of Theorems 1.3 and 1.4 by proving Assertions (∗) and (∗∗), respectively.

Given an étale algebra E/F of degree n, we define $X_{E/F,\,p}$ as the degree p hypersurface in $\mathbb{P}(E) = \mathbb{P}^{n-1}_F$, given by $\mathrm{tr}_{E/F}(x^p) = 0$ and

$$Y_{E/F,\,p} := X_{E/F,p} \cap H\,,$$

where $H \simeq \mathbb{P}^{n-2}_F$ is the hyperplane $\mathrm{tr}_{E/F}(x) = 0$ in $\mathbb{P}(E)$. Let $\Delta_{E/F}$ be the discriminant locus in $\mathbb{P}(E)$, i.e., the closed subvariety of $\mathbb{P}(E)$ defined by the condition $1, a, \dots, a^{n-1}$ are linearly dependent over F. (Here $a \in E$.) Clearly, $X_{E/F,p}$, $Y_{E/F,p}$ and $\Delta_{E/F}$ are F-forms of the varieties $X_{n,p}$, $Y_{n,p}$ and Δ_n defined in Section 2.

We now proceed with the proof of Assertion (∗) of Theorem 1.3. Assume (3) holds. Our goal is to show that a can be chosen so that (i) $E = F[a]$ in part (1), and (ii) $E' = F'[a]$ in part (2). In fact, only (i) needs to be proved; (ii) follows from (i) by Remark 3.2. Thus assertion (∗) can be restated as follows: if $X_{E/F,\,p}$ has an F-point, then $X_{E/F,\,p}$ has an F-point away from $\Delta_{E/F}$. By Lemma 2.1(e), $X_{n,p}$ is not contained in Δ_n; hence, $X_{E/F,\,p}$ is not contained in $\Delta_{E/F}$. Thus in order to prove Assertion (∗) of Theorem 1.3, it suffices to establish the following lemma.

LEMMA 8.1. *Suppose p is a prime, F is a p-closed field, and E/F is an étale algebra of degree $n \geqslant 3$. If $X_{E/F,\,p}$ has an F-point, then F-points are dense in $X_{E/F,\,p}$.*

Now observe that Lemma 8.1 is, in fact, a special case of Proposition 7.1. This completes the proof of Assertion (∗) of Theorem 1.3.

Let us now turn our attention to proving Assertion (∗∗) of Theorem 1.3. Since $Y_{E/F,\,p}$ is an F-form of $Y_{n,p}$, Lemma 2.1(f) tells us that under the assumptions of Theorem 1.4(∗∗), $Y_{E/F,\,p}$ is not contained in $\Delta_{E/F}$. Assuming that condition (3) of Theorem 1.4 holds, we have constructed an F-point of $Y_{E/F,\,p}(F)$; see (6.1). That is, (6.1) gives an F-point $0 \neq a \in E$ whose image $[a]$ in $\mathbb{P}(E)$ lies on $Y_{E/F,\,p}(F)$. We claim that under the assumptions of Theorem 1.4(∗∗), $Y_{E/F}$ is not a cone over $[a]$.

Indeed, if condition (i) of Theorem 1.4(∗∗) holds, i.e., $(y_1 : \ldots : y_m) \neq (1 : \ldots : 1)$ then formula (6.1) tells us that a is not a scalar (i.e., $a \notin F \cdot 1_E$). If $Y_{E/F,p}$ were a cone over $[a]$, then it would remain a cone over $[a]$ after passing to the algebraic closure $\overline{F}$ of F. When we pass from F to $\overline{F}$, E becomes split, i.e., isomorphic to F^n, $\mathbb{P}(E)$ reduces to $\mathbb{P}(F^n) \simeq \mathbb{P}^{n-1}$, $Y_{E/F,p}$ reduces to $Y_{n,p}$, and the condition that a is not a scalar in E translates into $[a] \neq (1 : \ldots : 1)$ in $\mathbb{P}^{n-1}$. By Lemma 2.1(g), $Y_{n,p}$ cannot be a cone over $[a]$, a contradiction. We conclude that $Y_{E/F,p}$ is not a cone over $[a]$, as claimed.

On the other hand, if conditon (ii) of Theorem 1.4(**) holds, i.e., if $p > 2$ and $\mathrm{char}(F_0) \neq 2$, then by Lemma 2.1(h), $Y_{n,p}$ is not a cone over any of its points and hence, neither is $Y_{E/F,p}$. This proves the claim.

Thus in order to prove Assertion (**), it suffices to establish the following lemma.

LEMMA 8.2. *Suppose p is a prime, F is a p-closed field, and E/F is an étale algebra of degree $n \geqslant 3$. If $Y_{E/F,p}$ has an F-point, and $Y_{E/F,p}$ is not a cone over this point, then F-points are dense in $Y_{E/F,p}$.*

Once again, Lemma 8.2 is a special case of Proposition 7.1. This completes the proof of Assertion (**) of Theorem 1.4. □

REMARK 8.3. Since $X_{n,p}$ is smooth and absolutely irreducible (see Lemma 2.1(a) and (c)), so is $X_{E/F,p}$. Hence, we can deduce Lemma 8.1 directly from the result of Colliot-Thélène's mentioned at the beginning of Section 7, without appealing to Proposition 7.1. We do need Proposition 7.1 to deduce Lemma 8.2 though, since $Y_{E/F,p}$ is not smooth in general; see Lemma 2.1(b).

9. Remarks on Theorems 1.3 and 1.4

REMARK 9.1. The requirement that $\mathrm{char}(F_0) \neq p$ is harmless. In characteristic p, $\mathrm{tr}(a^p) = \mathrm{tr}(a)^p$. In this setting the Hermite-Joubert Problem 1.1 amounts to finding an element $0 \neq a \in E$ of trace zero, which is always possible (assuming $n \geqslant 2$).

REMARK 9.2. Condition (1) in either theorem holds for F_0 if and only if it holds after we replace F_0 by $F_0^{(p)}$ (or by any finite extension F_1 such that $[F_1 : F_0]$ is prime to p). In particular, if F_0 does not contain ζ_p, we are free to replace F by $F(\zeta_p)$. Similarly for condition (2). Consequently, the assumption that $\zeta_p \in F_0$ in both theorems can be dropped if we ask that $y_1, \ldots, y_m$ lie in $F_0(\zeta_p)$, rather than F_0, in part (3).

REMARK 9.3. As we pointed out after the statement of Theorem 1.4 in the Introduction, conditions (i) and (ii) are very mild. Nevertheless, these conditions cannot be dropped entirely.

To illustrate this point, we will consider the following example: $n = 6 = 3^1 + 3^1$, $p = 3$ and $\mathrm{char}(F_0) = 2$. In this case the system (1.2) reduces to

$$\begin{cases} 3y_1 + 3y_2 = 0 \\ 3y_1^3 + 3y_2^3 = 0. \end{cases}$$

Conditions (1), (2) and (3) of Theorem 1.4 are satisfied in this example; we can take $a = 1_F$ in part (1), $F' := F_6$ and $a = 1_{F_6}$ in part (2), and $y = (1 : 1)$ in part (3). On the other hand, it is shown in [**Re**3] that no element $a \in E' - F'$ satisfies

$$\mathrm{tr}_{E'/F'}(a) = \mathrm{tr}_{E'/F'}(a^3) = 0$$

in part (2), for any finite extension F'/F_6 of degree prime to 3; see [**Re**$_3$, Theorem 2 and Remark (3) in Section 8]. Thus we cannot choose F' and $a \in E'$ in part (2), so that $E' = F'[a]$. Note that conditions (i) and (ii) of Theorem 1.4(**) both fail here. □

REMARK 9.4. Recall that the symmetric group S_n acts on

$$Y_{n,p} := \{(x_1 : \ldots : x_n) \,|\, x_1 + \ldots + x_n = x_1^p + \ldots + x_n^p = 0\} \subset \mathbb{P}^{n-2}$$

by permuting the variables $x_1, \ldots, x_n$. It turns out that the Hermite-Joubert Problem is related to versality of this action. We will not use this connection in the present paper, but will state it and outline a proof in this remark, for the interested reader. For the definition of various types of versality for group actions on algebraic varieties, see [**DR**, Introduction].

(a) The Hermite-Joubert Problem 1.1 has a positive solution if and only if the S_n-action on $Y_{n,p}$ weakly versal.

(b) The Local Hermite-Joubert Problem 1.2 has a positive solution if and only if the S_n-action on $Y_{n,p}$ is weakly p-versal.

Moreover, assume that $Y_{n,p}$ is not a cone. (This is a mild assumption on n, p, and F_0; see Lemma 2.1(g) and (h).) Then

(c) "weakly p-versal" can be replaced by "p-versal" in part (b).

(d) Furthermore, if $n \geqslant 5$, $p = 3$ and $\mathrm{char}(F_0) \neq 2$, then "weakly versal" can be replaced by "versal" in part (a).

To prove part (a), note that by [**DR**, Theorem 1.1] the S_n-action on $Y_{n,p}$ is weakly versal if and only if the twist ${}^{\tau}Y_{n,p}$ has an F-point for every field extension F/F_0 and every S_n-torsor $\tau\colon T \to \mathrm{Spec}(F)$. It is easy to see that ${}^{\tau}Y_{n,p}$ is precisely, $Y_{E/F,\,p}$, where E/F is the étale algebra corresponding to τ. Thus the S_n-action on $Y_{n,p}$ is weakly versal if and only if $Y_{E/F,p}$ has an F-point for every field extension F/F_0 and every étale algebra E/F, i.e., if and and only if the Hermite-Joubert Problem 1.1 has a positive solution, as claimed. Part (b) is proved by a similar argument, with [**DR**, Lemma 8.2] used in place of [**DR**, Theorem 1.1]. Part (c) is a consequence of Proposition 7.1. Part (d) is a consequence of Lemma 7.2(b). □

10. The Hermite-Joubert problem for $p = 2$

For $p = 2$, Theorems 1.3 and 1.4 can be strengthened to give the following answer to the Hermite-Joubert Problem 1.1.

COROLLARY 10.1. *Let F_0 be a field of characteristic $\neq 2$, and $n = 2^{k_1} + \ldots + 2^{k_m} \geqslant 3$, where the exponents $k_1, \ldots, k_m \geqslant 0$ are distinct integers.*

(a) Conditions (1), (2) and (3) of Theorem 1.3 (with $p = 2$) are equivalent to:

(4) For every field F containing F_0 and every n-dimensional étale algebra E/F, there exists an element $0 \neq a \in E$ such that $\mathrm{tr}_{E/F}(a^2) = 0$.

(b) Moreover, if (4) holds, and F is an infinite field, then the element $a \in E$ in (4) can be chosen so that $E = F[a]$.

(c) Conditions (1), (2) and (3) of Theorem 1.4 (with $p = 2$) are equivalent to:

(4′) For every field F containing F_0 and every n-dimensional étale algebra E/F, there exists an element $0 \neq a \in E$ such that $\mathrm{tr}_{E/F}(a) = \mathrm{tr}_{E/F}(a^2) = 0$.

(d) Moreover, if (4′) holds, $\mathrm{char}(F_0)$ *does not divide* n*, and* F *is an infinite field, then the element* $a \in E$ *in (4′) can be chosen so that* $E = F[a]$*.*

PROOF. By a theorem of Springer, a quadric hypersurface in $\mathbb{P}^l$ defined over a field F of characteristic $\neq 2$, has an F-point if and only if it has an $F^{(2)}$-point; see, e.g., [**Lam**, Theorem VII.2.7] or [**Pf**, Theorem 6.1.12]. Applying this to the hypersurfaces $X_{E/F,2} \subset \mathbb{P}_F^{n-1}$ and $Y_{E/F,2} \subset \mathbb{P}_F^{n-2}$ given by $\mathrm{tr}(x^2) = 0$ and $\mathrm{tr}_{E/F}(x) = \mathrm{tr}_{E/F}(x^2) = 0$, respectively, we see that (1) $\Longleftrightarrow$ (4) in part (a) and (1) $\Longleftrightarrow$ (4′) in part (c).

Proof of part (b). We begin by establishing the following claim. Let E/F be an étale algebra of degree n. Assume $X_{E/F,2}$ has an F-point. Then F-points are dense in $X_{E/F,2}$.

Indeed, by Lemma 2.1(a), $X_{n,2}$ is a smooth quadric hypersurface in $\mathbb{P}^{n-1}$, and hence, so is $X_{E/F,2}$. By Lemma 7.2(a) the existence of an F-point on $X_{E/F,2}$ implies that $X_{E/F,2}$ is rational over F. Since we are assuming that F is infinite, F-points are dense in $X_{E/F,2}$. This proves the claim.

Now observe that by Lemma 2.1(e), $X_{n,2}$ is not contained in Δ_n and thus $X_{E/F,2}$ is not contained in the discriminant locus $\Delta_{E/F}$. The claim tells us that there exists an F-point of $X_{E/F,2}$ away from $\Delta_{E/F}$. This F-point is represented by an element $a \in E$ such that $\mathrm{tr}_{E/F}(a^2) = 0$ and $F[a] = E$, as desired.

Finally, we turn to the proof of part (d). Since we are assuming that $\mathrm{char}(F_0) \neq 2$ and does not divide n, Lemma 2.1(b) tells us that $Y_{n,2}$ is a smooth quadric hypersurface in $\mathbb{P}^{n-2}$, and hence, so is any of its twisted forms $Y_{E/F,2}$. Moreover, by Lemma 2.1(f), $Y_{n,2}$ is not contained in Δ_n and hence, $Y_{E/F,2}$ is not contained in $\Delta_{E/F}$.

Now, arguing as in the proof of part (b) above, we see that if (4′) holds, then $Y_{E/F,2}$ is rational over F and hence, F-points are dense in $Y_{E/F,2}$ (recall that F is assumed to be an infinite field). In particular, there there exists an F-point of $Y_{E/F,2}$ away from the discriminant locus $\Delta_{E/F}$, and part (d) follows. This completes the proof of Corollary 10.1. □

REMARK 10.2. Suppose $p = 2$. Let us arrange the exponents $k_1, \dots, k_m$ in Corollary 10.1 so that $k_1, \dots, k_s$ are even and $k_{s+1}, \dots, k_m$ are odd. (Here $k_1, \dots, k_m$ are distinct non-negative integers; we do not require that $k_1 > \dots > k_m$.) The quadratic form $2^{k_1}y_1^2 + \dots + 2^{k_m}y_m^2$ is then equivalent to $q(z_1, \dots, z_n) = z_1^2 + \dots + z_s^2 + 2(z_{s+1}^2 + \dots + z_m^2)$. Condition (3) of Theorem 1.3 amounts to requiring q to be isotropic over F_0. Condition (3) of Theorem 1.4 is equivalent to saying that q has an isotropic vector in the hyperplane given by

$$2^{k_1/2}z_1 + \dots + 2^{k_s/2}z_s + 2^{(k_{s+1}-1)/2}z_1 + \dots + 2^{(k_m-1)/2}z_m = 0 \qquad (10.1)$$

in $\mathbb{P}^{m-1}$. Note that condition (3) of Theorem 1.3 fails if F_0 is formally real. On the other hand, condition (3) of Theorem 1.4 holds if the Witt index of q is ≥ 2. Indeed, in this case the quadric hypersurface in $\mathbb{P}^{m-1}$ given by $q = 0$ has a line defined over F_0; see [**Lam**, Theorem II.4.3]. Intersecting this line with the hyperplane (10.1) we obtain a desired isotropic vector defined over F_0. □

11. The Hermite-Joubert problem for $p = 3$

Springer's theorem has the following conjectural analogue for $p = 3$.

CONJECTURE 11.1 (J. W. S. Cassels, P. Swinnerton-Dyer [**Co**$_1$, p. 267]). *Let X be a cubic hypersurface in $\mathbb{P}^l$ defined over a field F. If $X(F_1) \neq 0$ for some finite extension F_1/F and $[F_1 : F]$ is prime to 3, then $X(F) \neq \emptyset$. In other words, if X has an $F^{(3)}$-point, then X has an F-point.*

REMARK 11.2. This long-standing conjecture remains largely open. To the best of our knowledge, the partial results proved in the 1976 paper of D. Coray [**Co**$_1$] remain state of the art. One special case, where the conjecture is known (and easy to prove) is the following:

Let X be a cubic hypersurface in $\mathbb{P}^l$ defined over a field F. If $[F_1 : F] = 2$ and $X(F_1) \neq \emptyset$, then $X(F) \neq \emptyset$; see [**Co**$_1$, Proposition 2.2].

REMARK 11.3. If $p = 3$, then the assumption that $\zeta_p \in F_0$ in the statements of Theorems 1.3 and 1.4 can be dropped.

To prove this, let us assume that $\zeta_3 \notin F_0$ and see what happens if we replace F_0 by $F_0(\zeta_3)$. As we explained in Remark 9.2, the validity of conditions (1) and (2) will not change. Since $[F_0(\zeta_3) : F_0] \leqslant 2$, Remark 11.2 tells us that the validity of condition (3) will not change either. □

In view of Corollary 10.1 and Conjecture 11.1, it is natural to expect the following answer to the Hermite-Joubert Problem 1.1 for $p = 3$.

CONJECTURE 11.4. *Let F_0 be a field of characteristic $\neq 3$, $n \geqslant 3$ be an integer, and $n = 3^{k_1} + \ldots + 3^{k_m}$ be the base 3 expansion of n.*

(a) Conditions (1), (2) and (3) of Theorem 1.3 (with $p = 3$) are equivalent to:

(4) For every field F containing F_0 and every n-dimensional étale algebra E/F, there exists an element $0 \neq a \in E$ such that $\mathrm{tr}_{E/F}(a^3) = 0$.

(b) Moreover, if (4) holds, $n \geqslant 4$, and F is an infinite field, then the element $a \in E$ in (4) can be chosen so that $E = F[a]$.

(c) Conditions (1), (2) and (3) of Theorem 1.4 (with $p = 3$) are equivalent to:

(4′) For every field F containing F_0 and every n-dimensional étale algebra E/F, there exists an element $0 \neq a \in E$ such that $\mathrm{tr}_{E/F}(a) = \mathrm{tr}_{E/F}(a^3) = 0$.

(d) Suppose (4′) holds, $n \geqslant 5$, $\mathrm{char}(F_0) \neq 2$, and F is an infinite field. Then the element a in (4′) can be chosen so that $E = F[a]$.

PROPOSITION 11.5. *Conjecture 11.4 follows from Conjecture 11.1.*

PROOF. Recall that by Remark 11.3, for $p = 3$, Theorems 1.3 and 1.4 remain valid even if F_0 does not contain ζ_3. The proof of the equivalences (1) $\Longleftrightarrow$ (4) in part (a) and (1) $\Longleftrightarrow$ (4′) in part (c) is now exactly the same as in Corollary 10.1, with Conjecture 11.1 used in place of Springer's theorem.

To prove part (b), let E/F be an étale algebra of degree n. By (4), $X_{E/F,3}$ has an F-point. By Lemma 2.1, $X_{E/F,3}$ is a smooth absolutely irreducible cubic hypersurface of dimension $n - 2 \geqslant 2$. Since it has an F-point, Lemma 7.2(b) tells us that $X_{E/F,3}$ is unirational over F. Since we are assuming that F is infinite, this implies that F-points are dense in $X_{E/F,3}$. On the other hand, by Lemma 2.1(e), $X_{n,3}$ is not contained in Δ_n. Hence, $X_{E/F,3}$ is not contained in $\Delta_{E/F}$. Therefore, we can find an F-point of $X_{E/F,3}$ away from $\Delta_{E/F}$, and part (b) follows.

We now turn to part (d). Since $n \geqslant 5$, Lemma 2.1(f) tells us that $Y_{n,3}$ is not contained in Δ_n. Hence, $Y_{E/F,3}$ is not contained in $\Delta_{E/F}$. By (4′), $Y_{E/F,3}$ has an F-point. Thus it suffices to show that F-points are dense in $Y_{E/F,3}$.

Since we are assuming that $n \geqslant 5$, Lemma 2.1(e) tells us that $Y_{n,3}$ is an absolutely irreducible cubic hypersurface of dimension $\geqslant 2$, and hence, so is $Y_{E/F,3}$. Moreover, since $\mathrm{char}(F_0) \neq 2$, $Y_{n,3}$ is not a cone by Lemma 2.1(h). Hence, neither is $Y_{E/F}$. Once again, by Lemma 7.2(b), the existence of an F-point on $Y_{E/F,3}$ implies that $Y_{E/F,3}$ is unirational. In particular, F-points are dense in $Y_{E/F,3}$. This completes the proof of Proposition 11.5. □

REMARK 11.6. Let $Z_{m,p}$ and $W_{m,p}$ be the degree p hypersurfaces cut out by the equation (1.1) and the system (1.2) in $\mathbb{P}^{m-1}$ and $\mathbb{P}^{m-2}$, respectively.

If $\zeta_p \in F_0$, it follows from Theorem 1.3 (respectively, Theorem 1.4) that $Z_{m,p}$ (respectively, $W_{m,p}$) has an F_0-point if and only if it has an $F_0^{(p)}$-point. Indeed, as we noted in Remark 9.2, the validity of conditions (1) and (2) does not change when we replace F_0 by $F_0^{(p)}$. Hence, neither does the validity of (3).

In particular, this shows that Conjecture 11.1 is true for the cubic hypersurfaces $Z_{m,3}$ and $W_{m,3}$ defined over F_0. Note also that for $p = 3$ the requirement that $\zeta_3 \in F_0$ can be dropped; see Remark 11.3.

12. When are there solutions to (1.1) and (1.2)?

LEMMA 12.1. *Let F_0 be a field of characteristic $\neq p$. Equation* (1.1) *has a solution $y = (y_1, \dots, y_m) \in \mathbb{P}^{m-1}(F_0)$ if one of the following conditions holds:*

(a) $\sqrt[p]{-p^{k_i-k_j}}$ lies in F_0, for some $1 \leqslant i < j \leqslant m$.

(b) $k_i \equiv k_j \pmod p$ for some $1 \leqslant i < j \leqslant m$ and either p is odd or $p = 2$ and $\sqrt{-1} \in F_0$.

(c) $m \geqslant p+1$, and either p is odd or $p = 2$ and $\sqrt{-1} \in F_0$.

PROOF. (a) Set $y_i := 1$, $y_j := \sqrt[p]{-p^{k_i-k_j}}$, and $y_h = 0$ for every $h \neq i, j$. Then $y = (y_1 : \ldots : y_m)$ is a solution to (1.1).

(b) If $k_i \equiv k_j \pmod p$, then $\sqrt[p]{-p^{k_i-k_j}} \in F_0$.

(c) If $m \geqslant p+1$, then $k_1, \dots, k_m$ cannot all be distinct modulo p, and part (b) applies. □

We will now prove the converse to Lemma 12.1(b) in the case, where $F_0 = \mathbb{Q}(\zeta_p)$ and p is odd.

PROPOSITION 12.2. *Let p be an odd prime and $F_0 = \mathbb{Q}(\zeta_p)$. Then the following conditions are equivalent.*

(a) Equation (1.1) *has no solutions in $\mathbb{P}^{m-1}(F_0)$.*

(b) The integers $k_1, \dots, k_m$ are distinct modulo p.

PROOF. The implication (a) $\Longrightarrow$ (b) follows from Lemma 12.1(b).

(b) $\Longrightarrow$ (a): Assume $(y_1 : \ldots : y_m) \in \mathbb{P}^{m-1}(F_0)$ is a solution to (1.1), i.e.,

$$p^{k_1} y_1^p + \ldots + p^{k_m} y_m^p = 0 \tag{12.1}$$

The p-adic valuation $\nu_p \colon \mathbb{Q}^* \to \mathbb{Z}$ can be extended to $\nu_p \colon \mathbb{Q}(\zeta_p)^* \to \Gamma$, where Γ is an additive subgroup of $\mathbb{Q}$ such that $[\Gamma : \mathbb{Z}] \leqslant p-1$; see [**Lang**, Theorem XII4.1

and Proposition XII.4.2]. In fact, we can take $\Gamma = \dfrac{1}{p-1}\mathbb{Z}$, but we will not need this in the sequel. From (12.1) we see that

$$\nu_p(p^{k_i} y_i^p) = \nu_p(p^{k_j} y_j^p)$$

for some $1 \leqslant i < j \leqslant m$ such that $y_i, y_j \neq 0$. It remains to show that $k_i - k_j$ is divisible by p. Indeed, assume the contrary. Then $k_i + p\nu_p(y_i) = k_j + p\nu_p(y_j)$, and

$$\frac{k_i - k_j}{p} = \nu_p(y_j) - \nu_p(y_i) \in \Gamma\,.$$

Thus $[\Gamma : Z] \geqslant [\frac{1}{p}\mathbb{Z} : \mathbb{Z}] = p$, a contradiction. □

LEMMA 12.3. *Let F_0 be a field of characteristic $\neq p$. The system (1.2) has a solution in $\mathbb{P}^{m-1}(F_0)$ if one of the following conditions holds:*

(a) p is odd and $k_i = k_j$ for some $i \neq j$,

(b) $\sqrt[p]{-p^{k_i - k_j}}$ and $\sqrt[p]{-p^{k_{i'} - k_{j'}}}$ both lie in F_0, for some 4-tuple of distinct integers i, j, i', j' between 1 and m,

(c) $m \geqslant p + 3$, and either p is odd or $p = 2$ and $\sqrt{-1} \in F_0$,

(d) $m \geqslant p + 2$ and $\mathrm{char}(F_0) > 0$.

PROOF. (a) Set $y_i := 1$, $y_j := -1$, and $y_h = 0$ for any $h \neq i, j$. Then

$$y = (y_1 : \ldots : y_m)$$

is a solution to (1.2).

(b) The hypersurface $Z_{m,p} \subset \mathbb{P}^{m-1}$ given by $p^{k_1} y_1^p + \ldots + p^{k_m} y_m^p = 0$, contains the line through $y := (y_1 : \ldots : y_m)$ and $y' := (y'_1 : \ldots : y'_m)$, where $y_i := 1$, $y_j := \sqrt[p]{-p^{k_i - k_j}}$ and $y_h = 0$ for every $h \neq i, j$, and similarly $y'_{i'} := 1$, $y'_{j'} := \sqrt[p]{-p^{k_{i'} - k_{j'}}}$ and $y_{h'} = 0$ for every $h' \neq i', j'$. Intersecting this line with the hyperplane $p^{k_1} y_1 + \ldots + p^{k_m} y_m = 0$, we obtain a solution to (1.2).

(c) Assume $m \geqslant p + 3$. Then there exist $1 \leqslant i < j \leqslant m$ such that $k_i \equiv k_j \pmod p$. Since $m - 2 \geqslant p + 1$, after removing k_i and k_j from the sequence $k_1, \ldots, k_m$, we will find two other distinct subscripts i' and j' such that $k_{i'} \equiv k_{j'} \pmod p$. The desired conclusion now follows from part (b).

(d) Let $\mathbb{F}$ be the prime subfield of F_0. By Chevalley's theorem $\mathbb{F}$ is a C_1-field; see [**Pf**, Theorem 5.2.1]. Note that the coefficients p^{k_i} of the system (1.2) all lie in $\mathbb{F}$. Since we are assuming that $m - 1 > p$, the C_1 property of $\mathbb{F}$ guarantees that the system (1.2) has a solution in $\mathbb{P}^{m-2}(\mathbb{F})$ and hence, in $\mathbb{P}^{m-2}(F_0)$. □

13. Proof of Theorem 1.5

By Theorem 1.4 it suffices to show that the system (1.2) has no non-trivial solutions in $\mathbb{Q}$. (Recall that by Remark 11.3, for $p = 3$, Theorem 1.4 is valid for $F_0 = \mathbb{Q}$, even though $\zeta_3 \notin \mathbb{Q}$.)

We will say that two triples, (k_1, k_2, k_3) and $(k'_1, k'_2, k'_3) \in \mathbb{Z}^3$, are *equivalent* if

$$(k'_1, k'_2, k'_3) = (k_{\sigma(1)} + c, k_{\sigma(2)} + c, k_{\sigma(3)} + c)\,,$$

for some $\sigma \in \mathrm{S}_3$ and $c \in \mathbb{Z}$. For each triple of integers (k_1, k_2, k_3), we would like to know whether or not the system

$$\begin{cases} 3^{k_1}y_1 + 3^{k_2}y_2 + 3^{k_3}y_3 = 0 \\ 3^{k_1}y_1^3 + 3^{k_2}y_2^3 + 3^{k_3}y_3^3 = 0 \end{cases} \tag{13.1}$$

has a non-trivial solution in $\mathbb{Q}$. For the purpose of proving Theorem 1.5, we may replace (k_1, k_2, k_3) by an equivalent triple (k_1', k_2', k_3'). This will cause the system (13.1) to be replaced by an equivalent system. Moreover, $k_1 + k_2 + k_3 \equiv k_1' + k_2' + k_3' \pmod 3$ and if k_1, k_2, k_3 are distinct, then so are k_1', k_2', k_3'.

One easily checks that any triple (k_1, k_2, k_3) with $k_1 + k_2 + k_3 \equiv 0$ or $1 \pmod 3$, is equivalent to some (k_1', k_2', k_3'), where $(k_1', k_2', k_3') \equiv (0,0,0)$, $(0,1,2)$ or $(0,0,1)$ $\pmod 3$. Thus it suffices to show that our system has no non-zero solutions over $\mathbb{Q}$ in each of these three cases.

Case 1: $k_1 = 3e_1$, $k_2 = 3e_2$, $k_3 = 3e_3$, where e_1, e_2, and e_3 are distinct integers. Substituting $z_i := 3^{e_i}y_i$, we obtain

$$\begin{cases} 3^{2e_1}z_1 + 3^{2e_2}z_2 + 3^{2e_3}z_3 = 0 \\ z_1^3 + z_2^3 + z_3^3 = 0. \end{cases}$$

By Fermat's last theorem, the only solutions to the second equation in $\mathbb{P}^2(\mathbb{Q})$ are

$$(1 : -1 : 0),\ (1 : 0 : -1) \text{ and } (0 : 1 : -1).$$

None of them satisfy the first equation.

Case 2: $k_1 = 3e_1$, $k_2 = 3e_2 + 1$, $k_3 = 3e_3 + 2$. In this case equation (1.1) has no non-trivial solutions over $\mathbb{Q}$ by Proposition 12.2. Hence, neither does the system (1.2).

Case 3: $k_1 = 3e_1$, $k_2 = 3e_2$, and $k_3 = 3e_3 + 1$, where $e_1 \neq e_2$. Once again, setting $z_i := 3^{e_i}y_i$, we reduce our system to

$$\begin{cases} 3^{2e_1}z_1 + 3^{2e_2}z_2 + 3^{2e_3+1}z_3 = 0 \\ z_1^3 + z_2^3 + 3z_3^3 = 0. \end{cases}$$

By [**Sel**, Theorem VIII, p. 301], the only solution $(z_1 : z_2 : z_3) \in \mathbb{P}^2(\mathbb{Q})$ to the second equation is $(1 : -1 : 0)$. Since $e_1 \neq e_2$, this point does not satisfy the first equation. This completes the proof of Theorem 1.5. □

14. Beyond Theorem 1.5

CONJECTURE 14.1. *Theorem* 1.5 *remains true for all triples* k_1, k_2, k_3 *of distinct non-negative integers.*

We offer the following partial result in support of Conjecture 14.1.

PROPOSITION 14.2. *Theorem* 1.5 *remains valid for any* $n = 3^{k_1} + 3^{k_2} + 3^{k_3}$ *such that* $k_1 > k_2 > k_3 \geqslant 0$ *and* $k_1 \not\equiv k_2 \pmod 3$.

In particular, the Hermite-Joubert Problem 1.1 (with $p = 3$ and $F_0 = \mathbb{Q}$) has a negative answer for $n = 3^{k_1} + 3^{k_2} + 3^{k_3}$, where $k_1 > k_2 > k_3 \geqslant 0$ and $k_1 \equiv k_3 \equiv 0 \pmod 3$, and $k_2 \equiv 2 \pmod 3$ or alternatively, if $k_2 \equiv k_3 \equiv 0 \pmod 3$ and $k_1 \equiv 2 \pmod 3$. These cases are not covered by Theorem 1.5. The smallest of these new examples is $n = 3^3 + 3^2 + 3^0 = 37$.

Proof of Proposition 14.2. By Theorem 1.4 it suffices to show that the system

$$\begin{cases} 3^{k_1}y_1 + 3^{k_2}y_2 + 3^{k_3}y_3 = 0 \\ 3^{k_1}y_1^3 + 3^{k_2}y_2^3 + 3^{k_3}y_3^3 = 0 \end{cases}$$

does not have a solution $(y_1, y_2, y_3) \neq (0, 0, 0)$ with $y_1, y_2, y_3 \in \mathbb{Q}$. Assume the contrary. After dividing both equations by 3^{k_3}, and replacing k_1, k_2 by $k_1 - k_3$ and $k_2 - k_3$ respectively, we may assume without loss of generality that $k_3 = 0$. Substituting $y_3 = -3^{k_1}y_1 - 3^{k_2}y_2$ into the second equation, we obtain

$$3^{k_1}y_1^3 + 3^{k_2}y_2^3 - 3^{3k_1}y_1^3 - 3^{2k_1+k_2+1}y_1^2y_2 - 3^{k_1+2k_2+1}y_1y_2^2 - 3^{3k_2}y_2^3 = 0\,. \tag{14.1}$$

Clearly $y_1, y_2 \neq 0$. Set

$$\begin{aligned} &M_1 := \nu_3(3^{k_1}y_1^3) = k_1 + 3\nu_3(y_1), \\ &M_2 := \nu_3(3^{k_2}y_2^3) = k_2 + 3\nu_3(y_2), \quad \text{and} \\ &M := \min(M_1, M_2)\,. \end{aligned}$$

Here ν_3 denotes the 3-adic valuation. Since $k_1 \not\equiv k_2 \pmod 3$, we have $M_1 \neq M_2$.

We claim that the 3-adic valuation of each of the last four terms on the left hand side of (14.1) is $> M$. If we manage to prove this claim, then we will be able to conclude that

$$\nu_3(3^{k_1}y_1^3 + 3^{k_2}y_2^3 - 3^{3k_1}y_2^3 - 3^{2k_1+k_2+1}y_1^3y_2 - 3^{k_1+2k_2+1}y_1y_2^2 - 3^{3k_2}y_2^3) = M,$$

contradicting (14.1), and Proposition 14.2 will follow.

To prove the claim, we will consider each term separately:

(i) $\nu_3(3^{3k_1}y_1^3) = 3k_1 + 3\nu_3(y_1) > M_1 \geqslant M$.

(ii) $\nu_3(3^{2k_1+k_2+1}y_1^2y_2) = 2k_1 + k_2 + 2\nu_3(y_1) + \nu_3(y_2) + 1 > \frac{2}{3}M_1 + \frac{1}{3}M_2 > \frac{2}{3}M + \frac{1}{3}M = M$.

(iii) $\nu_3(3^{k_1+2k_2+1}y_1y_2^2) = k_1 + 2k_2 + \nu_3(y_1) + 2\nu_3(y_2) + 1 > \frac{1}{3}M_1 + \frac{2}{3}M_2 > \frac{2}{3}M + \frac{1}{3}M = M$.

(iv) $\nu_3(3^{3k_2}y_2^3) = 3k_2 + 3\nu_3(y_2) > M_2 \geqslant M$.

This completes the proof of the claim and thus of Proposition 14.2. □

Using Proposition 14.2, one readily checks that Conjecture 14.1 follows from Conjecture 14.3 below.

Conjecture 14.3. *Let* $(q_1 : q_2 : q_3)$ *be a* $\mathbb{Q}$*-point of the curve* $C \subset \mathbb{P}^2$ *given by* $x_1^3 + x_2^3 + 9x_3^3 = 0$. *Then* $3^a q_1 + 3^b q_2 + q_3 \neq 0$ *for any integers* $a > b \geq 0$.

Note that if we view C as an elliptic curve with the origin at $(1 : -1 : 0)$, then the group $C(\mathbb{Q})$ of rational points is cyclic, generated by $(1 : 2 : -1)$; see [**Sel**, p. 357].

Acknowledgements

The authors are grateful to M. Bennett, N. Bruin, A. Duncan, S. Gaulhiac, D. Ghioca and B. Poonen for stimulating discussions.

Some of the results of this paper were announced during the Clifford Lectures at Tulane University in March 2015. The second author would like to thank M. Brion and M. Can for inviting him to take part in the Clifford Lectures and to submit a paper for the proceedings.

References

[CT] Jean-Louis Colliot-Thélène, *Rational connectedness and Galois covers of the projective line*, Ann. of Math. (2) **151** (2000), no. 1, 359–373, DOI 10.2307/121121. MR1745009

[EKM] Richard Elman, Nikita Karpenko, and Alexander Merkurjev, *The algebraic and geometric theory of quadratic forms*, American Mathematical Society Colloquium Publications, vol. 56, American Mathematical Society, Providence, RI, 2008. MR2427530

[Co_1] D. F. Coray, *Algebraic points on cubic hypersurfaces*, Acta Arith. **30** (1976), no. 3, 267–296. MR0429731

[Co_2] Daniel F. Coray, *Cubic hypersurfaces and a result of Hermite*, Duke Math. J. **54** (1987), no. 2, 657–670, DOI 10.1215/S0012-7094-87-05428-7. MR899410

[DR] Alexander Duncan and Zinovy Reichstein, *Versality of algebraic group actions and rational points on twisted varieties*, J. Algebraic Geom. **24** (2015), no. 3, 499–530, DOI 10.1090/S1056-3911-2015-00644-0. With an appendix containing a letter from J.-P. Serre. MR3344763

[He] Ch. Hermite, *Sur l'invariant du 18e ordre des formes du cinquième degré et sur le role qu'il joue dans la résolution de l'équation du cinquième degré, extrait de deux lettres de M. Hermite à l'éditeur* (French), J. Reine Angew. Math. **59** (1861), 304–305, DOI 10.1515/crll.1861.59.304. MR1579180

[Jo] P. Joubert, *Sur l'equation du sixième degré*, C-R. Acad. Sc. Paris **64** (1867), 1025-1029.

[Ko] János Kollár, *Unirationality of cubic hypersurfaces*, J. Inst. Math. Jussieu **1** (2002), no. 3, 467–476, DOI 10.1017/S1474748002000117. MR1956057

[Kr] Hanspeter Kraft, *A result of Hermite and equations of degree 5 and 6*, J. Algebra **297** (2006), no. 1, 234–253, DOI 10.1016/j.jalgebra.2005.04.015. MR2206857

[Lam] T. Y. Lam, *Introduction to quadratic forms over fields*, Graduate Studies in Mathematics, vol. 67, American Mathematical Society, Providence, RI, 2005. MR2104929

[Lang] Serge Lang, *Algebra*, 3rd ed., Graduate Texts in Mathematics, vol. 211, Springer-Verlag, New York, 2002. MR1878556

[Pf] Albrecht Pfister, *Quadratic forms with applications to algebraic geometry and topology*, London Mathematical Society Lecture Note Series, vol. 217, Cambridge University Press, Cambridge, 1995. MR1366652

[Re_1] Zinovy Reichstein, *On a theorem of Hermite and Joubert*, Canad. J. Math. **51** (1999), no. 1, 69–95, DOI 10.4153/CJM-1999-005-x. MR1692919

[Re_2] Zinovy Reichstein, *Essential dimension*, Proceedings of the International Congress of Mathematicians. Volume II, Hindustan Book Agency, New Delhi, 2010, pp. 162–188. MR2827790

[Re_3] Zinovy Reichstein, *Joubert's theorem fails in characteristic 2* (English, with English and French summaries), C. R. Math. Acad. Sci. Paris **352** (2014), no. 10, 773–777, DOI 10.1016/j.crma.2014.08.004. MR3262906

[RY_1] Zinovy Reichstein and Boris Youssin, *Essential dimensions of algebraic groups and a resolution theorem for G-varieties*, Canad. J. Math. **52** (2000), no. 5, 1018–1056, DOI 10.4153/CJM-2000-043-5. With an appendix by János Kollár and Endre Szabó. MR1782331

[RY_2] Z. Reichstein and B. Youssin, *Conditions satisfied by characteristic polynomials in fields and division algebras*, J. Pure Appl. Algebra **166** (2002), no. 1-2, 165–189, DOI 10.1016/S0022-4049(01)00009-3. MR1868544

[Sel] Ernst S. Selmer, *The Diophantine equation $ax^3 + by^3 + cz^3 = 0$*, Acta Math. **85** (1951), 203–362 (1 plate). MR0041871

Department of Mathematics, University of British Columbia, Vancouver, Canada
E-mail address: `mbrassil@math.ubc.ca`

Department of Mathematics, University of British Columbia, Vancouver, Canada
E-mail address: `reichst@math.ubc.ca`

Proceedings of Symposia in Pure Mathematics
Volume **94**, 2017
http://dx.doi.org/10.1090/pspum/094/01622

Some structure theorems for algebraic groups

Michel Brion

ABSTRACT. These are extended notes of a course given at Tulane University for the 2015 Clifford Lectures. Their aim is to present structure results for group schemes of finite type over a field, with applications to Picard varieties and automorphism groups.

CONTENTS

2010 *Mathematics Subject Classification.* Primary 14L15, 14L30, 14M17; Secondary 14K05, 14K30, 14M27, 20G15.

1. Introduction

The algebraic groups of the title are the group schemes of finite type over a field; they occur in many questions of algebraic geometry, number theory and representation theory. To analyze their structure, one seeks to build them up from algebraic groups of a specific geometric nature, such as smooth, connected, affine, proper... A first result in this direction asserts that *every algebraic group G has a largest connected normal subgroup scheme G^0, the quotient G/G^0 is finite and étale, and the formation of G^0 commutes with field extensions.* The main goal of this expository text is to prove two more advanced structure results:

THEOREM 1. *Every algebraic group G over a field k has a smallest normal subgroup scheme H such that the quotient G/H is affine. Moreover, H is smooth, connected and contained in the center of G^0; in particular, H is commutative. Also, $\mathcal{O}(H) = k$ and H is the largest subgroup scheme of G satisfying this property. The formation of H commutes with field extensions.*

THEOREM 2. *Every algebraic group G over k has a smallest normal subgroup scheme N such that G/N is proper. Moreover, N is affine and connected. If k is perfect and G is smooth, then N is smooth as well, and its formation commutes with field extensions.*

In particular, every smooth connected algebraic group over a perfect field is an extension of an *abelian variety* (i.e., a smooth connected proper algebraic group) by a smooth connected algebraic group which is affine, or equivalently *linear.* Both building blocks, abelian varieties and linear algebraic groups, have been extensively studied; see e.g. the books [**45**] for the former, and [**8**, **60**] for the latter.

Also, every algebraic group over a field is an extension of a linear algebraic group by an *anti-affine* algebraic group H, i.e., every global regular function on H is constant. Clearly, every abelian variety is anti-affine; but the converse turns out to be incorrect, unless k is algebraic over a finite field (see §5.5). Still, the structure of anti-affine groups over an arbitrary field can be reduced to that of abelian varieties; see [**11**, **56**] and also §5.5 again.

As a consequence, taking for G an anti-affine group which is not an abelian variety, one sees that the natural maps $H \to G/N$ and $N \to G/H$ are generally not isomorphisms with the notation of the above theorems. But when G is smooth and connected, one may combine these theorems to obtain more information on its structure, see §5.1.

The above theorems have a long history. Theorem 1 was first obtained by Rosenlicht in 1956 for smooth connected algebraic groups, see [**52**, Sec. 5]. The version presented here is due to Demazure and Gabriel, see [**23**, III.3.8]. In the setting of smooth connected algebraic groups again, Theorem 2 was announced by

Chevalley in the early 1950's. But he published his proof in 1960 only (see [**18**]), as he had first to build up a theory of Picard and Albanese varieties. Meanwhile, proofs of Chevalley's theorem had been published by Barsotti and Rosenlicht (see [**4**], and [**52**, Sec. 5] again). The present version of Theorem 2 is a variant of a result of Raynaud (see [**51**, IX.2.7]).

The terminology and methods of algebraic geometry have much evolved since the 1950's; this makes the arguments of Barsotti, Chevalley and Rosenlicht rather hard to follow. For this reason, modern proofs of the above results have been made available recently: first, a scheme-theoretic version of Chevalley's proof of his structure theorem by Conrad (see [**19**]); then a version of Rosenlicht's proof for smooth connected algebraic groups over algebraically closed fields (see [**15**, Chap. 2] and also [**44**]).

In this text, we present scheme-theoretic proofs of Theorems 1 and 2, with (hopefully) modest prerequisites. More specifically, we assume familiarity with the contents of Chapters 2 to 5 of the book [**39**], which will be our standard reference for algebraic geometry over an arbitrary field. Also, we do not make an explicit use of sheaves for the fpqc or fppf topology, even if these notions are in the background of several arguments.

To make the exposition more self-contained, we have gathered basic notions and results on group schemes over a field in Section 2, referring to the books [**23**] and [**24**] for most proofs. Section 3 is devoted to the proof of Theorem 1, and Section 4 to that of Theorem 2. Although the statements of both theorems are very similar, the first one is actually much easier. Its proof only needs a few preliminary results: some criteria for an algebraic group to be affine (§3.1), the notion of affinization of a scheme (§3.2) and a version of the rigidity lemma for "anti-affine" schemes (§3.3). In contrast, the proof of Theorem 2 is based on quite a few results on abelian varieties. Some of them are taken from [**45**], which will be our standard reference on that topic; less classical results are presented in §§4.1 and 4.2.

Section 5 contains applications and developments of the above structure theorems, in several directions. We begin with the Rosenlicht decomposition, which reduces somehow the structure of smooth connected algebraic groups to the linear and anti-affine cases (§5.1). We then show in §5.2 that every homogeneous space admits a projective equivariant compactification. §5.3 gathers some known results on the structure of commutative algebraic groups. In §5.4, we provide details on semi-abelian varieties, i.e., algebraic groups obtained as extensions of an abelian variety by a torus; these play an important rôle in various aspects of algebraic and arithmetic geometry. §5.5 is devoted to the classification of anti-affine algebraic groups, based on results from §§5.3 and 5.4. The final §5.6 contains developments on algebraic groups in positive characteristics, including a recent result of Totaro (see [**61**, §2]).

Further applications, of a geometric nature, are presented in Sections 6 and 7. We give a brief overview of the Picard schemes of proper schemes in §6.1, referring to [**35**] for a detailed exposition. §6.2 is devoted to structure results for the Picard variety of a proper variety X, in terms of the geometry of X. Likewise, §7.1 surveys the automorphism group schemes of proper schemes. §7.2 presents a useful descent property for actions of connected algebraic groups. In the final §7.3, based on [**12**], we show that every smooth connected algebraic group over a perfect field is the connected automorphism group of some normal projective variety.

Each section ends with a paragraph of notes and references, which also contains brief presentations of recent work, and some open questions. A general problem, which falls out of the scope of these notes, asks for a version of Theorem 2 in the setting of group schemes over (say) discrete valuation rings. A remarkable analogue of Theorem 1 has been obtained by Raynaud in that setting (see [**24**, VIB.12.10]). But Chevalley's structure theorem admits no direct generalization, as abelian varieties degenerate to tori. So finding a meaningful analogue of that theorem over a ring of formal power series is already an interesting challenge.

Notation and conventions. Throughout this text, we fix a ground field k with algebraic closure $\bar{k}$; the characteristic of k is denoted by $\mathrm{char}(k)$.

We denote by k_s the separable closure of k in $\bar{k}$ and by Γ the Galois group of k_s over k. Also, we denote by k_i the perfect closure of k in $\bar{k}$, i.e., the largest subfield of $\bar{k}$ that is purely inseparable over k. If $\mathrm{char}(k) = 0$ then $k_s = \bar{k}$ and $k_i = k$; if $\mathrm{char}(k) = p > 0$ then $k_i = \bigcup_{n \geq 0} k^{1/p^n}$.

We consider separated schemes over $\mathrm{Spec}(k)$ unless otherwise stated; we will call them k-schemes, or just schemes if this creates no confusion. Morphisms and products of schemes are understood to be over $\mathrm{Spec}(k)$. For any k-scheme X, we denote by $\mathcal{O}(X)$ the k-algebra of global sections of the structure sheaf $\mathcal{O}_X$. Given a field extension K of k, we denote the K-scheme $X \times \mathrm{Spec}(K)$ by X_K.

We identify a scheme X with its *functor of points* that assigns to any scheme S the set $X(S)$ of morphisms $f : S \to X$. When S is affine, i.e., $S = \mathrm{Spec}(R)$ for an algebra R, we also use the notation $X(R)$ for $X(S)$. In particular, we have the set $X(k)$ of *k-rational* points.

A *variety* is a geometrically integral scheme of finite type. The function field of a variety X will be denoted by $k(X)$.

2. Basic notions and results

2.1. Group schemes.

DEFINITION 2.1.1. A *group scheme* is a scheme G equipped with morphisms $m : G \times G \to G$, $i : G \to G$ and with a k-rational point e, which satisfy the following condition:

For any scheme S, the set $G(S)$ is a group with multiplication map $m(S)$, inverse map $i(S)$ and neutral element e.

This condition is equivalent to the commutativity of the following diagrams:

$$\begin{array}{ccc} G \times G \times G & \xrightarrow{m \times \mathrm{id}} & G \times G \\ \Big\downarrow{\scriptstyle \mathrm{id} \times m} & & \Big\downarrow{\scriptstyle m} \\ G \times G & \xrightarrow{\quad m \quad} & G \end{array}$$

(i.e., m is associative),

$$\begin{array}{ccccc} G & \xrightarrow{e \times \mathrm{id}} & G \times G & \xleftarrow{\mathrm{id} \times e} & G \\ & \searrow^{\mathrm{id}} & \Big\downarrow{\scriptstyle m} & \swarrow_{\mathrm{id}} & \\ & & G & & \end{array}$$

(i.e., e is the neutral element), and

$$\begin{array}{ccccc} G & \xrightarrow{\mathrm{id}\times i} & G\times G & \xleftarrow{i\times\mathrm{id}} & G \\ & {\scriptstyle e\circ f}\searrow & \downarrow{\scriptstyle m} & \swarrow{\scriptstyle e\circ f} & \\ & & G & & \end{array}$$

(i.e., i is the inverse map). Here $f : G \to \mathrm{Spec}(k)$ denotes the structure map.

We will write for simplicity $m(x, y) = xy$ and $i(x) = x^{-1}$ for any scheme S and points $x, y \in G(S)$.

REMARKS 2.1.2. (i) For any k-group scheme G, the base change under a field extension K of k yields a K-group scheme G_K.

(ii) The assignement $S \mapsto G(S)$ defines a *group functor*, i.e., a contravariant functor from the category of schemes to that of groups. In fact, the group schemes are exactly those group functors that are representable (by a scheme).

(iii) Some natural group functors are not representable. For example, consider the functor that assigns to any scheme S the group $\mathrm{Pic}(S)$ of isomorphism classes of invertible sheaves on S, and to any morphism of schemes $f : S' \to S$, the pull-back map $f^* : \mathrm{Pic}(S) \to \mathrm{Pic}(S')$. This yields a commutative group functor that we still denote by Pic. For any local ring R, we have $\mathrm{Pic}(\mathrm{Spec}(R)) = 0$. If Pic is represented by a scheme X, then every morphism $\mathrm{Spec}(R) \to X$ is constant for R local; hence every morphism $S \to X$ is locally constant. As a consequence, $\mathrm{Pic}(\mathbb{P}^1) = \mathrm{Hom}(\mathbb{P}^1, X) = 0$, a contradiction.

DEFINITION 2.1.3. Let G be a group scheme. A *subgroup scheme* of G is a (locally closed) subscheme H such that $H(S)$ is a subgroup of $G(S)$ for any scheme S. We say that H is *normal in* G, if $H(S)$ is a normal subgroup of $G(S)$ for any scheme S. We then write $H \trianglelefteq G$.

DEFINITION 2.1.4. Let G, H be group schemes. A morphism $f : G \to H$ is called a *homomorphism* if $f(S) : G(S) \to H(S)$ is a group homomorphism for any scheme S.

The *kernel* of the homomorphism f is the group functor $\mathrm{Ker}(f)$ such that $\mathrm{Ker}(f)(S) = \mathrm{Ker}(f(S) : G(S) \to H(S))$. It is represented by a closed normal subgroup scheme of G, the fiber of f at the neutral element of H.

DEFINITION 2.1.5. An *algebraic group* over k is a k-group scheme of finite type.

This notion of algebraic group is somewhat more general than the classical one. More specifically, the "algebraic groups defined over k" in the sense of [**8**, **60**] are the *geometrically reduced* k-group schemes of finite type. Yet both notions coincide in characteristic 0, as a consequence of the following result of Cartier:

THEOREM 2.1.6. *When* $\mathrm{char}(k) = 0$, *every algebraic group over* k *is reduced.*

PROOF. See [**23**, II.6.1.1] or [**24**, VIB.1.6.1]. A self-contained proof is given in [**45**, p. 101]. □

EXAMPLE 2.1.7. The *additive group* $\mathbb{G}_a$ is the affine line $\mathbb{A}^1$ equipped with the addition. More specifically, we have $m(x, y) = x + y$ and $i(x) = -x$ identically, and $e = 0$.

Consider a subgroup scheme $H \subseteq \mathbb{G}_a$. If $H \neq \mathbb{G}_a$, then H is the zero scheme $V(P)$ for some non-constant polynomial $P \in \mathcal{O}(\mathbb{G}_a) = k[x]$; we may assume that P has leading coefficient 1. We claim that P is an *additive polynomial*, i.e.,

$$P(x+y) = P(x) + P(y)$$

in the polynomial ring $k[x, y]$.

To see this, note that $P(0) = 0$ as $0 \in H(k)$, and

$$P(x+y) \in (P(x), P(y))$$

(the ideal of $k[x, y]$ generated by $P(x)$ and $P(y)$), as the addition $\mathbb{G}_a \times \mathbb{G}_a \to \mathbb{G}_a$ sends $H \times H$ to H. Thus, there exist $A(x, y)$, $B(x, y) \in k[x, y]$ such that

$$P(x+y) - P(x) - P(y) = A(x,y)\, P(x) + B(x,y)\, P(y).$$

Dividing $A(x, y)$ by $P(y)$, we may assume that $\deg_y A(x, y) < \deg(P)$ with an obvious notation. Since $\deg_y(P(x+y) - P(x) - P(y)) < \deg(P)$, it follows that $B = 0$. Likewise, we obtain $A = 0$; this yields the claim.

We now determine the additive polynomials. The derivative of any such polynomial P satisfies $P'(x+y) = P'(x)$, hence P' is constant. When $\mathrm{char}(k) = 0$, we obtain $P(x) = ax$ for some $a \in k$, hence H is just the (reduced) point 0. Alternatively, this follows from Theorem 2.1.6, since $H(\bar{k})$ is a finite subgroup of $(\bar{k}, +)$, and hence is trivial.

When $\mathrm{char}(k) = p > 0$, we obtain $P(x) = a_0 x + P_1(x^p)$, where P_1 is again an additive polynomial. By induction on $\deg(P)$, it follows that

$$P(x) = a_0 x + a_1 x^p + \cdots + a_n x^{p^n}$$

for some positive integer n and $a_0, \ldots, a_n \in k$. As a consequence, $\mathbb{G}_a$ has many subgroup schemes in positive characteristics; for example,

$$\alpha_{p^n} := V(x^{p^n})$$

is a non-reduced subgroup scheme supported at 0.

Note finally that the additive polynomials are exactly the endomorphisms of $\mathbb{G}_a$, and their kernels yield all subgroup schemes of that group scheme (in arbitrary characteristics).

EXAMPLE 2.1.8. The *multiplicative group* $\mathbb{G}_m$ is the punctured affine line $\mathbb{A}^1 \setminus 0$ equipped with the multiplication: we have $m(x, y) = xy$ and $i(x) = x^{-1}$ identically, and $e = 1$.

The subgroup schemes of $\mathbb{G}_m$ turn out to be $\mathbb{G}_m$ and the subschemes

$$\mu_n := V(x^n - 1)$$

of nth roots of unity, where n is a positive integer; these are the kernels of the endomorphisms $x \mapsto x^n$ of $\mathbb{G}_m$. Moreover, μ_n is reduced if and only if n is prime to $\mathrm{char}(k)$.

EXAMPLE 2.1.9. Given a vector space V, the *general linear group* $\mathrm{GL}(V)$ is the group functor that assigns to any scheme S, the automorphism group of the sheaf of $\mathcal{O}_S$-modules $\mathcal{O}_S \otimes_k V$. When V is of finite dimension n, the choice of a basis identifies V with k^n and $\mathrm{GL}(V)(S)$ with $\mathrm{GL}_n(\mathcal{O}(S))$, the group of invertible $n \times n$ matrices with coefficients in the algebra $\mathcal{O}(S)$. It follows that $\mathrm{GL}(V)$ is represented by an open affine subscheme of the affine scheme $\mathbb{A}^{n^2}$ (associated with the linear space of $n \times n$ matrices), the complement of the zero scheme of the determinant. This defines a group scheme GL_n, which is smooth, connected, affine and algebraic.

DEFINITION 2.1.10. A group scheme is *linear* if it is isomorphic to a closed subgroup scheme of GL_n for some positive integer n.

Clearly, every linear group scheme is algebraic and affine. The converse also holds, see Proposition 3.1.1 below.

Some natural classes of group schemes arising from geometry, such as automorphism group schemes and Picard schemes of proper schemes, are generally not algebraic. Yet they turn out to be locally of finite type; this motivates the following:

DEFINITION 2.1.11. A *locally algebraic group* over k is a k-group scheme, locally of finite type.

PROPOSITION 2.1.12. *The following conditions are equivalent for a locally algebraic group G with neutral element e:*

(1) *G is smooth.*
(2) *G is geometrically reduced.*
(3) *$G_{\bar{k}}$ is reduced at e.*

PROOF. Clearly, (1)$\Rightarrow$(2)$\Rightarrow$(3). We now show that (3)$\Rightarrow$(1). For this, we may replace G with $G_{\bar{k}}$ and hence assume that k is algebraically closed.

Observe that for any $g \in G(k)$, the local ring $\mathcal{O}_{G,g}$ is isomorphic to $\mathcal{O}_{G,e}$ as the left multiplication by g in G is an automorphism that sends e to g. It follows that $\mathcal{O}_{G,g}$ is reduced; hence every open subscheme of finite type of G is reduced as well. Since G is locally of finite type, it must be reduced, too. Thus, G contains a smooth k-rational point g. By arguing as above, we conclude that G is smooth. $\square$

2.2. Actions of group schemes.

DEFINITION 2.2.1. An *action* of a group scheme G on a scheme X is a morphism $a : G \times X \to X$ such that the map $a(S)$ yields an action of the group $G(S)$ on the set $X(S)$, for any scheme S.

This condition is equivalent to the commutativity of the following diagrams:

$$\begin{array}{ccc} G \times G \times X & \xrightarrow{m \times \mathrm{id}} & G \times X \\ {\scriptstyle \mathrm{id} \times a}\downarrow & & \downarrow{\scriptstyle a} \\ G \times X & \xrightarrow{\quad a \quad} & X \end{array}$$

(i.e., a is "associative"), and

$$\begin{array}{ccc} X & \xrightarrow{e \times \mathrm{id}} & G \times X \\ & {\scriptstyle \mathrm{id}}\searrow & \downarrow{\scriptstyle a} \\ & & X \end{array}$$

(i.e., the neutral element acts via the identity).

We may view a G-action on X as a homomorphism of group functors

$$a : G \longrightarrow \mathrm{Aut}_X,$$

where Aut_X denotes the *automorphism group functor* that assigns to any scheme S, the group of automorphisms of the S-scheme $X \times S$. The S-points of Aut_X are those morphisms $f : X \times S \to X$ such that the map

$$f \times p_2 : X \times S \longrightarrow X \times S, \quad (x, s) \longmapsto (f(x, s), s)$$

is an automorphism of $X \times S$; they may be viewed as families of automorphisms of X parameterized by S.

DEFINITION 2.2.2. A scheme X equipped with an action a of G will be called a *G-scheme*; we then write for simplicity $a(g, x) = g \cdot x$ for any scheme S and $g \in G(S)$, $x \in X(S)$.

The action is *trivial* if a is the second projection $p_2 : G \times X \to X$; equivalently, $g \cdot x = x$ identically.

REMARK 2.2.3. For an arbitrary action a, we have a commutative triangle

$$\begin{array}{ccc} G \times X & \xrightarrow{u} & G \times X \\ & \searrow^{a} & \downarrow{\scriptstyle p_2} \\ & & X, \end{array}$$

where $u(g, x) := (g, a(g, x))$. Since u is an automorphism (with inverse the map $(g, x) \mapsto (g, a(g^{-1}, x))$), it follows that the morphism a shares many properties of the scheme G. For example, a is always faithfully flat; it is smooth if and only if G is smooth.

In particular, the multiplication $m : G \times G \to G$ is faithfully flat.

DEFINITION 2.2.4. Let X, Y be G-schemes with actions a, b. A *morphism of G-schemes* $\varphi : X \to Y$ is a morphism of schemes such that the following square commutes:

$$\begin{array}{ccc} G \times X & \xrightarrow{a} & X \\ {\scriptstyle \mathrm{id} \times \varphi}\downarrow & & \downarrow{\scriptstyle \varphi} \\ G \times Y & \xrightarrow{b} & Y. \end{array}$$

In other words, $\varphi(g \cdot x) = g \cdot \varphi(x)$ identically; we then say that φ is *G-equivariant*. When Y is equipped with the trivial action of G, we say that φ is *G-invariant*.

DEFINITION 2.2.5. Let X be a G-scheme with action a, and Y a closed subscheme of X.

The *normalizer* (resp. *centralizer*) of Y in G is the group functor $\mathrm{N}_G(Y)$ (resp. $\mathrm{C}_G(Y)$) that associates with any scheme S, the set of those $g \in G(S)$ which induce an automorphism of $Y \times S$ (resp. the identity of $Y \times S$).

The *kernel* of a is the centralizer of X in G, or equivalently, the kernel of the corresponding homomorphism of group functors.

The action a is *faithful* if its kernel is trivial; equivalently, for any scheme S, every non-trivial element of $G(S)$ acts non-trivially on $X \times S$.

The *fixed point functor* of X is the subfunctor X^G that associates with any scheme S, the set of all $x \in X(S)$ such that for any S-scheme S' and any $g \in G(S')$, we have $g \cdot x = x$.

THEOREM 2.2.6. *Let G be a group scheme acting on a scheme X.*

(1) *The normalizer and centralizer of any closed subscheme $Y \subseteq X$ are represented by closed subgroup schemes of G.*
(2) *The functor of fixed points is represented by a closed subscheme of X.*

PROOF. See [**23**, II.1.3.6] or [**24**, VIB.6.2.4]. □

In particular, $\mathrm{N}_G(Y)$ is the largest subgroup scheme of G that acts on Y, and $\mathrm{C}_G(Y)$ is the kernel of this action. Moreover, X^G is the largest subscheme of X on which G acts trivially. We also say that $\mathrm{N}_G(Y)$ *stabilizes* Y, and $\mathrm{C}_G(Y)$ *fixes* Y *pointwise*.

When Y is just a k-rational point x, we have $\mathrm{N}_G(Y) = \mathrm{C}_G(Y) =: \mathrm{C}_G(x)$. This is the *stabilizer* of x in G, which is clearly represented by a closed subgroup scheme of G: the fiber at x of the *orbit map*

$$a_x : G \longrightarrow X, \quad g \longmapsto g \cdot x.$$

We postpone the definition of orbits to §2.7, where homogeneous spaces are introduced; we now record classical properties of the orbit map:

PROPOSITION 2.2.7. *Let G be an algebraic group acting on a scheme of finite type X via a.*

(1) *The image of the orbit map a_x is locally closed for any closed point $x \in X$.*
(2) *If k is algebraically closed and G is smooth, then there exists $x \in X(k)$ such that the image of a_x is closed.*

PROOF. (1) Consider the natural map $\pi : X_{\bar{k}} \to X$. Since π is faithfully flat and quasi-compact, it suffices to show that $\pi^{-1}(a_x(G))$ is locally closed (see e.g. [**31**, IV.2.3.12]). As $\pi^{-1}(a_x(G))$ is the image of the orbit map $(a_x)_{\bar{k}}$, we may assume k algebraically closed. Then $a_x(G)$ is constructible, and hence contains a dense open subset U of its closure. The pull-back $a_x^{-1}(U)$ is a non-empty open subset of the underlying topological space of G; hence that space is covered by the translates $g a_x^{-1}(U)$, where $g \in G(k)$. It follows that $a_x(G)$ is covered by the translates gU, and hence is open in its closure.

(2) Choose a closed G-stable subscheme $Y \subseteq X$ of minimal dimension and let $x \in Y(k)$. If $a_x(G)$ is not closed, then $Z := \overline{a_x(G)} \setminus a_x(G)$ (equipped with its reduced subscheme structure) is a closed subscheme of Y, stable by $G(k)$. Since the normalizer of Z is representable and $G(k)$ is dense in G, it follows that Z is stable by G. But $\dim(Z) < \dim(a_x(G)) \leq \dim(Y)$, a contradiction. □

EXAMPLE 2.2.8. Every group scheme G acts on itself by left multiplication, via

$$\lambda : G \times G \longrightarrow G, \quad (x, y) \longmapsto xy.$$

It also acts by right multiplication, via

$$\rho : G \times G \longrightarrow G, \quad (x, y) \longmapsto yx^{-1}$$

and by conjugation, via

$$\mathrm{Int} : G \times G \longrightarrow G, \quad (x, y) \longmapsto xyx^{-1}.$$

The actions λ and ρ are both faithful. The kernel of Int is the *center of* G.

DEFINITION 2.2.9. Let G, H be two group schemes and $a : G \times H \to H$ an action by group automorphisms, i.e., we have $g \cdot (h_1 h_2) = (g \cdot h_1)(g \cdot h_2)$ identically.

The *semi-direct product* $G \ltimes H$ is the scheme $G \times H$ equipped with the multiplication such that

$$(g,h) \cdot (g',h') = (gg', (g'^{-1} \cdot h)h'),$$

the neutral element $e_G \times e_H$, and the inverse such that $(g,h)^{-1} = (g^{-1}, g \cdot h^{-1})$.

By using the Yoneda lemma, one may readily check that $G \ltimes H$ is a group scheme. Moreover, H (identified with its image in $G \ltimes H$ under the closed immersion $h \mapsto (e_G, h)$) is a closed normal subgroup scheme, and G (identified with its image under the closed immersion $g \mapsto (g, e_H)$) is a closed subgroup scheme having a retraction

$$r : G \ltimes H \longrightarrow G, \quad (g,h) \longmapsto g$$

with kernel H. The given action of G on H is identified with the action by conjugation in $G \ltimes H$.

REMARKS 2.2.10. (i) With the above notation, G is a normal subgroup scheme of $G \ltimes H$ if and only if G acts trivially on H.

(ii) Conversely, consider a group scheme G and two closed subgroup schemes N, H of G such that H normalizes N and the inclusion of H in G admits a retraction $r : G \to H$ which is a homomorphism with kernel N. Form the semi-direct product $H \ltimes N$, where H acts on N by conjugation. Then one may check that the multiplication map

$$H \ltimes N \longrightarrow G, \quad (x,y) \longmapsto xy$$

is an isomorphism of group schemes, with inverse being the morphism

$$G \longrightarrow H \ltimes N, \quad z \longmapsto (r(z), r(z)^{-1}z).$$

2.3. Linear representations.

DEFINITION 2.3.1. Let G be a group scheme and V a vector space. A *linear representation* ρ *of* G *in* V is a homomorphism of group functors $\rho : G \to \mathrm{GL}(V)$. We then say that V is a *G-module.*

More specifically, ρ assigns to any scheme S and any $g \in G(S)$, an automorphism $\rho(g)$ of the sheaf of $\mathcal{O}_S$-modules $\mathcal{O}_S \otimes_k V$, functorially on S. Note that $\rho(g)$ is uniquely determined by its restriction to V (identified with $1 \otimes_k V \subseteq \mathcal{O}(S) \otimes_k V$, where 1 denotes the unit element of the algebra $\mathcal{O}(S)$), which yields a linear map $V \to \mathcal{O}(S) \otimes_k V$.

A linear subspace $W \subseteq V$ is a *G-submodule* if each $\rho(g)$ normalizes $\mathcal{O}_S \otimes_k W$. More generally, the notions of quotients, exact sequences, tensor operations of linear representations of abstract groups extend readily to the setting of group schemes.

EXAMPLES 2.3.2. (i) When $V = k^n$ for some positive integer n, a linear representation of G in V is a homomorphism of group schemes $\rho : G \to \mathrm{GL}_n$ or equivalently, a linear action of G on the affine space $\mathbb{A}^n$.

(ii) Let X be an affine G-scheme with action a. For any scheme S and $g \in G(S)$, we define an automorphism $\rho(g)$ of the $\mathcal{O}_S$-algebra $\mathcal{O}_S \otimes_k \mathcal{O}(X)$ by setting

$$\rho(g)(f) := f \circ a(g^{-1})$$

for any $f \in \mathcal{O}(X)$. This yields a representation ρ of G in $\mathcal{O}(X)$, which uniquely determines the action in view of the anti-equivalence of categories between affine schemes and algebras.

For instance, if G acts linearly on a finite-dimensional vector space V, then $\mathcal{O}(V) \cong \mathrm{Sym}(V^*)$ (the symmetric algebra of the dual vector space) as G-modules.

(iii) More generally, given any G-scheme X, one may define a representation ρ of G in $\mathcal{O}(X)$ as above. But in general, the G-action on X is not uniquely determined by ρ. For instance, if X is a proper G-variety, then $\mathcal{O}(X) = k$ and hence ρ is trivial.

LEMMA 2.3.3. *Let X, Y be quasi-compact schemes. Then the map*

$$\mathcal{O}(X) \otimes_k \mathcal{O}(Y) \longrightarrow \mathcal{O}(X \times Y), \quad f \otimes g \longmapsto ((x,y) \mapsto f(x)\, g(y))$$

is an isomorphism of algebras. In particular, we have a canonical isomorphism

$$\mathcal{O}(X) \otimes_k R \xrightarrow{\cong} \mathcal{O}(X_R)$$

for any quasi-compact scheme X and any algebra R.

PROOF. The assertion is well-known when X and Y are affine.

When X is affine and Y is arbitrary, we may choose a finite open covering $(V_i)_{1 \leq i \leq n}$ of Y; then the intersections $V_i \cap V_j$ are affine as well. Also, we have an exact sequence

$$0 \longrightarrow \mathcal{O}(Y) \longrightarrow \prod_i \mathcal{O}(V_i) \xrightarrow{d_Y} \prod_{i,j} \mathcal{O}(V_i \cap V_j),$$

where $d_Y((f_i)_i) := (f_i|_{V_i \cap V_j} - f_j|_{V_i \cap V_j})_{i,j}$. Tensoring with $\mathcal{O}(X)$ yields an exact sequence

$$0 \longrightarrow \mathcal{O}(X) \otimes_k \mathcal{O}(Y) \longrightarrow \prod_i \mathcal{O}(X \times V_i) \xrightarrow{d_{X,Y}} \prod_{i,j} \mathcal{O}(X \times (V_i \cap V_j)),$$

where $d_{X,Y}$ is defined similarly. Since the $X \times V_i$ form an open covering of $X \times Y$, the kernel of $d_{X,Y}$ is $\mathcal{O}(X \times Y)$; this proves the assertion in this case.

In the general case, we choose a finite open affine covering $(U_i)_{1 \leq i \leq m}$ of X and obtain an exact sequence

$$0 \longrightarrow \mathcal{O}(X) \otimes_k \mathcal{O}(Y) \longrightarrow \prod_i \mathcal{O}(U_i \times Y) \longrightarrow \prod_{i,j} \mathcal{O}((U_i \cap U_j) \times Y),$$

by using the above step. The assertion follows similarly. □

The quasi-compactness assumption in the above lemma is a mild finiteness condition, which is satisfied e.g. for affine or noetherian schemes.

PROPOSITION 2.3.4. *Let G be an algebraic group and X a G-scheme of finite type. Then the G-module $\mathcal{O}(X)$ is the union of its finite-dimensional submodules.*

PROOF. The action map $a : G \times X \to X$ yields a homomorphism of algebras $a^{\#} : \mathcal{O}(X) \to \mathcal{O}(G \times X)$. In view of Lemma 2.3.3, we may view $a^{\#}$ as a homomorphism $\mathcal{O}(X) \to \mathcal{O}(G) \otimes_k \mathcal{O}(X)$. Choose a basis (φ_i) of the vector space $\mathcal{O}(G)$. Then for any $f \in \mathcal{O}(X)$, there exists a family (f_i) of elements of $\mathcal{O}(X)$ such that $f_i \neq 0$ for only finitely many i's, and

$$a^{\#}(f) = \sum_i \varphi_i \otimes f_i.$$

Thus, we have identically

$$\rho(g)(f) = \sum_i \varphi_i(g^{-1})\, f_i.$$

Applying this to the action of G on itself by left multiplication, we obtain the existence of families $(\gamma_{ij})_j$, $(\psi_{ij})_j$ of elements of $\mathcal{O}(G)$ such that $\gamma_{ij} \neq 0$ for only finitely many i's, and

$$\varphi_i(h^{-1}g^{-1}) = \sum_j \gamma_{ij}(g^{-1})\,\psi_{ij}(h^{-1})$$

identically on $G \times G$. It follows that

$$\rho(g)\rho(h)(f) = \sum_{i,j} \gamma_{ij}(g^{-1})\,\psi_{ij}(h^{-1})\,f_i.$$

As a consequence, the span of the f_i's in $\mathcal{O}(G)$ is a finite-dimensional G-submodule, which contains $f = \sum_i \varphi_i(e)\,f_i$. □

PROPOSITION 2.3.5. *Let G be an algebraic group and X an affine G-scheme of finite type. Then there exists a finite-dimensional G-module V and a closed G-equivariant immersion $\iota : X \to V$.*

PROOF. We may choose finitely many generators $f_1, \ldots, f_n$ of the algebra $\mathcal{O}(X)$. By Proposition 2.3.4, each f_i is contained in some finite-dimensional G-submodule $W_i \subseteq \mathcal{O}(X)$. Thus, $W := W_1 + \cdots + W_n$ is a finite-dimensional G-submodule of $\mathcal{O}(X)$, which generates that algebra. This defines a surjective homomorphism of algebras $\mathrm{Sym}(W) \to \mathcal{O}(X)$, equivariant for the natural action of G on $\mathrm{Sym}(W)$. In turn, this yields the desired closed equivariant immersion. □

Examples of linear representations arise from the action of the stabilizer of a k-rational point on its infinitesimal neighborhoods, which we now introduce.

EXAMPLE 2.3.6. Let G be an algebraic group acting on a scheme X via a and let $Y \subseteq X$ be a closed subscheme. For any non-negative integer n, consider the *nth infinitesimal neighborhood $Y_{(n)}$ of Y in X*; this is the closed subscheme of X with ideal sheaf $\mathcal{I}_Y^{n+1}$, where $\mathcal{I}_Y \subseteq \mathcal{O}_X$ denotes the ideal sheaf of Y. The subschemes $Y_{(n)}$ form an increasing sequence, starting with $Y_{(0)} = Y$.

Next, assume that G stabilizes Y. Then $a^{-1}(Y) = p_2^{-1}(Y)$, and hence

$$a^{-1}(\mathcal{I}_Y)\mathcal{O}_{G\times X} = p_2^{-1}(\mathcal{I}_Y)\mathcal{O}_{G\times X}.$$

It follows that

$$a^{-1}(\mathcal{I}_Y^{n+1})\mathcal{O}_{G\times X} = p_2^{-1}(\mathcal{I}_Y^{n+1})\mathcal{O}_{G\times X}.$$

Thus, $a^{-1}(Y_{(n)}) = p_2^{-1}(Y_{(n)})$, i.e., G stabilizes $Y_{(n)}$ as well.

As a consequence, given a (say) locally noetherian G-scheme X equipped with a k-rational point $x = \mathrm{Spec}(\mathcal{O}_{X,x}/\mathfrak{m}_x)$, the algebraic group $\mathrm{C}_G(x)$ acts on each infinitesimal neighborhood $x_{(n)} = \mathrm{Spec}(\mathcal{O}_{X,x}/\mathfrak{m}_x^{n+1})$, which is a finite scheme supported at x. This yields a linear representation ρ_n of G on $\mathcal{O}_{X,x}/\mathfrak{m}_x^{n+1}$ by algebra automorphisms. In particular, $\mathrm{C}_G(x)$ acts linearly on $\mathfrak{m}_x/\mathfrak{m}_x^2$ and hence on the Zariski tangent space, $T_x(X) = (\mathfrak{m}_x/\mathfrak{m}_x^2)^*$.

Applying the above construction to the action of G on itself by conjugation, which fixes the point e, we obtain a linear representation of G in $\mathfrak{g} := T_e(G)$, called the *adjoint representation* and denoted by

$$\mathrm{Ad} : G \longrightarrow \mathrm{GL}(\mathfrak{g}).$$

This yields in turn a linear map

$$\mathrm{ad} := d\mathrm{Ad}_e : \mathfrak{g} \longrightarrow \mathrm{End}(\mathfrak{g})$$

(where the right-hand side denotes the space of endomorphisms of the vector space $\mathfrak{g}$), and hence a bilinear map

$$[\,,\,] : \mathfrak{g} \times \mathfrak{g} \longrightarrow \mathfrak{g}, \quad (x, y) \longmapsto \mathrm{ad}(x)(y).$$

One readily checks that $[x, x] = 0$ identically; also, $[\,,\,]$ satisfies the Jacobi identity (see e.g. [**23**, II.4.4.5]). Thus, $(\mathfrak{g}, [\,,\,])$ is a Lie algebra, called the *Lie algebra of* G; we denote it by $\mathrm{Lie}(G)$.

Denote by T_G the tangent sheaf of G, i.e., the sheaf of derivations of $\mathcal{O}_G$. By [**23**, II.4.4.6], we may also view $\mathrm{Lie}(G)$ as the Lie algebra $H^0(G, T_G)^G = \mathrm{Der}^G(\mathcal{O}_G)$ consisting of those global derivations of $\mathcal{O}_G$ that are invariant under the G-action via right multiplication; this induces an isomorphism

$$T_G \cong \mathcal{O}_G \otimes_k \mathrm{Lie}(G).$$

We have $\dim(G) \leq \dim \mathrm{Lie}(G)$ with equality if and only if G is smooth, as follows from Proposition 2.1.12. Also, every homomorphism of algebraic groups $f : G \to H$ differentiates to a homomorphism of Lie algebras

$$\mathrm{Lie}(f) := df_{e_G} : \mathrm{Lie}(G) \longrightarrow \mathrm{Lie}(H).$$

More generally, every action a of G on a scheme X yields a homomorphism of Lie algebras

$$\mathrm{Lie}(a) : \mathrm{Lie}(G) \longrightarrow H^0(X, T_X) = \mathrm{Der}(\mathcal{O}_X)$$

(see [**23**, II.4.4]).

When $\mathrm{char}(k) = p > 0$, the pth power of any derivation is a derivation; this equips $\mathrm{Lie}(G) = \mathrm{Der}^G(\mathcal{O}_G)$ with an additional structure of *p-Lie algebra*, also called *restricted Lie algebra* (see [**23**, II.7.3]). This structure is preserved by the above homomorphisms.

2.4. The neutral component. Recall that a scheme X is étale (over $\mathrm{Spec}(k)$) if and only if its underlying topological space is discrete and the local rings of X are finite separable extensions of k (see e.g. [**23**, I.4.6.1]). In particular, every étale scheme is locally of finite type. Also, X is étale if and only if the k_s-scheme X_{k_s} is constant; moreover, the assignement $X \mapsto X(k_s)$ yields an equivalence from the category of étale schemes (and morphisms of schemes) to that of discrete topological spaces equipped with a continuous action of the Galois group Γ (and Γ-equivariant maps); see [**23**, I.4.6.2, I.4.6.4].

Next, let X be a scheme, locally of finite type. By [**23**, I.4.6.5], there exists an étale scheme $\pi_0(X)$ and a morphism

$$\gamma = \gamma_X : X \longrightarrow \pi_0(X)$$

such that every morphism of schemes $f : X \to Y$, where Y is étale, factors uniquely through γ. Moreover, γ is faithfully flat and its fibers are exactly the connected components of X. The formation of the *scheme of connected components* $\pi_0(X)$ commutes with field extensions in view of [**23**, I.4.6.7].

As a consequence, given a morphism of schemes $f : X \to Y$, where X and Y are locally of finite type, we obtain a commutative diagram

$$\begin{array}{ccc} X & \xrightarrow{f} & Y \\ {\scriptstyle \gamma_X}\downarrow & & \downarrow{\scriptstyle \gamma_Y} \\ \pi_0(X) & \xrightarrow{\pi_0(f)} & \pi_0(Y), \end{array}$$

where $\pi_0(f)$ is uniquely determined. Applying this construction to the two projections $p_1 : X \times Y \to X$, $p_2 : X \times Y \to Y$, we obtain a canonical morphism $\pi_0(X \times Y) \to \pi_0(X) \times \pi_0(Y)$, which is in fact an isomorphism (see [**23**, I.4.6.10]). In particular, the formation of the scheme of connected components commutes with finite products.

It follows easily that for any locally algebraic group scheme G, there is a unique group scheme structure on $\pi_0(G)$ such that γ is a homomorphism. Moreover, given an action a of G on a scheme X, locally of finite type, we have a compatible action $\pi_0(a)$ of $\pi_0(G)$ on $\pi_0(X)$.

THEOREM 2.4.1. *Let G be a locally algebraic group and denote by G^0 the connected component of e in G.*

(1) *G^0 is the kernel of $\gamma : G \to \pi_0(G)$.*
(2) *The formation of G^0 commutes with field extensions.*
(3) *G^0 is a geometrically irreducible algebraic group.*
(4) *The connected components of G are irreducible, of finite type and of the same dimension.*

PROOF. (1) This holds as the fibers of γ are the connected components of G.

(2) This follows from the fact that the formation of γ commutes with field extensions.

(3) Consider first the case of an algebraically closed field k. Then the reduced neutral component G^0_{red} is smooth by Proposition 2.1.12, and hence locally irreducible. Since G^0_{red} is connected, it is irreducible.

Returning to an arbitrary ground field, G^0 is geometrically irreducible by (2) and the above step. We now show that G^0 is of finite type. Choose a non-empty open subscheme of finite type $U \subseteq G^0$; then U is dense in G^0. Consider the multiplication map of G^0, and its pull-back

$$n : U \times U \longrightarrow G^0.$$

We claim that n is faithfully flat.

Indeed, n is flat by Remark 2.2.3. To show that n is surjective, let $g \in G^0(K)$ for some field extension K of k. Then $U_K \cap g\,i(U_K)$ is non-empty, since G^0_K is irreducible. Thus, there exists a field extension L of k and $x, y \in U(L)$ such that $g = xy^{-1}$. This yields the claim.

By that claim and the quasi-compactness of $U \times U$, we see that G^0 is quasi-compact as well. But G^0 is also locally of finite type; hence it is of finite type.

(4) Let $X \subseteq G$ be a connected component. Since G is locally of finite type, we may choose a closed point $x \in X$; then the residue field $\kappa(x)$ is a finite extension of k. Thus, we may choose a field extension K of $\kappa(x)$, which is finite and stable under $\mathrm{Aut}_k(\kappa(x))$. The structure map $\pi : X_K \to X$ is finite and faithfully flat, hence open and closed; moreover, every point x' of $\pi^{-1}(x)$ is K-rational (as $\kappa(x')$

is a quotient field of $K \otimes_k \kappa(x)$). Thus, the fiber of γ_K at x' is the translate $x'G_K^0$ (since $x'^{-1}\gamma_K^{-1}\gamma_K(x')$ is a connected component of G_K and contains e). As a consequence, $\pi(x'G_K^0)$ is irreducible, open and closed in G, and contains $\pi(x') = x$; so $\pi(x'G_K^0) = X$. This shows that X is irreducible of dimension $\dim(G^0)$. To check that X is of finite type, observe that $X_K = \bigcup_{x' \in \pi^{-1}(x)} x'G_K^0$ is of finite type, and apply descent theory (see [**31**, IV.2.7.1]).

□

With the notation and assumptions of the above theorem, G^0 is called the *neutral component* of G. Note that G is equidimensional of dimension $\dim(G^0)$.

REMARKS 2.4.2. (i) Let G be a locally algebraic group acting on a scheme X, locally of finite type. If k is separably closed, then every connected component of X is stable by G^0.

(ii) A locally algebraic group G is algebraic if and only if $\pi_0(G)$ is finite.

(iii) By [**23**, II.5.1.8], the category of étale group schemes is equivalent to that of discrete topological groups equipped with a continuous action of Γ by group automorphisms, via the assignement $G \mapsto G(k_s)$. Under this equivalence, the finite étale group schemes correspond to the (abstract) finite groups equipped with a Γ-action by group automorphisms.

These results reduce the structure of locally algebraic groups to that of algebraic groups; we will concentrate on the latter in the sequel.

2.5. Reduced subschemes. Recall that every scheme X has a largest reduced subscheme $X_{\rm red}$; moreover, $X_{\rm red}$ is closed in X and has the same underlying topological space. Every morphism of schemes $f : X \to Y$ sends $X_{\rm red}$ to $Y_{\rm red}$.

PROPOSITION 2.5.1. *Let G be a smooth algebraic group acting on a scheme of finite type X.*

1. *$X_{\rm red}$ is stable by G.*
2. *Let $\eta : \tilde{X} \to X_{\rm red}$ denote the normalization. Then there is a unique action of G on $\tilde{X}$ such that η is equivariant.*
3. *When k is separably closed, every irreducible component of $X_{\rm red}$ is stable by G^0.*

PROOF. (1) As G is geometrically reduced, $G \times X_{\rm red}$ is reduced by [**31**, IV.6.8.5]. Thus, $G \times X_{\rm red} = (G \times X)_{\rm red}$ is sent to $X_{\rm red}$ by a.

(2) Likewise, as G is geometrically normal, $G \times \tilde{X}$ is normal by [**31**, IV.6.8.5] again. So the map $\mathrm{id} \times \eta : G \times \tilde{X} \to G \times X$ is the normalization. This yields a morphism $\tilde{a} : G \times \tilde{X} \to \tilde{X}$ such that the square

$$\begin{array}{ccc} G \times \tilde{X} & \xrightarrow{\tilde{a}} & \tilde{X} \\ {\scriptstyle \mathrm{id}\times\eta}\downarrow & & \downarrow{\scriptstyle \eta} \\ G \times X_{\rm red} & \xrightarrow{a} & X_{\rm red}, \end{array} \tag{2.5.1}$$

commutes, where a denotes the G-action. Since η induces an isomorphism on a dense open subscheme of $\tilde{X}$, we have $\tilde{a}(e, \tilde{x}) = \tilde{x}$ identically on $\tilde{X}$. Likewise, $\tilde{a}(g, \tilde{a}(h, \tilde{x})) = \tilde{a}(gh, \tilde{x})$ identically on $G \times G \times \tilde{X}$, i.e., $\tilde{a}$ is an action.

(3) Let Y be an irreducible component of $X_{\rm red}$. Then the normalization $\tilde{Y}$ is a connected component of $\tilde{X}$, and hence is stable by G^0 (Remark 2.4.2 (i)). Using the

surjectivity of the normalization map $\tilde{Y} \to Y$ and the commutative square (2.5.1), it follows that Y is stable by G^0. □

When the field k is perfect, the product of any two reduced schemes is reduced (see [**23**, I.2.4.13]). It follows that the natural map $(X \times Y)_{\mathrm{red}} \to X_{\mathrm{red}} \times Y_{\mathrm{red}}$ is an isomorphism; in particular, the formation of X_{red} commutes with field extensions. This implies easily the following:

PROPOSITION 2.5.2. *Let G be a group scheme over a perfect field k.*

(1) *Any action of G on a scheme X restricts to an action of the closed subgroup scheme G_{red} on X_{red}.*
(2) *If G is locally algebraic, then G_{red} is the largest smooth subgroup scheme of G.*

Note that G_{red} is not necessarily normal in G, as shown by the following:

EXAMPLE 2.5.3. Consider the $\mathbb{G}_m$-action on $\mathbb{G}_a$ by multiplication. If $\mathrm{char}(k) = p$, then every subgroup scheme $\alpha_{p^n} = V(x^{p^n}) \subset \mathbb{G}_a$ is normalized by this action (since x^{p^n} is homogeneous), but not centralized (since $\mathbb{G}_m$ acts non-trivially on $\mathcal{O}(\alpha_{p^n}) = k[x]/(x^{p^n})$). Thus, we may form the corresponding semi-direct product $G := \mathbb{G}_m \ltimes \alpha_{p^n}$. Then G is an algebraic group; moreover, $G_{\mathrm{red}} = \mathbb{G}_m$ is not normal in G by Remark 2.2.10 (i).

To obtain a similar example with G finite, just replace $\mathbb{G}_m$ with its subgroup scheme μ_ℓ of ℓ-th roots of unity, where ℓ is prime to p.

We now obtain a structure result for finite group schemes:

PROPOSITION 2.5.4. *Let G be a finite group scheme over a perfect field k. Then the multiplication map induces an isomorphism $G_{\mathrm{red}} \ltimes G^0 \xrightarrow{\cong} G$.*

PROOF. Consider, more generally, a finite scheme X. We claim that the morphism $\gamma : X \to \pi_0(X)$ restricts to an isomorphism $X_{\mathrm{red}} \cong \pi_0(X)$. To check this, we may assume that X is irreducible; then $X = \mathrm{Spec}(R)$ for some local artinian k-algebra R with residue field K being a finite extension of k. Since k is perfect, K lifts uniquely to a subfield of R, which is clearly the largest subfield of that algebra. Then γ_X is the associated morphism $\mathrm{Spec}(R) \to \mathrm{Spec}(K)$; this yields our claim.

Returning to our finite group scheme G, we obtain an isomorphism of group schemes $G_{\mathrm{red}} \xrightarrow{\cong} \pi_0(G)$ via γ. This yields in turn a retraction of G to G_{red} with kernel G^0. So the desired statement follows from Remark 2.2.10 (ii). □

With the notation and assumptions of the above proposition, G_{red} is a finite étale group scheme, which corresponds to the finite group $G(\bar{k})$ equipped with the action of the Galois group Γ. Also, G^0 is finite and its underlying topological space is just the point e; such a group scheme is called *infinitesimal*. Examples of infinitesimal group schemes include α_{p^n} and μ_{p^n} in characteristic $p > 0$. When $\mathrm{char}(k) = 0$, every infinitesimal group scheme is trivial by Theorem 2.1.6.

Proposition 2.5.4 can be extended to the setting of algebraic groups over perfect fields, see Corollary 2.8.7. But it fails over any imperfect field, as shown by the following example of a finite group scheme G such that G_{red} is not a subgroup scheme:

EXAMPLE 2.5.5. Let k be an imperfect field, i.e., $\mathrm{char}(k) = p > 0$ and $k \neq k^p$. Choose $a \in k \setminus k^p$ and consider the finite subgroup scheme $G \subset \mathbb{G}_a$ defined as the kernel of the additive polynomial $x^{p^2} - ax^p$. Then $G_{\mathrm{red}} = V(x(x^{p(p-1)} - a))$ is smooth at 0 but not everywhere, since $x^{p(p-1)} - a = (x^{p-1} - a^{1/p})^p$ over k_i. So G_{red} admits no group scheme structure in view of Proposition 2.1.12.

2.6. Torsors.

DEFINITION 2.6.1. Let X be a scheme equipped with an action a of a group scheme G, and $f : X \to Y$ a G-invariant morphism of schemes.

We say that f is a *G-torsor over Y* (or a *principal G-bundle over Y*) if it satisfies the following conditions:

(1) f is faithfully flat and quasi-compact.
(2) The square

(2.6.1)
$$\begin{array}{ccc} G \times X & \xrightarrow{a} & X \\ {\scriptstyle p_2}\downarrow & & \downarrow{\scriptstyle f} \\ X & \xrightarrow{f} & Y \end{array}$$

is cartesian.

REMARKS 2.6.2. (i) The condition (2) may be rephrased as follows: for any scheme S and any points $x, y \in X(S)$, we have $f(x) = f(y)$ if and only if there exists $g \in G(S)$ such that $y = g \cdot x$; moreover, such a g is unique. This is the scheme-theoretic version of the notion of principal bundle in topology.

(ii) Consider a group scheme G and a scheme Y. Let G act on $G \times Y$ via left multiplication on itself. Then the projection $p_2 : G \times Y \to Y$ is a G-torsor, called the *trivial G-torsor over Y*.

(iii) One easily checks that a G-torsor $f : X \to Y$ is trivial if and only if f has a section. In particular, a G-torsor X over $\mathrm{Spec}(k)$ is trivial if and only if X has a k-rational point. When G is algebraic, this holds of course if k is algebraically closed, but generally not over an arbitrary field k. Assume for instance that k contains some element t which is not a square, and consider the scheme $X := V(x^2 - t) \subset \mathbb{A}^1$. Then X is normalized by the action of μ_2 on $\mathbb{A}^1$ via multiplication; this yields a non-trivial μ_2-torsor over $\mathrm{Spec}(k)$.

(iv) For any G-torsor $f : X \to Y$, the topology of Y is the quotient of the topology of X by the equivalence relation defined by f (see [**31**, IV.2.3.12]). As a consequence, the assignement $Z \mapsto f^{-1}(Z)$ yields a bijection from the open (resp. closed) subschemes of Y to the open (resp. closed) G-stable subschemes of X.

DEFINITION 2.6.3. Let G be a group scheme acting on a scheme X. A morphism of schemes $f : X \to Y$ is a *categorical quotient* of X by G, if f is G-invariant and every G-invariant morphism of schemes $\varphi : X \to Z$ factors uniquely through f.

In view of its universal property, a categorical quotient is unique up to unique isomorphism.

PROPOSITION 2.6.4. *Let G be an algebraic group, and $f : X \to Y$ be a G-torsor. Then f is a categorical quotient by G.*

PROOF. Consider a G-invariant morphism $\varphi : X \to Z$. Then $\varphi^{-1}(U)$ is an open G-stable subscheme for any open subscheme U of Z. Thus, f restricts to a G-torsor $f_U : \varphi^{-1}(U) \to V$ for some open subscheme $V = V(U)$ of Y. To show that φ factors uniquely through f, it suffices to show that $\varphi_U : \varphi^{-1}(U) \to U$ factors uniquely through f_U for any affine U. Thus, we may assume that Z is affine. Then φ corresponds to a G-invariant homomorphism $\mathcal{O}(Z) \to \mathcal{O}(X)$, i.e., to a homomorphism $\mathcal{O}(Z) \to \mathcal{O}(X)^G$ (the subalgebra of G-invariants in $\mathcal{O}(X)$). So it suffices to check that the map

$$f^{\#} : \mathcal{O}_Y \longrightarrow f_*(\mathcal{O}_X)^G$$

is an isomorphism.

Since f is faithfully flat, it suffices in turn to show that the natural map

$$\mathcal{O}_X = f^*(\mathcal{O}_Y) \to f^*(f_*(\mathcal{O}_X)^G)$$

is an isomorphism. We have canonical isomorphisms

$$f^*(f_*(\mathcal{O}_X)) \cong p_{2*}(a^*(\mathcal{O}_X)) \cong p_{2*}(\mathcal{O}_{G\times X}) \cong \mathcal{O}(G) \otimes_k \mathcal{O}_X,$$

where the first isomorphism follows from the cartesian square (2.6.1) and the faithful flatness of f, and the third isomorphism follows from Lemma 2.3.3. Moreover, the composition of these isomorphisms identifies the G-action on $f^*(f_*(\mathcal{O}_X))$ with that on $\mathcal{O}(G) \otimes_k \mathcal{O}_X$ via left multiplication on $\mathcal{O}(G)$. Thus, taking G-invariants yields the desired isomorphism. □

PROPOSITION 2.6.5. *Let $f : X \to Y$ be a G-torsor.*

(1) *The morphism f is finite (resp. affine, proper, of finite presentation) if and only if so is the scheme G.*
(2) *When Y is of finite type, f is smooth if and only if G is smooth.*

PROOF. (1) This follows from the cartesian diagram (2.6.1) together with descent theory (see [**31**, IV.2.7.1]).

Likewise, (2) follows from [**31**, IV.6.8.3]. □

REMARKS 2.6.6. (i) As a consequence of the above proposition, every torsor $f : X \to Y$ under an algebraic group G is of finite presentation. In particular, f is surjective on $\bar{k}$-rational points, i.e., the induced map $X(\bar{k}) \to Y(\bar{k})$ is surjective. But f is generally not surjective on S-points for an arbitrary scheme S (already for $S = \mathrm{Spec}(k)$). Still, f satisfies the following weaker version of surjectivity:

For any scheme S and any point $y \in Y(S)$, there exists a faithfully flat morphism of finite presentation $\varphi : S' \to S$ and a point $x \in X(S')$ such that $f(x) = y$.

Indeed, viewing y as a morphism $S \to Y$, we may take $S' := X \times_Y S$, $\varphi := p_2$ and $x := p_1$.

(ii) Consider a G-scheme X, a G-invariant morphism of schemes $f : X \to Y$ and a faithfully flat quasi-compact morphism of schemes $v : Y' \to Y$. Form the cartesian square

$$\begin{array}{ccc} X' & \xrightarrow{f'} & Y' \\ {\scriptstyle u}\downarrow & & \downarrow{\scriptstyle v} \\ X & \xrightarrow{f} & Y. \end{array}$$

Then there is a unique action of G on X' such that u is equivariant and f' is invariant. Moreover, f is a G-torsor if and only if f' is a G-torsor. Indeed, this follows again from descent theory, more specifically from [**31**, IV.2.6.4] for the condition (2), and [**31**, IV.2.7.1] for (3).

(iii) In the above setting, f is a G-torsor if and only if the base change f_K is a G_K-torsor for some field extension K of k.

(iv) Consider two G-torsors $f : X \to Y$, $f' : X' \to Y$ and a G-equivariant morphism $\varphi : X \to X'$ of schemes over Y. Then φ is an isomorphism: to check this, one may reduce by descent to the case where f and f' are trivial. Then φ is identified with an endomorphism of the trivial torsor. But every such endomorphism is of the form $(g, y) \mapsto (g\psi(y), y)$ for a unique morphism $\psi : Y \to G$, and hence is an automorphism with inverse $(g, y) \mapsto (g\psi(y)^{-1}, y)$.

EXAMPLE 2.6.7. Let G be an algebraic group. Then $\gamma : G \to \pi_0(G)$ is a G^0-torsor.

Indeed, recall from §2.4 that the formation of γ commutes with field extensions. By Remark 2.6.6 (iii), we may thus assume k algebraically closed. Then the finite étale scheme $\pi_0(G)$ just consists of finitely many k-rational points, say $x_1, \ldots, x_n$, and the fiber F_i of γ at x_i contains a k-rational point, say g_i. Recall that F_i is a connected component of G; thus, the translate $g_i^{-1}F_i$ is a connected component of G through e, and hence equals G^0. It follows that G is the disjoint union of the translates g_iG^0, which are the fibers of γ; this yields our assertion.

2.7. Homogeneous spaces and quotients.

PROPOSITION 2.7.1. *Let $f : G \to H$ be a homomorphism of algebraic groups.*

(1) *The image $f(G)$ is closed in H.*
(2) *f is a closed immersion if and only if its kernel is trivial.*

PROOF. As in the proof of Proposition 2.2.7, we may assume that k is algebraically closed.

(1) Consider the action a of G on H given by $g \cdot h := f(g)h$. By Proposition 2.2.7 again, there exists $h \in H(k)$ such that the image of the orbit map a_h is closed. But $a_h(G) = a_e(G)h$ and hence $a_e(G) = f(G)$ is closed.

(2) Clearly, $\mathrm{Ker}(f)$ is trivial if f is a closed immersion. Conversely, if $\mathrm{Ker}(f)$ is trivial then the fiber of f at any point $x \in X$ consists of that point; in particular, f is quasi-finite. By Zariski's Main Theorem (see [**31**, IV.8.12.6]), f factors as an immersion followed by a finite morphism. As a consequence, there exists a dense open subscheme U of $f(G)$ such that the restriction $f^{-1}(U) \to U$ is finite. Since the translates of $U_{\bar{k}}$ by $G(\bar{k})$ cover $f(G_{\bar{k}})$, it follows that $f_{\bar{k}}$ is finite; hence f is finite as well. Choose an open affine subscheme V of $f(G)$; then so is $f^{-1}(V)$, and $\mathcal{O}(f^{-1}(V))$ is a finite module over $\mathcal{O}(V)$ via $f^{\#}$. Moreover, the natural map

$$\mathcal{O}(V)/\mathfrak{m} \longrightarrow \mathcal{O}(f^{-1}(V)/\mathfrak{m}\mathcal{O}(f^{-1}(V)) = \mathcal{O}(f^{-1}(\mathrm{Spec}\,\mathcal{O}(V)/\mathfrak{m}))$$

is an isomorphism for any maximal ideal $\mathfrak{m}$ of $\mathcal{O}(V)$. By Nakayama's lemma, it follows that $f^{\#}$ is surjective; this yields the assertion. □

As a consequence of the above proposition, *every subgroup scheme of an algebraic group is closed.*

We now come to an important existence result:

THEOREM 2.7.2. *Let G be an algebraic group and $H \subseteq G$ a subgroup scheme.*

(1) *There exists a G-scheme G/H equipped with a G-equivariant morphism*

$$q : G \longrightarrow G/H,$$

which is an H-torsor for the action of H on G by right multiplication.

(2) *The scheme G/H is of finite type. It is smooth if G is smooth.*
(3) *If H is normal in G, then G/H has a unique structure of algebraic group such that q is a homomorphism.*

PROOF. See [**24**, VIA.3.2]. ☐

REMARKS 2.7.3. (i) With the notation and assumptions of the above theorem, q is the categorical quotient of G by H, in view of Proposition 2.6.4. In particular, q is unique up to unique isomorphism; it is called the *quotient morphism.* The *homogeneous space* G/H is equipped with a k-rational point $x := q(e)$, the *base point.* The stabilizer $C_G(x)$ equals H, since it is the fiber of q at x.

(ii) By the universal property of categorical quotients, the homomorphism of algebras $q^{\#} : \mathcal{O}(G/H) \to \mathcal{O}(G)^H$ is an isomorphism.

(iii) The morphism q is faithfully flat and lies in a cartesian diagram

$$\begin{array}{ccc} G \times H & \xrightarrow{n} & G \\ {\scriptstyle p_1}\downarrow & & \downarrow{\scriptstyle q} \\ G & \xrightarrow{q} & G/H, \end{array}$$

where n denotes the restriction of the multiplication $m : G \times G \to G$. Also, q is of finite presentation in view of Proposition 2.6.5.

(iv) Since q is flat and G, H are equidimensional, we see that G/H is equidimensional of dimension $\dim(G) - \dim(H)$.

(v) We have $(G/H)(\bar{k}) = G(\bar{k})/H(\bar{k})$ as follows e.g. from Remark 2.6.6 (i). In particular, if k is perfect (so that G_{red} is a subgroup scheme of G), then the scheme G/G_{red} has a unique $\bar{k}$-rational point. Since that scheme is of finite type, it is finite and local; its base point is its unique k-rational point.

Next, we obtain two further factorization properties of quotient morphisms:

PROPOSITION 2.7.4. *Let $f : G \to H$ be a homomorphism of algebraic groups, $N := \mathrm{Ker}(f)$ and $q : G \to G/N$ the quotient homomorphism. Then there is a unique homomorphism $\iota : G/N \to H$ such that the triangle*

$$\begin{array}{ccc} G & \xrightarrow{f} & H \\ {\scriptstyle q}\downarrow & \nearrow{\scriptstyle \iota} & \\ G/N & & \end{array}$$

commutes. Moreover, ι is an isomorphism onto a subgroup scheme of H.

PROOF. Clearly, f is N-invariant; thus, it factors through a unique morphism $\iota : G/N \to H$ by Theorem 2.7.2. We check that ι is a homomorphism: let S be a scheme and $x, y \in (G/N)(S)$. By Remark 2.6.6, there exist morphisms of schemes $\varphi : T \to S$, $\psi : U \to S$ and points $x_T \in G(T)$, $y_U \in G(U)$ such that $q(x_T) = x$, $q(y_U) = y$. Using the fibered product $S' := T \times_S U$, we thus obtain a morphism $f : S' \to S$ and points $x', y' \in G(S')$ such that $q(x') = x$, $q(y') = y$; then $q(x'y') = xy$. Since $f(x'y') = f(x')f(y')$, we have $\iota(xy) = \iota(x)\iota(y)$. One may check likewise that $\mathrm{Ker}(\iota)$ is trivial. Thus, ι is a closed immersion; hence its image is a subgroup scheme in view of Proposition 2.7.1. ☐

PROPOSITION 2.7.5. *Let G be an algebraic group, X a G-scheme of finite type and $x \in X(k)$. Then the orbit map $a_x : G \to X$, $g \mapsto g \cdot x$ factors through a unique immersion $j_x : G/\mathrm{C}_G(x) \to X$.*

PROOF. See [**23**, III.3.5.2] or [**24**, V.10.1.2]. □

With the above notation and assumptions, we may define the *orbit* of x as the locally closed subscheme of X corresponding to the immersion j_x.

2.8. Exact sequences, isomorphism theorems.

DEFINITION 2.8.1. Let $j : N \to G$ and $q : G \to Q$ be homomorphisms of group schemes. We have an *exact sequence*

$$1 \longrightarrow N \xrightarrow{j} G \xrightarrow{q} Q \longrightarrow 1 \tag{2.8.1}$$

if the following conditions hold:

(1) j induces an isomorphism of N with $\mathrm{Ker}(q)$.
(2) For any scheme S and any $y \in Q(S)$, there exists a faithfully flat morphism $f : S' \to S$ of finite presentation and $x \in G(S')$ such that $q(x) = y$.

Then G is called an *extension* of Q by N.
We say that q is an *isogeny* if N is finite.

REMARKS 2.8.2. (i) The condition (1) holds if and only if the sequence of groups

$$1 \longrightarrow N(S) \xrightarrow{j(S)} G(S) \xrightarrow{q(S)} Q(S)$$

is exact for any scheme S.

(ii) The condition (2) holds whenever q is faithfully flat of finite presentation, as already noted in Remark 2.6.6(i).

(iii) As for exact sequences of abstract groups, one may define the push-forward of the exact sequence (2.8.1) under any homomorphism $N \to N'$, and the pull-back under any homomorphism $Q' \to Q$. Also, exactness is preserved under field extensions.

Next, consider an algebraic group G and a normal subgroup scheme N; then we have an exact sequence

$$1 \longrightarrow N \longrightarrow G \xrightarrow{q} G/N \longrightarrow 1 \tag{2.8.2}$$

by Theorem 2.7.2 and the above remarks. Conversely, given an exact sequence (2.8.1) of algebraic groups, j is a closed immersion and q factors through a closed immersion $\iota : G/N \to Q$ by Proposition 2.7.4. Since q is surjective, ι is an isomorphism; this identifies the exact sequences (2.8.1) and (2.8.2).

As another consequence of Proposition 2.7.4, *the category of commutative algebraic groups is abelian.* Moreover, the above notion of exact sequence coincides with the categorical notion. In this setting, the set of isomorphism classes of extensions of Q by N has a natural structure of commutative group, that we denote by $\mathrm{Ext}^1(Q, N)$.

We now extend some classical isomorphism theorems for abstract groups to the setting of group schemes, in a series of propositions:

PROPOSITION 2.8.3. *Let G be an algebraic group and $N \trianglelefteq G$ a normal subgroup scheme with quotient $q : G \to G/N$. Then the assignement $H \mapsto H/N$ yields a bijective correspondence between the subgroup schemes of G containing N and the subgroup schemes of G/N, with inverse the pull-back. Under this correspondence, the normal subgroup schemes of G containing N correspond to the normal subgroup schemes of G/N.*

PROOF. See [**24**, VIA.5.3.1]. □

PROPOSITION 2.8.4. *Let G be an algebraic group and $N \subseteq H \subseteq G$ subgroup schemes with quotient maps $q_N : G \to G/N$, $q_H : G \to G/H$.*

(1) *There exists a unique morphism $f : G/N \to G/H$ such that the triangle*

$$\begin{array}{ccc} G & \xrightarrow{q_H} & G/H \\ {\scriptstyle q_N}\downarrow & \nearrow_{f} & \\ G/N & & \end{array}$$

commutes. Moreover, f is G-equivariant and faithfully flat of finite presentation. The fiber of f at the base point of G/H is the homogeneous space H/N.

(2) *If N is normal in H, then the action of H on G by right multiplication factors through an action of H/N on G/N that centralizes the action of G. Moreover, f is an H/N-torsor.*

(3) *If H and N are normal in G, then we have an exact sequence*

$$1 \longrightarrow H/N \longrightarrow G/N \xrightarrow{f} G/H \longrightarrow 1.$$

PROOF. (1) The existence of f follows from the fact that q_N is a categorical quotient. To show that f is equivariant, let S be a scheme, $g \in G(S)$ and $y \in (G/N)(S)$. Then there exists a morphism $S' \to S$ and $y' \in G(S')$ such that $q_N(y') = y$. So

$$f(g \cdot y) = f(g \cdot q_N(y') = (f \circ q_N)(gy') = q_H(gy') = g \cdot q_H(y') = g \cdot y.$$

One checks similarly that the fiber of f at the base point x equals H/N.

Next, note that the multiplication map $n : G \times H \to H$ yields a morphism $r : G \times H/N \to G/N$. We claim that the square

$$(2.8.3) \qquad \begin{array}{ccc} G \times H/N & \xrightarrow{r} & G/N \\ {\scriptstyle p_1}\downarrow & & \downarrow{\scriptstyle f} \\ G & \xrightarrow{q_H} & G/H \end{array}$$

is cartesian. The commutativity of this square follows readily from the equivariance of the involved morphisms. Let S be a scheme and $g \in G(S)$, $y \in (G/N)(S)$. Then $q_H(g) = f(y)$ if and only if $f(g^{-1} \cdot y) = q_H(e) = f(x)$, i.e., $g^{-1}y \in (H/N)(S)$. It follows that the map $G \times H/N \to G \times_{G/H} G/N$ is bijective on S-points; this yields the claim.

Since q and p_1 are faithfully flat of finite presentation, the same holds for f in view of the cartesian square (2.8.3).

(2) The existence of the action $G/N \times H/N \to G/N$ follows similarly from the universal property of the quotient $G \times H \to G/N \times H/N$. One may check by lifting points as in the proof of (1) that this action centralizes the G-action. Finally, f is a G-torsor in view of the cartesian square (2.8.3) again.

(3) This follows readily from (1) together with Proposition 2.7.4 (or argue by lifting points to check that f is a homomorphism). □

PROPOSITION 2.8.5. *Let G be an algebraic group, $H \subseteq G$ a subgroup scheme and $N \trianglelefteq G$ a normal subgroup scheme. Consider the semi-direct product $H \ltimes N$, where H acts on N by conjugation.*

(1) *The map*

$$f : H \ltimes N \longrightarrow G, \quad (x, y) \longmapsto xy$$

is a homomorphism with kernel $H \cap N$ identified with a subgroup scheme of $H \ltimes N$ via $x \mapsto (x^{-1}, x)$.

(2) *The image $H \cdot N$ of f is the smallest subgroup scheme of G containing H and N.*

(3) *The natural maps $H/H \cap N \to H \cdot N/N$ and $N/H \cap N \to H \cdot N/H$ are isomorphisms.*

(4) *If H is normal in G, then $H \cdot N$ is normal in G as well.*

PROOF. The assertions (1) and (2) are easily checked.

(3) We have a commutative diagram

$$\begin{array}{ccc} H & \longrightarrow & H \ltimes N/N \\ \downarrow & & \downarrow \\ H/H \cap N & \longrightarrow & H \cdot N/N, \end{array}$$

where the top horizontal arrow is an isomorphism and the vertical arrows are $H \cap N$-torsors. This yields the first isomorphism by using Proposition 2.6.4. The second isomorphism is obtained similarly.

(4) This may be checked as in the proof of Proposition 2.7.4. □

We also record a useful observation:

LEMMA 2.8.6. *Keep the notation and assumptions of the above proposition. If $G = H \cdot N$, then $G(\bar{k}) = H(\bar{k})\, N(\bar{k})$. The converse holds when G/N is smooth.*

PROOF. The first assertion follows e.g. from Remark 2.6.6 (i).

For the converse, consider the quotient homomorphism $q : G \to G/N$: it restricts to a homomorphism $H \to G/N$ with kernel $H \cap N$, and hence factors through a closed immersion $i : H/H \cap N \to G/N$ by Proposition 2.7.4. Since $G(\bar{k}) = H(\bar{k})N(\bar{k})$, we see that i is surjective on $\bar{k}$-rational points. As G/N is smooth, i must be an isomorphism. Thus, $H \cdot N/N = G/N$. By Proposition 2.8.3, we conclude that $H \cdot N = G$. □

We may now obtain the promised generalization of the structure of finite group schemes over a perfect field (Proposition 2.5.4):

COROLLARY 2.8.7. *Let G be an algebraic group over a perfect field k.*

(1) $G = G_{\mathrm{red}} \cdot G^0$.
(2) $G_{\mathrm{red}} \cap G^0 = G^0_{\mathrm{red}}$ *is the smallest subgroup scheme* H *of* G *such that* G/H *is finite.*

PROOF. (1) This follows from Lemma 2.8.6, since $G/G^0 \cong \pi_0(G)$ is smooth and $G(\bar{k}) = G_{\mathrm{red}}(\bar{k})$.

(2) Let $H \subseteq G$ be a subgroup scheme. Since G/H is of finite type, the finiteness of G/H is equivalent to the finiteness of $(G/H)(\bar{k}) = G(\bar{k})/H(\bar{k}) = G(\bar{k})/H_{\mathrm{red}}(\bar{k})$. Thus, G/H_{red} is finite if and only if so is G/H. Likewise, using the finiteness of H/H^0, one may check that G/H is finite if and only if so is G/H^0_{red}. Under these conditions, the homogeneous space $G^0_{\mathrm{red}}/H^0_{\mathrm{red}}$ is finite as well; since it is also smooth and connected, it follows that $G^0_{\mathrm{red}} = H^0_{\mathrm{red}}$, i.e., $G^0_{\mathrm{red}} \subseteq H$.

To complete the proof, it suffices to check that G/G^0_{red} is finite, or equivalently that $G(\bar{k})/G^0(\bar{k})$ is finite. But this follows from the finiteness of G/G^0. □

DEFINITION 2.8.8. An exact sequence of group schemes (2.8.1) is called *split* if $q : G \to Q$ has a section which is a homomorphism.

Any such section s yields an endomorphism $r := s \circ q$ of the group scheme G with kernel N; moreover, r may be viewed as a retraction of G to the image of s, isomorphic to H. By Remark 2.2.10 (ii), this identifies (2.8.1) with the exact sequence

$$1 \longrightarrow N \xrightarrow{\ i\ } H \ltimes N \xrightarrow{\ r\ } H \longrightarrow 1.$$

2.9. The relative Frobenius morphism. Throughout this subsection, we assume that the ground field k has characteristic $p > 0$.

Let X be a k-scheme and n a positive integer. The *nth absolute Frobenius morphism of* X is the endomorphism

$$F^n_X : X \longrightarrow X$$

which is the identity on the underlying topological space and such that the homomorphism of sheaves of algebras $(F^n_X)^{\#} : \mathcal{O}_X \to (F^n_X)_*(\mathcal{O}_X) = \mathcal{O}_X$ is the p^nth power map, $f \mapsto f^{p^n}$.

Clearly, every morphism of k-schemes $f : X \to Y$ lies in a commutative square

$$\begin{array}{ccc} X & \xrightarrow{\ f\ } & Y \\ {\scriptstyle F^n_X}\downarrow & & \downarrow{\scriptstyle F^n_Y} \\ X & \xrightarrow{\ f\ } & Y. \end{array}$$

Note that F^n_X is generally not a morphism of k-schemes, since the p^nth power map is generally not k-linear. To address this, define a k-scheme $X^{(n)}$ by the cartesian square

$$\begin{array}{ccc} X^{(n)} & \longrightarrow & X \\ \downarrow & & \downarrow{\scriptstyle \pi} \\ \mathrm{Spec}(k) & \xrightarrow{\ F^n_k\ } & \mathrm{Spec}(k), \end{array}$$

where π denotes the structure map and $F^n_k := F^n_{\mathrm{Spec}(k)}$ corresponds to the p^nth power map of k. Then F^n_X factors through a unique morphism of k-schemes

$$F^n_{X/k} : X \longrightarrow X^{(n)},$$

the *nth relative Frobenius morphism*. Equivalently, the above cartesian square extends to a commutative diagram

$$\begin{array}{ccc} X & & \\ {\scriptstyle F^n_{X/k}}\downarrow & \searrow^{F^n_X} & \\ X^{(n)} & \longrightarrow & X \\ \downarrow & & \downarrow{\scriptstyle \pi} \\ \operatorname{Spec}(k) & \xrightarrow{F^n_k} & \operatorname{Spec}(k). \end{array}$$

The underlying topological space of $X^{(n)}$ is X again, and the structure sheaf is given by

$$\mathcal{O}_{X^{(n)}}(U) = \mathcal{O}_X(U) \otimes_{F^n} k$$

for any open subset $U \subseteq X$, where the right-hand side denotes the tensor product of $\mathcal{O}_X(U)$ and k over k acting on $\mathcal{O}_X(U)$ via scalar multiplication, and on k via the p^nth power map. Thus, we have in $\mathcal{O}_X(U) \otimes_{F^n} k$

$$tf \otimes u = f \otimes t^{p^n} u$$

for any $f \in \mathcal{O}_X(U)$ and $t, u \in k$. The k-algebra structure on $\mathcal{O}_X(U) \otimes_{F^n} k$ is defined by

$$t(f \otimes u) = f \otimes tu$$

for any such f, t and u. The morphism $F^n_{X/k}$ is again the identity on the underlying topological spaces; the associated homomorphism of sheaves of algebras is the map

$$(F^n_{X/k})^{\#} : \mathcal{O}_X(U) \otimes_{F^n} k \longrightarrow \mathcal{O}_X(U), \quad f \otimes t \longmapsto t f^{p^n}. \tag{2.9.1}$$

Using this description, one readily checks that *the formation of the nth relative Frobenius morphism commutes with field extensions*. Moreover, for any positive integers m, n, we have an isomorphism of schemes

$$(X^{(m)})^{(n)} \cong X^{(m+n)}$$

that identifies the composition $F^n_{X^{(m)}/k} \circ F^m_{X/k}$ with $F^{m+n}_{X/k}$. In particular, $F^n_{X/k}$ may be seen as the nth iterate of the relative Frobenius morphism $F_{X/k}$.

Also, note that *the formation of $F^n_{X/k}$ is compatible with closed subschemes and commutes with finite products*. Specifically, any morphism of k-schemes $f : X \to Y$ induces a morphism of k-schemes $f^{(n)} : X^{(n)} \to Y^{(n)}$ such that the square

$$\begin{array}{ccc} X & \xrightarrow{f} & Y \\ {\scriptstyle F^n_{X/k}}\downarrow & & \downarrow{\scriptstyle F^n_{Y/k}} \\ X^{(n)} & \xrightarrow{f^{(n)}} & Y^{(n)} \end{array}$$

commutes. If f is a closed immersion, then so is $f^{(n)}$. Moreover, for any two schemes X, Y, the map

$$p_1^{(n)} \times p_2^{(n)} : (X \times Y)^{(n)} \longrightarrow X^{(n)} \times Y^{(n)}$$

is an isomorphism (where $p_1 : X \times Y \to X$, $p_2 : X \times Y \to Y$ denote the projections), and the triangle

$$\begin{array}{ccc} X \times Y & \xrightarrow{F^n_{X\times Y/k}} & (X \times Y)^{(n)} \\ & \searrow^{F^n_{X/k} \times F^n_{Y/k}} & \downarrow \cong \\ & & X^{(n)} \times Y^{(n)} \end{array}$$

commutes.

We now record some geometric properties of the relative Frobenius morphism:

LEMMA 2.9.1. *Let X be a scheme of finite type and n a positive integer.*

(1) *The morphism $F^n_{X/k}$ is finite and purely inseparable.*
(2) *The scheme-theoretic image of $F^n_{X/k}$ is geometrically reduced for $n \gg 0$.*

PROOF. (1) Since $F^n_{X/k}$ is the identity on the underlying topological spaces, we may assume that X is affine. Let $R := \mathcal{O}(X)$, then the image of the homomorphism $(F^n_{X/k})^\# : R \otimes_{F^n} k \to R$ is the k-subalgebra kR^{p^n} generated by the p^nth powers. Thus, $F^n_{X/k}$ is integral, and hence finite since R is of finite type. Also, $F^n_{X/k}$ is clearly purely inseparable.

(2) Let $I \subset R$ denote the ideal consisting of nilpotent elements. Since the algebra R is of finite type, there exists a positive integer n_0 such that $f^n = 0$ for all $f \in I$ and all $n \geq n_0$. Choose n_1 such that $p^{n_1} \geq n_0$, then $(F^n_{X/k})^\#$ sends I to 0 for any $n \geq n_1$. Thus, the image of $F^n_{X/k}$ is reduced for $n \gg 0$. Since the formation of $F^n_{X/k}$ commutes with field extensions, this completes the proof. □

PROPOSITION 2.9.2. *Let G be a k-group scheme.*

(1) *There is a unique structure of k-group scheme on $G^{(n)}$ such that $F^n_{G/k}$ is a homomorphism.*
(2) *If G is algebraic, then $\mathrm{Ker}(F^n_{G/k})$ is infinitesimal. Moreover, $G/\mathrm{Ker}(F^n_{G/k})$ is smooth for $n \gg 0$.*

PROOF. (1) This follows from the fact that the formation of the relative Frobenius morphism commutes with finite products.

(2) This is a consequence of the above lemma together with Proposition 2.1.12. □

Notes and references.

Most of the notions and results presented in this section can be found in [**23**] and [**24**] in a much greater generality. We provide some specific references:

Proposition 2.1.12 is taken from [**24**, VIA.1.3.1]; Proposition 2.2.7 follows from results in [**23**, II.5.3]; Lemma 2.3.3 is a special case of [**23**, I.2.2.6]; Theorem 2.4.1 follows from [**23**, II.5.1.1, II.5.1.8]; Proposition 2.5.4 holds more generally for locally algebraic groups, see [**23**, II.2.2.4]; Example 2.5.5 is in [**24**, VIA.1.3.2].

Our definition of torsors is somewhat ad hoc: what we call G-torsors over Y should be called G_Y-torsors, where G_Y denotes the group scheme $p_2 : G \times Y \to Y$ (see [**23**, III.4.1] for general notions and results on torsors).

Proposition 2.6.4 is a special case of a result of Mumford, see [**46**, Prop. 0.1]; Proposition 2.7.1 is a consequence of [**23**, II.5.5.1]; Proposition 2.7.4 is a special case of [**24**, VIA.5.4.1].

Theorem 2.7.2 (on the existence of homogeneous spaces) is a deep result, since no direct construction of these spaces is known in this generality. In the setting of affine algebraic groups, homogeneous spaces may be constructed by a method of Chevalley; this is developed in [**23**, III.3.5].

Propositions 2.8.4 and 2.8.5 are closely related to results in [**24**, VIA.5.3]. We have provided additional details to be used later.

Proposition 2.9.2 (2) holds more generally for locally algebraic groups, see [**24**, VII.8.3].

Many interesting extensions of algebraic groups are not split, but quite a few of them turn out to be *quasi-split*, i.e., split after pull-back by some isogeny. For example, the extension

$$1 \longrightarrow G^0 \longrightarrow G \longrightarrow \pi_0(G) \longrightarrow 1$$

is quasi-split for any algebraic group G (see [**9**, Lem. 5.11] when G is smooth and k is algebraically closed of characteristic 0; the general case follows from [**14**, Thm. 1.1]). Further instances of quasi-split extensions will be obtained in Theorems 4.2.5, 5.3.1 and 5.6.3 below. On the other hand, the group G of upper triangular unipotent 3×3 matrices lies in an extension

$$1 \longrightarrow \mathbb{G}_a \longrightarrow G \longrightarrow \mathbb{G}_a^2 \longrightarrow 1,$$

which is not quasi-split. It would be interesting to determine those classes of algebraic groups that yield quasi-split extensions.

3. Proof of Theorem 1

3.1. Affine algebraic groups. In this subsection, we obtain several criteria for an algebraic group to be affine, which will be used throughout the sequel. We begin with a classical result:

PROPOSITION 3.1.1. *Every affine algebraic group is linear.*

PROOF. Let G be an affine algebraic group. By Proposition 2.3.5, there exist a finite-dimensional G-module V and a closed G-equivariant immersion $\iota : G \to V$, where G acts on itself by left multiplication. Since the latter action is faithful, the G-action on V is faithful as well. In other words, the corresponding homomorphism $\rho : G \to \mathrm{GL}(V)$ has a trivial kernel. By Proposition 2.7.1, it follows that ρ is a closed immersion. □

Next, we relate the affineness of algebraic groups with that of subgroup schemes and quotients:

PROPOSITION 3.1.2. *Let H be a subgroup scheme of an algebraic group G.*

(1) *If H and G/H are both affine, then G is affine as well.*
(2) *If G is affine, then H is affine. If in addition $H \trianglelefteq G$, then G/H is affine.*

PROOF. (1) Since H is affine, the quotient morphism $q : G \to G/H$ is affine as well, in view of Proposition 2.6.5 and Theorem 2.7.2 (3). This yields the statement.

(2) The first assertion follows from the closedness of H in G (Proposition 2.7.1). The second assertion is proved in [**23**, III.3.7.3], see also [**24**, VIB.11.7]. □

REMARK 3.1.3. With the notation and assumptions of the above proposition, G is smooth (resp. proper, finite) if H and G/H are both smooth (resp. proper, finite), as follows from the same argument. Also, G is connected if H and G/H are both connected; since all these schemes have k-rational points, this is equivalent to geometric connectedness.

The above proposition yields that every algebraic group has an "affine radical":

LEMMA 3.1.4. *Let G be an algebraic group.*

(1) *G has a largest smooth connected normal affine subgroup scheme, $L(G)$.*
(2) *$L(G/L(G))$ is trivial.*
(3) *The formation of $L(G)$ commutes with separable algebraic field extensions.*

PROOF. (1) Let L_1, L_2 be two smooth connected normal affine subgroup schemes of G. Then the product $L_1 \cdot L_2 \subseteq G$ is a normal subgroup scheme by Proposition 2.8.5. Since $L_1 \cdot L_2$ is a quotient of $L_1 \ltimes L_2$, it is smooth and connected. Also, by using the isomorphism $L_1 \cdot L_2/L_1 \cong L_2/L_1 \cap L_2$ together with Proposition 3.1.2, we see that $L_1 \cdot L_2$ is affine.

Next, take L_1 as above and of maximal dimension. Then $\dim(L_1 \cdot L_2/L_1) = 0$ by Proposition 2.7.4. Since $L_1 \cdot L_2/L_1$ is smooth and connected, it must be trivial. It follows that $L_2 \subseteq L_1$; this proves the assertion.

(2) Denote by $M \subseteq G$ the pull-back of $L(G/L(G))$ under the quotient map $G \to G/L(G)$. By Proposition 2.8.3, M is a normal subgroup scheme of G containing $L(G)$. Moreover, M is affine, smooth and connected, since so are $L(G)$ and $M/L(G)$. Thus, $M = L(G)$; this yields the assertion by Proposition 2.8.3 again.

(3) This follows from a classical argument of Galois descent, see [**57**, V.22]. More specifically, it suffices to check that the formation of $L(G)$ commutes with Galois extensions. Let K be such an extension of k, and $\mathcal{G}$ the Galois group. Then $\mathcal{G}$ acts on $G_K = G \times \operatorname{Spec}(K)$ via its action on K. Let $L' := L(G_K)$; then for any $\gamma \in \mathcal{G}$, the image $\gamma(L')$ is also a smooth connected affine normal K-subgroup scheme of G_K. Thus, $\gamma(L') \subseteq L'$. Since this also holds for γ^{-1}, we obtain $\gamma(L') = L'$. As G_K is covered by $\mathcal{G}$-stable affine open subschemes, it follows (by arguing as in [**57**, V.20]) that there exists a unique subscheme $M \subseteq G$ such that $L' = M_K$. Then M is again a smooth connected affine normal subgroup scheme of G, and hence $M \subseteq L(G)$. On the other hand, we clearly have $L(G)_K \subseteq L'$; we conclude that $M = L(G)$. □

REMARK 3.1.5. In fact, the formation of L commutes with separable field extensions that are not necessarily algebraic. This can be shown by adapting the proof of [**21**, 1.1.9], which asserts that the formation of the unipotent radical commutes with all separable field extensions. That proof involves methods of group schemes over rings, which go beyond the scope of this text.

Our final criterion for affineness is of geometric origin:

PROPOSITION 3.1.6. *Let $a : G \times X \to X$ be an action of an algebraic group on an irreducible locally noetherian scheme and let $x \in X(k)$. Then the quotient group scheme $\mathrm{C}_G(x)/\operatorname{Ker}(a)$ is affine.*

PROOF. We may replace G with $\mathrm{C}_G(x)$, and hence assume that G fixes x. Consider the nth infinitesimal neighborhoods, $x_{(n)} := \operatorname{Spec}(\mathcal{O}_{X,x}/\mathfrak{m}_x^{n+1})$, where n runs over the positive integers; these form an increasing sequence of finite subschemes of X supported at x. As seen in Example 2.3.6, each $x_{(n)}$ is stabilized by G; this yields a linear representation ρ_n of G in $\mathcal{O}_{X,x}/\mathfrak{m}_x^{n+1} =: V_n$, a finite-dimensional vector space. Denote by N_n the kernel of ρ_n; then N_n contains $\operatorname{Ker}(a)$. As ρ_n is a quotient of ρ_{n+1}, we have $N_{n+1} \subseteq N_n$. Since G is of finite type, it follows that there exists n_0 such that $N_n = N_{n_0} =: N$ for all $n \geq n_0$. Then N acts trivially on each subscheme $x_{(n)}$. As X is locally noetherian and irreducible, the union of these subschemes is dense in X; it follows that N acts trivially on X, by using the representability of the fixed point functor X^G (Theorem 2.2.6). Thus, $N = \operatorname{Ker}(a)$. So $\rho_{n_0} : G \to \operatorname{GL}(V_{n_0})$ factors through a closed immersion $j : G/\operatorname{Ker}(a) \to \operatorname{GL}(V_{n_0})$ by Proposition 2.7.1. □

COROLLARY 3.1.7. *Let G be a connected algebraic group and Z its center. Then G/Z is affine.*

PROOF. Consider the action of G on itself by inner automorphisms. Then the kernel of this action is Z and the neutral element is fixed. So the assertion follows from Proposition 3.1.6. □

The connectedness assumption in the above corollary cannot be removed in view of Example 4.2.2 below.

3.2. The affinization theorem. Every scheme X is equipped with a morphism to an affine scheme, namely, the canonical morphism

$$\varphi = \varphi_X : X \to \operatorname{Spec} \mathcal{O}(X).$$

The restriction of φ_X to any affine open subscheme $U \subseteq X$ is the morphism $U \to \operatorname{Spec} \mathcal{O}(X)$ associated with the restriction homomorphism $\mathcal{O}(X) \to \mathcal{O}(U)$. Moreover, φ satisfies the following universal property: every morphism $f : X \to Y$, where Y is an affine scheme, factors uniquely through φ. We say that φ is the *affinization morphism* of X, and denote $\operatorname{Spec} \mathcal{O}(X)$ by $\operatorname{Aff}(X)$. When X is of finite type, $\operatorname{Aff}(X)$ is not necessarily of finite type; equivalently, the algebra $\mathcal{O}(X)$ is not necessarily finitely generated (even when X is a quasi-projective variety, see Example 3.2.3 below).

Also, every morphism of schemes $f : X \to Y$ lies in a commutative diagram

$$\begin{array}{ccc} X & \xrightarrow{\ f\ } & Y \\ {\scriptstyle \varphi_X}\downarrow & & \downarrow{\scriptstyle \varphi_Y} \\ \operatorname{Aff}(X) & \xrightarrow{\operatorname{Aff}(f)} & \operatorname{Aff}(Y), \end{array}$$

where $\operatorname{Aff}(f)$ is the morphism of affine schemes associated with the ring homomorphism $f^\# : \mathcal{O}(Y) \to \mathcal{O}(X)$.

For quasi-compact schemes, the formation of the affinization morphism commutes with field extensions and finite products, as a consequence of Lemma 2.3.3. It follows that for any algebraic group G, there is a canonical group scheme structure on $\operatorname{Aff}(G)$ such that φ_G is a homomorphism. Moreover, given an action a of G on a quasi-compact scheme X, the map $\operatorname{Aff}(a)$ is an action of $\operatorname{Aff}(G)$ on $\operatorname{Aff}(X)$, compatibly with a.

With these observations at hand, we may make an important step in the proof of Theorem 1:

THEOREM 3.2.1. *Let G be an algebraic group, $\varphi : G \to \operatorname{Aff}(G)$ its affinization morphism and $H := \operatorname{Ker}(\varphi)$. Then H is the smallest normal subgroup scheme of G such that G/H is affine. Moreover, $\mathcal{O}(H) = k$ and $\operatorname{Aff}(G) = G/H$. In particular, $\mathcal{O}(G) = \mathcal{O}(G/H)$; thus, the algebra $\mathcal{O}(G)$ is finitely generated.*

PROOF. Consider a normal subgroup scheme N of G such that G/N is affine. Then we have a commutative diagram of homomorphisms

$$\begin{array}{ccc} G & \xrightarrow{\quad q \quad} & G/N \\ \Big\downarrow{\scriptstyle \varphi_G} & & \Big\downarrow{\scriptstyle \varphi_{G/N}} \\ \mathrm{Aff}(G) & \xrightarrow{\mathrm{Aff}(q)} & \mathrm{Aff}(G/N), \end{array}$$

where q is the quotient morphism and $\varphi_{G/N}$ is an isomorphism. Since H is the fiber of φ_G at the neutral element e_G, it follows that $H \subseteq N$.

We now claim that H is the kernel of the action of G on $\mathcal{O}(G)$ via left multiplication. Denote by K the latter kernel; we check that $H(R) = K(R)$ for any algebra R. Note that $H(R)$ consists of those $x \in G(R)$ such that $f(x) = f(e)$ for all $f \in \mathcal{O}(G)$ (since $\mathcal{O}(G \times \mathrm{Spec}(R)) = \mathcal{O}(G) \otimes_k R$). Also, $K(R)$ consists of those $x \in G(R)$ such that $f(xy) = f(y)$ for all $f \in \mathcal{O}(G \times \mathrm{Spec}(R'))$ and $y \in G(R')$, where R' runs over all R-algebras. In particular, $f(x) = f(e)$ for all $f \in \mathcal{O}(G)$, and hence $K(R) \subseteq H(R)$.

To show the opposite inclusion, choose a basis $(\varphi_i)_{i \in I}$ of the k-vector space $\mathcal{O}(G)$; then the R'-module $\mathcal{O}(G \times \mathrm{Spec}(R')) = \mathcal{O}(G) \otimes_k R'$ is free with basis $(\varphi_i)_{i \in I}$. Thus, for any $f \in \mathcal{O}(G) \otimes_k R'$, there exists a unique family $(\psi_i = \psi_i(f))_{i \in I}$ in $\mathcal{O}(G) \otimes_k R'$ such that $f(xy) = \sum_i \psi_i(x)\, \varphi_i(y)$ identically. So the equalities $f(xy) = f(y)$ for all $y \in G(R')$ are equivalent to the equalities

$$\sum_i (\psi_i(x) - \psi_i(e))\, \varphi_i(y) = 0$$

for all such y. Since the latter equalities are satisfied for any $x \in H(R)$, this yields the inclusion $H(R) \subseteq K(R)$, and completes the proof of the claim.

By Proposition 2.3.4, there exists an increasing family of finite-dimensional G-submodules $(V_i)_{i \in I}$ of $\mathcal{O}(G)$ such that $\mathcal{O}(G) = \bigcup_i V_i$. Denoting by K_i the kernel of the corresponding homomorphism $G \to \mathrm{GL}(V_i)$, we see that $H = K$ is the decreasing intersection of the K_i. Since the topological space underlying G is noetherian and each K_i is closed in G, there exists $i \in I$ such that $H = K_i$. It follows that G/H is affine. We have proved that H is the smallest normal subgroup scheme of G having an affine quotient.

The affinization morphism φ_G factors through a unique morphism of affine schemes $\iota : G/H \to \mathrm{Aff}(G)$. The associated homomorphism

$$\iota^{\#} : \mathcal{O}(\mathrm{Aff}(G)) = \mathcal{O}(G) \to \mathcal{O}(G/H) = \mathcal{O}(G)^H$$

is an isomorphism; thus, so is ι. This shows that $\mathrm{Aff}(G) = G/H$.

Next, consider the kernel N of the affinization morphism φ_H. Then $N \trianglelefteq H$ and the quotient group H/N is affine. Since G/H is affine as well, it follows by Proposition 3.1.2 that the homogeneous space G/N is affine. Thus, the quotient morphism $G \to G/N$ factors through a unique morphism $\mathrm{Aff}(G) \to G/N$. Taking fibers at e yields that $H \subseteq N$; thus, $H = N$. Hence the action on H on itself via left multiplication yields a trivial action on $\mathcal{O}(H)$. As $\mathcal{O}(H)^H = k$, we conclude that $\mathcal{O}(H) = k$. □

COROLLARY 3.2.2. *Let G be an algebraic group acting faithfully on an affine scheme X. Then G is affine.*

PROOF. The action of G on X factors through an action of $\mathrm{Aff}(G)$ on $\mathrm{Aff}(X) = X$. Thus, the subgroup scheme H of Theorem 3.2.1 acts trivially on X. Hence H is trivial; this yields the assertion. □

EXAMPLE 3.2.3. Let E be an elliptic curve equipped with an invertible sheaf $\mathcal{L}$ such that $\deg(\mathcal{L}) = 0$ and $\mathcal{L}$ has infinite order in $\mathrm{Pic}(E)$. (Such a pair $(E, \mathcal{L})$ exists unless k is algebraic over a finite field, as follows from [**59**]; see also [**63**]). Choose an invertible sheaf $\mathcal{M}$ on E such that $\deg(\mathcal{M}) > 0$. Denote by L, M the line bundles on E associated with $\mathcal{L}, \mathcal{M}$ and consider their direct sum,

$$\pi : X := L \oplus M \longrightarrow E.$$

Then X is a quasi-projective variety and

$$\pi_*(\mathcal{O}_X) \cong \bigoplus_{\ell,m} \mathcal{L}^{\otimes \ell} \otimes_{\mathcal{O}_E} \mathcal{M}^{\otimes m},$$

where the sum runs over all pairs of non-negative integers. Thus,

$$\mathcal{O}(X) \cong \bigoplus_{\ell,m} H^0(E, \mathcal{L}^{\otimes \ell} \otimes_{\mathcal{O}_E} \mathcal{M}^{\otimes m}).$$

In particular, the algebra $\mathcal{O}(X)$ is equipped with a bi-grading. If this algebra is finitely generated, then the pairs (ℓ, m) such that $\mathcal{O}(X)_{\ell,m} \neq 0$ form a finitely generated monoid under componentwise addition; as a consequence, the convex cone $C \subset \mathbb{R}^2$ generated by these pairs is closed. But we have $\mathcal{O}(X)_{\ell,0} = 0$ for any $\ell \geq 1$, since $\mathcal{L}^{\otimes \ell}$ is non-trivial and has degree 0. Also, given any positive rational number t, we have $\mathcal{O}(X)_{n,tn} \neq 0$ for any positive integer n such that tn is integer, since $\deg(\mathcal{L}^{\otimes n} \otimes_{\mathcal{O}_E} \mathcal{M}^{\otimes tn}) > 0$. Thus, C is not closed, a contradiction. We conclude that *the algebra $\mathcal{O}(X)$ is not finitely generated.*

3.3. Anti-affine algebraic groups.

DEFINITION 3.3.1. An algebraic group G over k is *anti-affine* if $\mathcal{O}(G) = k$.

By Lemma 2.3.3, G is anti-affine if and only if G_K is anti-affine for some field extension K of k.

LEMMA 3.3.2. *Every anti-affine algebraic group is smooth and connected.*

PROOF. Let G be an algebraic group. Recall that the group of connected components $\pi_0(G) \cong G/G^0$ is finite and étale. Also, $\mathcal{O}(\pi_0(G)) \cong \mathcal{O}(G)^{G^0}$ by Remark 2.7.3 (ii). If G is anti-affine, then it follows that $\mathcal{O}(\pi_0(G)) = k$. Thus, $\pi_0(G)$ is trivial, i.e., G is connected.

To show that G is smooth, we may assume that k is algebraically closed. Then G_{red} is a smooth subgroup scheme of G; moreover, the homogeneous space G/G_{red} is finite by Remark 2.7.3(v). As above, it follows that $G = G_{\mathrm{red}}$. □

We now obtain a generalization of a classical rigidity lemma (see [**45**, p. 43]):

LEMMA 3.3.3. *Let X, Y, Z be schemes such that X is quasi-compact, $\mathcal{O}(X) = k$ and Y is locally noetherian and irreducible. Let $f : X \times Y \to Z$ be a morphism. Assume that there exist k-rational points $x_0 \in X$, $y_0 \in Y$ such that $f(x, y_0) = f(x_0, y_0)$ identically. Then $f(x, y) = f(x_0, y)$ identically.*

PROOF. Let $z_0 := f(x_0, y_0)$; this is a k-rational point of Z. As in Example 2.3.6, consider the nth infinitesimal neighborhoods of this point,

$$z_{0,(n)} := \operatorname{Spec}(\mathcal{O}_{Z,z_0}/\mathfrak{m}_{z_0}^{n+1}),$$

where n runs over the positive integers. These form an increasing sequence of finite subschemes of Z supported at z_0, and one checks as in the above example that $X \times y_{0,(n)}$ is contained in the fiber of f at $z_{0,(n)}$, where $y_{0,(n)} := \operatorname{Spec}(\mathcal{O}_{Y,y_0}/\mathfrak{m}_{y_0}^{n+1})$. In other words, f restricts to a morphism $f_n : X \times y_{0,(n)} \to z_{0,(n)}$. Consider the associated homomorphism of algebras $f_n^{\#} : \mathcal{O}(z_{0,(n)}) \to \mathcal{O}(X \times y_{0,(n)})$. By Lemma 2.3.3 and the assumptions on X, we have $\mathcal{O}(X \times y_{0,(n)}) = \mathcal{O}(X) \otimes_k \mathcal{O}(y_{0,(n)}) = \mathcal{O}(y_{0,(n)})$. Since $z_{0,(n)}$ is affine, it follows that f_n factors through a morphism $g_n : y_{0,(n)} \to Z$, i.e., $f_n(x, y) = g_n(y)$ identically. In particular, $f(x, y) = f(x_0, y)$ on $X \times y_{0,(n)}$.

Next, consider the largest closed subscheme $W \subseteq X \times Y$ on which $f(x, y) = f(x_0, y)$, i.e., W is the pull-back of the diagonal in $Z \times Z$ under the morphism $(x, y) \mapsto (f(x, y), f(x_0, y))$. Then W contains $X \times y_{0,(n)}$ for all n. Since Y is locally noetherian and irreducible, the union of the $y_{0,(n)}$ is dense in Y. It follows that the union of the $X \times y_{0,(n)}$ is dense in $X \times Y$; we conclude that $W = X \times Y$. □

PROPOSITION 3.3.4. *Let H be an anti-affine algebraic group, G an algebraic group and $f : H \to G$ a morphism of schemes such that $f(e_H) = e_G$. Then f is a homomorphism and factors through the center of G^0.*

PROOF. Since H is connected by Lemma 3.3.2, we see that f factors through G^0. Thus, we may assume that G is connected.

Consider the morphism

$$\varphi : H \times H \longrightarrow G, \quad (x, y) \longmapsto f(xy)f(y)^{-1}f(x)^{-1}.$$

Then $\varphi(x, e_H) = e_G = \varphi(e_H, e_H)$ identically; also, H is irreducible in view of Lemma 3.3.2. Thus, the rigidity lemma applies, and yields $\varphi(x, y) = \varphi(e_H, y) = e_G$ identically. This shows that f is a homomorphism.

The assertion that f factors through the center of G is proved similarly by considering the morphism

$$\psi : H \times G \longrightarrow G, \quad (x, y) \longmapsto f(x)yf(x)^{-1}y^{-1}.$$

□

In particular, every anti-affine group G is commutative. Also, note that G/H is anti-affine for any subgroup scheme $H \subseteq G$ (since $\mathcal{O}(G/H) = \mathcal{O}(G)^H$).

We may now complete the proof of Theorem 1 with the following:

PROPOSITION 3.3.5. *Let G be an algebraic group and H the kernel of the affinization morphism of G.*

(1) *H is contained in the center of G^0.*

(2) *H is the largest anti-affine subgroup of G.*

PROOF. (1) By Theorem 3.2.1, H is anti-affine. So the assertion follows from Lemma 3.3.2 and Proposition 3.3.4, or alternatively, from Corollary 3.1.7.

(2) Consider another anti-affine subgroup $N \subseteq G$. Then the quotient group $N/N \cap H$ is anti-affine, and also affine (since $N/N \cap H$ is isomorphic to a subgroup of G/H, and the latter is affine). As a consequence, $N/N \cap H$ is trivial, that is, N is contained in H. □

We will denote the largest anti-affine subgroup of an algebraic group G by G_{ant}. For later use, we record the following observations:

LEMMA 3.3.6. *Let G be an algebraic group and $N \trianglelefteq G$ a normal subgroup scheme. Then the quotient map $G \to G/N$ yields an isomorphism*

$$G_{\mathrm{ant}}/G_{\mathrm{ant}} \cap N \xrightarrow{\cong} (G/N)_{\mathrm{ant}}.$$

PROOF. By Proposition 2.7.4, we have a closed immersion of algebraic groups $G_{\mathrm{ant}}/G_{\mathrm{ant}} \cap N \to G/N$; moreover, $G_{\mathrm{ant}}/G_{\mathrm{ant}} \cap N$ is anti-affine. So we obtain a closed immersion of commutative algebraic groups $j : G_{\mathrm{ant}}/G_{\mathrm{ant}} \cap N \to (G/N)_{\mathrm{ant}}$. Denote by C the cokernel of j; then C is anti-affine as a quotient of $(G/N)_{\mathrm{ant}}$. Also, C is a subgroup of $(G/N)/(G_{\mathrm{ant}}/G_{\mathrm{ant}} \cap N)$, which is a quotient group of G/G_{ant}. Since the latter group is affine, it follows that C is affine as well, by using Proposition 3.1.2. Thus, C is trivial, i.e., j is an isomorphism. □

LEMMA 3.3.7. *The following conditions are equivalent for an algebraic group G:*

(1) *G is proper.*
(2) *G_{ant} is an abelian variety and G/G_{ant} is finite.*

Under these conditions, we have $G_{\mathrm{ant}} = G^0_{\mathrm{red}}$; in particular, G^0_{red} is a smooth connected algebraic group and its formation commutes with field extensions.

PROOF. (1)⇒(2) As G_{ant} is smooth, connected and proper, it is an abelian variety. Also, G/G_{ant} is proper and affine, hence finite.

(2)⇒(1) This follows from Remark 3.1.3.

For the final assumption, note that the quotient group scheme G^0/G_{ant} is finite and connected, hence infinitesimal. So the algebra $\mathcal{O}(G^0/G_{\mathrm{ant}})$ is local with residue field k (via evaluation at e). It follows that $(G^0/G_{\mathrm{ant}})_{\mathrm{red}} = e$ and hence that $G^0_{\mathrm{red}} \subseteq G_{\mathrm{ant}}$; this yields the assertion. □

Notes and references.

Some of the main results of this section originate in Rosenlicht's article [**52**]. More specifically, Corollary 3.1.7 is a scheme-theoretic version of [**52**, Thm. 13], and Theorem 3.2.1, of [**52**, Cor. 3, p. 431].

Also, Theorem 3.2.1, Lemma 3.3.2 and Proposition 3.3.4 are variants of results from [**23**, III.3.8].

The rigidity lemma 3.3.3 is a version of [**56**, Thm. 1.7].

4. Proof of Theorem 2

4.1. The Albanese morphism. Throughout this subsection, A denotes an abelian variety, i.e., a smooth connected proper algebraic group. Then A is commutative by Corollary 3.1.7. Thus, we will denote the group law additively; in particular, the neutral element will be denoted by 0. Also, the variety A is projective (see [**45**, p. 62]).

LEMMA 4.1.1. *Every morphism $f : \mathbb{P}^1 \to A$ is constant.*

PROOF. We may assume that k is algebraically closed. Suppose that f is non-constant and denote by $C \subseteq A$ its image, with normalization $\eta : \tilde{C} \to C$. Then f factors through a surjective morphism $\mathbb{P}^1 \to \tilde{C}$. By Lüroth's theorem, it follows that $\tilde{C} \cong \mathbb{P}^1$. Thus, f factors through the normalization $\eta : \mathbb{P}^1 \to C$ and hence it suffices to show that η is constant. In other words, we may assume that f is birational to its image. Then the differential

$$df : T_{\mathbb{P}^1} \longrightarrow f^*(T_A)$$

is non-zero at the generic point of $\mathbb{P}^1$ and hence is injective. Since the tangent sheaf T_A is trivial and $T_{\mathbb{P}^1} \cong \mathcal{O}_{\mathbb{P}^1}(2)$, we obtain an injective map $\mathcal{O}_{\mathbb{P}^1} \to \mathcal{O}_{\mathbb{P}^1}(-2)^{\oplus n}$, where $n := \dim(A)$. This yields a contradiction, since $H^0(\mathbb{P}^1, \mathcal{O}_{\mathbb{P}^1}) = k$ while $H^0(\mathbb{P}^1, \mathcal{O}_{\mathbb{P}^1}(-2)) = 0$. □

THEOREM 4.1.2. *Let X be a smooth variety and $f : X \dashrightarrow A$ a rational map. Then f is a morphism.*

PROOF. Again, we may assume k algebraically closed. View f as a morphism $U \to A$, where $U \subseteq X$ is a non-empty open subvariety. Denote by $Y \subseteq X \times A$ the closure of the graph of f, with projections $p_1 : Y \to X$, $p_2 : Y \to A$. Then p_1 is proper (since so is A) and birational (since it restricts to an isomorphism over U). Assume that p_1 is not an isomorphism. Then p_1 contracts some rational curve in Y, i.e., there exists a non-constant morphism $g : \mathbb{P}^1 \to Y$ such that $p_1 \circ g$ is constant (see [**22**, Prop. 1.43]). It follows that $p_2 \circ g : \mathbb{P}^1 \to A$ is non-constant; but this contradicts Lemma 4.1.1. □

LEMMA 4.1.3. *Let X, Y be varieties equipped with k-rational points x_0, y_0 and let $f : X \times Y \to A$ be a morphism. Then we have identically*

$$f(x,y) - f(x_0,y) - f(x,y_0) + f(x_0,y_0) = 0.$$

PROOF. By a result of Nagata (see [**48, 49**], and [**40**] for a modern proof), we may choose a compactification of X, i.e., an open immersion $X \to \bar{X}$, where $\bar{X}$ is a proper variety. Replacing $\bar{X}$ with its normalization, we may further assume that $\bar{X}$ is normal. Also, we may assume k algebraically closed from the start.

Denote by U the smooth locus of $\bar{X}$ and by V the smooth locus of Y. By Theorem 4.1.2, the rational map $f : \bar{X} \times Y \dashrightarrow A$ yields a morphism $g : U \times V \to A$. Also, note that the complement $\bar{X} \setminus U$ has codimension at least 2, since $\bar{X}$ is normal and $k = \bar{k}$. It follows that $\mathcal{O}(U) = \mathcal{O}(\bar{X}) = k$. Using again the assumption that $k = \bar{k}$, we may choose points $x_1 \in U(k)$, $y_1 \in V(k)$. Consider the morphism

$$\varphi : U \times V \longrightarrow A, \quad (x,y) \longmapsto g(x,y) - g(x_1,y) - g(x,y_1) + g(x_1,y_1).$$

Then $\varphi(x,y_1) = 0$ identically, and hence $\varphi = 0$ by rigidity (Lemma 3.3.3). It follows that $f(x,y) - f(x_1,y) - f(x,y_1) + f(x_1,y_1) = 0$ identically on $X \times Y$. This readily yields the desired equation. □

PROPOSITION 4.1.4. *Let G be a smooth connected algebraic group.*

(1) *Let $f : G \to A$ be a morphism to an abelian variety sending e to 0. Then f is a homomorphism.*
(2) *There exists a smallest normal subgroup scheme N of G such that the quotient G/N is an abelian variety. Moreover, N is connected.*
(3) *For any abelian variety A, every morphism $f : G \to A$ sending e to 0 factors uniquely through the quotient homomorphism $\alpha : G \to G/N$.*
(4) *The formation of N commutes with separable algebraic field extensions.*

PROOF. (1) This follows from Lemma 4.1.3 applied to the map $G \times G \to A$, $(x,y) \mapsto f(xy)$ and to $x_0 = y_0 = e$.

(2) Consider two normal subgroup schemes N_1, N_2 of G such that G/N_1, G/N_2 are abelian varieties. Then $N_1 \cap N_2$ is normal in G and $G/N_1 \cap N_2$ is proper, since the natural map

$$G/N_1 \cap N_2 \longrightarrow G/N_1 \times G/N_2$$

is a closed immersion of algebraic groups (Proposition 2.7.1). Since $G/N_1 \cap N_2$ is smooth and connected, it is an abelian variety as well. It follows that there exists a smallest such subgroup scheme, say N.

We claim that the neutral component N^0 is normal in G. To check this, we may assume k algebraically closed. Then $G(k)$ is dense in G and normalizes N^0; this yields the claim.

The natural homomorphism $G/N^0 \to G/N$ is finite, since it is a torsor under the finite group N/N^0 (Propositions 2.6.5 and 2.8.4). As a consequence, G/N^0 is proper; hence $N = N^0$ by the minimality assumption. Thus, N is connected.

(3) By (1), f is a homomorphism; denote its kernel by H. Then G/H is an abelian variety in view of Proposition 2.7.4. Thus, H contains N, and the assertion follows from Proposition 2.8.4.

(4) This is checked by a standard argument of Galois descent, as in the proof of Lemma 3.1.4. □

REMARK 4.1.5. In fact, the formation of N commutes with all separable field extensions. This may be proved as sketched in Remark 3.1.5.

With the notation and assumptions of Proposition 4.1.4, we say that the quotient morphism

$$\alpha : G \longrightarrow G/N$$

is the *Albanese homomorphism of* G, and $G/N =: \mathrm{Alb}(G)$ the *Albanese variety.*

Actually, Proposition 4.1.4 (3) extends to any *pointed variety*, i.e., a variety equipped with a k-rational point:

THEOREM 4.1.6. *Let (X, x) be a pointed variety. Then there exists an abelian variety A and a morphism $\alpha = \alpha_X : X \to A$ sending x to 0, such that for any abelian variety B, every morphism $X \to B$ sending x to 0 factors uniquely through α.*

PROOF. See [**58**, Thm. 5] when k is algebraically closed; the general case is obtained in [**66**, Thm. A1]. □

The morphism $\alpha : X \to A$ in the above theorem is uniquely determined by (X, x); it is called the *Albanese morphism*, and A is again the *Albanese variety*, denoted by $\mathrm{Alb}(X)$. Combining that theorem with Theorem 4.1.2 and Lemma 4.1.3, we obtain readily:

COROLLARY 4.1.7. (1) *For any smooth pointed variety (X, x) and any open subvariety $U \subseteq X$ containing x, we have a commutative square*

$$\begin{array}{ccc} U & \xrightarrow{\alpha_U} & \mathrm{Alb}(U) \\ {\scriptstyle j}\downarrow & & \downarrow{\scriptstyle \mathrm{Alb}(j)} \\ X & \xrightarrow{\alpha_X} & \mathrm{Alb}(X), \end{array}$$

where $j : U \to X$ denotes the inclusion and $\mathrm{Alb}(j)$ *is an isomorphism.*

(2) *For any pointed varieties* (X,x), (Y,y), *we have* $\alpha_{X\times Y} = \alpha_X \times \alpha_Y$. *In particular, the natural map*

$$\mathrm{Alb}(X \times Y) \longrightarrow \mathrm{Alb}(X) \times \mathrm{Alb}(Y)$$

is an isomorphism.

(3) *Any action* a *of a smooth connected algebraic group* G *on a pointed variety* X *yields a unique homomorphism* $f = f_a : \mathrm{Alb}(G) \to \mathrm{Alb}(X)$ *such that the square*

$$\begin{array}{ccc} G \times X & \xrightarrow{\quad a \quad} & X \\ \Big\downarrow{\scriptstyle \alpha_G \times \alpha_X} & & \Big\downarrow{\scriptstyle \alpha_X} \\ \mathrm{Alb}(G) \times \mathrm{Alb}(X) & \xrightarrow{f+\mathrm{id}} & \mathrm{Alb}(X) \end{array}$$

commutes.

The formation of the Albanese morphism commutes with separable algebraic field extensions, by Galois descent again. But it does not commute with arbitrary field extensions, as shown by Example 4.2.7 below.

4.2. Abelian torsors.

LEMMA 4.2.1. *Let* A *an abelian variety and* n *a non-zero integer. Then the multiplication map*

$$n_A : A \longrightarrow A, \quad x \longmapsto nx$$

is an isogeny.

PROOF. See [**45**, p. 62]. □

EXAMPLE 4.2.2. With the above notation, consider the semi-direct product $G := \mathbb{Z}/2 \ltimes A$, where $\mathbb{Z}/2$ (viewed as a constant group scheme) acts on A via $x \mapsto \pm x$. One may check that the center Z of G is the kernel of the multiplication map 2_A and hence is finite; thus, G/Z is not affine. So Corollary 3.1.7 does not extend to disconnected algebraic groups.

Also, note that G is not contained in any connected algebraic group, as follows from Proposition 3.3.4.

LEMMA 4.2.3. *Let* G *be a smooth connected commutative algebraic group, with group law denoted additively, and let* $f : X \to \mathrm{Spec}(k)$ *be a* G*-torsor. Then there exists a positive integer* n *and a morphism* $\varphi : X \to G$ *such that*

$$\varphi(g \cdot x) = \varphi(x) + ng$$

identically on $G \times X$.

PROOF. If X has a k-rational point x, then the orbit map $a_x : G \to X$, $g \mapsto g \cdot x$ is a G-equivariant isomorphism, where G acts on itself by translation. So we may just take $\varphi = a_x^{-1}$ and $n = 1$ (i.e., φ is G-equivariant).

In the general case, since X is a smooth variety, it has a K-rational point x_1 for some finite Galois extension of fields K/k. Denote by $\mathcal{G}$ the corresponding Galois group and by $x_1, \ldots, x_n$ the distinct conjugates of x_1 under $\mathcal{G}$. Then the first step yields G_K-equivariant isomorphisms $g_i : X_K \to G_K$ for $i = 1, \ldots, n$, such that $x = g_i(x) \cdot x_i$ identically. Consider the morphism

$$\phi : X_K \longrightarrow G_K, \quad x \longmapsto g_1(x) + \cdots + g_n(x).$$

Then ϕ is equivariant under $\mathcal{G}$, since that group permutes the x_i's and hence the g_i's. Also, we have $\phi(g \cdot x) = \phi(x) + ng$ identically. So ϕ descends to the desired morphism $X \to G$. □

PROPOSITION 4.2.4. *Let A be an abelian variety and $f : X \to Y$ an A-torsor, where X, Y are smooth varieties. Then there exists a positive integer n and a morphism $\varphi : X \to A$ such that $\varphi(a \cdot x) = \varphi(x) + na$ identically on $A \times X$.*

PROOF. Let $\eta_Y : \operatorname{Spec} k(Y) \to Y$ be the generic point. Then the base change $X \times_Y \operatorname{Spec} k(Y) \to \operatorname{Spec} k(Y)$ is an $A_{k(Y)}$-torsor. Using Lemma 4.2.3, we obtain a map $\psi : X \times_Y \operatorname{Spec} k(Y) \to A_{k(Y)}$ satisfying the required covariance property. Composing ψ with the natural maps $\eta_X \times f : \operatorname{Spec} k(X) \to X \times_Y \operatorname{Spec} k(Y)$ and $\pi : A_{k(Y)} \to A$ yields a map $\operatorname{Spec} k(X) \to A$, which may be viewed as a rational map $X \dashrightarrow A$ and hence (by Theorem 4.1.2) as a morphism $\varphi : X \to A$. Clearly, φ satisfies the same covariance property as ψ. □

THEOREM 4.2.5. *Let G be a smooth connected algebraic group and $A \subseteq G$ an abelian subvariety. Then $A \subseteq Z(G)$ and there exists a connected normal subgroup scheme $H \subseteq G$ such that $G = A \cdot H$ and $A \cap H$ is finite. If k is perfect, then we may take H smooth.*

PROOF. By Proposition 3.3.4, A is contained in the center of G. The quotient map $q : G \to G/A$ is an A-torsor; also, G/A is smooth, since so is G. Thus, Proposition 4.2.4 yields a map $\varphi : G \to A$ such that $\varphi(ag) = \varphi(g) + na$ identically, for some integer $n > 0$. Composing φ with a translation of A, we may assume that $\varphi(e_G) = 0$. Then φ is a homomorphism in view of Proposition 4.1.4; its restriction to A is the multiplication n_A.

We claim that $G = A \cdot \operatorname{Ker}(\varphi)$. Since G is smooth, it suffices by Lemma 2.8.6 to show the equality $G(\bar{k}) = A(\bar{k}) \operatorname{Ker}(\varphi)(\bar{k})$. Let $g \in G(\bar{k})$; by Lemma 4.2.1, there exists $a \in A(\bar{k})$ such that $\varphi(g) = na$. Thus, $\varphi(a^{-1}g) = 0$; this yields the claim.

Let $H := \operatorname{Ker}(\varphi)^0$. Then $A \cap H$ is finite, since it is contained in $A \cap \operatorname{Ker}(\varphi) = \operatorname{Ker}(n_A)$. We now show that $G = A \cdot H$. The homogeneous space $G/A \cdot H$ is smooth and connected, since so is G. On the other hand, $G/A \cdot H = A \cdot \operatorname{Ker}(\varphi)/A \cdot H$ is homogeneous under $\operatorname{Ker}(\varphi)/H = \pi_0(\operatorname{Ker}(\varphi))$, and hence is finite. Thus, $G/A \cdot H$ is trivial; this yields the assertion. Since H centralizes A, it follows that H is normal in G.

Finally, if k is perfect, then we may replace H with H_{red} in the above argument. □

COROLLARY 4.2.6. *Let A be an abelian variety and $B \subseteq A$ an abelian subvariety. Then there exists an abelian subvariety $C \subseteq A$ such that $A = B + C$ and $B \cap C$ is finite.*

PROOF. By the above theorem, there exists a connected subgroup scheme H of A such that $A = B + H$ and $B \cap H$ is finite. Now replace H with H_{red}, which is an abelian subvariety by Lemma 3.3.7. □

The following example displays several specific features of algebraic groups over imperfect fields. It is based on Weil restriction; we refer to Appendix A of [**21**] for the definition and main properties of this notion.

EXAMPLE 4.2.7. Let k be an imperfect field. Choose a non-trivial finite purely inseparable extension K of k. Let A' be a non-trivial abelian variety over K; then the Weil restriction $\mathrm{R}_{K/k}(A') =: G$ is a smooth connected commutative algebraic group over k, of dimension $[K : k] \dim(A')$. Moreover, there is an exact sequence of algebraic groups over K

$$(4.2.1) \qquad 1 \longrightarrow U' \longrightarrow G_K \xrightarrow{q} A' \longrightarrow 1,$$

where U' is non-trivial, smooth, connected and unipotent (see [**21**, A.5.11]). It follows readily that q is the Albanese homomorphism of G_K.

Let H be a smooth connected affine algebraic group over k. We claim that every homomorphism $f : H \to G$ is constant. Indeed, the morphisms of k-schemes $f : H \to G$ correspond bijectively to the morphisms of K-schemes $f' : H_K \to A'$, via the assignement $f \mapsto f' := q \circ f_K$ (see [**21**, A.5.7]). Since q is a homomorphism, this bijection sends homomorphisms to homomorphisms. As every homomorphism $H_K \to A'$ is constant, this proves the claim.

Next, consider the Albanese homomorphism $\alpha : G \to \mathrm{Alb}(G)$. If $\alpha_K = q$, then $\mathrm{Ker}(\alpha)_K = \mathrm{Ker}(\alpha_K) = U'$; as a consequence, $\mathrm{Ker}(\alpha)$ is smooth, connected and affine. By the claim, it follows that $\mathrm{Ker}(\alpha)$ is trivial, i.e., G is an abelian variety. Then so is G_K, but this contradicts the non-triviality of U'. So *the formation of the Albanese morphism does not commute with arbitrary field extensions.*

Note that $U' = L(G_K)$ (the largest smooth connected normal affine subgroup scheme of G_K, introduced in Lemma 3.1.4). Also, $L(G)$ is trivial by the claim. Thus, *the formation of $L(G)$ does not commute with arbitrary field extensions.*

Likewise, *G is not an extension of an abelian variety by a smooth connected affine algebraic group.* We will see in Theorem 4.3.2 that every smooth connected algebraic group over a *perfect* field lies in a unique such extension.

For later use, we analyze the structure of G in more detail. We claim that *there is a unique exact sequence*

$$1 \longrightarrow A \longrightarrow G \longrightarrow U \longrightarrow 1,$$

where A is an abelian variety and U is unipotent. This is equivalent to the assertion that G_{ant} is an abelian variety, and G/G_{ant} is unipotent. So it suffices to show the corresponding assertion for G_K. We may choose a positive integer n such that $p^n_{U'} = 0$. Thus, the extension (4.2.1) is trivialized by push-out via $p^n_{U'}$. It follows that (4.2.1) is also trivialized by pull-back via $p^n_{A'}$, and hence $G_K \cong (U' \times A')/F$ for some finite group scheme F (isomorphic to the kernel of $p^n_{A'}$). So the image of A' in G_K is an abelian variety with unipotent quotient; this proves the claim.

Next, we claim that *there exists a connected unipotent subgroup scheme $V \subseteq G$ such that $G = A \cdot V$ and $A \cap V$ is finite.* This follows from Theorem 4.2.5, or directly by taking for V the neutral component of $\mathrm{Ker}(p^m_G)$, where m is chosen so that $p^m_U = 0$. Yet *there exists no smooth connected subgroup scheme $H \subseteq G$ such that $G = A \cdot H$ and $A \cap H$ is finite.* Otherwise, the quotient homomorphism $G \to U$ restricts to a homomorphism $H \to U$ with finite kernel. Thus, H is affine; also, G is an extension of the abelian variety $A/A \cap H$ by H. This yields a contradiction.

4.3. Completion of the proof of Theorem 2. The final ingredient in Rosenlicht's proof of the Chevalley structure theorem is the following:

LEMMA 4.3.1. *Every non-proper algebraic group over an algebraically closed field contains an affine subgroup scheme of positive dimension.*

We now give a brief outline of the proof of this lemma, which is presented in detail in [**15**, Sec. 2.3]; see also [**44**, Sec. 4]. Let G be a non-proper algebraic group over $k = \bar{k}$. Then G^0_{red} is non-proper as well, and hence we may assume that G is smooth and connected. By [**48, 49**], there exists a compactification X of G, i.e., X is a proper variety containing G as an open subvariety; then the boundary $X \setminus G$ is non-empty. The action of G by left multiplication on itself induces a faithful rational action

$$a : G \times X \dashrightarrow X.$$

One shows (this is the main step of the proof) that there exists a proper variety X' and a birational morphism $\varphi : X' \to X$ such that the induced birational action $a' : G \times X' \dashrightarrow X'$ normalizes some irreducible divisor $D \subset X'$, i.e., a' induces a rational action $G \times D \dashrightarrow D$. Then one considers the "orbit map" a'_x for a general point $x \in D$, and the corresponding "stabilizer" $C_G(x)$ (these have to be defined appropriately). By adapting the argument of Proposition 3.1.6, one shows that $\mathrm{C}_G(x)$ is affine; it has positive dimension, since $\dim(G/C_G(x)) \leq \dim(D) = \dim(G) - 1$. So $\mathrm{C}_G(x)$ is the desired subgroup scheme.

We now show how to derive Theorem 2 from Lemma 4.3.1, under the assumptions that k is perfect and G is smooth and connected. We then have the following more precise result, which is a version of Chevalley's structure theorem:

THEOREM 4.3.2. *Let G be a smooth connected algebraic group over a perfect field k, and $L = L(G)$ the largest smooth connected affine normal subgroup scheme.*

(1) *L is the kernel of the Albanese homomorphism of G.*
(2) *The formation of L commutes with field extensions.*

PROOF. (1) Recall that the existence of $L(G)$ has been obtained in Lemma 3.1.4, as well as the triviality of $L(G/L(G))$. We may thus replace G with G/L and assume that L is trivial. Also, since the formations of $L(G)$ and of the Albanese homomorphism commute with algebraic field extensions (see Lemma 3.1.4 again, and Proposition 4.1.4), we may assume that k is algebraically closed.

Consider the center Z of G. If Z is proper, then its reduced neutral component Z^0_{red} is an abelian variety, say A. Since G/Z is affine (Corollary 3.1.7) and Z/A is finite, G/A is affine as well by Proposition 3.1.2. Also, by Theorem 4.2.5, there exists a smooth connected normal subgroup scheme $H \trianglelefteq G$ such that $G = A \cdot H$ and $A \cap H$ is finite. Then $H/A \cap H \cong G/A$ is affine, and hence H is affine by Proposition 3.1.2 again. It follows that H is trivial, and $G = A$ is an abelian variety.

On the other hand, if Z is not proper, then it contains an affine subgroup scheme N of positive dimension by Lemma 4.3.1. Thus, N^0_{red} is a non-trivial smooth connected central subgroup scheme of G. But this contradicts the assumption that $L(G)$ is trivial.

(2) Let K be a field extension of k. Then the exact sequence

$$1 \longrightarrow L \longrightarrow G \longrightarrow A \longrightarrow 1$$

yields an exact sequence of smooth connected algebraic groups over K

$$1 \longrightarrow L_K \longrightarrow G_K \longrightarrow A_K \longrightarrow 1,$$

where L_K is affine and A_K is an abelian variety. It follows readily that L_K equals $L(G_K)$ and is the kernel of the Albanese homomorphism of G_K. □

REMARK 4.3.3. With the notation and assumptions of the above theorem, every smooth connected affine subgroup scheme of G (not necessarily normal) has a trivial image in the abelian variety A. Thus, L is the largest smooth connected affine subgroup scheme of G.

We now return to an arbitrary field and obtain the following:

THEOREM 4.3.4. *Let G be a smooth algebraic group and denote by N the kernel of the Albanese homomorphism of G^0. Then N is the smallest normal subgroup scheme of G such that the quotient is proper. Moreover, N is affine and connected.*

PROOF. By Proposition 4.1.4, N is connected. Also, as G^0/N is proper and G/G^0 is finite, G/N is proper.

We now show that N is normal in G. For this, we may assume that k is separably closed, since the formation of N commutes with separable algebraic field extensions (Proposition 4.1.4 again). Then $G(k)$ is dense in G by smoothness, and normalizes N by the uniqueness of the Albanese homomorphism. Thus, G normalizes N.

Next, we show that N is contained in every normal subgroup scheme $H \trianglelefteq G$ such that G/H is proper. Indeed, one sees as above that $H^0 \trianglelefteq G$ and G/H^0 is proper as well; hence G^0/H^0 is an abelian variety. Thus, $H^0 \supseteq N$ as desired.

Finally, we show that N is affine. For this, we may assume that G is connected. By Theorem 4.3.2, there exists an exact sequence of algebraic groups over the perfect closure k_i,

$$1 \longrightarrow L_i \longrightarrow G_{k_i} \longrightarrow A_i \longrightarrow 1,$$

where L_i is smooth, connected and affine, and A_i is an abelian variety. This exact sequence is defined over some subfield $K \subseteq k_i$, finite over k. In other words, there exists a finite purely inseparable field extension K of k and a smooth connected affine normal subgroup scheme $L' \trianglelefteq G_K$ such that G_K/L' is an abelian variety.

By Lemma 4.3.5 below, we may choose a subgroup scheme $L \subseteq G$ such that $L_K \supseteq L'$ and L_K/L' is finite. As a consequence, $L_K \trianglelefteq G_K$ and hence $L \trianglelefteq G$. Also, L_K is affine and hence L is affine. Finally, $(G/L)_K \cong G_K/L_K \cong (G_K/L')/(L_K/L')$ is an abelian variety, and hence G/L is an abelian variety. Thus, $N \subseteq L$ is affine. □

LEMMA 4.3.5. *Let G be an algebraic group over k. Let K be a finite purely inseparable field extension of k, and $H' \subseteq G_K$ a K-subgroup scheme. Then there exists a k-subgroup scheme $H \subseteq G$ such that $H' \subseteq H_K$ and H_K/H' is finite.*

PROOF. We may choose a positive integer n such that $K^{p^n} \subseteq k$. Consider the nth relative Frobenius homomorphism

$$F^n := F^n_{G_K/K} : G_K \longrightarrow G_K^{(n)}$$

as in §2.9. Denote by H'_n the pull-back under F^n of the subgroup scheme $H'^{(n)}$ of $G_K^{(n)}$. Then $H' \subseteq H'_n$ and the quotient H'_n/H' is finite, since F^n is the identity on the underlying topological spaces and remains so over $\bar{k}$. Denote by $\mathcal{I}' \subset \mathcal{O}_{G_K}$ the sheaf of ideals of H'; then the sheaf of ideals $\mathcal{I}'_n$ of H'_n is generated by the p^nth powers of local sections of $\mathcal{I}'$, as follows from (2.9.1). Since $K^{p^n} \subseteq k$, every such power lies in $\mathcal{O}_G$. Thus, $\mathcal{I}'_n = \mathcal{I}_K$ for a unique sheaf of ideals $\mathcal{I} \subset \mathcal{O}_G$. The corresponding closed k-subscheme $H \subseteq G$ satisfies $H_K = H'_n$, and hence is the desired subgroup scheme. □

REMARK 4.3.6. Let G be a smooth connected algebraic group over k with Albanese homomorphism $\alpha : G \to \mathrm{Alb}(G)$ and consider a field extension K of k. Then the homomorphism $\alpha_K : G_K \to \mathrm{Alb}(G)_K$ factors through a unique homomorphism

$$f : \mathrm{Alb}(G_K) \longrightarrow \mathrm{Alb}(G)_K.$$

Since α is surjective, f is surjective as well. Moreover, $\mathrm{Ker}(f)$ is infinitesimal: indeed, denoting by $\alpha' : G_K \to \mathrm{Alb}(G_K)$ the Albanese homomorphism, we have $\mathrm{Ker}(\alpha') \subseteq \mathrm{Ker}(\alpha_K)$ and $\mathrm{Ker}(f) \cong \mathrm{Ker}(\alpha_K)/\mathrm{Ker}(\alpha')$. In particular, $\mathrm{Ker}(f)$ is affine and connected. But $\mathrm{Ker}(f)$ is also a subgroup scheme of the abelian variety $\mathrm{Alb}(G_K)$; hence it must be finite and local.

In loose terms, the formation of the Albanese homomorphism commutes with field extensions up to purely inseparable isogenies.

Since every algebraic group is an extension of a smooth algebraic group by an infinitesimal one (Proposition 2.9.2), Theorem 4.3.4 implies readily:

COROLLARY 4.3.7. *Every connected algebraic group G is an extension of an abelian variety by a connected affine algebraic group.*

Yet G may contain no smallest connected affine subgroup scheme with quotient an abelian variety, as shown by the following:

EXAMPLE 4.3.8. Assume that $\mathrm{char}(k) = p > 0$ and let A be a non-trivial abelian variety. Then A contains two non-trivial infinitesimal subgroup schemes I, J such that $J \subsetneq I$: we may take $I = \mathrm{Ker}(F_{A/k})$ and $J = \mathrm{Ker}(F^2_{A/k})$, where $F^n_{A/k}$ denotes the nth relative Frobenius morphism as in §2.9. Thus,

$$G := (A \times I)/\mathrm{diag}(J)$$

is a connected commutative algebraic group, where $\mathrm{diag}(x) := (x, x)$. Consider the infinitesimal subgroup schemes $N_1 := (J \times I)/\mathrm{diag}(J)$ and $N_2 := \mathrm{diag}(I)/\mathrm{diag}(J)$ of G. Then $G/N_1 \cong A/J$ and $G/N_2 \cong A$ are abelian varieties. Also, $N_1 \cap N_2$ is trivial, and G is not an abelian variety.

To complete the proof of Theorem 2, it remains to treat the general case, where G is an arbitrary algebraic group over an arbitrary field. We then have to prove:

LEMMA 4.3.9. *Any algebraic group G has a smallest normal subgroup scheme N such that G/N is proper. Moreover, N is affine and connected.*

PROOF. The existence of N is obtained as in the proof of Proposition 4.1.4.

We now claim that there exists a connected affine normal subgroup $H \trianglelefteq G$ such that G/H is proper. For this, we may reduce to the case where G is smooth by using Proposition 2.9.2 if $\mathrm{char}(k) > 0$. Then we may take for H the kernel of the Albanese homomorphism of G^0 (Theorem 4.3.4); this proves the claim.

Given H as in that claim, N is a normal subgroup scheme of H, and hence is affine. Moreover, H/N is affine, connected, and proper since G/N is proper. Thus, H/N is infinitesimal; it follows that N is connected. □

Notes and references.

Theorem 4.1.2 is called *Weil's extension theorem.* The proof presented here is taken from that of [**22**, Cor. 1.44].

Many results of this section are due to Rosenlicht. More specifically, Lemma 4.2.3 is a version of [**52**, Thm. 14], and Theorem 4.2.5, of [**52**, Cor. , p. 434]; Lemma 4.3.1 is [**52**, Lem. 1, p. 437].

Corollary 4.2.6 is called *Poincaré's complete reducibility theorem*; it is proved e.g. in [**45**, p. 173] over an algebraically closed field, and in [**20**, Cor. 3.20] over an arbitrary field. It implies that every abelian variety is isogenous to a product of simple abelian varieties, and these are uniquely determined up to isogeny and reordering.

Example 4.2.7 develops a construction of Raynaud, see [**24**, XVII.C.5]. The proof of Lemma 4.3.5 is taken from [**10**, 9.2 Thm. 1]. Corollary 4.3.7 is due to Raynaud, see [**51**, IX.2.7].

5. Some further developments

5.1. The Rosenlicht decomposition. Throughout this section, G denotes a smooth connected algebraic group. By Theorem 2, G has a smallest connected normal affine subgroup scheme $G_{\rm aff}$ with quotient being an abelian variety; also, recall that $G_{\rm aff}$ is the kernel of the Albanese homomorphism $\alpha : G \to \mathrm{Alb}(G)$. On the other hand, by Theorem 1, every algebraic group H has a largest anti-affine subgroup scheme that we will denote by $H_{\rm ant}$; moreover, $H_{\rm ant}$ is smooth, connected, and contained in the center of H^0. Also, $H_{\rm ant}$ is the smallest normal subgroup scheme of H having an affine quotient.

We will analyze the structure of G in terms of those of $G_{\rm aff}$ and $G_{\rm ant}$. Note that $(G_{\rm aff})_{\rm ant}$ is trivial (since $G_{\rm aff}$ is affine), but $(G_{\rm ant})_{\rm aff}$ may have positive dimension as we will see in §5.5.

THEOREM 5.1.1. *Keep the above notation and assumptions.*

(1) $G = G_{\rm aff} \cdot G_{\rm ant}$.
(2) $G_{\rm aff} \cap G_{\rm ant}$ *contains* $(G_{\rm ant})_{\rm aff}$.
(3) *The quotient* $(G_{\rm aff} \cap G_{\rm ant})/(G_{\rm ant})_{\rm aff}$ *is finite.*

PROOF. (1) By Proposition 2.8.5, $G_{\rm aff} \cdot G_{\rm ant}$ is a normal subgroup scheme of G. Moreover, the quotient $G \to G/G_{\rm aff} \cdot G_{\rm ant}$ factors through a homomorphism $G/G_{\rm aff} \to G/G_{\rm aff} \cdot G_{\rm ant}$ and also through a homomorphism $G/G_{\rm ant} \to G/G_{\rm aff} \cdot G_{\rm ant}$, in view of Proposition 2.8.4. In particular, $G/G_{\rm aff} \cdot G_{\rm ant}$ is a quotient of an abelian variety, and hence is an abelian variety as well. Also, $G/G_{\rm aff} \cdot G_{\rm ant}$ is a quotient of an affine algebraic group, and hence is affine as well (Proposition 3.1.2). Thus, $G/G_{\rm aff} \cdot G_{\rm ant}$ is trivial; this proves the assertion.

(2) Proposition 2.8.5 yields an isomorphism $G_{\rm ant}/G_{\rm ant} \cap G_{\rm aff} \cong G/G_{\rm aff}$. In particular, $G_{\rm ant}/G_{\rm ant} \cap G_{\rm aff}$ is an abelian variety. Since $G_{\rm ant} \cap G_{\rm aff}$ is affine, the assertion follows from Theorem 2.

(3) Consider the quotient $\bar{G} := G/(G_{\rm ant})_{\rm aff}$. Then $\bar{G} = \bar{G}_{\rm aff} \cdot \bar{G}_{\rm ant}$, where $\bar{G}_{\rm aff} := G_{\rm aff}/(G_{\rm ant})_{\rm aff}$ and $\bar{G}_{\rm ant} := G_{\rm ant}/(G_{\rm ant})_{\rm aff}$. Moreover, $\bar{G}_{\rm aff}$ is affine (as a quotient of $G_{\rm aff}$) and $\bar{G}_{\rm ant}$ is an abelian variety. Thus, $\bar{G}_{\rm ant} \cap \bar{G}_{\rm ant}$ is finite. This yields the assertion in view of Proposition 2.8.3. □

REMARK 5.1.2. The above theorem is equivalent to the assertion that *the multiplication map of G induces an isogeny*

$$(G_{\rm aff} \times G_{\rm ant})/(G_{\rm ant})_{\rm aff} \longrightarrow G,$$

where $(G_{\rm ant})_{\rm aff}$ is viewed as a subgroup scheme of $G_{\rm aff} \times G_{\rm ant}$ via $x \mapsto (x, x^{-1})$.

Theorem 5.1.1 may also be reformulated in terms of the two Albanese varieties $\mathrm{Alb}(G) = G/G_{\mathrm{aff}}$ and $\mathrm{Alb}(G_{\mathrm{ant}}) = G_{\mathrm{ant}}/(G_{\mathrm{ant}})_{\mathrm{aff}}$. Namely, the inclusion $\iota : G_{\mathrm{ant}} \to G$ yields a homomorphism $\mathrm{Alb}(\iota) : \mathrm{Alb}(G_{\mathrm{ant}}) \to \mathrm{Alb}(G)$ with kernel isomorphic to $(G_{\mathrm{aff}} \cap G_{\mathrm{ant}})/(G_{\mathrm{ant}})_{\mathrm{aff}}$. The assertion (1) is equivalent to the surjectivity of $\mathrm{Alb}(\iota)$, and (3) amounts to the finiteness of its kernel. So Theorem 5.1.1 just means that $\mathrm{Alb}(\iota)$ *is an isogeny.*

PROPOSITION 5.1.3. *With the above notation and assumptions, G_{ant} is the smallest subgroup scheme $H \subseteq G$ such that $G = H \cdot G_{\mathrm{aff}}$.*

PROOF. Let $H \subseteq G$ be a subgroup scheme. Then $H_{\mathrm{ant}} \cdot G_{\mathrm{aff}} \subseteq G$ is a normal subgroup scheme, since H_{ant} is central in G; moreover, $G/H_{\mathrm{ant}} \cdot G_{\mathrm{aff}}$ is a quotient of G/G_{aff}, hence an abelian variety. If $G = H \cdot G_{\mathrm{aff}}$, then

$$G/H_{\mathrm{ant}} \cdot G_{\mathrm{aff}} \cong H/H \cap (H_{\mathrm{ant}} \cdot G_{\mathrm{aff}}).$$

Moreover, $H \cap (H_{\mathrm{ant}} \cdot G_{\mathrm{aff}})$ is a normal subgroup scheme of H containing H_{ant}; thus, the quotient $H/H \cap (H_{\mathrm{ant}} \cdot G_{\mathrm{aff}})$ is affine. So $G/H_{\mathrm{ant}} \cdot G_{\mathrm{aff}}$ is trivial, i.e., $G = H_{\mathrm{ant}} \cdot G_{\mathrm{aff}}$. By Proposition 2.8.4, it follows that $G/H_{\mathrm{ant}} \cong G_{\mathrm{aff}}/G_{\mathrm{aff}} \cap H_{\mathrm{ant}}$. The right-hand side is the quotient of an affine algebraic group by a normal subgroup scheme, and hence is affine. By Theorem 1, it follows that $H_{\mathrm{ant}} \supseteq G_{\mathrm{ant}}$. □

REMARK 5.1.4. By the above proposition, the extension

$$1 \longrightarrow G_{\mathrm{aff}} \longrightarrow G \longrightarrow \mathrm{Alb}(G) \longrightarrow 1$$

is split if and only if $G_{\mathrm{aff}} \cap G_{\mathrm{ant}}$ is trivial. But this fails in general; in fact, $G_{\mathrm{aff}} \cap G_{\mathrm{ant}}$ is generally of positive dimension, and hence the above extension does not split after pull-back by any isogeny (see Remark 5.5.6 below for specific examples).

We now present two applications of Theorem 5.1.1; first, to the derived subgroup $\mathcal{D}(G)$. Recall from [**23**, II.5.4.8] (see also [**24**, VIB.7.8]) that $\mathcal{D}(G)$ is the subgroup functor of G that assigns to any scheme S, the set of those $g \in G(S)$ such that g lies in the commutator subgroup of $G(S')$ for some scheme S', faithfully flat and of finite presentation over S. Moreover, the group functor $\mathcal{D}(G)$ is represented by a smooth connected subgroup scheme of G, and $\mathcal{D}(G)(\bar{k})$ is the commutator subgroup of $G(\bar{k})$.

COROLLARY 5.1.5. *With the above notation and assumptions, we have $\mathcal{D}(G) = \mathcal{D}(G_{\mathrm{aff}})$. In particular, $\mathcal{D}(G)$ is affine. Also, G is commutative if and only if G_{aff} is commutative.*

Our second application of Theorem 5.1.1 characterizes Lie algebras of algebraic groups in characteristic 0:

COROLLARY 5.1.6. *Assume that* $\mathrm{char}(k) = 0$ *and consider a finite-dimensional Lie algebra $\mathfrak{g}$ over k, with center $\mathfrak{z}$. Then the following conditions are equivalent:*

(1) $\mathfrak{g} = \mathrm{Lie}(G)$ *for some algebraic group G.*
(2) $\mathfrak{g} = \mathrm{Lie}(L)$ *for some linear algebraic group L.*
(3) $\mathfrak{g}/\mathfrak{z}$ *(viewed as a Lie subalgebra of* $\mathrm{Lie}(\mathrm{GL}(\mathfrak{g}))$ *via the adjoint representation) is the Lie algebra of some algebraic subgroup of* $\mathrm{GL}(\mathfrak{g})$.

PROOF. Since (2)⇔(3) follows from [**17**, V.5.3] and (2)⇒(1) is obvious, it suffices to show that (1)⇒(2).

Let G be an algebraic group such that $\mathfrak{g} = \mathrm{Lie}(G)$. We may assume that G is connected. Then $G = G_{\mathrm{aff}} \cdot G_{\mathrm{ant}} = G_{\mathrm{aff}} \cdot Z$ and hence $\mathfrak{g} = \mathfrak{g}_{\mathrm{aff}} + \mathfrak{z}$ with an obvious notation. Thus, we have a decomposition of Lie algebras $\mathfrak{g} = \mathfrak{g}_{\mathrm{aff}} \oplus \mathfrak{a}$ for some linear subspace $\mathfrak{a} \subseteq \mathfrak{z}$, viewed as an abelian Lie algebra. Let $n := \dim(\mathfrak{a})$, then $L := G_{\mathrm{aff}} \times \mathbb{G}_a^n$ is a connected linear algebraic group with Lie algebra isomorphic to $\mathfrak{g}$. □

The Lie algebras satisfying the condition (2) above are called *algebraic*.

5.2. Equivariant compactification of homogeneous spaces.

DEFINITION 5.2.1. Let G be an algebraic group and $H \subseteq G$ a subgroup scheme. An *equivariant compactification* of the homogeneous space G/H is a proper G-scheme X equipped with an open equivariant immersion $G/H \to X$ with schematically dense image.

Equivalently, X is a G-scheme equipped with a base point $x \in X(k)$ such that the G-orbit of x is schematically dense in X and the stabilizer $\mathrm{C}_G(x)$ equals H.

THEOREM 5.2.2. *Let G be an algebraic group and $H \subseteq G$ a subgroup scheme. Then G/H has an equivariant compactification by a projective scheme.*

PROOF. If G is affine, then H is the stabilizer of a line ℓ in some finite-dimensional G-module V (see e.g. [**23**, II.2.3.5]). Then the closure of the G-orbit of ℓ in the projective space of lines of V yields the desired projective equivariant compactification X, in view of Proposition 2.7.5. Note that X is equipped with an ample G-linearized invertible sheaf in the sense of [**46**, I.3 Def. 1.6].

If G is proper, then the homogeneous space G/H is proper as well. So it suffices to check that G/H is projective. For this, we may assume k algebraically closed by using [**31**, IV.9.1.5]. Note that $H \subseteq H \cdot G^0 \subseteq G$ and the quotient scheme $G/H \cdot G^0$ is finite and étale, hence constant. So the scheme G/H is a finite disjoint union of copies of $H \cdot G^0/G^0 \cong G^0/G^0 \cap H$. Thus, we may assume that G is connected. If $\mathrm{char}(k) = 0$, then G, and hence G/H, are abelian varieties and thus projective. If $\mathrm{char}(k) > 0$, then Proposition 2.9.2 yields that $G/\mathrm{Ker}(F^n_{G/k})$ is an abelian variety for $n \gg 0$; thus, $G/H \cdot \mathrm{Ker}(F^n_{G/k})$ is an abelian variety as well, and hence is projective. Moreover, the natural map $f : G/H \to G/H \cdot \mathrm{Ker}(F^n_{G/k})$ is the quotient by the action of the infinitesimal group scheme $\mathrm{Ker}(F^n_{G/k})$, and hence is finite. It follows that G/H is projective.

For an arbitrary algebraic group G, Theorem 2 yields an affine normal subgroup scheme $N \subseteq G$ such that G/N is proper. Then $H \cdot N$ is a subgroup scheme of G and $G/H \cdot N \cong (G/N)/(H \cdot N/N)$ is proper as well, hence projective.

We now claim that it suffices to show the existence of a projective $H \cdot N$-equivariant compactification Y of $H \cdot N/H$ having an $H \cdot N$-linearized ample line bundle. Indeed, by [**46**, Prop. 7.1] applied to the projection $p_1 : G \times Y \to G$ and to the $H \cdot N$-torsor $q : G \to G/H \cdot N$, there exists a unique cartesian square

$$\begin{array}{ccc} G \times Y & \xrightarrow{p_1} & G \\ {\scriptstyle r}\downarrow & & \downarrow{\scriptstyle q} \\ X & \xrightarrow{f} & G/H \cdot N, \end{array}$$

where X is a G-scheme, r and f are G-equivariant, and r is an $H \cdot N$ torsor for the action defined by $x \cdot (g, y) := (gx^{-1}, xy)$, where $x \in H \cdot N$, $g \in G$ and $y \in Y$. Moreover, f is projective, and hence so is X. (We may view $f : X \to G/H \cdot N$ as the homogeneous fiber bundle with fiber Y associated with the principal bundle $q : G \to G/H \cdot N$ and the $H \cdot N$-scheme Y). Also, as in the proof of Proposition 2.8.4, we have a cartesian square

$$\begin{array}{ccc} G \times H \cdot N/H & \xrightarrow{p_1} & G \\ \downarrow n & & \downarrow q \\ G/H & \xrightarrow{f} & G/H \cdot N, \end{array}$$

where n is obtained from the multiplication map $G \times H \cdot N \to G$ (indeed, this square is commutative and the horizontal arrows are $H \cdot N/H$-torsors). Since Y is an equivariant compactification of $H \cdot N/H$, it follows that X is the desired equivariant compactification of G/H. This proves the claim.

In view of the first step of the proof (the case of an affine group G), it suffices in turn to check that $H \cdot N$ acts on $H \cdot N/H$ via an affine quotient group.

By Lemma 3.3.6, $(H \cdot N)_{\mathrm{ant}}$ is a quotient of $(H \ltimes N)_{\mathrm{ant}}$. The latter is the fiber at the neutral element of the affinization morphism $H \ltimes N \to \operatorname{Spec} \mathcal{O}(H \ltimes N)$. Also, recall that $H \ltimes N \cong H \times N$ as schemes, N is affine, and the affinization morphism commutes with finite products; thus, $(H \ltimes N)_{\mathrm{ant}} = H_{\mathrm{ant}}$. As a consequence, we have $(H \cdot N)_{\mathrm{ant}} = H_{\mathrm{ant}}$. Since $H \cdot N/H \cong (H \cdot N/H_{\mathrm{ant}})/(H/H_{\mathrm{ant}})$, and $H \cdot N/H_{\mathrm{ant}} = H \cdot N/(H \cdot N)_{\mathrm{ant}}$ is affine, this completes the proof. □

REMARKS 5.2.3. (i) With the notation and assumptions of the above theorem, the homogeneous space G/H is quasi-projective. In particular, every algebraic group is quasi-projective.

(ii) Assume in addition that G is smooth. Then G/H has an equivariant compactification by a *normal* projective scheme, as follows from Proposition 2.5.1.

(iii) If $\mathrm{char}(k) = 0$ then every homogeneous space has a *smooth* projective equivariant compactification. This follows indeed from the existence of an equivariant resolution of singularities; see [**38**, Prop. 3.9.1, Thm. 3.36].

(iv) Over any imperfect field k, there exist smooth connected algebraic groups G having no smooth compactification (equivariant or not). For example, choose $a \in k \setminus k^p$, where $p := \mathrm{char}(k)$, and consider the subgroup scheme $G \subseteq \mathbb{G}_a^2$ defined as the kernel of the homomorphism

$$\mathbb{G}_a^2 \longrightarrow \mathbb{G}_a, \quad (x, y) \longmapsto y^p - x - ax^p.$$

Then G_{k_i} is the kernel of the homomorphism $(x, y) \mapsto (y - a^{1/p}x)^p - x$; thus, $G_{k_i} \cong (\mathbb{G}_a)_{k_i}$ via the map $(x, y) \mapsto y - a^{1/p}x$. As a consequence, G is smooth, connected and unipotent. Moreover, G is equipped with a compactification X, the zero subscheme of $y^p - xz^{p-1} - ax^p$ in $\mathbb{P}^2$; the complement $X \setminus G$ consists of a unique k_i-rational point P with homogeneous coordinates $(1, a^{1/p}, 0)$. One may check that P is a regular, non-smooth point of X. Since G is a regular curve, it follows that X is its unique regular compactification.

5.3. Commutative algebraic groups. This section gives a brief survey of some structure results for the groups of the title. We will see how to build them from two classes: the (commutative) unipotent groups and the groups of multiplicative type, i.e., those algebraic groups M such that $M_{\bar{k}}$ is isomorphic to a subgroup scheme of some torus $\mathbb{G}^n_{m,\bar{k}}$. Both classes are stable under taking subgroup schemes, quotients, commutative group extensions, and extensions of the base field (see [**23**, IV.2.2.3, IV.2.2.6] for the unipotent groups, and [**23**, IV.1.2.4, IV.1.4.5] for those of multiplicative type). Also, every homomorphism between groups of different classes is constant (see [**23**, IV.2.2.4]).

The structure of commutative unipotent algebraic groups is very simple if $\mathrm{char}(k) = 0$: every such group G is isomorphic to its Lie algebra via the exponential map (see [**23**, IV.2.4]). In particular, G is a *vector group*, i.e., the additive group of a finite-dimensional vector space, uniquely determined up to isomorphism of vector spaces. In characteristic $p > 0$, the commutative unipotent groups are much more involved (see [**23**, V.1] for structure results over a perfect field); we just recall that for any such group G, the multiplication map p^n_G is zero for $n \gg 0$.

Next, we consider algebraic groups of multiplicative type. Any such group M is uniquely determined by its *character group* $X^*(M)$, consisting of the homomorphisms of k_s-group schemes $G_{k_s} \to (\mathbb{G}_m)_{k_s}$. Also, $X^*(M)$ is a finitely generated abelian group equipped with an action of the Galois group Γ. Moreover, the assignement $M \mapsto X^*(M)$ yields an anti-equivalence from the category of algebraic groups of multiplicative type (and homomorphisms) to that of finitely generated abelian groups equipped with a Γ-action (and Γ-equivariant homomorphisms). Also, M is a torus (resp. smooth) if and only if the group $X^*(M)$ is torsion-free (resp. p-torsion free if $\mathrm{char}(k) = p > 0$).

It follows readily that every algebraic group of multiplicative type M has a largest subtorus, namely, M^0_{red}; its character group is the quotient of $X^*(M)$ by the torsion subgroup. Moreover, the formation of the largest subtorus commutes with arbitrary field extensions (see [**23**, IV.1.3] for these results).

THEOREM 5.3.1. *Let G be a commutative affine algebraic group.*

(1) *G has a largest subgroup scheme of multiplicative type M, and the quotient G/M is unipotent.*
(2) *When k is perfect, G also has a largest unipotent subgroup scheme U, and $G = M \times U$.*
(3) *Returning to an arbitrary field k, there exists a subgroup scheme $H \subseteq G$ such that $G = M \cdot H$ and $M \cap H$ is finite.*
(4) *Any exact sequence of algebraic groups*

$$1 \longrightarrow G_1 \longrightarrow G \longrightarrow G_2 \longrightarrow 1$$

induces exact sequences

$$1 \to M_1 \to M \to M_2 \to 1, \quad 1 \to U_1 \to U \to U_2 \to 1$$

with an obvious notation.

PROOF. The assertions (1) and (2) are proved in [**23**, IV.3.1.1] and [**24**, XVII.7.2.1].

(3) We argue as in the proof of Theorem 4.3.4. By (2), we have $G_{k_i} = M_{k_i} \times U_i$ for a unique unipotent subgroup scheme $U_i \subseteq G_{k_i}$. Thus, U_i is defined over some subfield $K \subseteq k_i$, finite over k, and $G_K = M_K \times U'$ with an obvious notation. By Lemma 4.3.5, there exists a subgroup scheme $H \subseteq G$ such that $H_K \supseteq U'$ and H_K/U' is finite. Then $G_K = M_K \cdot H_K$ and $M_K \cap H_K$ is finite. This yields the assertion.

(4) We have $M_1 = M \cap G_1$ by construction. It follows that the quotient map $q : G \to G_2$ induces a closed immersion of group schemes $\iota : M/M_1 \to M_2$. Let N be the scheme-theoretic image of ι; then N is of multiplicative type and G_2/N is a quotient of G/M, hence is unipotent. Thus, $N = M_2$; this yields the first exact sequence, and in turn the second one. □

The assertion (2) of the above proposition fails over any imperfect field, as shown by the following variant of Example 4.2.7:

EXAMPLE 5.3.2. Let k be an imperfect field and choose a non-trivial finite purely inseparable field extension K of k. Then $G := \mathrm{R}_{K/k}(\mathbb{G}_{m,K})$ is a smooth connected commutative algebraic group of dimension $[K : k]$. Moreover, there is a canonical exact sequence

$$1 \longrightarrow U' \longrightarrow G_K \xrightarrow{q} \mathbb{G}_{m,K} \longrightarrow 1, \tag{5.3.1}$$

where U' is a non-trivial smooth connected unipotent group (see [**21**, A.5.11]). Also, we have a closed immersion of algebraic groups

$$j : \mathbb{G}_{m,k} \longrightarrow \mathrm{R}_{K/k}(\mathbb{G}_{m,K}) = G$$

in view of [**21**, A.5.7]. Thus, $j_K : \mathbb{G}_{m,K} \to G_K$ a closed immersion of algebraic groups as well, which yields a splitting of the exact sequence (5.3.1). Moreover, we have an exact sequence

$$1 \longrightarrow \mathbb{G}_{m,k} \xrightarrow{j} G \longrightarrow U \longrightarrow 1, \tag{5.3.2}$$

where U is smooth, connected and unipotent (since U_K is isomorphic to U').

We claim that (5.3.2) is not split. Indeed, it suffices to show that every homomorphism $f : H \to G$ is constant, where H is a smooth connected unipotent group. But as in Example 4.2.7, these homomorphisms correspond bijectively to the homomorphisms $f' : H_K \to \mathbb{G}_{m,K}$ via the assignement $f \mapsto f' := q \circ f$. Moreover, every such homomorphism is constant; this completes the proof of the claim.

Next, consider a smooth connected commutative algebraic group (not necessarily affine) over a perfect field k. By combining Theorems 4.3.2 and 5.3.1, we obtain an exact sequence

$$1 \longrightarrow T \times U \longrightarrow G \xrightarrow{q} A \longrightarrow 1, \tag{5.3.3}$$

where $T \subseteq G$ is the largest subtorus, and $U \subseteq G$ the largest smooth connected unipotent subgroup scheme. This yields in turn two exact sequences

$$1 \longrightarrow T \longrightarrow G/U \longrightarrow A \longrightarrow 1, \quad 1 \longrightarrow U \longrightarrow G/T \longrightarrow A \longrightarrow 1$$

and a homomorphism

$$f : G \longrightarrow G/U \times_A G/T \tag{5.3.4}$$

which is readily seen to be an isomorphism.

Thus, the structure of G is reduced to those of G/U and G/T. The former will be described in the next section. We now describe the latter under the assumption that $\mathrm{char}(k) = 0$; we then consider extensions

$$1 \longrightarrow U \longrightarrow G \longrightarrow A \longrightarrow 1, \tag{5.3.5}$$

where U is unipotent and A is a prescribed abelian variety. As U is a vector group, such an extension is called a *vector extension* of A.

When $U \cong \mathbb{G}_a$, every extension (5.3.5) yields a $\mathbb{G}_a$-torsor over A; this defines a map

$$\mathrm{Ext}^1(A, \mathbb{G}_a) \to H^1(A, \mathcal{O}_A)$$

which is in fact an isomorphism (see e.g. [**50**, III.17]). For an arbitrary vector group U, we obtain an isomorphism

$$\mathrm{Ext}^1(A, U) \cong H^1(A, \mathcal{O}_A) \otimes_k U.$$

Viewing the right-hand side as $\mathrm{Hom}_{\mathrm{gp.sch.}}(H^1(A, \mathcal{O}_A)^*, U)$, it follows that there is a *universal vector extension*

$$1 \longrightarrow H^1(A, \mathcal{O}_A)^* \longrightarrow E(A) \longrightarrow A \longrightarrow 1, \tag{5.3.6}$$

from which every extension (5.3.5) is obtained via push-out by a unique linear map $H^1(A, \mathcal{O}_A)^* \to U$.

REMARKS 5.3.3. (i) When $\mathrm{char}(k) > 0$, every abelian variety A still has a universal vector extension $E(A)$, as shown by the above arguments. Yet note that $E(A)$ classifies extensions of A by vector groups, which form a very special class of smooth connected unipotent commutative groups in this setting.

(ii) Little is known on the structure of commutative algebraic groups over imperfect fields, due to the failure of Chevalley's structure theorem. We will present a partial remedy to that failure in Corollary 5.6.8, as a consequence of a structure result for commutative algebraic groups in positive characteristics (Theorem 5.6.3).

5.4. Semi-abelian varieties.

DEFINITION 5.4.1. A *semi-abelian variety* is an algebraic group obtained as an extension

$$1 \longrightarrow T \longrightarrow G \stackrel{q}{\longrightarrow} A \longrightarrow 1, \tag{5.4.1}$$

where T is a torus and A an abelian variety.

REMARKS 5.4.2. (i) With the above notation, q is the Albanese homomorphism. It follows that T and A are uniquely determined by G.

(ii) Every semi-abelian variety is smooth and connected; it is also commutative in view of Corollary 5.1.5. Moreover, the multiplication map n_G is an isogeny for any non-zero integer n, since this holds for abelian varieties (by Lemma 4.2.1) and for tori.

(iii) Given a semi-abelian variety G over k and an extension of fields K of k, the K-algebraic group G_K is a semi-abelian variety.

In the opposite direction, we will need:

LEMMA 5.4.3. *Let G be an algebraic group. If $G_{\bar{k}}$ is a semi-abelian variety, then G is a semi-abelian variety.*

PROOF. Clearly, G is smooth, connected and commutative. Also, arguing as at the end of the proof of Theorem 4.3.4, we obtain a finite extension K of k and an exact sequence of algebraic groups over K

$$1 \longrightarrow T' \longrightarrow G_K \stackrel{q'}{\longrightarrow} A' \longrightarrow 1,$$

where T' is a torus and A' an abelian variety; here q' is the Albanese homomorphism. On the other hand, Theorem 4.3.4 yields an exact sequence

$$1 \longrightarrow N \longrightarrow G \stackrel{q}{\longrightarrow} A \longrightarrow 1,$$

where N is connected and affine, and A is an abelian variety; also, q is the Albanese homomorphism. By Remark 4.3.6, N_K contains T' and the quotient N_K/T' is infinitesimal; as a consequence, T' is the largest subtorus of N_K. Consider the largest subgroup scheme $M \subseteq N$ of multiplicative type; then N/M is unipotent and connected. Since the formation of M commutes with field extensions, we have $(N/M)_K = N_K/M_K$. The latter is a quotient of N_K/T'; hence N_K/M_K is infinitesimal, and so is N/M. Let $T := M^0_{\mathrm{red}}$; this is the largest subtorus of N and M/T is infinitesimal, hence so is N/T. It follows that $G/T =: A$ is proper; since G is smooth and connected, A is an abelian variety. Thus, G is a semi-abelian variety; moreover, $T_K = T'$ and $A_K = A'$ in view of Remark 5.4.2 (i). □

REMARK 5.4.4. More generally, if G_K is a semi-abelian variety for some field extension K of k, then G is a semi-abelian variety as well. This may be checked as in Remarks 3.1.5 and 4.1.5, by adapting the proof of [**21**, 1.1.9].

Next, we obtain a geometric characterization of semi-abelian varieties:

PROPOSITION 5.4.5. *Let G be a smooth connected algebraic group over a perfect field k. Then G is a semi-abelian variety if and only if every morphism (of schemes) $f : \mathbb{A}^1 \to G$ is constant.*

PROOF. Assume that G is semi-abelian and consider the exact sequence (5.4.1). Then $q \circ f : \mathbb{A}^1 \to A$ is constant by Lemma 4.1.1. Thus, there exists $g \in G(\bar{k})$ such that $f_{\bar{k}}$ factors through the translate $gT_{\bar{k}} \subset G_{\bar{k}}$. So we may view $f_{\bar{k}}$ as a morphism $\mathbb{A}^1_{\bar{k}} \to (\mathbb{A}^1_{\bar{k}} \setminus 0)^n$ for some positive integer n. Since every morphism $\mathbb{A}^1_{\bar{k}} \to \mathbb{A}^1_{\bar{k}} \setminus 0$ is constant, we see that f is constant.

Conversely, assume that every morphism $\mathbb{A}^1 \to G$ is constant. By [**60**, 14.3.10], it follows that every smooth connected unipotent subgroup of G is trivial. In particular, the unipotent radical of G_{aff} is trivial, i.e., G_{aff} is reductive. Also, G_{aff} has no root subgroups and hence is a torus. So G is a semi-abelian variety. □

As a consequence, semi-abelian varieties are stable under group extensions in a strong sense:

COROLLARY 5.4.6. *Let G be a smooth connected algebraic group and $N \subseteq G$ a subgroup scheme.*

(1) *If G is a semi-abelian variety, then so is G/N. If in addition N is smooth and connected, then N is a semi-abelian variety as well.*
(2) *If N and G/N are semi-abelian varieties, then so is G.*
(3) *Let $f : G \to H$ be an isogeny, where H is a semi-abelian variety. Then G is a semi-abelian variety.*

PROOF. We may assume k algebraically closed by Lemma 5.4.3.

(1) We have a commutative diagram of extensions

$$\begin{array}{ccccccccc} 1 & \longrightarrow & T \cap N & \longrightarrow & N & \longrightarrow & B & \longrightarrow & 1 \\ & & \downarrow & & \downarrow & & \downarrow & & \\ 1 & \longrightarrow & T & \longrightarrow & G & \longrightarrow & A & \longrightarrow & 1, \end{array}$$

where T is a torus, A an abelian variety, and B a subgroup scheme of A. This yields an exact sequence

$$1 \longrightarrow T/T \cap N \longrightarrow G/N \longrightarrow A/B \longrightarrow 1,$$

where $T/T \cap N$ is a torus and A/B is an abelian variety. Thus, G/N is a semi-abelian variety. The assertion on N follows from Proposition 5.4.5.

(2) Denote by $q : G \to G/N$ the quotient morphism. Let $f : \mathbb{A}^1 \to G$ be a morphism; then $q \circ f$ is constant by Proposition 5.4.5 again. Translating by some $g \in G(k)$, we may thus assume that f factors through N. Then f is constant; this yields the assertion.

(3) Consider the Albanese homomorphism $\alpha_H : H \to \mathrm{Alb}(H)$ and the kernel N of the composition $\alpha_H \circ f$. Then N is an extension of $\mathrm{Ker}(\alpha_H)$ (a torus) by $\mathrm{Ker}(f)$ (a finite group scheme). As a consequence, N is affine and N/T is finite, where $T \subseteq N$ denotes the largest subtorus. Since $G/N \cong \mathrm{Alb}(H)$ is an abelian variety, it follows that G/T is an abelian variety as well. Thus, G is a semi-abelian variety. □

REMARKS 5.4.7. (i) Proposition 5.4.5 fails over any imperfect field k, Indeed, as in Remark 5.2.3 (iv), consider the subgroup scheme $G \subset \mathbb{G}_a^2$ defined by $y^p = x + ax^p$, where $p := \mathrm{char}(k)$ and $a \in k \setminus k^p$. Then every morphism $f : \mathbb{A}^1 \to G$ is just given by $x(t), y(t) \in k[t]$ such that

$$y(t)^p = x(t) + a\,x(t)^p.$$

Thus, $x(t) = z(t)^p$ for a unique $z(t) \in k^{1/p}[t]$ such that

$$y(t) = z(t) + a^{1/p}\,z(t)^p.$$

Consider the monomial of highest degree in $z(t)$, say $a_n t^n$. If $n \geq 1$, then the monomial of highest degree in $y(t)$ is $a^{1/p}\,a_n^p\,t^{np}$. Since $a^{1/p} \notin k$ and $a_n^p \in k$, this contradicts the fact that $y(t) \in k[t]$. So $n = 0$, i.e., $z(t)$ is constant; then so are $x(t)$ and $y(t)$.

(ii) Consider a semi-abelian variety G and a semi-abelian subvariety $H \subseteq G$. Then the induced homomorphism between Albanese varieties, $\mathrm{Alb}(H) \to \mathrm{Alb}(G)$, has a finite kernel that may be arbitrary large. For example, let H be a non-trivial abelian variety over an algebraically closed field; then H contains a copy of the constant group scheme $\mathbb{Z}/\ell$ for any prime number $\ell \neq \mathrm{char}(k)$ (see [**45**, p. 64]). Choosing a root of unity of order ℓ in k, we obtain a closed immersion $j : \mathbb{Z}/\ell \to \mathbb{G}_m$ and, in turn, a commutative diagram of extensions

$$\begin{array}{ccccccccc}
1 & \longrightarrow & \mathbb{Z}/\ell & \longrightarrow & H & \longrightarrow & A & \longrightarrow & 1 \\
 & & \Big\downarrow{\scriptstyle j} & & \Big\downarrow & & \Big\downarrow{\scriptstyle \mathrm{id}} & & \\
1 & \longrightarrow & \mathbb{G}_m & \longrightarrow & G & \longrightarrow & A & \longrightarrow & 1,
\end{array}$$

where A is an abelian variety. So G is a semi-abelian variety containing H, and the kernel of the homomorphism $H = \mathrm{Alb}(H) \to \mathrm{Alb}(G) = A$ is $\mathbb{Z}/\ell$.

(iii) Consider again a semi-abelian variety G and let $H \subseteq G$ be a subgroup scheme. Then H^0_{red} is a semi-abelian variety. To see this, we may replace G, H with G/H_{ant}, H/H_{ant}, and hence assume that H is affine. Let T be the largest torus of G, then $H/T \cap H$ is affine and isomorphic to a subgroup scheme of the abelian variety G/T. Thus, $H/T \cap H$ is finite; it follows that $(H/T \cap H)^0_{\mathrm{red}}$ is trivial, and hence $H^0_{\mathrm{red}} \subseteq T$. So $H^0_{\mathrm{red}} = (T \cap H)^0_{\mathrm{red}}$ is a torus.

Next, we extend Lemma 4.1.3 to morphisms to semi-abelian varieties:

LEMMA 5.4.8. *Let X, Y be varieties equipped with k-rational points x_0, y_0 and let $f : X \times Y \to G$ be a morphism to a semi-abelian variety. Then we have identically*

$$f(x,y)\, f(x,y_0)^{-1}\, f(x_0,y)^{-1}\, f(x_0,y_0) = e.$$

PROOF. We may assume that $f(x_0,y_0) = e$. Let $q : G \to A$ denote the Albanese homomorphism. By Lemma 4.1.3, we have identically

$$(q \circ f)(x,y) = (q \circ f)(x,y_0)\,(q \circ f)(x_0,y).$$

Thus, the morphism

$$\varphi : X \times Y \longrightarrow G, \quad (x,y) \longmapsto f(x,y)\, f(x,y_0)^{-1}\, f(x_0,y)^{-1}$$

factors through the torus $T := \mathrm{Ker}(q)$. Also, we have identically $\varphi(x,y_0) = \varphi(x_0,y) = e$. We now show that φ is constant. For this, we may assume k algebraically closed; then $T \cong \mathbb{G}_m^n$ and accordingly, $\varphi = \varphi_1 \times \cdots \times \varphi_n$ for some $\varphi_i : X \times Y \to \mathbb{G}_m$. Equivalently, $\varphi \in \mathcal{O}(X \times Y)^*$ (the unit group of the algebra $\mathcal{O}(X \times Y)$). By [**25**, Lem. 2.1], there exists $u_i \in \mathcal{O}(X)^*$ and $v_i \in \mathcal{O}(Y)^*$ such that $\varphi_i(x,y) = u_i(x)\, v_i(y)$ identically. As $\varphi_i(x,y_0) = \varphi_i(x_0,y) = 1$, it follows that $\varphi_i = 1$; thus, φ factors through e. □

REMARKS 5.4.9. (i) In view of Corollary 5.4.6 and Lemma 5.4.8, *every smooth connected algebraic group G admits a universal homomorphism to a semi-abelian variety, which satisfies the assertions of Proposition* 4.1.4. In particular, the kernel N of this homomorphism is connected. If k is perfect, then one may check that $N = \mathcal{R}_u(G_{\mathrm{aff}}) \cdot \mathcal{D}(G_{\mathrm{aff}})$, where $\mathcal{R}_u$ denotes the unipotent radical (and $\mathcal{D}$ the derived subgroup); as a consequence, N is smooth. This fails over an arbitrary field k, as shown by Example 4.2.7 again (then N is the kernel of the Albanese homomorphism).

(ii) More generally, *every pointed variety admits a universal morphism to a semi-abelian variety*, as follows from [**57**, Thm. 7] over an algebraically closed field, and from [**66**, Thm. A1] over an arbitrary field. This universal morphism, which gives back that of Theorem 4.1.6, is still called the Albanese morphism.

Finally, we discuss the structure of semi-abelian varieties by adapting the approach to the classification of vector extensions (§5.3). Consider first the extensions

$$1 \longrightarrow \mathbb{G}_m \longrightarrow G \xrightarrow{\;q\;} A \longrightarrow 1, \tag{5.4.2}$$

where A is a prescribed abelian variety. Any such extension yields a $\mathbb{G}_m$-torsor over A, or equivalently a line bundle over that variety. This defines a map

$$\mathrm{Ext}^1(A, \mathbb{G}_m) \longrightarrow \mathrm{Pic}(A).$$

By the Weil-Barsotti formula (see e.g. [**50**, III.17, III.18]), this map is injective and its image is the subgroup of translation-invariant line bundles over A; this is the group of k-rational points of the dual abelian variety $\widehat{A}$. We may thus identify $\mathrm{Ext}^1(A, \mathbb{G}_m)$ with $\widehat{A}(k)$. Also, using a Poincaré sheaf, we will view the points of $\widehat{A}(k)$ as the algebraically trivial invertible sheaves $\mathcal{L}$ on A equipped with a rigidification at 0, i.e., an isomorphism $k \xrightarrow{\cong} \mathcal{L}_0$.

Denoting by $\mathcal{L} \in \widehat{A}(k)$ the invertible sheaf that corresponds to the extension (5.4.2), we obtain an isomorphism of sheaves of algebras over A

$$q_*(\mathcal{O}_G) \cong \bigoplus_{n=-\infty}^{\infty} \mathcal{L}^{\otimes n},$$

since G is the $\mathbb{G}_m$-torsor over A associated with $\mathcal{L}$. The multiplication in the right-hand side is given by the natural isomorphisms

$$\mathcal{L}^{\otimes m} \otimes_{\mathcal{O}_A} \mathcal{L}^{\otimes n} \xrightarrow{\cong} \mathcal{L}^{\otimes(m+n)}.$$

For an arbitrary extension (5.4.1), consider the associated extension

$$(5.4.3) \qquad 1 \longrightarrow T_{k_s} \longrightarrow G_{k_s} \xrightarrow{q_{k_s}} A_{k_s} \longrightarrow 1$$

and denote by Λ the character group of T. Then the split torus T_{k_s} is canonically isomorphic to $\mathrm{Hom}_{\mathrm{gp.sch.}}(\Lambda, (\mathbb{G}_m)_{k_s})$. Moreover, every $\lambda \in \Lambda$ yields an extension of the form (5.4.2) via pushout by $\lambda : T_{k_s} \to (\mathbb{G}_m)_{k_s}$. This defines a map

$$\gamma_{k_s} : \mathrm{Ext}^1(A_{k_s}, T_{k_s}) \to \mathrm{Hom}_{\mathrm{gp.}}(\Lambda, \widehat{A}(k_s))$$

which is readily seen to be an isomorphism by identifying T_{k_s} with $(\mathbb{G}_m)^n_{k_s}$ and accordingly Λ with $\mathbb{Z}^n$. Likewise, we obtain an isomorphism of sheaves of algebras over A_{k_s}:

$$(5.4.4) \qquad q_*(\mathcal{O}_G)_{k_s} \cong \bigoplus_{\lambda \in \Lambda} c(\lambda),$$

where $c : \Lambda \to \widehat{A}(k_s)$ denotes the map classifying the extension. Here again, the multiplication of the right-hand side is given by the natural isomorphisms

$$c(\lambda) \otimes_{\mathcal{O}_A} c(\mu) \xrightarrow{\cong} c(\lambda + \mu).$$

By construction, γ_{k_s} is equivariant for the natural actions of Γ. Composing γ_{k_s} with the base change map $\mathrm{Ext}^1(A, T) \to \mathrm{Ext}^1(A_{k_s}, T_{k_s})$ (which is Γ-invariant), we thus obtain a map

$$(5.4.5) \qquad \gamma : \mathrm{Ext}^1(A, T) \to \mathrm{Hom}^{\Gamma}_{\mathrm{gp.}}(\Lambda, \widehat{A}(k_s)).$$

By Galois descent, this yields:

PROPOSITION 5.4.10. *Let A be an abelian variety and T a torus. Then the map (5.4.5) is an isomorphism.*

5.5. Structure of anti-affine groups. Let G be an anti-affine group. Recall from Lemma 3.3.2 and Proposition 3.3.4 that G is smooth, connected and commutative; also, every quotient of G is anti-affine.

PROPOSITION 5.5.1. *When* $\mathrm{char}(k) = p > 0$, *every anti-affine group over k is a semi-abelian variety.*

PROOF. By Lemma 5.4.3, we may assume k algebraically closed. Using the isomorphism (5.3.4), we may further assume that G lies in an exact sequence

$$1 \longrightarrow U \longrightarrow G \xrightarrow{q} A \longrightarrow 1,$$

where U is unipotent and A is an abelian variety. We may then choose a positive integer n such that the power map p_U^n is zero. Then p_G^n factors through a morphism $f : G/U = A \to G$. Since f is surjective and the square

$$\begin{array}{ccc} G & \xrightarrow{p_G^n} & G \\ {\scriptstyle q}\downarrow & & \downarrow{\scriptstyle q} \\ A & \xrightarrow{p_A^n} & A \end{array}$$

commutes, the composition $q \circ f : A \to A$ equals p_A^n and hence is an isogeny. Let B be the image of f; then B is an abelian subvariety of G, isogenous to A via q. It follows that $G = U \cdot B$ and $U \cap B$ is finite. Thus, $U/U \cap B \cong G/B$ is affine (as a quotient of U) and anti-affine (as a quotient of G), hence trivial. Since U is smooth and connected, it must be trivial as well. □

Returning to an arbitrary field, we now characterize those semi-abelian varieties that are anti-affine:

PROPOSITION 5.5.2. *Let G be a semi-abelian variety, extension of an abelian variety A by a torus T with character group Λ, and let $c : \Lambda \to \widehat{A}(k_s)$ be the map classifying this extension. Then G is anti-affine if and only if c is injective.*

PROOF. Since G is anti-affine if and only if $G_{\bar{k}}$ is anti-affine, we may assume k algebraically closed. Then the isomorphism (5.4.4) yields

$$\mathcal{O}(G) = H^0(A, q_*(\mathcal{O}_G)) \cong \bigoplus_{\lambda \in \Lambda} H^0(A, c(\lambda)).$$

Thus, G is anti-affine if and only if $H^0(A, c(\lambda)) = 0$ for all non-zero λ. So it suffices to check that $H^0(A, \mathcal{L}) = 0$ for any non-zero $\mathcal{L} \in \widehat{A}(k)$.

If $H^0(A, \mathcal{L}) \neq 0$, then $\mathcal{L} \cong \mathcal{O}_A(D)$ for some non-zero effective divisor D on A. Since $\mathcal{L}$ is algebraically trivial, the intersection number $D \cdot C$ is zero for any irreducible curve C on A. Now choose a smooth point $x \in D_{\mathrm{red}}(k)$; then there exists an irreducible curve C through x which intersects D_{red} transversally at that point. Then $D \cdot C > 0$, a contradiction. □

COROLLARY 5.5.3. *The anti-affine semi-abelian varieties over a field k are classified by the pairs (A, Λ), where A is an abelian variety over k and $\Lambda \subset \widehat{A}(k_s)$ is a Γ-stable free abelian subgroup of finite rank.*

If the ground field k is finite, then the group $\widehat{A}(k_s) = \widehat{A}(\bar{k})$ is the union of the subgroups $\widehat{A}(K)$, where K runs over all finite extensions of k. Since all these subgroups are finite, $\widehat{A}(k_s)$ is a torsion group. In view of Proposition 5.5.1 and Corollary 5.5.3, this readily yields:

COROLLARY 5.5.4. *Any anti-affine group over a finite field is an abelian variety.*

Using the Rosenlicht decomposition (Theorem 5.1.1), this yields in turn:

COROLLARY 5.5.5. *Let G be a smooth connected algebraic group over a finite field k. Then there is a unique decomposition $G = L \cdot A$, where $L \trianglelefteq G$ is smooth, connected and affine, and $A \subseteq G$ is an abelian variety. Moreover, $L \cap A$ is finite.*

REMARK 5.5.6. Returning to an abelian variety A over an arbitrary field k, assume that there exists an invertible sheaf $\mathcal{L} \in \widehat{A}(k)$ of infinite order (such a pair $(A, \mathcal{L})$ exists unless k is algebraic over a finite field, as seen in Example 3.2.3). Consider the associated extension

$$1 \longrightarrow \mathbb{G}_m \longrightarrow G \longrightarrow A \longrightarrow 1.$$

Then G is anti-affine by Proposition 5.5.2, and hence the above extension is not split. In fact, it does not split after pull-back by any isogeny $B \to A$: otherwise, it would split after pull-back by $n_A : A \to A$ for some positive integer n, or equivalently, after push-forward under the nth power map of $\mathbb{G}_m$. But this push-forward amounts to replacing $\mathcal{L}$ with $\mathcal{L}^{\otimes n}$, which is still of infinite order.

Next, we turn to the classification of anti-affine groups in characteristic 0. As a first step, we obtain:

LEMMA 5.5.7. *Let G be a connected commutative algebraic group over a field of characteristic 0. Let $T \subseteq G$ be the largest torus and $U \subseteq G$ the largest unipotent subgroup scheme. Then G is anti-affine if and only if G/U and G/T are anti-affine.*

PROOF. If G is anti-affine, then so are its quotients G/U and G/T. Conversely, assume that G/U and G/T are anti-affine. Then $G/G_{\mathrm{ant}} \cdot U$ is anti-affine, and also affine as a quotient of G/G_{ant}. Thus, $G = G_{\mathrm{ant}} \cdot U$ and hence $G/G_{\mathrm{ant}} \cong U/U \cap G_{\mathrm{ant}}$ is unipotent. Likewise, one shows that G/G_{ant} is a torus. Thus, G/G_{ant} is trivial. □

With the notation and assumptions of the above lemma, G/U is a semi-abelian variety and G/T is a vector extension of A. So to complete the classification, it remains to characterize those vector extensions that are anti-affine:

PROPOSITION 5.5.8. *Assume that* $\mathrm{char}(k) = 0$. *Let G be an extension of an abelian variety A by a vector group U and denote by $\gamma : H^1(A, \mathcal{O}_A)^* \to U$ the linear map classifying the extension. Then G is anti-affine if and only if γ is surjective.*

PROOF. We have to show that the universal vector extension $E(A)$ is anti-affine, and every anti-affine vector extension (5.3.5) is a quotient of $E(A)$.

Let $V := H^1(A, \mathcal{O}_A)^*$; then $V = E(A)_{\mathrm{aff}}$ with the notation of the Rosenlicht decomposition. By that decomposition, $(E(A)_{\mathrm{ant}})_{\mathrm{aff}} \subseteq V \cap E(A)_{\mathrm{aff}}$ and the quotient is finite. Since $V \cap E(A)_{\mathrm{aff}}$ is a vector group, we obtain the equality $(E(A)_{\mathrm{ant}})_{\mathrm{aff}} = V \cap E(A)_{\mathrm{aff}} =: W$. In view of Remark 5.1.2, this yields a commutative diagram of extensions

$$\begin{array}{ccccccccc} 1 & \longrightarrow & W & \longrightarrow & E(A)_{\mathrm{ant}} & \longrightarrow & A & \longrightarrow & 1 \\ & & \downarrow{\scriptstyle\iota} & & \downarrow & & \downarrow{\scriptstyle\mathrm{id}} & & \\ 1 & \longrightarrow & V & \longrightarrow & E(A) & \longrightarrow & A & \longrightarrow & 1, \end{array}$$

where ι is injective. As a consequence, every extension of A by $\mathbb{G}_a$ is obtained by pushout from the top extension, i.e., the resulting map $W^* \to \mathrm{Ext}^1(A, \mathbb{G}_a)$ is surjective. Since the bottom extension is universal, it follows that $W = V$ and $E(A)_{\mathrm{ant}} = E(A)$.

Next, let G be a vector extension of A. If G is anti-affine, then the classifying map $\gamma : E(A) \to G$ is surjective by Lemma 3.3.6. The converse assertion follows from the fact that every quotient of an anti-affine group is anti-affine. □

As a consequence, *the anti-affine vector extensions of A are classified by the linear subspaces $V \subseteq H^1(A, \mathcal{O}_A)$* by assigning to any such extension, the image of the transpose of its classifying map. Combining this result with the isomorphism (5.3.4), Lemma 5.5.7 and Proposition 5.5.2, we obtain the desired classification:

THEOREM 5.5.9. *When* $\mathrm{char}(k) > 0$, *the anti-affine groups are classified by the pairs (A, Λ), where A is an abelian variety over k and $\Lambda \subseteq \widehat{A}(k_s)$ is a Γ-stable free abelian subgroup of finite rank.*

When $\mathrm{char}(k) = 0$, *the anti-affine groups are classified by the triples (A, Λ, V), where (A, Λ) is as above and $V \subseteq H^1(A, \mathcal{O}_A)$ is a linear subspace.*

5.6. Commutative algebraic groups (continued). We first show that every group as in the title has a "semi-abelian radical":

LEMMA 5.6.1. *Let G be a commutative algebraic group.*

(1) *G has a largest semi-abelian subvariety that we will denote by G_{sab}. Moreover, $(G/G_{\mathrm{sab}})_{\mathrm{sab}}$ is trivial.*
(2) *The formation of G_{sab} commutes with algebraic field extensions.*

PROOF. (1) This follows from the stability of semi-abelian varieties under taking quotients and extensions (Corollary 5.4.6) as in the proof of Lemma 3.1.4.

(2) When $\mathrm{char}(k) = 0$, the statement is obtained by Galois descent as in the proof of Lemma 3.1.4 again. When $\mathrm{char}(k) = p > 0$, we have $G_{\mathrm{ant}} \subseteq G_{\mathrm{sab}}$ by Proposition 5.5.1; also, $G_{\mathrm{sab}}/G_{\mathrm{ant}} = (G/G_{\mathrm{ant}})_{\mathrm{sab}}$ as follows from Corollary 5.4.6 again. Since G/G_{ant} is affine, $(G/G_{\mathrm{ant}})_{\mathrm{sab}}$ is just its largest subtorus. As the formations of G_{ant} and of the largest subtorus commute with field extensions, the assertion follows. □

Next, we characterize those commutative algebraic groups that have a trivial semi-abelian radical:

LEMMA 5.6.2. *If* $\mathrm{char}(k) = p > 0$, *then the following conditions are equivalent for a commutative algebraic group G:*

(1) *G_{sab} is trivial.*
(2) *G is affine and its largest subgroup of multiplicative type is finite.*
(3) *The multiplication map n_G is zero for some positive integer n.*

If in addition G is smooth and connected, these conditions are equivalent to G being unipotent.

PROOF. (1)⇒(2) Note that G_{ant} is trivial in view of Proposition 5.5.1. Thus, G is affine. Also, G contains no non-trivial torus; this yields the assertion.

(2)⇒(3) By Theorem 5.3.1, we have an exact sequence

$$1 \longrightarrow M \longrightarrow G \longrightarrow U \longrightarrow 1,$$

where M is of multiplicative type and U is unipotent. Then M is finite, and hence killed by n_M for some positive integer n. Also, recall that U is killed by p_U^m for some m. It follows that $np_G^m = 0$.

(3)⇒(1) This follows from the fact that $n_H \neq 0$ for any non-trivial semi-abelian variety H and any $n \neq 0$.

When G is smooth and connected, the condition (2) implies that G is unipotent, in view of Theorem 5.3.1 again. Conversely, if G is unipotent, then it clearly satisfies the condition (2). □

We say that a commutative algebraic group G *has finite exponent* if it satisfies the above condition (3). When $\mathrm{char}(k) = 0$, this just means that G is finite; when $\mathrm{char}(k) > 0$, this amounts to G being an extension of a unipotent algebraic group by a finite group scheme of multiplicative type.

THEOREM 5.6.3. *Assume that* $\mathrm{char}(k) = p > 0$ *and consider a commutative algebraic group* G *and its largest semi-abelian subvariety* G_{sab}.

(1) *The quotient* G/G_{sab} *has finite exponent.*
(2) *There exists a subgroup scheme* $H \subseteq G$ *such that* $G = G_{\mathrm{sab}} \cdot H$ *and* $G_{\mathrm{sab}} \cap H$ *is finite.*
(3) *If* G *is smooth and connected, then* G/G_{sab} *is unipotent. If in addition* k *is perfect, then we may take for* H *the largest smooth connected unipotent subgroup scheme of* G.

PROOF. (1) This follows from Lemmas 5.6.1 and 5.6.2.

(2) By (1), we may choose $n > 0$ such that $n_{G/G_{\mathrm{sab}}} = 0$. Then n_G factors through G_{sab}; also, recall that $n_{G_{\mathrm{sab}}}$ is an isogeny. It follows that $G = G_{\mathrm{sab}} \cdot \mathrm{Ker}(n_G)$ and $G_{\mathrm{sab}} \cap \mathrm{Ker}(n_G)$ is finite.

(3) The first assertion is a consequence of Lemma 5.6.2 again.

Assume k perfect and consider the Rosenlicht decomposition $G = G_{\mathrm{aff}} \cdot G_{\mathrm{ant}}$. Then $G_{\mathrm{aff}} = T \times U$, where T is a torus and U a smooth connected unipotent group. Thus, $G = T \cdot U \cdot G_{\mathrm{ant}}$. Also, $T \cdot G_{\mathrm{ant}}$ is a semi-abelian subvariety of G; moreover, $G/T \cdot G_{\mathrm{ant}}$ is isomorphic to a quotient of U, and hence contains no semi-abelian variety. Thus, $T \cdot G_{\mathrm{ant}} = G_{\mathrm{sab}}$ and hence $G = G_{\mathrm{sab}} \cdot U$. Moreover, $G_{\mathrm{sab}} \cap U$ is clearly finite. □

In analogy with the structure of commutative affine algebraic groups (Theorem 5.3.1), note that the class of commutative algebraic groups of finite exponent is stable under taking subgroup schemes, quotients, commutative group extensions, and extensions of the ground field. Moreover, when G is a semi-abelian variety and H a commutative algebraic group of finite exponent, every homomorphism $\varphi : G \to H$ is constant, and every homomorphism $\psi : H \to G$ factors through a finite subgroup scheme of G.

Also, note the following analogue of Lemma 3.3.6, which yields that the assignement $G \mapsto G_{\mathrm{sab}}$ is close to being exact:

LEMMA 5.6.4. *Assume that* $\mathrm{char}(k) = p$. *Let* G *be a commutative algebraic group and* $H \subseteq G$ *a subgroup scheme.*

(1) $H_{\mathrm{sab}} \subseteq G_{\mathrm{sab}} \cap H$ *and the quotient is finite.*
(2) *The quotient map* $q : G \to G/H$ *yields an isomorphism*

$$G_{\mathrm{sab}}/G_{\mathrm{sab}} \cap H \xrightarrow{\cong} (G/H)_{\mathrm{sab}}.$$

PROOF. (1) This follows readily from Corollary 5.4.6 and Lemma 5.6.1.

(2) Observe that q restricts to a closed immersion of group schemes

$$\iota : G_{\mathrm{sab}}/G_{\mathrm{sab}} \cap H \longrightarrow G/H$$

that we will regard as an inclusion. Also, by Theorem 5.6.3, the quotient $G/G_{\mathrm{sab}} \cdot H$ has finite exponent. Since $G/G_{\mathrm{sab}} \cdot H \cong (G/H)/(G_{\mathrm{sab}} \cdot H/H)$ is isomorphic to the cokernel of ι, it follows that $G_{\mathrm{sab}}/G_{\mathrm{sab}} \cap H \supseteq (G/H)_{\mathrm{sab}}$. But the opposite inclusion holds by Corollary 5.4.6 again; this yields the assertion. □

REMARK 5.6.5. When $\mathrm{char}(k) = 0$, the quotient G/G_{sab} is not necessarily unipotent for a smooth connected commutative group G. This happens for example when G is the universal vector extension of a non-trivial abelian variety; then G_{sab} is trivial.

Still, the structure of such a group G reduces somehow to that of its anti-affine part. Specifically, one may check that there exists a subtorus $T \subseteq G$ and a vector subgroup $U \subseteq G$ such that the multiplication map $T \times U \times G_{\mathrm{ant}} \to G$ is an isogeny; moreover, T is uniquely determined up to isogeny, and U is uniquely determined.

As an application of Theorem 5.6.3, we present a remedy to (or a measure of) the failure of Chevalley's structure theorem over imperfect fields. To state it, we need the following:

DEFINITION 5.6.6. A smooth connected algebraic group G is called a *pseudo-abelian variety* if it does not contain any non-trivial smooth connected affine normal subgroup scheme.

REMARKS 5.6.7. (i) By Lemma 3.1.4, every smooth connected algebraic group G lies in a unique exact sequence

$$1 \longrightarrow L \longrightarrow G \longrightarrow Q \longrightarrow 1,$$

where L is smooth, connected and linear, and Q is a pseudo-abelian variety.

(ii) If k is perfect, then every pseudo-abelian variety is just an abelian variety, as follows from Theorem 4.3.2. But there exist non-proper pseudo-abelian varieties over any imperfect field, as shown by Example 4.2.7.

COROLLARY 5.6.8. *Let G be a pseudo-abelian variety. Then G is commutative and lies in a unique exact sequence*

$$1 \longrightarrow A \longrightarrow G \longrightarrow U \longrightarrow 1, \tag{5.6.1}$$

where A is an abelian variety and U is unipotent.

PROOF. The commutativity of G follows from Corollary 5.1.5.

For the remaining assertion, we may assume that $\mathrm{char}(k) = p > 0$. By Theorem 5.6.3, G lies in a unique extension

$$1 \longrightarrow G_{\mathrm{sab}} \longrightarrow G \longrightarrow U \longrightarrow 1,$$

where G_{sab} is a semi-abelian variety and U is unipotent. Since G contains no non-trivial torus, G_{sab} is an abelian variety. This shows the existence of the exact sequence (5.6.1); for its uniqueness, just observe that $A = G_{\mathrm{ant}}$. □

Notes and references.

Theorem 5.1.1 is due to Rosenlicht (see [**52**, Cor. 5, p. 140]). This result is very useful for reducing questions about general algebraic groups to the linear and anti-affine cases; see [**43**] for a recent application.

When $\mathrm{char}(k) = p > 0$, characterizing Lie algebras of *smooth* algebraic groups among p-Lie algebras seems to be an open problem. It is well-known that every finite-dimensional p-Lie algebra is the Lie algebra of some *infinitesimal* group scheme; see [**23**, II.7.3, II.7.4] for this result and further developments.

The proof of Theorem 5.2.2 is adapted from [**14**, Thm. 4.13]. The quasi-projectivity of homogeneous spaces is a classical result, see e.g. [**51**, Cor. VI.2.6].

Also, the existence of equivariant compactifications of certain homogeneous spaces having no separable point at infinity has attracted recent interest, see [**26**, **27**].

In positive characteristics, the existence of (equivariant or not) *regular* projective compactifications of homogeneous spaces is an open question.

Example 5.3.2 is due to Raynaud, see [**24**, XVII.C.5].

The algebraic group considered in Remark 5.4.7 (i) is an example of a *k-wound* unipotent group G, i.e., every morphism $\mathbb{A}^1 \to G$ is constant. This notion plays an important rôle in the structure of smooth connected unipotent groups over imperfect fields, see [**21**, B.2].

The notion of Albanese morphism extends to non-pointed varieties by replacing semi-abelian varieties with semi-abelian torsors. More specifically, for any variety X, there exists a semi-abelian variety Alb^0_X, an Alb^0_X-torsor Alb^1_X (over $\mathrm{Spec}(k)$) and a morphism

$$u_X : X \longrightarrow \mathrm{Alb}^1_X$$

such that for any semi-abelian variety A^0, any torsor A^1 under A^0, and any morphism $f : X \to A^1$, there exists a unique morphism of varieties

$$g^1 : \mathrm{Alb}^1_X \to A^1$$

such that $g^1 \circ u_X = f$ and there exists a unique morphism of algebraic groups

$$g^0 : \mathrm{Alb}^0_X \to A^0$$

such that g_1 is g_0-equivariant (see [**66**, App. A]).

The structure of anti-affine algebraic groups has been obtained in [**11**] and [**56**] independently. Our exposition follows that of [**11**] with some simplifications.

The notion of a pseudo-abelian variety is due to Totaro in [**61**], as well as Corollary 5.6.8 and further results about these varieties. In particular, it is shown that every smooth connected commutative group of exponent p occurs as the unipotent quotient of some pseudo-abelian variety, see [**61**, Cor. 6.5, Cor. 7.3]). This yields many more examples of pseudo-abelian varieties than those constructed in Example 4.2.7. Yet a full description of pseudo-abelian varieties is an open problem.

6. The Picard scheme

6.1. Definitions and basic properties.

DEFINITION 6.1.1. Let X be a scheme. The *relative Picard functor*, denoted by $\mathrm{Pic}_{X/k}$, is the commutative group functor that assigns to any scheme S the group $\mathrm{Pic}(X \times S)/p_2^* \mathrm{Pic}(S)$, where $p_2 : X \times S \to S$ denotes the projection, and to any morphism of schemes $f : S' \to S$, the homomorphism induced by pull-back.

If X is equipped with a k-rational point x, then for any scheme S, the map $x \times \mathrm{id} : S \to X \times S$ is a section of $p_2 : X \times S \to S$. Thus,

$$(x \times \mathrm{id})^* : \mathrm{Pic}(X \times S) \to \mathrm{Pic}(S)$$

is a retraction of

$$p_2^* : \mathrm{Pic}(S) \to \mathrm{Pic}(X \times S).$$

So we may view $\mathrm{Pic}_{X/k}(S)$ as the group of isomorphism classes of invertible sheaves on $X \times S$, trivial along $x \times S$. If in addition S is equipped with a k-rational point s, then we obtain a pull-back map $s^* : \mathrm{Pic}_{X/k}(S) \to \mathrm{Pic}_{X/k}(s) = \mathrm{Pic}(X)$ with kernel isomorphic to $\mathrm{Pic}(X \times S)/p_1^* \mathrm{Pic}(X) \times p_2^* \mathrm{Pic}(S)$. Indeed, the map

$$s^* \times x^* : \mathrm{Pic}(X \times S) \longrightarrow \mathrm{Pic}(X) \times \mathrm{Pic}(S)$$

is a retraction of

$$p_1^* \times p_2^* : \mathrm{Pic}(X) \times \mathrm{Pic}(S) \longrightarrow \mathrm{Pic}(X \times S).$$

The kernel of $s^* \times x^*$ is called the group of *divisorial correspondences.*

THEOREM 6.1.2. *Let X be a proper scheme having a k-rational point.*

(1) $\mathrm{Pic}_{X/k}$ *is represented by a locally algebraic group (that will still be denoted by* $\mathrm{Pic}_{X/k}$*).*
(2) $\mathrm{Lie}(\mathrm{Pic}_{X/k}) = H^1(X, \mathcal{O}_X)$.
(3) *If* $H^2(X, \mathcal{O}_X) = 0$ *then* $\mathrm{Pic}_{X/k}$ *is smooth.*

PROOF. (1) This is proved in [**47**, II.15] via a characterization of commutative locally algebraic groups among commutative group functors, in terms of seven axioms. When X is projective, there is an alternative proof via the Hilbert scheme, see [**35**, Thm. 4.8].

The assertion (2) follows from [**35**, Thm. 5.11], and (3) from [**35**, Prop. 5.19]. □

With the notation and assumptions of the above theorem, the neutral component, $\mathrm{Pic}^0_{X/k}$, is a connected algebraic group by Theorem 2.4.1. The group of connected components, $\pi_0(\mathrm{Pic}_{X/k})$, is called the *Néron-Severi group*; we denote it by $\mathrm{NS}_{X/k}$. The formation of $\mathrm{Pic}_{X/k}$ commutes with field extensions; hence the same holds for $\mathrm{Pic}^0_{X/k}$ and $\mathrm{NS}_{X/k}$. Also, the commutative group $\mathrm{NS}_{X/k}(\bar{k})$ is finitely generated in view of [**6**, XIII.5.1].

REMARKS 6.1.3. (i) For any pointed scheme (S, s), we have a natural isomorphism of groups

$$\mathrm{Hom}_{\mathrm{pt.sch.}}(S, \mathrm{Pic}_{X/k}) \cong \mathrm{Pic}(X \times S)/p_1^* \mathrm{Pic}(X) \times p_2^* \mathrm{Pic}(S), \tag{6.1.1}$$

where the left-hand side denotes the subgroup of $\mathrm{Hom}(S, \mathrm{Pic}_{X/k})$ consisting of those f such that $f(s) = 0$, i.e., the kernel of $s^* : \mathrm{Pic}_{X/k}(S) \to \mathrm{Pic}(X)$.

(ii) If X is an abelian variety, then $\mathrm{Pic}^0_{X/k}$ is the dual abelian variety, $\widehat{X}$.

DEFINITION 6.1.4. Let X be a proper scheme over a perfect field k, having a k-rational point. The *Picard variety* $\mathrm{Pic}^0(X)$ is the reduced scheme $(\mathrm{Pic}^0_{X/k})_{\mathrm{red}}$.

With the above notation and assumptions, $\mathrm{Pic}^0(X)$ is a smooth connected commutative algebraic group; its formation commutes with field extensions.

Returning to a proper scheme X over an arbitrary ground field k, having a k-rational point, we say that *the Picard variety of X exists* if $\mathrm{Pic}^0(X) := (\mathrm{Pic}^0_{X/k})_{\mathrm{red}}$ is a subgroup scheme of the Picard scheme. We will see that this holds under mild assumptions on the singularities of X.

6.2. Structure of Picard varieties. Throughout this subsection, we consider a proper variety X equipped with a k-rational point x.

PROPOSITION 6.2.1. *The Albanese variety of X is canonically isomorphic to the dual of the largest abelian subvariety of* $\mathrm{Pic}^0_{X/k}$.

PROOF. Let A be an abelian variety. Then we have

$$\mathrm{Hom}_{\mathrm{gp.sch.}}(A, \mathrm{Pic}^0_{X/k}) = \mathrm{Hom}_{\mathrm{pt.sch.}}(A, \mathrm{Pic}^0_{X/k})$$

in view of Proposition 3.3.4. Moreover,

$$\mathrm{Hom}_{\mathrm{pt.sch.}}(A, \mathrm{Pic}^0_{X/k}) = \mathrm{Hom}_{\mathrm{pt.sch.}}(A, \mathrm{Pic}_{X/k}).$$

Using the isomorphism (6.1.1), this yields an isomorphism

$$\mathrm{Hom}_{\mathrm{gp.sch.}}(A, \mathrm{Pic}^0_{X/k}) \cong \mathrm{Pic}(A \times X)/p_1^* \mathrm{Pic}(A) \times p_2^* \mathrm{Pic}(X),$$

which is contravariant in A and X. Exchanging the roles of A, X and using (6.1.1) again, we obtain an isomorphism

$$\mathrm{Hom}_{\mathrm{gp.sch.}}(A, \mathrm{Pic}^0_{X/k}) \cong \mathrm{Hom}_{\mathrm{pt.sch.}}(X, \mathrm{Pic}^0_A),$$

contravariant in A and X again. This readily yields the assertion (and reproves the existence of the Albanese morphism in this setting). $\square$

PROPOSITION 6.2.2. *When X is geometrically normal, its Picard variety exists and is an abelian variety.*

PROOF. By Lemma 3.3.7, it suffices to show that $\mathrm{Pic}^0_{X/k}$ is proper. For this, we may assume k algebraically closed. In view of Theorem 4.3.2, it suffices in turn to show that $\mathrm{Pic}^0_{X/k}$ contains no non-trivial smooth connected affine subgroup. As any such subgroup contains a copy of $\mathbb{G}_a$ or $\mathbb{G}_m$ (see e.g. [**60**, 3.4.9, 6.2.5, 6.3.4]), we are reduced to checking that every homomorphism from $\mathbb{G}_a$ or $\mathbb{G}_m$ to $\mathrm{Pic}^0_{X/k}$ is constant. But we clearly have

$$\mathrm{Hom}_{\mathrm{gp.sch.}}(\mathbb{G}_a, \mathrm{Pic}^0_{X/k}) \subseteq \mathrm{Hom}_{\mathrm{pt.sch.}}(\mathbb{A}^1, \mathrm{Pic}_{X/k}).$$

Moreover,

$$\mathrm{Hom}_{\mathrm{pt.sch.}}(\mathbb{A}^1, \mathrm{Pic}_{X/k}) \cong \mathrm{Pic}(X \times \mathbb{A}^1)/p_1^* \mathrm{Pic}(X)$$

in view of (6.1.1) and the triviality of $\mathrm{Pic}(\mathbb{A}^1)$. Since X is normal, the divisor class group $\mathrm{Cl}(X \times \mathbb{A}^1)$ is isomorphic to $\mathrm{Cl}(X)$ via p_1^*; moreover, for any Weil divisor D on X, the pull-back $p_1^*(D)$ is Cartier if and only if D is Cartier. Thus, the map $p_1^* : \mathrm{Pic}(X) \to \mathrm{Pic}(X \times \mathbb{A}^1)$ is an isomorphism. As a consequence, every homomorphism $\mathbb{G}_a \to \mathrm{Pic}^0_{X/k}$ is constant.

Arguing similarly with $\mathbb{G}_a$ replaced by $\mathbb{G}_m$, we obtain

$$\mathrm{Hom}_{\mathrm{gp.sch.}}(\mathbb{G}_m, \mathrm{Pic}^0_{X/k}) \subseteq \mathrm{Pic}(X \times (\mathbb{A}^1 \setminus 0))/p_1^* \mathrm{Pic}(X).$$

Also, the pull-back map $\mathrm{Cl}(X \times \mathbb{A}^1) \to \mathrm{Cl}(X \times (\mathbb{A}^1 \setminus 0))$ is surjective. It follows that $p_1^* : \mathrm{Cl}(X) \to \mathrm{Cl}(X \times (\mathbb{A}^1 \setminus 0))$ is an isomorphism and restricts to an isomorphism on Picard groups. Thus, every homomorphism $\mathbb{G}_m \to \mathrm{Pic}^0_{X/k}$ is constant as well. $\square$

DEFINITION 6.2.3. A scheme X is *semi-normal* if X is reduced and every finite bijective morphism of schemes $f : Y \to X$ that induces an isomorphism on all residue fields is an isomorphism.

Examples of semi-normal schemes include of course normal varieties, and also divisors with smooth normal crossings. Nodal curves are semi-normal; cuspidal curves are not. By [**29**, Cor. 5.7], semi-normality is preserved under separable field extensions. But it is not preserved under arbitrary field extensions, as shown by Example 6.2.5 below. We say that a scheme X is *geometrically semi-normal* if $X_{\bar{k}}$ is semi-normal.

PROPOSITION 6.2.4. *When X is geometrically semi-normal, its Picard variety exists and is a semi-abelian variety.*

PROOF. Using Lemma 5.6.1, we may assume k perfect. By Proposition 5.4.5, it suffices to show that every morphism $\mathbb{A}^1 \to \mathrm{Pic}^0_{X/k}$ is constant. By (6.1.1), we have

$$\mathrm{Hom}_{\mathrm{pt.sch.}}(\mathbb{A}^1, \mathrm{Pic}_{X/k}) \subseteq \mathrm{Pic}(X \times \mathbb{A}^1)/p_1^* \mathrm{Pic}(X).$$

So it suffices in turn to show that the map

$$p_1^* : \mathrm{Pic}(X) \longrightarrow \mathrm{Pic}(X \times \mathbb{A}^1)$$

is an isomorphism. When X is affine, this follows from [**62**, Thm. 3.6]. For an arbitrary variety X, we consider the first terms of the Leray spectral sequence associated with the morphism $p_1 : X \times \mathbb{A}^1 \to X$ and the sheaf $\mathcal{O}^*_{X \times \mathbb{A}^1}$ consisting of the units of the structure sheaf. This yields an exact sequence

$$0 \longrightarrow H^1(X, p_{1*}(\mathcal{O}^*_{X\times\mathbb{A}^1})) \xrightarrow{p_1^*} H^1(X \times \mathbb{A}^1, \mathcal{O}^*_{X\times\mathbb{A}^1}) \longrightarrow H^0(X, R^1p_{1*}(\mathcal{O}^*_{X\times\mathbb{A}^1})).$$

Also, the natural map $\mathcal{O}^*_X \to p_{1*}(\mathcal{O}^*_{X\times\mathbb{A}^1})$ is an isomorphism, since $R[t]^* = R^*$ for any integral domain R. Thus,

$$H^1(X, p_{1*}(\mathcal{O}^*_{X\times\mathbb{A}^1})) = H^1(X, \mathcal{O}^*_X) = \mathrm{Pic}(X).$$

Moreover, $H^1(X \times \mathbb{A}^1, \mathcal{O}^*_{X\times\mathbb{A}^1}) = \mathrm{Pic}(X \times \mathbb{A}^1)$. Thus, it suffices to show that $R^1p_{1*}(\mathcal{O}^*_{X\times\mathbb{A}^1}) = 0$. Recall that $R^1p_{1*}(\mathcal{O}^*_{X\times\mathbb{A}^1})$ is the sheaf on X associated with the presheaf $U \mapsto \mathrm{Pic}(U \times \mathbb{A}^1)$. When U is affine, we already saw that the map $p_1^* : \mathrm{Pic}(U) \to \mathrm{Pic}(U \times \mathbb{A}^1)$ is an isomorphism. Moreover, for any $x \in X$ and $L \in \mathrm{Pic}(U)$, there exists an open affine neighborhood V of x in U such that $L|_V$ is trivial. This yields the desired vanishing. □

The above proposition does not extend to semi-normal schemes, as shown by the following:

EXAMPLE 6.2.5. Let k be an imperfect field. Set $p := \mathrm{char}(k)$ and choose $a \in k \setminus k^p$. Like in Remark 5.2.3 (iv), consider the regular projective curve X defined as the zero scheme of $y^p - xz^{p-1} - ax^p$ in $\mathbb{P}^2$. Then X is not geometrically semi-normal, since $X_{\bar{k}}$ is a cuspidal curve: the zero scheme of $(y - a^{1/p}x)^p - xz^{p-1}$. We now show that $\mathrm{Pic}^0_{X/k}$ is a smooth connected unipotent group, non-trivial if $p \geq 3$.

The smoothness of $\mathrm{Pic}^0_{X/k}$ follows from Theorem 6.1.2 (3). Also, the group $\mathrm{Pic}^0_{X/k}(k_s)$ is non-trivial and killed by p when $p \geq 3$, see [**36**, Thm. 6.10.1, Lem. 6.11.1]. It follows that $\mathrm{Pic}^0_{X/k}$ is non-trivial and killed by p as well. By using the structure of commutative algebraic groups, e.g., Lemma 5.6.2, this implies that $\mathrm{Pic}^0_{X/k}$ is unipotent.

Next, we present a classical example of a smooth projective surface for which the Picard scheme is not smooth:

EXAMPLE 6.2.6. Assume that k is algebraically closed of characteristic 2. Let E be an ordinary elliptic curve, i.e., it has a (unique) k-rational point z_0 of order 2. Let F be another elliptic curve and consider the automorphism σ of $E \times F$ such that

$$\sigma(z, w) = (z + z_0, -w)$$

identically. Then σ has order 2 and fixes no point of $E \times F$. Thus, there exists a quotient morphism

$$q : E \times F \longrightarrow X$$

for the group $\langle\sigma\rangle$ generated by σ; moreover, q is finite and étale, and X is a smooth projective surface. Denote by $E/\langle z_0\rangle$ the quotient of E by the subgroup generated by z_0. Then the projection $E \times F \to E$ descends to a morphism

$$\alpha : X \longrightarrow E/\langle z_0\rangle.$$

We claim that α is the Albanese morphism of the pointed variety (X, x), where $x := q(0,0)$. Consider indeed a morphism

$$f : X \longrightarrow A,$$

where A is an abelian variety and $f(x) = 0$. Composing f with q yields a σ-invariant morphism

$$g : E \times F \longrightarrow A, \quad (0,0) \longmapsto 0.$$

By Lemma 4.1.3, we have $g(z,w) = g(z,0) + g(0,w)$ identically. Moreover, the map $h : F \to A$, $w \mapsto g(0,w)$ is a homomorphism by Proposition 4.1.4. In particular, $h(-w) = -h(w)$. But $h(-w) = h(w)$ by σ-invariance. Thus, h factors through the kernel of 2_F, a finite group scheme. Since F is smooth and connected, it follows that h is constant. Thus, $g(z,w) = g(z,0) = g(z+z_0,0)$; this yields our claim.

Combining that claim with Propositions 6.2.1 and 6.2.2, we see that

$$\mathrm{Pic}^0(X) \cong \widehat{E/\langle z_0\rangle} \cong E/\langle z_0\rangle.$$

Next, we claim that *the tangent sheaf* T_X *is trivial.* Indeed, since q is étale, we have

$$q^*(T_X) \cong T_{E\times F} \cong \mathcal{O}_{E\times F} \otimes_k (\mathrm{Lie}(E) \oplus \mathrm{Lie}(F)). \tag{6.2.1}$$

These isomorphisms are equivariant for the natural action of σ on $\mathcal{O}_{E\times F}$ and its trivial action on $\mathrm{Lie}(E) \oplus \mathrm{Lie}(F)$ (recall that σ acts on E by a translation, and on F by -1). Thus, we obtain

$$T_X \cong q_*(q^*(T_X))^{\langle\sigma\rangle} \cong q_*(\mathcal{O}_{E\times F})^{\langle\sigma\rangle} \otimes_k (\mathrm{Lie}(E) \oplus \mathrm{Lie}(F)) \cong \mathcal{O}_X \otimes_k (\mathrm{Lie}(E) \oplus \mathrm{Lie}(F)).$$

This yields the claim.

By that claim and the Riemann-Roch theorem, we obtain $\chi(\mathcal{O}_X) = 0$. Since $h^0(\mathcal{O}_X) = 1$ and $h^2(\mathcal{O}_X) = h^0(\omega_X) = h^0(\mathcal{O}_X) = 1$, this yields $h^1(\mathcal{O}_X) = 2$. In other words, the Lie algebra of $\mathrm{Pic}_{X/k}$ has dimension 2; thus, $\mathrm{Pic}_{X/k}$ *is not smooth.*

Notes and references.

As mentioned in the introduction, a detailed reference for Picard schemes is Kleiman's article [**35**]; see also [**10**], especially Chapter 8 for general results on the Picard functor, and Chapter 9 for applications to relative curves.

Proposition 6.2.2 is well-known, see e.g. [**35**, Thm. 5.4].

Proposition 6.2.4 is due to Alexeev when k is algebraically closed, see [**2**, Thm. 4.1.7].

Example 6.2.6 is due to Igusa, see [**33**].

Assume that k is perfect and consider a proper scheme X having a k-rational point. Then the Picard variety of X lies in a unique exact sequence of the form (5.3.3),

$$1 \longrightarrow T_X \times U_X \longrightarrow \mathrm{Pic}^0(X) \longrightarrow A_X \longrightarrow 1,$$

where T_X is a torus, U_X a smooth connected commutative algebraic group, and A_X an abelian variety. The affine part $T_X \times U_X$ has been described by Geisser in [**28**]; in particular, the dual of the character group of T_X is isomorphic to the étale cohomology group $H^1_{\text{ét}}(X_{\bar{k}}, \mathbb{Z})$ as a Galois module (see [**28**, Thm. 1], and [**64**, Thm. 7.1], [**2**, Cor. 4.2.5] for closely related results). Also, when X is reduced, U_X is the kernel of the pull-back map $f^* : \mathrm{Pic}_{X/k} \to \mathrm{Pic}_{X^+/k}$, where $f : X^+ \to X$ denotes the semi-normalization (see [**28**, Thm. 3]).

It is an open problem whether every smooth connected algebraic group G over a perfect field k is the Picard variety of some proper scheme X. When k is algebraically closed of characteristic 0, one can construct an appropriate scheme X by using the structure of G described in §5.3, see [**13**, Thm. 1.1]. In fact, X may be taken projective, and normal except at finitely many points. Yet when $\mathrm{char}(k) > 0$, the unipotent part of the Picard variety of a scheme satisfying these conditions is not arbitrary, see [**13**, Thm. 1.2].

In fact, it is not even known whether every torus is the largest subtorus of some Picard variety; equivalently, whether every free abelian group equipped with an action of the Galois group Γ is obtained as $H^1_{\text{ét}}(X_{\bar{k}}, \mathbb{Z})$ for some proper scheme X (which can be assumed semi-normal).

In another direction, the structure of Picard schemes over imperfect fields is largely unknown; see [**61**, Ex. 3.1] for a remarkable example.

7. The automorphism group scheme

7.1. Basic results and examples. Recall from §2.2 the automorphism group functor of a scheme X, i.e., the group functor Aut_X that assigns to any scheme S the automorphism group of the S-scheme $X \times S$. We have the following representability result for Aut_X, analogous to that for the Picard scheme (Theorem 6.1.2):

THEOREM 7.1.1. *Let X be a proper scheme.*

(1) *The group functor* Aut_X *is represented by a locally algebraic group (that will be still denoted by* Aut_X*).*
(2) $\mathrm{Lie}(\mathrm{Aut}_X) = H^0(X, T_X)$*, where T_X denotes the sheaf of derivations of the structure sheaf $\mathcal{O}_X$.*
(3) *If $H^1(X, T_X) = 0$, then* Aut_X *is smooth.*

PROOF. (1) This is obtained in [**41**, Thm. 3.7] via an axiomatic characterization of locally algebraic groups among group functors, which generalizes that of [**47**]. When X is projective, the result follows from the existence of the Hilbert scheme. More specifically, the functor of endomorphisms is represented by an open subscheme End_X of the Hilbert scheme $\mathrm{Hilb}_{X \times X}$, by sending each endomorphism u to its graph (the image of $\mathrm{id} \times u : X \to X \times X$), see [**37**, Thm. I.10]. Moreover, Aut_X is represented by an open subscheme of End_X in view of [**37**, Lem. I.10.1].

(2) See [**41**, Lem. 3.4] and also [**23**, II.4.2.4].

(3) This follows from [**32**, III.5.9]. □

With the above notation and assumptions, we say that Aut_X is the *automorphism group scheme* of X; its neutral component, Aut^0_X, is a connected algebraic group. The formations of Aut_X and Aut^0_X commute with field extensions. For any group scheme G, the datum of a G-action on X is equivalent to a homomorphism $G \to \mathrm{Aut}_X$.

EXAMPLE 7.1.2. Let C be a smooth, projective, geometrically irreducible curve of genus $g \geq 2$. Then T_C is the dual of the canonical sheaf ω_C, and hence $\deg(T_C) =$

$2-2g<0$; it follows that $H^0(C,T_C)=0$. Thus, Aut_C is étale; equivalently, Aut^0_C is trivial.

In fact, Aut_C *is finite*. To see this, it suffices to show that $\mathrm{Aut}_C(\bar{k})$ is finite; thus, we may assume k algebraically closed. Let $f \in \mathrm{Aut}_C(k)$ and consider its graph $\Gamma_f \subset C \times C$. Choose a point $x \in C(k)$; then $\mathcal{O}_C(x)$ is an ample invertible sheaf on C, and hence $\mathcal{L} := \mathcal{O}_C(x) \boxtimes \mathcal{O}_C(x)$ is an ample invertible sheaf on $C \times C$. Moreover, the pull-back of $\mathcal{L}$ to $\Gamma_f \cong C$ is isomorphic to $\mathcal{O}_C(x+f(x))$; by the Riemann-Roch theorem, the Hilbert polynomial $P : n \mapsto \chi(C, \mathcal{O}_C(n(x+f(x))))$ is independent of f. As a consequence, Aut_C is equipped with an immersion into the Hilbert scheme $\mathrm{Hilb}^P_{C\times C}$. Since the latter is projective, this yields the assertion.

By the above argument, Aut_C is an algebraic group for any geometrically irreducible curve C (take for x a smooth closed point of C).

EXAMPLE 7.1.3. Let A be an abelian variety. Then the action of A on itself by translation yields a homomorphism

$$\tau : A \longrightarrow \mathrm{Aut}^0_A .$$

Clearly, $\mathrm{Ker}(\tau)$ is trivial and hence τ is a closed immersion. Since A is smooth and Aut^0_A is an irreducible scheme of finite type, of dimension at most

$$h^0(T_A) = h^0(\mathcal{O}_A \otimes_k \mathrm{Lie}(A)) = \dim \mathrm{Lie}(A) = \dim(A),$$

it follows that τ *is an isomorphism.*

Also, one readily checks that

$$\mathrm{Aut}_A \cong \mathrm{Aut}_{A,0} \ltimes A,$$

where $\mathrm{Aut}_{A,0} \subseteq \mathrm{Aut}_A$ denotes the stabilizer of the origin. It follows that $\mathrm{Aut}_{A,0} \cong \pi_0(\mathrm{Aut}_A)$ is étale.

If A is an elliptic curve, then $\mathrm{Aut}_{A,0}$ is finite, as may be seen by arguing as in the preceding example. But this fails for abelian varieties of higher dimension, for example, when $A = B \times B$ for a non-trivial abelian variety B: then $\mathrm{Aut}_{A,0}$ contains the constant group scheme $\mathrm{GL}_2(\mathbb{Z})$ acting by linear combinations of entries.

Next, we present an application of Theorem 2 to the structure of connected automorphism group schemes. Recall that a variety X is *uniruled* if there exist an integral scheme of finite type Y and a dominant rational map $\mathbb{P}^1 \times Y \dashrightarrow X$ such that the induced map $\mathbb{P}^1_y \dashrightarrow X$ is non-constant for some $y \in Y$. Also, X is uniruled if and only if $X_{\bar{k}}$ is uniruled, see [**37**, IV.1.3].

PROPOSITION 7.1.4. *Let X be a proper variety. If X is not uniruled, then Aut^0_X is proper.*

PROOF. We may assume k algebraically closed. If Aut^0_X is not proper, then it contains a connected affine normal subgroup scheme N of positive dimension, as follows from Theorem 2. In turn, the reduced subgroup scheme N_{red} contains a subgroup scheme H isomorphic to $\mathbb{G}_a$ or $\mathbb{G}_m$. Since H acts faithfully on X, the action morphism $H \times X \to X$ yields a uniruling. □

EXAMPLE 7.1.5. Assume that k is algebraically closed of characteristic 2. Let X be the smooth projective surface constructed in Example 6.2.6. We claim that Aut_X *is not smooth.*

To see this, recall that the tangent sheaf T_X is trivial; hence $\mathrm{Lie}(\mathrm{Aut}_X)$ has dimension 2. On the other hand, X is equipped with an action of the elliptic curve

E, via its action on $E \times F$ by translations on itself (which commutes with the involution σ). This yields a homomorphism $E \to \mathrm{Aut}_X$ that factors through

$$f : E \longrightarrow (\mathrm{Aut}^0_X)_{\mathrm{red}} =: G.$$

Since X contains an E-stable curve isomorphic to E (the image of $E \times y$, where $y \in F(k)$ and $y \neq -y$), the action of E on X is faithful, and hence f is a closed immersion. We now show that f *is an isomorphism*; this implies the claim for dimension reasons.

First, note that every morphism $\mathbb{P}^1 \to X$ is constant, since the Albanese morphism $\alpha : X \to E/\langle z_0 \rangle$ has its fibers at all closed points isomorphic to the elliptic curve F, and $E/\langle z_0 \rangle$ is an elliptic curve. In particular, X is not uniruled. By the above proposition, it follows that G is an abelian variety.

Combining this with Proposition 3.1.6, we see that $C_G(x)$ is finite for any $x \in X(k)$. Thus, the G-orbits of k-rational points of X are abelian varieties, isogenous to G. Since X is not an abelian variety (as its Albanese morphism is not an isomorphism), it follows that $\dim(G) \leq 1$. Thus, G must be the image of f.

7.2. Blanchard's lemma. Consider a group scheme G acting on a scheme X, and a morphism of schemes $f : X \to Y$. In general, the G-action on X does not descend to an action on Y; for example, when $G = X$ is an elliptic curve acting on itself by translations, and f is the quotient by $\mathbb{Z}/2$ acting via $x \mapsto \pm x$. Yet we will obtain a descent result under the assumption that f is proper and $f_*(\mathcal{O}_X) = \mathcal{O}_Y$; then f is surjective and its fibers are connected by [**31**, III.4.3.2, III.4.3.4]. Such a descent result was first proved by Blanchard in the setting of holomorphic transformation groups, see [**7**, I.1].

THEOREM 7.2.1. *Let G be a connected algebraic group, X a G-scheme of finite type, Y a scheme of finite type and $f : X \to Y$ a proper morphism such that $f^\# : \mathcal{O}_Y \to f_*(\mathcal{O}_X)$ is an isomorphism. Then there exists a unique action of G on Y such that f is equivariant.*

PROOF. Let $a : G \times X \to X$ denote the action. We show that there is a unique morphism $b : G \times Y \to Y$ such that the square

$$\begin{array}{ccc} G \times X & \xrightarrow{a} & X \\ {\scriptstyle \mathrm{id} \times f} \downarrow & & \downarrow {\scriptstyle f} \\ G \times Y & \xrightarrow{b} & Y \end{array}$$

commutes. By [**31**, II.8.11], it suffices to check that the morphism

$$\mathrm{id} \times f : G \times X \longrightarrow G \times Y$$

is proper, the map

$$(\mathrm{id} \times f)^\# : \mathcal{O}_{G \times Y} \longrightarrow (\mathrm{id} \times f)_*(\mathcal{O}_{G \times X})$$

is an isomorphism, and the composition

$$f \circ a : G \times X \longrightarrow Y$$

is constant on the fibers of $\mathrm{id} \times f$.

Since f is proper, $\mathrm{id} \times f$ is proper as well. Consider the square

$$\begin{array}{ccc} G \times X & \xrightarrow{p_X} & X \\ {\scriptstyle \mathrm{id}\times f}\downarrow & & \downarrow{\scriptstyle f} \\ G \times Y & \xrightarrow{p_Y} & Y, \end{array}$$

where p_X, p_Y denote the projections. As this square is cartesian and both horizontal arrows are flat, the natural map

$$p_Y^* f_*(\mathcal{O}_X) \longrightarrow (\mathrm{id} \times f)_* p_X^*(\mathcal{O}_X),$$

is an isomorphism, and hence so is the natural map $\mathcal{O}_{G\times Y} \to (\mathrm{id}\times f)_*(\mathcal{O}_{G\times X})$. We now check that $f \circ a$ is constant on the fibers of $\mathrm{id}\times f$. Let K be a field extension of k and let $g \in G(K)$, $y \in Y(K)$. Then the fiber of $\mathrm{id} \times f$ at (g, y) is just $g \times f^{-1}(y)$. So it suffices to show that the morphism

$$h : G_K \times f^{-1}(y) \longrightarrow Y_K, \quad (g, x) \longmapsto f(g \cdot x)$$

is constant on $g \times f^{-1}(y)$. But this follows e.g. from the rigidity lemma 3.3.3 applied to the irreducible components of $f^{-1}(y)$, since h is constant on $e \times f^{-1}(y)$, and $f^{-1}(y)$ is connected.

It remains to show that a is an action of the group scheme G. Note that e acts on X via the identity; moreover, the composite morphism of sheaves

$$\mathcal{O}_Y \xrightarrow{b^\#} b_*(\mathcal{O}_{G\times Y}) \xrightarrow{(e\times \mathrm{id})^\#} b_*(\mathcal{O}_{e\times Y}) \cong \mathcal{O}_Y$$

is the identity, since so is the analogous morphism

$$\mathcal{O}_X \xrightarrow{a^\#} a_*(\mathcal{O}_{G\times X}) \xrightarrow{(e\times \mathrm{id})^\#} a_*(\mathcal{O}_{e\times X}) \cong \mathcal{O}_X$$

and $f_*(\mathcal{O}_X) = \mathcal{O}_Y$. Likewise, the square

$$\begin{array}{ccc} G \times G \times Y & \xrightarrow{\mathrm{id}\times b} & G \times Y \\ {\scriptstyle m\times \mathrm{id}}\downarrow & & \downarrow{\scriptstyle b} \\ G \times Y & \xrightarrow{b} & Y \end{array}$$

commutes on closed points, and the corresponding square of morphisms of sheaves commutes as well, since the analogous square with Y replaced by X commutes. □

COROLLARY 7.2.2. *Let $f : X \to Y$ be a morphism of proper schemes such that $f^\# : \mathcal{O}_Y \to f_*(\mathcal{O}_X)$ is an isomorphism.*

(1) *f induces a homomorphism*

$$f_* : \mathrm{Aut}_X^0 \longrightarrow \mathrm{Aut}_Y^0 .$$

(2) *If f is birational, then f_* is a closed immersion.*

PROOF. (1) This follows readily from the above theorem applied to the action of Aut_X^0 on X.

(2) By Proposition 2.7.1, it suffices to check that $\mathrm{Ker}(f_*)$ is trivial. Let S be a scheme and $u \in \mathrm{Aut}_X^0(S)$ such that $f_*(u) = \mathrm{id}$. As f is birational, there exists a dense open subscheme $V \subseteq Y$ such that f pulls back to an isomorphism $f^-(V) \to V$. Then u pulls back to the identity on $f^{-1}(V) \times S$. Since the latter is dense in $X \times S$, we obtain $u = \mathrm{id}$. □

The above corollary applies to a birational morphism $f : X \to Y$, where X and Y are proper varieties and Y is normal: then $f^{\#} : \mathcal{O}_Y \to f_*(\mathcal{O}_X)$ is an isomorphism by Zariski's Main Theorem.

Corollary 7.2.2(1) also applies to the two projections $p_1 : X \times Y \to X$ and $p_2 : X \times Y \to Y$, where X, Y are proper varieties: then $\mathcal{O}(X) = k = \mathcal{O}(Y)$ and hence $p_{1*}(\mathcal{O}_{X\times Y}) = \mathcal{O}_X$, $p_{2*}(\mathcal{O}_{X\times Y}) = \mathcal{O}_Y$ in view of Lemma 2.3.3. This implies readily the following:

Corollary 7.2.3. *Let X, Y be proper varieties. Then the homomorphism*

$$p_{1*} \times p_{2*} : \mathrm{Aut}^0_{X\times Y} \longrightarrow \mathrm{Aut}^0_X \times \mathrm{Aut}^0_Y$$

is an isomorphism with inverse the natural homomorphism

$$\mathrm{Aut}^0_X \times \mathrm{Aut}^0_Y \longrightarrow \mathrm{Aut}^0_{X\times Y}, \quad (u,v) \longmapsto ((x,y) \mapsto (u(x), v(y))).$$

7.3. Varieties with prescribed connected automorphism group.

Theorem 7.3.1. *Let G be a smooth connected algebraic group of dimension n over a perfect field k. Then there exists a normal projective variety X such that $G \cong \mathrm{Aut}^0_X$ and* $\dim(X) \leq 2n+1$.

The proof will occupy the rest of this subsection. We will use the action of $G\times G$ on G via left and right multiplication; this identifies G with the homogeneous space $(G \times G)/\mathrm{diag}(G)$, and e to the base point of that homogeneous space. By Theorem 5.2.2, we may choose a projective compactification Y of G which is $G\times G$-equivariant, i.e., Y is equipped with two commuting G-actions, on the left and on the right. We may further assume that Y is normal in view of Proposition 2.5.1.

Denote by Aut^G_Y the centralizer of G in Aut_Y relative to the right G-action. Then Aut^G_Y is a closed subgroup scheme of Aut_Y by Theorem 2.2.6. Also, the left G-action on Y yields a homomorphism

$$\varphi : G \longrightarrow \mathrm{Aut}^G_Y .$$

Lemma 7.3.2. *The above map φ is an isomorphism.*

Proof. Since the left G-action on itself is faithful, the kernel of φ is trivial and hence φ is a closed immersion. We show that φ is surjective on $\bar{k}$-points: if $u \in \mathrm{Aut}^G_Y(\bar{k})$, then u stabilizes the open orbit of the right $G_{\bar{k}}$-action, i.e., $G_{\bar{k}}$. As u commutes with that action, it follows that the pull-back of u to $G_{\bar{k}} \subseteq Y_{\bar{k}}$ is the left multiplication by some $g \in G(\bar{k})$. Since $G_{\bar{k}}$ is dense in $Y_{\bar{k}}$, we conclude that $u = \varphi(g)$.

As G is smooth, it suffices to check that $\mathrm{Lie}(\varphi)$ is an isomorphism to complete the proof. We have $\mathrm{Lie}(\mathrm{Aut}^G_Y) = H^0(Y, T_Y)^G = \mathrm{Der}^G(\mathcal{O}_Y)$, where $\mathrm{Der}(\mathcal{O}_Y)$ denotes the Lie algebra of derivations of $\mathcal{O}_Y$, and $\mathrm{Der}^G(\mathcal{O}_Y)$ the Lie subalgebra of invariants under the right G-action. Moreover, the pull-back $j : \mathrm{Der}(\mathcal{O}_Y) \to \mathrm{Der}(\mathcal{O}_G)$ is injective by the density of G in Y. Also, recall that $\mathrm{Der}^G(\mathcal{O}_G) \cong \mathrm{Lie}(G)$ and this identifies the composition $j \circ \mathrm{Lie}(\varphi)$ with the identity of $\mathrm{Lie}(G)$. This yields the desired statement. □

For any closed subscheme $F \subseteq G$, we denote by Aut^F_Y the centralizer of F in Aut_Y, where F is identified with $e \times F \subseteq e \times G$. Then again, Aut^F_Y is a closed subgroup scheme of Aut_Y; we denote its neutral component by $\mathrm{Aut}^{F,0}_Y$.

LEMMA 7.3.3. *With the above notation, there exists a finite étale subscheme* $F \subset G$ *such that* $\mathrm{Aut}_Y^{F,0} = \mathrm{Aut}_Y^{G,0}$.

PROOF. Since G is smooth, $G(\bar{k})$ is dense in $G_{\bar{k}}$ and hence the above lemma yields an isomorphism $G_{\bar{k}} \cong \mathrm{Aut}^{G(\bar{k})}(Y_{\bar{k}})$ with an obvious notation. Consider the family of (closed) subgroup schemes $\mathrm{Aut}_{Y_{\bar{k}}}^{F,0} \subseteq \mathrm{Aut}_{Y_{\bar{k}}}^{0}$, where F runs over the finite subsets $F \subseteq G(\bar{k})$, stable by the Galois group Γ. Since $\mathrm{Aut}_{Y_{\bar{k}}}^{0}$ is of finite type, we may choose F so that $\mathrm{Aut}_{Y_{\bar{k}}}^{F,0}$ is minimal in this family. Let Ω be a Γ-orbit in $G(\bar{k})$, then $\mathrm{Aut}_{Y_{\bar{k}}}^{F\cup\Omega,0} \subseteq \mathrm{Aut}_{Y_{\bar{k}}}^{F,0}$ and hence equality holds. Thus, $\mathrm{Aut}_{Y_{\bar{k}}}^{F,0} = \mathrm{Aut}_{Y_{\bar{k}}}^{G(\bar{k}),0} = G_{\bar{k}}$. This yields the assertion by using the bijective correspondence between finite étale subschemes of G and finite Γ-stable subsets of $G(\bar{k})$. □

Next, consider the diagonal homomorphism $\mathrm{Aut}_Y^0 \to \mathrm{Aut}_Y^0 \times \mathrm{Aut}_Y^0$. By Corollary 7.2.3, we may identify the right-hand side with $\mathrm{Aut}_{Y\times Y}^0$; this yields a closed immersion of algebraic groups

$$\Delta : \mathrm{Aut}_Y^0 \longrightarrow \mathrm{Aut}_{Y\times Y}^0 .$$

Also, for any finite étale subscheme $F \subseteq G$, the morphism

$$\Gamma : F \times Y \longrightarrow Y \times Y, \quad (g, x) \longmapsto (x, g \cdot x)$$

is finite, and hence we may view its image Z as a closed reduced subscheme of $Y \times Y$. Note that Z is stable by the G-action via $\Delta \circ \varphi$; also, $Z_{\bar{k}}$ is the union of the graphs of the automorphisms $g \in F(\bar{k})$ of Y.

LEMMA 7.3.4. *Keep the above notation.*

(1) Δ *identifies* Aut_Y^0 *with* $\mathrm{Aut}_{Y\times Y,\mathrm{diag}(Y)}^0$ *(the neutral component of the stabilizer of the diagonal in* $\mathrm{Aut}_{Y\times Y}$*).*

(2) *If* F *is a finite étale subscheme of* G *that satisfies the assertion of Lemma 7.3.3 and contains* e*, then* Δ *identifies* $\varphi(G)$ *with* $\mathrm{Aut}_{Y\times Y,Z,\mathrm{red}}^0$ *(the reduced neutral component of the stabilizer of* Z *in* $\mathrm{Aut}_{Y\times Y}$*).*

PROOF. (1) Just observe that for any scheme S and for any $u, v \in \mathrm{Aut}_X(S)$, the automorphism $u \times v$ of $(X \times S) \times_S (X \times S)$ stabilizes the diagonal if and only if $u = v$.

(2) We may assume k algebraically closed; then we may view F as a finite subset of $G(k)$. Also, by Proposition 2.5.1, the reduced subgroup $\mathrm{Aut}_{Y\times Y,Z,\mathrm{red}}^0$ stabilizes the graph Γ_f of any $f \in F$, since these graphs form the irreducible components of $Z_{\bar{k}}$. Now observe that for any scheme S and for any $f, g \in \mathrm{Aut}_X(S)$, the automorphism $g \times g$ of $(X \times S) \times_S (X \times S)$ stabilizes Γ_f if and only if g commutes with f. Thus, Δ identifies $\varphi(G)$ with $\mathrm{Aut}_{Y\times Y,Z,\mathrm{red}}^0$. □

We now choose a finite étale subscheme $F \subseteq G$ that satisfies the assumption of Lemma 7.3.3 and contains e. Denote by

$$f : X \longrightarrow Y \times Y$$

the normalization of the blowing-up of $Y \times Y$ along Z. Then the G-action on $Y \times Y$ via $\Delta \circ \varphi$ lifts to a unique action on X; in other words, we have a homomorphism

$$f^* : G \longrightarrow \mathrm{Aut}_X^0 .$$

As f is birational, f^* is a closed immersion by the argument of Corollary 7.2.2 (ii). Also, $Y \times Y$ is normal, since Y is normal and k is perfect. Thus, the above corollary yields a closed immersion

$$f_* : \mathrm{Aut}^0_X \longrightarrow \mathrm{Aut}^0_{Y\times Y}.$$

Moreover, the composition $f_* \circ f^*$ is an isomorphism of G to its image $(\Delta \circ \varphi)(G)$, since f is birational. We will identify G with its image, and hence $f_* \circ f^*$ with id.

Consider first the case where $\mathrm{char}(k) = 0$. Then Aut^0_X is smooth and hence stabilizes the exceptional locus $E \subset X$ of the birational morphism f. As a consequence, the action of Aut^0_X on $Y \times Y$ via f_* stabilizes the image $f(E) \subset Y \times Y$. If in addition $n \geq 2$, then $f(E) = Z$ and hence f_* sends Aut^0_X to $\mathrm{Aut}^0_{Y\times Y,Z}$, i.e. to G by Lemma 7.3.4. It follows that f^* is an isomorphism with inverse f_*.

If $n = 1$, then E is empty. We now reduce to the case where $n = 2$ as follows: we choose a smooth, projective, geometrically integral curve C of genus ≥ 1, equipped with a k-rational point c. Then $\mathrm{Aut}^0_{C,c}$ is trivial, as seen in Examples 7.1.2 and 7.1.3. Using Corollary 7.2.3, it follows that the natural map

$$\mathrm{Aut}^0_{Y\times Y,Z} \longrightarrow \mathrm{Aut}^0_{Y\times Y\times C,Z\times c}$$

is an isomorphism; by Lemma 7.3.4, this yields an isomorphism

$$G \cong \mathrm{Aut}^0_{Y\times Y\times C,Z\times c}.$$

Also, note that $Y \times Y \times C$ is a normal projective variety. We now consider the morphism $f' : X' \to Y \times Y \times C$ obtained as the normalization of the blowing-up along $Z \times c$. Since the latter has codimension 2 in $Y \times Y \times C$, the exceptional locus $E' \subset X'$ satisfies $f'(E') = Z \times c$. So the above argument yields again that f'^* is an isomorphism with inverse f'_*. This completes the proof of Theorem 7.3.1 in characteristic 0.

Next, assume that $\mathrm{char}(k) = p > 0$. We will use the following additional result:

LEMMA 7.3.5. *Let $f : X \to Y \times Y$ be as above and assume that $n - 1$ is not a multiple of p (in particular, $n \geq 2$). Then the differential*

$$\mathrm{Lie}(f^*) : \mathrm{Lie}(G) \longrightarrow \mathrm{Lie}(\mathrm{Aut}^0_X)$$

is an isomorphism.

PROOF. As $\mathrm{Lie}(f_*) \circ \mathrm{Lie}(f^*) = \mathrm{id}$, it suffices to show that the image of $\mathrm{Lie}(f_*)$ is contained in $\mathrm{Lie}(G)$. For this, we may assume that $k = \bar{k}$. We may thus view F as a finite subset of $G(k)$ containing e.

We will use the action of $\mathrm{Lie}(\mathrm{Aut}^0_X) = \mathrm{Der}(\mathcal{O}_X)$ on the "jacobian ideal" of f, defined as follows. Recall that the sheaf of differentials, $\Omega^1_X = \mathcal{I}_{\mathrm{diag}(X)}/\mathcal{I}^2_{\mathrm{diag}(X)}$ is equipped with a linearization for Aut_X (see [**24**, I.6] for background on linearized sheaves). Likewise, $\Omega^1_{Y\times Y}$ is equipped with a linearization for $\mathrm{Aut}_{Y\times Y}$ and hence for Aut^0_X acting via f_*. Moreover, the canonical map

$$f^*(\Omega^1_{Y\times Y}) \longrightarrow \Omega^1_X$$

is a morphism of Aut^0_X-linearized sheaves, since it arises from the canonical map $f^{-1}(\mathcal{I}_{\mathrm{diag}(Y\times Y)}) \to \mathcal{I}_{\mathrm{diag}(X)}$. Denoting by $\Omega^{2n}_{Y\times Y}$ the $2n$th exterior power of $\Omega^1_{Y\times Y}$ and defining Ω^{2n}_X likewise, this yields a morphism of Aut^0_X-linearized sheaves

$$f^*(\Omega^{2n}_{Y\times Y}) \longrightarrow \Omega^{2n}_X$$

and in turn, a morphism of Aut^0_X-linearized sheaves

$$Hom(\Omega^{2n}_X, f^*(\Omega^{2n}_{Y\times Y})) \longrightarrow End(\Omega^{2n}_X).$$

The image $\mathcal{J}_f$ of this morphism is an Aut^0_X-linearized subsheaf of the sheaf of algebras $End(\Omega^{2n}_X)$. In particular, for any open subvariety U of X, the Lie algebra $\mathrm{Der}(\mathcal{O}_X)$ acts on the algebra $\mathrm{End}(\Omega^{2n}_U)$ by derivations that stabilize $\Gamma(U, \mathcal{J}_f)$.

Denote by $Y_{\mathrm{reg}} \subseteq Y$ the regular (or smooth) locus and consider the open subvariety $V \subseteq Y_{\mathrm{reg}} \times Y_{\mathrm{reg}}$ consisting of those points that lie in at most one graph Γ_g, where $g \in F$. Then $Z \cap V$ is a disjoint union of smooth varieties of dimension n, and is dense in Z. Let $U := f^{-1}(V)$; then the pull-back

$$f_U : U \longrightarrow V$$

is the blowing-up along $Z \cap V$, and hence U is smooth. Thus, the sheaf Ω^{2n}_U is invertible and $\mathcal{J}_{f_U}$ is just a sheaf of ideals of $\mathcal{O}_U$. A classical computation in local coordinates shows that $\mathcal{J}_{f_U} = \mathcal{O}_U(-(n-1)E)$, where E denotes the exceptional divisor of f_U. Hence we obtain an injective map

$$\mathrm{Der}(\mathcal{O}_X) \to \mathrm{Der}(\mathcal{O}_U, \mathcal{J}_{f_U}) = \mathrm{Der}(\mathcal{O}_U, \mathcal{O}_U(-(n-1)E))$$

with an obvious notation. Since $n-1$ is not a multiple of p, we have

$$\mathrm{Der}(\mathcal{O}_U, \mathcal{O}_U(-(n-1)E)) = \mathrm{Der}(\mathcal{O}_U, \mathcal{O}_U(-E)).$$

(Indeed, if $D \in \mathrm{Der}(\mathcal{O}_U, \mathcal{O}_U(-(n-1)E)))$ and z is a local generator of $\mathcal{O}_U(E)$ at $x \in X$, then $D(z^{n-1}) = (n-1)z^{n-2}D(z) \in z^{n-1}\mathcal{O}_{X,x}$ and hence $D(z) \in z\mathcal{O}_{X,x}$). Also, the natural map

$$\mathrm{Der}(\mathcal{O}_U) \longrightarrow \mathrm{Der}(f_{U*}(\mathcal{O}_U)) = \mathrm{Der}(\mathcal{O}_V)$$

is injective and sends $\mathrm{Der}(\mathcal{O}_U, \mathcal{O}_U(-E))$ to $\mathrm{Der}(\mathcal{O}_V, f_{U*}(\mathcal{O}_U(-E)))$. Moreover, $f_{U*}(\mathcal{O}_U(-E))$ is the ideal sheaf of $Z \cap V$, and hence is stable by $\mathrm{Der}(\mathcal{O}_X)$ acting via the composition

$$\mathrm{Der}(\mathcal{O}_X) \longrightarrow \mathrm{Der}(f_*(\mathcal{O}_X)) = \mathrm{Der}(\mathcal{O}_{Y\times Y}) \longrightarrow \mathrm{Der}(\mathcal{O}_V).$$

It follows that the image of $\mathrm{Lie}(f_*)$ stabilizes the ideal sheaf of the closure of $Z \cap V$ in $Y \times Y$, ie., of Z. By arguing as in the proof of Lemma 7.3.4 (2), we conclude that $\mathrm{Lie}(f_*)$ sends $\mathrm{Der}(\mathcal{O}_X)$ to $\mathrm{Lie}(G)$. □

As f^* is a closed immersion and G is smooth, the above lemma completes the proof of Theorem 7.3.1 when p does not divide $n-1$. Next, when p divides $n-1$, we replace $Y \times Y$ (resp. Z) with $Y \times Y \times C$ (resp. $Z \times c$) for a pointed curve (C, c) as above. This replaces the codimension n of Z in $Y \times Y$ with $n+1$, and hence we obtain a normal projective variety X' of dimension $2n+1$ such that $G \cong \mathrm{Aut}^0_{X'}$.

REMARKS 7.3.6. (i) For any smooth connected *linear* algebraic group G of dimension n, there exists a normal projective *unirational* variety of dimension at most $2n+2$ such that $G \cong \mathrm{Aut}^0_X$. Indeed, the variety G is unirational (see [**24**, XIV.6.10]) and hence so is $Y \times Y$ with the notation of the above proof. Also, in that proof, the pointed curve (C, c) may be replaced with a pair (S, C), where S is a smooth rational projective surface such that Aut^0_S is trivial, and $C \subset S$ is a smooth, geometrically irreducible curve. Such a pair is obtained by taking for S the blowing-up of $\mathbb{P}^2$ at 4 points in general position, and for C an exceptional curve.

(ii) If $\mathrm{char}(k) = 0$ then every connected algebraic group G is the connected automorphism group of some *smooth* projective variety, as follows from the existence of an equivariant resolution of singularities (see [**38**, Prop. 3.9.1, Thm. 3.36]).

(iii) Still assuming $\mathrm{char}(k) = 0$, consider a finite-dimensional Lie algebra $\mathfrak{g}$. Then $\mathfrak{g}$ is algebraic if and only if $\mathfrak{g} \cong \mathrm{Der}(\mathcal{O}_X)$ for some proper scheme X, as follows by combining Corollary 5.1.6, Theorem 7.1.1 and Theorem 7.3.1. Moreover, X may be chosen smooth, projective and unirational by the above remarks.

Notes and references.

Most results of §7.1 are taken from [**41**]. Those of §7.2 are algebraic analogues of classical results about holomorphic transformation groups, see [**1**, 2.4].

In the setting of complex analytic varieties, the problem of realizing a given Lie group as an automorphism group has been extensively studied. It is known that every finite group is the automorphism group of a smooth projective complex curve (see [**30**]); moreover, every compact connected real Lie group is the automorphism group of a bounded domain (satisfying additional conditions), see [**5, 55**]. Also, every connected real Lie group of dimension n is the automorphism group of a Stein complete hyperbolic manifold of dimension $2n$ (see [**34, 65**]). Theorem 7.3.1, obtained in [**12**, Thm. 1], may be viewed as an algebraic analogue of the latter result. The proof presented here is a streamlined version of that in [**12**].

There are still many open questions about automorphism group schemes. For instance, can one realize any algebraic group over an arbitrary field as the full automorphism group scheme of a proper scheme? Also, very little is known on the group of connected components $\pi_0(\mathrm{Aut}_X)$, where X is a proper scheme, or on the analogously defined group $\pi_0(\mathrm{Aut}(M))$, where M is a compact complex manifold (then $\mathrm{Aut}(M)$ is a complex Lie group, possibly with infinitely many components). As mentioned in [**16**], it is not known whether there exists a compact complex manifold M for which $\pi_0(\mathrm{Aut}(M))$ is not finitely generated.

Note added in proof: such an example (with M projective algebraic) has been constructed by John Lesieutre, see arXiv:1609.06391.

Acknowledgements. These are extended notes of a series of talks given at Tulane University for the 2015 Clifford Lectures. I am grateful to the organizer, Mahir Can, for his kind invitation, and to the other speakers and participants for their interest and stimulating discussions.

These notes are also based on a course given at Institut Camille Jordan, Lyon, during the 2014 special period on algebraic groups and representation theory. I also thank this institution for its hospitality, and the participants of the special period who made it so successful.

Last but not least, I warmly thank Raphaël Achet, Mahir Can, Bruno Laurent, Preena Samuel and an anonymous referee for their careful reading of preliminary versions of this text and their very helpful comments.

References

[1] Dmitri N. Akhiezer, *Lie group actions in complex analysis*, Aspects of Mathematics, E27, Friedr. Vieweg & Sohn, Braunschweig, 1995. MR1334091

[2] Valery Alexeev, *Complete moduli in the presence of semiabelian group action*, Ann. of Math. (2) **155** (2002), no. 3, 611–708, DOI 10.2307/3062130. MR1923963

[3] Satoshi Arima, *Commutative group varieties*, J. Math. Soc. Japan **12** (1960), 227–237. MR0136611

[4] Iacopo Barsotti, *Structure theorems for group-varieties*, Ann. Mat. Pura Appl. (4) **38** (1955), 77–119. MR0071849
[5] Eric Bedford and Jiri Dadok, *Bounded domains with prescribed group of automorphisms*, Comment. Math. Helv. **62** (1987), no. 4, 561–572, DOI 10.1007/BF02564462. MR920057
[6] P. Berthelot, A. Grothendieck, and L. Illusie, *Théorie des intersections et théorème de Riemann-Roch* (French), Lecture Notes in Mathematics, Vol. 225, Springer-Verlag, Berlin-New York, 1971. Séminaire de Géométrie Algébrique du Bois-Marie 1966–1967 (SGA 6); Dirigé par P. Berthelot, A. Grothendieck et L. Illusie. Avec la collaboration de D. Ferrand, J. P. Jouanolou, O. Jussila, S. Kleiman, M. Raynaud et J. P. Serre. MR0354655
[7] André Blanchard, *Sur les variétés analytiques complexes* (French), Ann. Sci. Ecole Norm. Sup. (3) **73** (1956), 157–202. MR0087184
[8] Armand Borel, *Linear algebraic groups*, 2nd ed., Graduate Texts in Mathematics, vol. 126, Springer-Verlag, New York, 1991. MR1102012
[9] A. Borel and J.-P. Serre, *Théorèmes de finitude en cohomologie galoisienne* (French), Comment. Math. Helv. **39** (1964), 111–164. MR0181643
[10] Siegfried Bosch, Werner Lütkebohmert, and Michel Raynaud, *Néron models*, Ergebnisse der Mathematik und ihrer Grenzgebiete (3) [Results in Mathematics and Related Areas (3)], vol. 21, Springer-Verlag, Berlin, 1990. MR1045822
[11] Michel Brion, *Anti-affine algebraic groups*, J. Algebra **321** (2009), no. 3, 934–952, DOI 10.1016/j.jalgebra.2008.09.034. MR2488561
[12] Michel Brion, *On automorphisms and endomorphisms of projective varieties*, Automorphisms in birational and affine geometry, Springer Proc. Math. Stat., vol. 79, Springer, Cham, 2014, pp. 59–81, DOI 10.1007/978-3-319-05681-4_4. MR3229345
[13] Michel Brion, *Which algebraic groups are Picard varieties?*, Sci. China Math. **58** (2015), no. 3, 461–478, DOI 10.1007/s11425-014-4882-3. MR3319917
[14] Michel Brion, *On extensions of algebraic groups with finite quotient*, Pacific J. Math. **279** (2015), no. 1-2, 135–153, DOI 10.2140/pjm.2015.279.135. MR3437773
[15] Michel Brion, Preena Samuel, and V. Uma, *Lectures on the structure of algebraic groups and geometric applications*, CMI Lecture Series in Mathematics, vol. 1, Hindustan Book Agency, New Delhi; Chennai Mathematical Institute (CMI), Chennai, 2013. MR3088271
[16] Serge Cantat and Abdelghani Zeghib, *Holomorphic actions, Kummer examples, and Zimmer program* (English, with English and French summaries), Ann. Sci. Éc. Norm. Supér. (4) **45** (2012), no. 3, 447–489. MR3014483
[17] Claude Chevalley, *Théorie des groupes de Lie. Tome III. Théorèmes généraux sur les algèbres de Lie* (French), Actualités Sci. Ind. no. 1226, Hermann & Cie, Paris, 1955. MR0068552
[18] C. Chevalley, *Une démonstration d'un théorème sur les groupes algébriques* (French), J. Math. Pures Appl. (9) **39** (1960), 307–317. MR0126447
[19] Brian Conrad, *A modern proof of Chevalley's theorem on algebraic groups*, J. Ramanujan Math. Soc. **17** (2002), no. 1, 1–18. MR1906417
[20] Brian Conrad, *Chow's K/k-image and K/k-trace, and the Lang-Néron theorem*, Enseign. Math. (2) **52** (2006), no. 1-2, 37–108. MR2255529
[21] Brian Conrad, Ofer Gabber, and Gopal Prasad, *Pseudo-reductive groups*, 2nd ed., New Mathematical Monographs, vol. 26, Cambridge University Press, Cambridge, 2015. MR3362817
[22] Olivier Debarre, *Higher-dimensional algebraic geometry*, Universitext, Springer-Verlag, New York, 2001. MR1841091
[23] M. Demazure, P. Gabriel, *Groupes algébriques*, Masson, Paris, 1970.
[24] M. Demazure, A. Grothendieck, *Schémas en groupes I, II, III (SGA 3)*, Springer Lecture Notes in Math. **151**, **152**, **153** (1970); revised version edited by P. Gille and P. Polo, vols. **I** and **III**, Soc. Math. de France, Paris, 2011.
[25] R. Fossum and B. Iversen, *On Picard groups of algebraic fibre spaces*, J. Pure Appl. Algebra **3** (1973), 269–280. MR0357396
[26] O. Gabber, *On pseudo-reductive groups and compactification theorems*, Oberwolfach Reports **9** (2012), 2371–2374.
[27] Ofer Gabber, Philippe Gille, and Laurent Moret-Bailly, *Fibrés principaux sur les corps valués henséliens* (French, with English and French summaries), Algebr. Geom. **1** (2014), no. 5, 573–612, DOI 10.14231/AG-2014-025. MR3296806
[28] Thomas Geisser, *The affine part of the Picard scheme*, Compos. Math. **145** (2009), no. 2, 415–422, DOI 10.1112/S0010437X08003710. MR2501424

[29] S. Greco and C. Traverso, *On seminormal schemes*, Compositio Math. **40** (1980), no. 3, 325–365. MR571055

[30] Leon Greenberg, *Maximal groups and signatures*, Discontinuous groups and Riemann surfaces (Proc. Conf., Univ. Maryland, College Park, Md., 1973), Princeton Univ. Press, Princeton, N.J., 1974, pp. 207–226. Ann. of Math. Studies, No. 79. MR0379835

[31] A. Grothendieck, *Éléments de géométrie algébrique (rédigés avec la collaboration de J. Dieudonné) : IV. Étude locale des schémas et des morphismes de schémas*, Publ. Math. IHÉS **20** (1964), 5–259, **24** (1965), 5–231, **28** (1966), 5–255, **32** (1967), 5–361.

[32] A. Grothendieck and M. Raynaud, *Revêtements étales et groupe fondamental (SGA 1)* (French), Documents Mathématiques (Paris) [Mathematical Documents (Paris)], 3, Société Mathématique de France, Paris, 2003. Séminaire de géométrie algébrique du Bois Marie 1960–61. [Algebraic Geometry Seminar of Bois Marie 1960-61]; Directed by A. Grothendieck; With two papers by M. Raynaud; Updated and annotated reprint of the 1971 original [Lecture Notes in Math., 224, Springer, Berlin; MR0354651 (50 #7129)]. MR2017446

[33] Jun-ichi Igusa, *On some problems in abstract algebraic geometry*, Proc. Nat. Acad. Sci. U. S. A. **41** (1955), 964–967. MR0074085

[34] Su-Jen Kan, *Complete hyperbolic Stein manifolds with prescribed automorphism groups*, Comment. Math. Helv. **82** (2007), no. 2, 371–383, DOI 10.4171/CMH/95. MR2319933

[35] Steven L. Kleiman, *The Picard scheme*, Fundamental algebraic geometry, Math. Surveys Monogr., vol. 123, Amer. Math. Soc., Providence, RI, 2005, pp. 235–321. MR2223410

[36] Tatsuji Kambayashi, Masayoshi Miyanishi, and Mitsuhiro Takeuchi, *Unipotent algebraic groups*, Lecture Notes in Mathematics, Vol. 414, Springer-Verlag, Berlin-New York, 1974. MR0376696

[37] János Kollár, *Rational curves on algebraic varieties*, Ergebnisse der Mathematik und ihrer Grenzgebiete. 3. Folge. A Series of Modern Surveys in Mathematics [Results in Mathematics and Related Areas. 3rd Series. A Series of Modern Surveys in Mathematics], vol. 32, Springer-Verlag, Berlin, 1996. MR1440180

[38] János Kollár, *Lectures on resolution of singularities*, Annals of Mathematics Studies, vol. 166, Princeton University Press, Princeton, NJ, 2007. MR2289519

[39] Qing Liu, *Algebraic geometry and arithmetic curves*, Oxford Graduate Texts in Mathematics, vol. 6, Oxford University Press, Oxford, 2002. Translated from the French by Reinie Erné; Oxford Science Publications. MR1917232

[40] W. Lütkebohmert, *On compactification of schemes*, Manuscripta Math. **80** (1993), no. 1, 95–111, DOI 10.1007/BF03026540. MR1226600

[41] Hideyuki Matsumura and Frans Oort, *Representability of group functors, and automorphisms of algebraic schemes*, Invent. Math. **4** (1967), 1–25. MR0217090

[42] B. Mazur and William Messing, *Universal extensions and one dimensional crystalline cohomology*, Lecture Notes in Mathematics, Vol. 370, Springer-Verlag, Berlin-New York, 1974. MR0374150

[43] S. Meng, D.-Q. Zhang, *Jordan property for non-linear algebraic groups and projective varieties*, arXiv:1507.02230.

[44] J. S. Milne, *A proof of the Barsotti-Chevalley theorem on algebraic groups*, preprint, arXiv:1311.6060v2.

[45] David Mumford, *Abelian varieties*, Tata Institute of Fundamental Research Studies in Mathematics, vol. 5, Published for the Tata Institute of Fundamental Research, Bombay; by Hindustan Book Agency, New Delhi, 2008. With appendices by C. P. Ramanujam and Yuri Manin; Corrected reprint of the second (1974) edition. MR2514037

[46] D. Mumford, J. Fogarty, and F. Kirwan, *Geometric invariant theory*, 3rd ed., Ergebnisse der Mathematik und ihrer Grenzgebiete (2) [Results in Mathematics and Related Areas (2)], vol. 34, Springer-Verlag, Berlin, 1994. MR1304906

[47] J. P. Murre, *On contravariant functors from the category of pre-schemes over a field into the category of abelian groups (with an application to the Picard functor)*, Inst. Hautes Études Sci. Publ. Math. **23** (1964), 5–43. MR0206011

[48] Masayoshi Nagata, *Imbedding of an abstract variety in a complete variety*, J. Math. Kyoto Univ. **2** (1962), 1–10. MR0142549

[49] Masayoshi Nagata, *A generalization of the imbedding problem of an abstract variety in a complete variety*, J. Math. Kyoto Univ. **3** (1963), 89–102. MR0158892

[50] F. Oort, *Commutative group schemes*, Lecture Notes in Mathematics, vol. 15, Springer-Verlag, Berlin-New York, 1966. MR0213365

[51] Michel Raynaud, *Faisceaux amples sur les schémas en groupes et les espaces homogènes* (French), Lecture Notes in Mathematics, Vol. 119, Springer-Verlag, Berlin-New York, 1970. MR0260758

[52] Maxwell Rosenlicht, *Some basic theorems on algebraic groups*, Amer. J. Math. **78** (1956), 401–443. MR0082183

[53] Maxwell Rosenlicht, *Extensions of vector groups by abelian varieties*, Amer. J. Math. **80** (1958), 685–714. MR0099340

[54] Maxwell Rosenlicht, *Toroidal algebraic groups*, Proc. Amer. Math. Soc. **12** (1961), 984–988. MR0133328

[55] Rita Saerens and William R. Zame, *The isometry groups of manifolds and the automorphism groups of domains*, Trans. Amer. Math. Soc. **301** (1987), no. 1, 413–429, DOI 10.2307/2000347. MR879582

[56] Carlos Sancho de Salas and Fernando Sancho de Salas, *Principal bundles, quasi-abelian varieties and structure of algebraic groups*, J. Algebra **322** (2009), no. 8, 2751–2772, DOI 10.1016/j.jalgebra.2009.08.001. MR2560900

[57] Jean-Pierre Serre, *Groupes algébriques et corps de classes* (French), Hermann, Paris, 1975. Deuxième édition; Publication de l'Institut de Mathématique de l'Université de Nancago, No. VII; Actualités Scientifiques et Industrielles, No. 1264. MR0466151

[58] Jean-Pierre Serre, *Exposés de séminaires (1950-1999)* (French), Documents Mathématiques (Paris) [Mathematical Documents (Paris)], 1, Société Mathématique de France, Paris, 2001. MR1942136

[59] D. T. Tèĭt and I. R. Šafarevič, *The rank of elliptic curves* (Russian), Dokl. Akad. Nauk SSSR **175** (1967), 770–773. MR0237508

[60] T. A. Springer, *Linear algebraic groups*, 2nd ed., Progress in Mathematics, vol. 9, Birkhäuser Boston, Inc., Boston, MA, 1998. MR1642713

[61] Burt Totaro, *Pseudo-abelian varieties* (English, with English and French summaries), Ann. Sci. Éc. Norm. Supér. (4) **46** (2013), no. 5, 693–721. MR3185350

[62] Carlo Traverso, *Seminormality and Picard group*, Ann. Scuola Norm. Sup. Pisa (3) **24** (1970), 585–595. MR0277542

[63] Douglas Ulmer, *Elliptic curves with large rank over function fields*, Ann. of Math. (2) **155** (2002), no. 1, 295–315, DOI 10.2307/3062158. MR1888802

[64] Charles A. Weibel, *Pic is a contracted functor*, Invent. Math. **103** (1991), no. 2, 351–377, DOI 10.1007/BF01239518. MR1085112

[65] Jörg Winkelmann, *Realizing connected Lie groups as automorphism groups of complex manifolds*, Comment. Math. Helv. **79** (2004), no. 2, 285–299, DOI 10.1007/s00014-003-0794-5. MR2059433

[66] Olivier Wittenberg, *On Albanese torsors and the elementary obstruction*, Math. Ann. **340** (2008), no. 4, 805–838, DOI 10.1007/s00208-007-0170-7. MR2372739

Institut Fourier, CS 40700, 38058 Grenoble Cedex 9, France
E-mail address: `Michel.Brion@univ-grenoble-alpes.fr`
URL: `https://www-fourier.ujf-grenoble.fr/~mbrion/`

Proceedings of Symposia in Pure Mathematics
Volume **94**, 2017
http://dx.doi.org/10.1090/pspum/094/01623

Structure and classification of pseudo-reductive groups

Brian Conrad and Gopal Prasad

Abstract. The theory of pseudo-reductive groups, developed by the authors jointly with Gabber, was motivated by applications to finiteness theorems over local and global function fields. Subsequent work by the authors aimed at a more comprehensive understanding of exceptional behavior in characteristic 2 yielded improvements to the general theory in *all* positive characteristics and a classification theorem in terms of a "generalized standard" construction over arbitrary fields (depending on many ingredients from the initial work by the authors and Gabber).

We provide an overview of the general theory from the vantage point of improvements found during the more recent work and survey the proof of the general classification theorem, including examples and applications illustrating many phenomena.

Contents

2010 *Mathematics Subject Classification.* Primary 20G15.

1. Introduction

1.1. Motivation. Let G be a smooth connected affine group over a field k. The theory of pseudo-reductive groups begins with the observation that the unipotent radical $\mathscr{R}_u(G_{\overline{k}})$ over an algebraic closure $\overline{k}$ of k generally does not arise from a k-subgroup of G when k is not perfect (though it always does when k is perfect, by Galois descent); see Example 1.2.3 for the most basic counterexamples over every imperfect field.

Such failure of k-descent arises almost immediately upon confronting several natural questions in the arithmetic of linear algebraic groups over global function fields over finite fields. This was the reason that the authors with Gabber first investigated pseudo-reductive groups, which we later learned had been studied by Borel and Tits as part of their work on rational conjugacy theorems for general smooth connected affine groups (announcing some results in [**BoTi3**] without proofs).

An important ingredient missing in the work of Borel and Tits is the "standard construction" (see §2.2). One of the main results emerging from [**CGP**] and [**CP**] is that all pseudo-reductive groups are "standard" away from characteristics 2 and 3 and that the"non-standard" possibilities in characteristics 2 and 3 can be described in a useful way via a "generalized standard" construction (see §10.2). The story of standardness and how it can fail in small characteristics underlies the most interesting arithmetic applications and provides a valuable guide to the *characteristic-free* general structure theory.

The proof of the ubiquity of the generalized standard construction in [**CP**] depends on many results in [**CGP**], in addition to requiring further new techniques. This survey is a user's guide to that proof.

REMARK 1.1.1. The nearly 900-page total length of [**CGP**] and [**CP**] is due to the treatment of many topics, some of which are not directly relevant to the study of generalized standardness. To provide a geodesic path through the proof that most pseudo-reductive groups are standard (in fact, away from characteristics 2 and 3 all pseudo-reductive groups are standard), and that every pseudo-reductive group (over an arbitrary field) has a canonical pseudo-reductive central quotient that is generalized standard, we use some of the newer ideas in [**CP**].

In this survey we aim to supply enough discussion of intermediate results, proofs, and examples to give the reader a sense of the scope and usefulness of the overall theory, but some topics are only touched upon here in an abbreviated form; e.g., the difficult construction (via birational group laws) of "split" pseudo-reductive groups with an irreducible *non-reduced* root system of any rank (over any imperfect field of characteristic 2) is only briefly described.

The robust theory of root groups and open cells in pseudo-reductive groups, as given in [**CGP**, Ch. 3], rests on scheme-theoretic dynamic techniques with 1-parameter subgroups described in [**CGP**, Ch. 2]. Although the theory of pseudo-reductive groups rests on the theory of reductive groups, the same dynamic methods can also be used to simplify the development of the theory of reductive groups (even over rings [**C3**]). In this article we will survey dynamic methods (and their applications to root groups and root systems), the general structure theory, and Galois-twisted forms. This includes a Tits-style classification of perfect pseudo-reductive groups G (see [**CP**, §6.3]) in the spirit of Tits' work [**Ti1**] (completed by Selbach [**Sel**]) in the connected semisimple case.

REMARK 1.1.2. It is surprising that a Tits-style classification theorem holds in the perfect pseudo-reductive case because a "Chevalley form" (i.e., k_s/k-form admitting a k-split maximal k-torus) generally *does not exist*, even over every local and global function field (see [**CP**, Ex. C.1.6]).

The main point of the theory of pseudo-reductive groups is two-fold:

(i) problems for arbitrary linear algebraic groups over imperfect fields k can often be reduced to the pseudo-reductive case, but there is generally no simple way to reduce rationality problems to the reductive case,

(ii) there is a rich structure theory for pseudo-reductive groups G in terms of root systems, root groups, and open cells similar to the reductive case.

In (ii) there are subtleties not encountered in the reductive case; e.g., there is no link to SL_2 with which one can develop the structure of root groups and open cells in the pseudo-reductive case, root groups can have very large dimension even if $k = k_s$, and in the perfect pseudo-reductive case the automorphism functor is represented by an affine k-group scheme that is generally non-smooth with maximal smooth closed k-subgroup whose identity component is larger than G/Z_G.

EXAMPLE 1.1.3. An interesting rationality question for which the difficulties alluded to in (i) block any easy inference from the reductive case is this: if G is a smooth connected affine group over a field k then is $G(k)$-conjugation transitive on the set of maximal split k-tori in G? The answer is well-known to be affirmative for reductive G [**Bo2**, 20.9(ii)], and was announced in general (without proof) by Borel and Tits in [**BoTi3**]; a complete proof is given in [**CGP**, Thm. C.2.3]. We will present this result as Theorem 4.2.9, its proof makes use of the structure of pseudo-reductive groups via Theorem 4.2.4.

One source of arithmetic motivation for the development of the general theory of pseudo-reductive groups was the desire to strengthen results in Oesterlé's beautiful work [**Oes**] on Tamagawa numbers for smooth connected affine groups over global function fields over finite fields. For example, parts of [**Oes**] were conditional on the finiteness of the degree-1 Tate–Shafarevich set

$$Ш^1_S(k,G) := \ker(\mathrm{H}^1(k,G) \longrightarrow \prod_{v \notin S} \mathrm{H}^1(k_v,G))$$

for any smooth connected affine group G over a global function field k and finite set S of places of k. (The analogous finiteness problem with k a number field was settled affirmatively by Borel and Serre [**BS**, Thm. 7.1] by using several ingredients: a finiteness result for adelic coset spaces [**Bo1**, Thm. 5.4], finiteness of degree-1 Galois cohomology of linear algebraic groups over local fields of characteristic 0, and class field theory. Their approach does not adapt to positive characteristic.)

A well-known application of the finiteness of $Ш^1_S(k,G)$, going back to [**BS**] over number fields, is to finiteness aspects of the failure of local-global principles for homogeneous spaces over k (see Example 1.1.4 below). But even for homogeneous spaces under "nice" groups, it is the Tate–Shafarevich sets of *stabilizer* subgroup schemes (at k-points) that are relevant. Since such subgroup schemes can be non-smooth when $\mathrm{char}(k) > 0$, one wants finiteness for $Ш^1_S(k,G)$ without smoothness hypotheses on G (in which case $\mathrm{H}^1(k,G)$ denotes the set of isomorphism classes of G-torsors for the fppf topology over k, and likewise over each k_v). The avoidance of smoothness is not a trivial matter, since if k is imperfect then in general G_{red} is not k-smooth, nor even a k-subgroup of G (see [**CGP**, Ex. A.8.3] for examples).

EXAMPLE 1.1.4. Let k be a global field, X a homogeneous space for an affine k-group scheme H of finite type, and S a finite set of places of k. For the equivalence relation on $X(k)$ of being in the same $H(k_v)$-orbit for all $v \notin S$, does each equivalence class consist of only finitely many $H(k)$-orbits? (More informally, is the failure of a local-to-global principle for the H-action on X governed by a finite set?) This problem for the equivalence class of a point $x_0 \in X(k)$ is very quickly reduced to the question of finiteness of $Ш^1_S(k,H_{x_0})$, where $H_{x_0} := \{h \in H \mid h.x_0 = x_0\}$ is the *scheme-theoretic* stabilizer of x_0 in X; see the beginning of [**C2**, §6] for this well-known reduction step.

Even if H is connected reductive, the stabilizer H_{x_0} may be arbitrarily bad (e.g., if $X = \mathrm{GL}_n/G$ for a closed k-subgroup scheme $G \subset \mathrm{GL}_n =: H$ and $x_0 = 1$ then $H_{x_0} = G$). By a trick with torsors when $\mathrm{char}(k) > 0$, the general question of finiteness of $Ш^1_S(k,G)$ for affine k-group schemes G of finite type can be reduced to the case of smooth connected affine G; see [**C2**, Lemma 6.1.1, §6.2].

The finiteness of $Ш^1_S(k,G)$ for connected semisimple G is a consequence of the Hasse Principle for simply connected semisimple k-groups (see [**Ha2**, Satz A] and [**BP**, App. B] when $\mathrm{char}(k) > 0$), and for commutative G it is a consequence of class field theory and the structure theory of Tits [**CGP**, App. B] for possibly non-split smooth connected unipotent groups over imperfect fields when $\mathrm{char}(k) > 0$ (see [**Oes**, IV, 2.6(a)]).

The analogous general finiteness problem over number fields is deduced from the settled semisimple and commutative cases in [**BS**] using that $\mathscr{R}_u(G_{\overline{k}})$ descends to a smooth connected unipotent normal k-subgroup $U \subset G$ (which moreover must be k-split) since k is perfect. But no such k-descent U generally exists when k is not perfect (see Examples 1.2.3 and 1.2.4). The structure theory of pseudo-reductive groups, especially the role of the "standard construction", makes possible what may presently appear to be impossible: to deduce the finiteness of $Ш^1_S(k, G)$ for general smooth connected affine k-groups G from the settled commutative case (over k) and semisimple case (over *finite extensions* of k).

1.2. Initial definitions and examples. Let G be a smooth connected affine group over a field k. In the absence of a descent of $\mathscr{R}_u(G_{\overline{k}}) \subset G_{\overline{k}}$ to a k-subgroup of G, the following notion is the best substitute:

DEFINITION 1.2.1. The *k-unipotent radical* $\mathscr{R}_{u,k}(G)$ is the maximal smooth connected unipotent normal k-subgroup of G.

If K/k is any extension field then obviously

$$\mathscr{R}_{u,k}(G)_K \subset \mathscr{R}_{u,K}(G_K). \tag{1.2.1.1}$$

Standard spreading-out and specialization arguments yield equality in (1.2.1.1) if K/k is separable [**CGP**, Prop. 1.1.9(1)], such as when $K = k_s$ or when k is the function field of a regular curve X (or higher-dimensional normal variety) over a field and K is the fraction field of the completed local ring $\mathscr{O}^{\wedge}_{X,x}$ at a point $x \in X$.

If K/k is not separable, such as $K = \overline{k}$ when k is imperfect, then the inclusion (1.2.1.1) is generally strict. To make examples with non-equality in (1.2.1.1) we shall use Weil restriction $\mathrm{R}_{k'/k}$ through a finite extension of fields k'/k that is not separable, so let us first review why $\mathrm{R}_{k'/k}$ is a very well-behaved operation when k'/k is separable. The key point is that the product decomposition $k' \otimes_k k_s = \prod_\sigma k_s$ defined by $a' \otimes b \mapsto (\sigma(a')b)$, with σ varying through the set of k-embeddings of k' into k_s, yields a direct product decomposition of k_s-schemes

$$\mathrm{R}_{k'/k}(X')_{k_s} = \mathrm{R}_{(k' \otimes_k k_s)/k_s}(X'_{k' \otimes_k k_s}) = \prod_\sigma (X' \otimes_{k',\sigma} k_s) \tag{1.2.1.2}$$

for any quasi-projective k'-scheme X'. For example, if G' is a connected reductive k'-group then $\mathrm{R}_{k'/k}(G')$ is a connected reductive k-group since $\mathrm{R}_{k'/k}(G')_{k_s}$ is a direct product of connected reductive groups.

REMARK 1.2.2. For a finite extension of fields k'/k, the Weil restriction functor $\mathrm{R}_{k'/k}$ on quasi-projective k'-schemes preserves smoothness (by the infinitesimal criterion) but if k'/k is not separable then the $\overline{k}$-algebra $k' \otimes_k \overline{k}$ is not a direct product of copies of $\overline{k}$ and consequently $\mathrm{R}_{k'/k}$ has bad properties (illustrated in [**CGP**, A.5]): it generally destroys properness, surjectivity, geometric connectedness, geometric irreducibility, and non-emptiness. Thus, the good behavior for separable k'/k (inspired by the classical idea of viewing a d-dimensional complex manifold as a $2d$-dimensional real-analytic manifold by using the $\mathbf{R}$-basis $\{1, i\}$ of $\mathbf{C}$ to convert local holomorphic coordinates into local real-analytic coordinates) is not generally a useful guide to the non-separable case.

Here are examples in which $\mathscr{R}_{u,k}(G) = 1$ but $\mathscr{R}_{u,\overline{k}}(G_{\overline{k}}) \neq 1$, with k any imperfect field, so (1.2.1.1) fails to be an equality with $K = \overline{k}$:

EXAMPLE 1.2.3. Let k be any imperfect field, with $p = \mathrm{char}(k) > 0$, and let k'/k be a nontrivial purely inseparable finite extension. The Weil restriction $G = \mathrm{R}_{k'/k}(\mathrm{GL}_1)$ (informally, "$k'^{\times}$ as a k-group") is a commutative smooth connected affine k-group. Explicitly, G is the Zariski-open subspace complementary to the hypersurface defined by vanishing of the norm polynomial $\mathrm{N}_{k'/k}$ in the affine space over k associated to the k-vector space k', so G is smooth and connected with dimension $[k':k]>1$.

The commutative unipotent smooth connected k-group $\mathscr{R}_{u,k}(G)$ is trivial because the p-torsion group $G(k_s)[p]$ is equal to ${k'_s}^{\times}[p]=1$. However, $\mathscr{R}_{u,\overline{k}}(G_{\overline{k}})\neq 1$ since G is not a torus (as G/GL_1 has dimension $[k':k]-1>0$ and is killed by the p-power $[k':k]$, so it is unipotent).

More generally, if k'/k is a finite extension of fields and G' is any connected reductive k'-group then consideration of the functorial meaning of Weil restriction (instead of using the crutch of commutativity as above) shows that $G:=\mathrm{R}_{k'/k}(G')$ satisfies $\mathscr{R}_{u,k}(G)=1$ [**CGP**, Prop. 1.1.10]. However, if k'/k is not separable and $G'\neq 1$ then necessarily $\mathscr{R}_{u,\overline{k}}(G_{\overline{k}})\neq 1$ (see [**CGP**, Ex. 1.1.12, Ex. 1.6.1]).

EXAMPLE 1.2.4. Let k'/k be a purely inseparable extension of degree $p=\mathrm{char}(k)$ and consider $G=\mathrm{R}_{k'/k}(\mathrm{SL}_p)/\mathrm{R}_{k'/k}(\mu_p)$. The inclusion $G\hookrightarrow \mathrm{R}_{k'/k}(\mathrm{PGL}_p)$ has codimension $\dim \mathrm{R}_{k'/k}(\mu_p)=\dim \mathrm{R}_{k'/k}(\mathrm{GL}_1)[p]=p-1>0$ with image $\mathscr{D}(\mathrm{R}_{k'/k}(\mathrm{PGL}_p))$ and $\mathscr{R}_{u,k}(G)=1$, but $\mathscr{R}_{u,\overline{k}}(G_{\overline{k}})\neq 1$ [**CGP**, Prop. 1.3.4, Ex. 1.3.5].

DEFINITION 1.2.5. A *pseudo-reductive* k-group is a smooth connected affine k-group G such that $\mathscr{R}_{u,k}(G)=1$. If also $G=\mathscr{D}(G)$ then G is *pseudo-semisimple.*

EXAMPLE 1.2.6. A mild but very useful generalization of Example 1.2.3 is given by direct products: the pseudo-reductive k-groups $\mathrm{R}_{k'/k}(G')$ for nonzero finite reduced k-algebras k' and smooth affine k'-groups G' with connected reductive fibers. (Concretely, if $k'=\prod k'_i$ for fields k'_i and if G'_i denotes the k'_i-fiber of G' then $\mathrm{R}_{k'/k}(G')=\prod \mathrm{R}_{k'_i/k}(G'_i)$.) This construction is far from exhaustive: the pseudo-reductive k-group G built in Example 1.2.4 is *not* a k-isogenous quotient of any k-group of the form $\mathrm{R}_{k'/k}(G')$ for a nonzero finite reduced k-algebra k' and smooth affine k'-group G' with connected reductive fibers [**CGP**, Ex. 1.4.7].

Over perfect k pseudo-reductivity coincides with reductivity (in the connected case), but Examples 1.2.3 and 1.2.4 provide many non-reductive pseudo-reductive groups over any imperfect field. If we define the *k-radical* $\mathscr{R}_k(G)$ similarly to $\mathscr{R}_{u,k}(G)$ by replacing "unipotent" with "solvable" then any pseudo-semisimple G satisfies $\mathscr{R}_k(G)=1$ (as $\mathscr{R}(G_{\overline{k}})=\mathscr{R}_u(G_{\overline{k}})$, since $G_{\overline{k}}$ is perfect) but the converse is false! More specifically, for any imperfect field k of characteristic $p>0$ and degree-p purely inseparable extension k'/k, the Weil restriction $G=\mathrm{R}_{k'/k}(\mathrm{PGL}_p)$ is pseudo-reductive (by Example 1.2.3) and satisfies $G\neq\mathscr{D}(G)$ (as we will explain in Example 2.2.3) but $\mathscr{R}_k(G)=1$; see [**CGP**, Ex. 11.2.1].

If G is a smooth connected affine group over a field k then $G/\mathscr{R}_{u,k}(G)$ is clearly pseudo-reductive, so every such G uniquely fits into a short exact sequence

$$1\longrightarrow U\longrightarrow G\longrightarrow G/U\longrightarrow 1 \tag{1.2.6.1}$$

expressing it as an extension of a pseudo-reductive k-group by a smooth connected unipotent k-group U. The usefulness of (1.2.6.1) rests on being able to analyze the outer terms. For the left term, this requires applying Tits' structure theory for smooth connected unipotent groups [**CGP**, App. B] because if k is not perfect then U is often not k-split (i.e., U may not admit a composition series consisting of smooth closed k-subgroups with successive quotients k-isomorphic to $\mathbf{G}_{\mathrm{a}}$):

EXAMPLE 1.2.7. If k'/k is a nontrivial purely inseparable finite extension in characteristic $p > 0$ and $G := \mathrm{R}_{k'/k}(\mathrm{GL}_n)/\mathrm{GL}_1$ with $n \geqslant 1$ then the central smooth connected k-subgroup

$$U := \mathrm{R}_{k'/k}(\mathrm{GL}_1)/\mathrm{GL}_1 = \ker(G \twoheadrightarrow \mathrm{R}_{k'/k}(\mathrm{PGL}_n))$$

of dimension $[k' : k] - 1 > 0$ is unipotent since it is killed by the p-power $[k' : k]$ yet U does not contain $\mathbf{G}_{\mathrm{a}}$ as a k-subgroup [**CGP**, Ex. B.2.8]. Thus, U is not k-split.

In §2.2 we will introduce the *standard construction* that builds many pseudo-reductive groups from Weil restrictions of connected reductive groups over finite extensions of k. (This construction also involves auxiliary commutative pseudo-reductive k-groups.) The ubiquity of the standard construction when $\mathrm{char}(k) \neq 2, 3$ leads to a useful general principle (requiring care in characteristics 2 and 3):

> to solve a problem for general smooth connected affine k-groups, the structure theory of pseudo-reductive k-groups should reduce the task to the commutative case over k and the connected semisimple case over all finite extensions k'/k.

The method by which one applies the structure theory to carry out such a reduction depends on the specific problem under consideration. Here are two examples:

EXAMPLE 1.2.8. Let k be a global function field over a finite field. In Example 1.1.4 we saw that the problem of finiteness of degree-1 Tate–Shafarevich sets $Ш^1_S(k, G)$ for arbitrary affine k-group schemes G of finite type and finite sets S of places of k naturally leads one to the study of pseudo-reductive groups. After reducing this problem to the case of smooth connected G, one can apply Galois-twisting to (1.2.6.1) to eventually reduce to pseudo-reductive G; see [**C2**, §6.3]. (This latter reduction is harder than its analogue over number fields because $\mathscr{R}_{u,k}(G)$ is generally not k-split.)

Pseudo-reductivity has not yet played a role beyond its definition. The structure theory of pseudo-reductive groups, especially the ubiquity of the "standard construction" away from characteristics 2 and 3 and a precise understanding of the "non-standard" possibilities in characteristics 2 and 3, is what allows one to reduce the pseudo-reductive case over k to the settled commutative case over k and the settled semisimple case over finite extensions of k to solve the general finiteness problem for $Ш^1_S(k, G)$ (see [**C2**, §6.4]).

EXAMPLE 1.2.9. For global function fields k, the finiteness of the Tamagawa number of any smooth connected affine k-group was settled by Harder [**Ha1**] and Oesterlé [**Oes**, IV, 1.3] in the semisimple and commutative cases respectively, and the general case is deduced from this via the structure theory of pseudo-reductive groups in [**C2**, §7.3–§7.4]. This deduction uses standardness (and control of non-standardness when $\mathrm{char}(k) = 2, 3$) very differently from how standardness (and its controlled failure in small characteristic) is used in the proof of finiteness of degree-1 Tate–Shafarevich sets.

The failure of the standard construction to be exhaustive in characteristics 2 and 3 is due to three sources (at least the first two of which below were known to Tits in an embryonic form [**Ti3**, Cours 1991-92, 5.3, 6.4]). Firstly, one can make "exotic" generalizations of the standard construction by using non-central Frobenius factorizations (see [**CGP**, §7.1, §7.4]) that exist in characteristic $p > 0$ if and only if the Dynkin diagram has an edge of multiplicity p (i.e., $p = 2, 3$). Here are constructions with $p = 2$:

EXAMPLE 1.2.10. Over a field k of characteristic 2, consider the exceptional isogeny between types B_n and C_n ($n \geqslant 1$, with B_1 and C_1 understood to denote A_1). This is given by $\pi_q : \mathrm{SO}(q) \to \mathrm{Sp}(\overline{B}_q)$ for any non-degenerate quadratic space (V, q) of dimension $2n+1$ over k and the associated symplectic space $(V/V^{\perp}, \overline{B}_q)$ of dimension $2n$, where $V^{\perp}$ is the defect line of q and $\overline{B}_q$ is induced by the bilinear form $B_q(v, v') = q(v+v') - q(v) - q(v')$. The kernel of the composite map

$$\mathrm{Spin}(q) \longrightarrow \mathrm{SO}(q) \xrightarrow{\pi_q} \mathrm{Sp}(\overline{B}_q)$$

is killed by the Frobenius isogeny $F_{\mathrm{Spin}(q)/k} : \mathrm{Spin}(q) \to \mathrm{Spin}(q^{(2)})$ (with $q^{(2)}$ the scalar extension of q by the squaring endomorphism of k), thereby yielding a non-central factorization of $F_{\mathrm{Spin}(q)/k}$:

$$\mathrm{Spin}(q) \longrightarrow \mathrm{Sp}(\overline{B}_q) \longrightarrow \mathrm{Spin}(q^{(2)}).$$

There is a not so widely known analogue of this Frobenius factorization for any simply connected k-group G of type C_n ($n \geqslant 2$) in place of $\mathrm{Spin}(q)$. This is fully explained in Example 7.5.5, and goes as follows. Among the minimal non-central k-subgroup schemes of $\ker F_{G/k}$ that are normal in G, there is a unique minimal one; call it N. For $\overline{G} := G/N$, the restriction to $V := \mathrm{im}(\mathrm{Lie}(G) \to \mathrm{Lie}(\overline{G}))$ of the quadratic map $X \mapsto X^{[2]}$ on $\mathrm{Lie}(\overline{G})$ is valued in a unique line L, and the resulting quadratic form $q : V \to L$ is $\overline{G}$-invariant and non-degenerate. The composite k-homomorphism $G \to \overline{G} \to \mathrm{SO}(q)$ uniquely factors through the central isogeny $\mathrm{Spin}(q) \to \mathrm{SO}(q)$ via a purely inseparable isogeny $G \to \mathrm{Spin}(q)$ through which $F_{G/k}$ uniquely factors.

In §7.5 such non-central Frobenius factorizations yield non-standard pseudo-semisimple k-groups H for which $H^{\mathrm{ss}}_{\overline{k}} := H_{\overline{k}}/\mathscr{R}(H_{\overline{k}})$ is simply connected of any type B_n or C_n ($n \geqslant 1$) that we wish (and adapts to F_4 for $p = 2$ and G_2 for $p = 3$).

To describe a second source of non-standard examples, we need to make an observation concerning root systems associated to standard pseudo-reductive groups. In §2.3 we will see that if G is a pseudo-reductive k-group and T is a split maximal k-torus in G (as exists when $k = k_s$) then the set $\Phi(G, T)$ of nontrivial T-weights on $\mathrm{Lie}(G)$ is a root system (spanning the $\mathbf{Q}$-vector space $\mathrm{X}(T')_{\mathbf{Q}}$ for the maximal k-torus $T' := T \cap \mathscr{D}(G)$ in $\mathscr{D}(G)$ that is an isogeny complement in T to the maximal central k-torus in G). Inspection of the standard construction shows that this root system is always *reduced* when G is standard [**CGP**, Cor. 4.1.6], and in fact without any standardness hypotheses $\Phi(G, T)$ is reduced whenever $\mathrm{char}(k) \neq 2$ (see Theorem 3.1.7) or k is perfect.

Now choose an imperfect field k with characteristic 2 and an integer $n \geqslant 1$. There exist pseudo-semisimple k-groups G with a split maximal k-torus T of dimension n such that $\Phi(G, T)$ is the unique *non-reduced* irreducible root system BC_n of rank n. The construction of such G (see §8) is rather delicate, involving birational group laws, in contrast with the preceding constructions that rest only on concrete operations with affine groups via Weil restrictions and fiber products. The existence of such k-groups G is ultimately due to the combinatorial fact that among all reduced and irreducible root systems, precisely type C_n ($n \geqslant 1$) admits a root that is divisible in the weight lattice (and moreover only divisible by ± 2).

REMARK 1.2.11. The pseudo-semisimple groups G with root system BC_n whose construction is reviewed in §8 admit a natural quotient map $f : G \twoheadrightarrow \overline{G} = \mathrm{R}_{k'/k}(\mathrm{Sp}_{2n})$ for a nontrivial finite extension k'/k contained inside $k^{1/2}$. If $[k : k^2]$ is finite then such G can be built for which the induced map $G(k) \to \overline{G}(k) = \mathrm{Sp}_{2n}(k')$ is bijective, and the structure of G ensures that the inverse bijection is a counterexample to a conjecture [**BoTi2**, 8.19] of Borel and Tits on the algebraicity of certain "abstract" homomorphisms between connected linear algebraic groups, even if we restrict to perfect connected linear algebraic groups. This affirms the expectation of Borel and Tits that restrictions on k (e.g., avoidance of imperfect fields of characteristic 2) may be needed in their conjecture.

A third source of non-standard pseudo-reductive groups occurs only over imperfect fields k of characteristic 2. In the rank-1 pseudo-semisimple case, one class of such k-groups arises from purely inseparable finite extensions K/k and a nonzero proper kK^2-subspaces $V \subset K$ such that the ratios v'/v of nonzero elements of V generate K as a k-algebra. Given such data, which exist over k if and only if $[k : k^2] > 2$, the k-subgroup $H_{V,K/k} \subset \mathrm{R}_{K/k}(\mathrm{SL}_2)$ generated by the points $\left(\begin{smallmatrix} 1 & v \\ 0 & 1 \end{smallmatrix}\right)$ and $\left(\begin{smallmatrix} 1 & 0 \\ v & 1 \end{smallmatrix}\right)$ for all $v \in V$ is perfect and pseudo-reductive. The k-isomorphism class of $H_{V,K/k}$ depends on V precisely up to $K^\times$-scaling, so this class of k-groups constitutes a "continuous family"; it plays a role in the above birational construction for BC_n, and admits as higher-rank generalizations certain "special orthogonal" groups attached to a distinguished class of degenerate quadratic forms in characteristic 2 (see §7.3). The properties of the k-groups $H_{V,K/k}$ are addressed in §7.2.

1.3. Terminology and notation. For a finite flat extension $B \to B'$ of noetherian rings, we denote by $\mathrm{R}_{B'/B}$ the Weil restriction functor assigning to any quasi-projective B'-scheme X' the quasi-projective B-scheme $\mathrm{R}_{B'/B}(X')$ representing the functor on B-algebras $A \rightsquigarrow X'(A \otimes_B B')$; we refer the reader to [**CGP**, A.5] for a discussion of the existence and basic properties of this functor (especially beyond the classical case when B' is finite étale over B).

For any scheme X, the underlying reduced closed subscheme (with the same topological space) is denoted X_{red}. For a group scheme H of finite type over a field k, H^{sm} denotes the maximal smooth closed k-subgroup; see [**CGP**, C.4.1–C.4.2] for its existence and basic properties and see [**CGP**, A.8.2] for the equality with H_{red} when H is of multiplicative type (but H^{sm} is usually much smaller than H_{red}; see [**CGP**, A.8.3, C.4.2] for examples).

For a smooth affine group G over a field k, $G_{\overline{k}}^{\mathrm{red}}$ denotes the quotient of $G_{\overline{k}}$ modulo its unipotent radical; we define $G_{\overline{k}}^{\mathrm{ss}}$ similarly using the radical. A finite-dimensional quadratic space (V, q) over a field k is *non-degenerate* if $q \neq 0$ and the projective hypersurface $(q = 0) \subset \mathbf{P}(V^*)$ is k-smooth.

For a group scheme G of finite type over a field k and smooth closed k-subgroup H, the *scheme-theoretic centralizer* $Z_G(H)$ is the closed k-subgroup scheme of G representing the functor assigning to any k-algebra A the group of points $g \in G(A)$ whose conjugation action on the A-group G_A is trivial on H_A; see [**CGP**, A.1.9ff.] for the existence of $Z_G(H)$. In the special case $H = G$ (with G smooth) it is called the *scheme-theoretic center* and is denoted by Z_G (e.g., $Z_{\mathrm{SL}_n} = \mu_n$ for all integers $n > 1$). The existence and basic properties of $Z_G(H)$ when H is of multiplicative type (but possibly not smooth) is addressed in [**CGP**, Prop. A.8.10].

Whenever we speak of "centralizer", "kernel" (for a homomorphism), and "intersection" (of closed subgroup schemes), it is always understood that we intend the scheme-theoretic notions (which may not be smooth). In [**CGP**, A.1] many basic definitions in the group scheme context are reviewed and the relationship with more classical definitions (when available) is discussed, such as quotients modulo closed subgroups.

1.4. Simplifications and corrections. In addition to surveying the combined works [**CGP**] and [**CP**], we have taken the opportunity to provide some simplifications and improvements, as well as a few corrections. For the convenience of the reader we highlight those items here, beginning with the simplifications.

(i) It is an important fact in the general theory that the Weil restriction $\mathrm{R}_{k'/k}(G')$ is perfect for any (possibly non-separable) finite extension of fields k'/k and connected semisimple k'-group G' that is simply connected. The original proof given in [**CGP**, Cor. A.7.11] relies on group scheme techniques over artinian rings; in Proposition 2.2.4 we provide a shorter and simpler proof using only smooth affine groups over fields.

(ii) In the study of pseudo-reductivity, an important notion is that of a *pseudo-parabolic* k-subgroup of a smooth connected affine k-group G. The definition of pseudo-parabolicity via a dynamic procedure (rather than by a geometric property of the associated coset space), given in Definition 2.3.6, may initially look ad hoc. However, it is a powerful concept (and is equivalent to parabolicity when G is reductive).

As for parabolic k-subgroups, every pseudo-parabolic k-subgroup of a smooth connected affine k-group G is its own scheme-theoretic normalizer; we give a proof of this fact in Theorem 4.3.6 that is substantially simpler than the proof given in [**CGP**, Prop. 3.5.7].

(iii) It is very useful that any pseudo-reductive k-group G admitting a split maximal k-torus T (such as whenever $k = k_s$) contains a *Levi k-subgroup* $L \supset T$ (i.e., $L_{\overline{k}} \to G_{\overline{k}}^{\mathrm{red}}$ is an isomorphism); moreover, one can control the position inside G of the simple positive root groups for L. The proof here as Theorem 5.4.4 is simpler than the one in [**CGP**, Thm. 3.4.6].

(iv) The first main classification theorem in the general theory of pseudo-reductive groups is that the standard construction is ubiquitous away from specific situations over imperfect fields of characteristics 2 and 3. This is made precise in the absolutely pseudo-simple case in Theorem 7.4.8, whose proof is much simpler than the one in [**CGP**, Cor. 6.3.5, Prop. 6.3.6].

One of the key facts that this rests upon, recorded here in Theorem 7.2.5(i), is that for any field k that is not imperfect of characteristic 2 and any absolutely pseudo-simple k-group G whose root system over k_s has rank 1 and whose minimal field of definition for its geometric unipotent radical is K/k, the natural map $i_G : G \to \mathrm{R}_{K/k}(G_K/\mathscr{R}_{u,K}(G_K))$ is an isomorphism. The proof here, based on ideas from [**CP**, §3.1], is a substantial simplification of the proof given in [**CGP**, Thm. 6.1.1].

(v) One of the main results in [**CP**] concerns the ubiquity of the "generalized standard" construction, a generalization of the standard construction that accounts for exceptional phenomena over fields k of characteristic 2 satisfying $[k : k^2] > 2$. A crucial step towards the proof of its ubiquity is that the "generalized standard" property is insensitive to passage to the derived group. We give a proof of this fact (in Proposition 10.2.5) that is significantly simpler than the proof in [**CP**, §9.1].

Now we mention four corrections. The first correction concerns the relationship between pseudo-parabolic k-subgroups of a pseudo-reductive k-group G and parabolic subgroups of $G' = G_{\overline{k}}^{\text{red}}$. The formulation of such a dictionary in [**CGP**, Prop. 3.5.4] omits the hypothesis that the chosen maximal k-torus T is split; this is needed to justify scalar extension to k_s at the start of the argument. We provide a much simpler proof of the corrected formulation in Proposition 4.3.3, moreover avoiding the passage to k_s.

That missing split hypothesis does not harm the proofs of results in [**CGP**] (or work in [**C2**] and [**CP**] relying on [**CGP**]) because every appeal to [**CGP**, Prop. 3.5.4] (e.g., in the proof of [**CP**, Prop. 8.1.4]) takes place over a separably closed field (where all tori are split) with two exceptions:

(a) [**CGP**, Cor. 3.5.11] has a formulation that permits its proof to begin by extending the ground field to its separable closure,
(b) [**CGP**, Prop. 11.4.4] concerns a pseudo-reductive k-group with a split maximal k-torus, and its proof works using the corrected formulation of [**CGP**, Prop. 3.5.4] because *every* pseudo-parabolic k-subgroup contains a split maximal k-torus (see Lemma 4.2.7, which has no logical dependence on anything in [**CGP**, Ch. 11]).

The second correction involves [**CGP**, Prop. 3.3.15] that provides three basic properties of minimal pseudo-parabolic k-subgroups P in a pseudo-reductive k-group G containing a split maximal k-torus. The formulation is correct but there is a gap in Step 1 of the proof in [**CGP**]: it was overlooked to show (as is needed in the proof) that every minimal pseudo-parabolic k-subgroup of such a G necessarily contains a split maximal k-torus. We establish [**CGP**, Prop. 3.3.15(1),(2)] by more direct means as Proposition 3.3.7, and establish part (3) as Proposition 4.2.8.

The third correction is at the end of the proof of [**CGP**, Lemma 7.1.2]. Replace the last sentence with: "By the Chevalley commutation relations [**SGA3**, XXIII, 3.3.1(iii), 3.4.1(iii)], if c is a positive root and c' is a short positive root such that $ic+c'$ is a long root then $r_{i,1} = p$ vanishes in k." (That c may be short was missed.)

The final correction is that [**CP**, Prop. 8.4.3] was not formulated in enough generality for later needs (in the proof of [**CP**, Thm. 9.2.1]), but its proof applies in the required additional generality. We record that result here in the appropriate generality as Proposition 10.1.15 and provide a proof; it implies that certain data entering into the "generalized standard" construction can be canonically recovered from the output of that construction (see Corollary 10.2.6). Galois descent then ensures that the "generalized standard" property over a field k is insensitive to scalar extension to k_s (Corollary 10.2.8), so when proving a given pseudo-reductive k-group is generalized standard (as in one of our main results, Theorem 10.2.13) it is sufficient to work over k_s. Passage to k_s is essential for accessing calculations with root groups and properties of the rank-1 case.

2. Standard groups and dynamic methods

2.1. Basic properties of pseudo-reductive groups. If K/k is a separable extension of fields (e.g., $K = k_s$) then a smooth connected affine k-group G is pseudo-reductive if and only if G_K is pseudo-reductive, since (1.2.1.1) is an equality in such cases. In particular, since $G(k_s)$ is Zariski-dense in G_{k_s}, it follows easily by

Galois descent from k_s that if G is pseudo-reductive and N is a smooth connected normal k-subgroup of G then N is pseudo-reductive. For example, the derived group $\mathscr{D}(G)$ of a pseudo-reductive group is always pseudo-reductive.

A consequence of the pseudo-reductivity of the derived group is that solvable pseudo-reductive groups G are always commutative [**CGP**, Prop. 1.2.3]. Indeed, solvability implies that $\mathscr{D}(G)$ is unipotent (as we may check over $\overline{k}$ by using the structure of solvable smooth connected affine $\overline{k}$-groups), yet $\mathscr{D}(G)$ inherits pseudo-reductivity from G and hence $\mathscr{D}(G) = 1$; i.e., G is commutative. However, in contrast with tori (which can be studied by means of Galois lattices), it is generally very difficult to say anything about the structure of commutative pseudo-reductive groups (e.g., they can admit nontrivial étale p-torsion in characteristic $p > 0$ [**CGP**, Ex. 1.6.3]). Hence, in structure theorems for pseudo-reductive groups we shall treat the commutative case as a black box.

Further similarities with the reductive case are given by the following result that often enables one to reduce general questions for pseudo-reductive groups to the separate consideration of commutative and pseudo-semisimple cases:

PROPOSITION 2.1.1. *Let G be a pseudo-reductive k-group and $T \subset G$ a k-torus.*

(i) *The scheme-theoretic centralizer $Z_G(T)$ is pseudo-reductive, and it is commutative when T is maximal in G.*
(ii) *Any Cartan k-subgroup C of G is commutative and pseudo-reductive and $G = C \cdot \mathscr{D}(G)$.*
(iii) *The derived group $\mathscr{D}(G)$ is perfect (i.e., pseudo-semisimple).*

PROOF. The proof of the first part of (i) entails using the analogue for $G_{\overline{k}}^{\mathrm{red}} := G_{\overline{k}}/\mathscr{R}_u(G_{\overline{k}})$ and the good behavior of torus centralizers under quotient maps (such as $G_{\overline{k}} \twoheadrightarrow G_{\overline{k}}^{\mathrm{red}}$) to show that $\mathscr{R}_{u,k}(Z_G(T)) \subset \mathscr{R}_{u,k}(G)$. Any Cartan k-subgroup C is certainly nilpotent, so the derived group $\mathscr{D}(C)$ is a smooth connected *unipotent* normal k-subgroup of C. But we have shown that C is pseudo-reductive, so $\mathscr{D}(C) = 1$; i.e., C is commutative. This proves (i).

We next claim that for any smooth connected affine k-group H and Cartan k-subgroup C of H, $H = C \cdot \mathscr{D}(H)$. Indeed, $H/\mathscr{D}(H)$ is commutative and hence is its own Cartan k-subgroup, so the Cartan subgroup C of H maps onto $H/\mathscr{D}(H)$. This yields (ii).

To prove (iii), let C be a Cartan k-subgroup of G. By [**CGP**, Lemma 1.2.5(ii)], $C \cap \mathscr{D}(G)$ is a Cartan subgroup of $\mathscr{D}(G)$. Hence, $\mathscr{D}(G) = (C \cap \mathscr{D}(G)) \cdot \mathscr{D}(\mathscr{D}(G)) \subset C \cdot \mathscr{D}(\mathscr{D}(G))$. Therefore, $G = C \cdot \mathscr{D}(G) = C \cdot \mathscr{D}(\mathscr{D}(G))$. But C is commutative, so $G/\mathscr{D}(\mathscr{D}(G))$ is commutative and thus $\mathscr{D}(G) \subset \mathscr{D}(\mathscr{D}(G))$. The reverse inclusion is obvious. □

The commutativity of any Cartan k-subgroup C of a pseudo-reductive k-group G implies immediately that C coincides with its own scheme-theoretic centralizer in G, so C contains the scheme-theoretic center Z_G of G (as in the reductive case). Another feature of pseudo-reductive groups reminiscent of the connected reductive case (and not shared by smooth connected affine groups in general) is that normality is *transitive* for smooth connected k-subgroups:

PROPOSITION 2.1.2. *If G is a pseudo-reductive k-group, H is a smooth connected normal k-subgroup of G, and N is a smooth connected normal k-subgroup of H then N is normal in G.*

The case of perfect N can be settled by using the known analogous result for $G^{\mathrm{red}}_{\overline{k}}$. The general case is deduced from this via considerations with root systems over k_s. See [**CGP**, Prop. 1.2.7, Rem. 3.1.10] for details.

Despite the preceding favorable basic properties, in general pseudo-reductivity is *not* an especially robust notion (in contrast with reductivity):

EXAMPLE 2.1.3. Over any imperfect field k of characteristic $p > 0$, pseudo-reductivity is usually not inherited by quotients modulo central k-subgroup schemes (e.g., $\mathrm{R}_{k'/k}(\mathrm{SL}_p)/\mu_p$ is not pseudo-reductive for any nontrivial purely inseparable finite extension k'/k in characteristic p [**CGP**, Ex. 1.3.5]) or modulo pseudo-semisimple normal k-subgroups (see [**CGP**, Ex. 1.6.4]).

Although central quotients G/Z of pseudo-reductive k-groups G can fail to be pseudo-reductive, such failure is governed by the commutative case: for any Cartan k-subgroup C of G, the k-smooth central quotient G/Z is pseudo-reductive if and only if its Cartan k-subgroup C/Z is pseudo-reductive [**CGP**, Lemma 9.4.1].

EXAMPLE 2.1.4. The failure of $\mathrm{R}_{k'/k}(\mathrm{SL}_p)/\mu_p$ to be pseudo-reductive for a nontrivial purely inseparable finite extension k'/k in characteristic p is explained by the failure of pseudo-reductivity of its Cartan k-subgroup $Q = \mathrm{R}_{k'/k}(\mathrm{GL}_1^{p-1})/\mu_p$, where μ_p is canonically included into the first factor $\mathrm{R}_{k'/k}(\mathrm{GL}_1)$. Indeed, for a degree-$p$ subextension k'_0/k of k'/k the quotient $\mathrm{R}_{k'_0/k}(\mu_p)/\mu_p$ is a k-subgroup of Q, and since ${k'_0}^p \subset k$ this k-subgroup coincides with the k-group $\mathrm{R}_{k'_0/k}(\mathrm{GL}_1)/\mathrm{GL}_1$ that is smooth, connected, and unipotent of dimension $p-1 > 0$.

Among the central quotients of a pseudo-reductive group G, the central quotient G/Z_G (with $Z_G \subset G$ the scheme-theoretic center) is especially useful. Fortunately, G/Z_G is *always* pseudo-reductive and has trivial scheme-theoretic center (but it might not be perfect, in contrast with the reductive case); see §6.1.

As a final illustration of the contrast between pseudo-reductive and reductive groups, recall that any connected reductive group H over a field k is unirational [**Bo2**, 18.2(ii)]. (The k-group H is generated by its perfect derived group and its maximal central k-torus, so alternatively one can appeal to the more general fact [**CGP**, Prop. A.2.11] that *every* perfect smooth connected affine k-group is generated by k-tori.) This unirationality property yields the important consequence that $H(k)$ is Zariski-dense in H when k is infinite. This fails badly in the pseudo-reductive case:

EXAMPLE 2.1.5. For every imperfect field k there exist pseudo-reductive k-groups G that are not unirational, and for rational function fields $k = \kappa(v)$ over fields κ of positive characteristic there exist nontrivial pseudo-reductive k-groups G such that $G(k)$ is not Zariski-dense in G. In view of the unirationality of the perfect $\mathscr{D}(G)$ [**CGP**, Prop. A.2.11] and the equality $G = C \cdot \mathscr{D}(G)$ for a (commutative pseudo-reductive) Cartan k-subgroup $C \subset G$, all obstructions arise in the commutative case.

To make a commutative pseudo-reductive k-group that is either not unirational or does not have a Zariski-dense locus of k-points, it suffices to construct a smooth connected unipotent k-group U with either of these properties and build a commutative pseudo-reductive extension of U by GL_1 over k.

Let $U = \{y^q = x - c^{p-1}x^p\}$ where $p = \mathrm{char}(k) > 0$, $c \in k - k^p$, and $q = p^r > 1$. In [**CGP**, Ex. 11.3.1] it is shown that: U admits a commutative pseudo-reductive extension by GL_1 over k, U is not unirational over k if $q > 2$, and when $k = \kappa(v)$ for a field κ of characteristic $p > 0$ the group $U(k)$ is finite if $q > 2$ and $c = v$.

2.2. The standard construction. A large class of pseudo-reductive groups can be built by using actions of commutative pseudo-reductive k-groups on Weil restrictions to k of connected reductive groups over finite (possibly inseparable) extensions of k. Before describing this construction, we address the preservation of pseudo-reductivity under certain central pushouts that "replace" a Cartan k-subgroup with another commutative pseudo-reductive k-group.

PROPOSITION 2.2.1. *Let $\mathscr{G}$ be a pseudo-reductive k-group and $\mathscr{C}$ a commutative pseudo-reductive k-subgroup satisfying $\mathscr{C} = Z_{\mathscr{G}}(\mathscr{C})$. Let C be another commutative pseudo-reductive k-group equipped with an action on $\mathscr{G}$ and with a k-homomorphism $\phi : \mathscr{C} \to C$ respecting the actions on $\mathscr{G}$. The cokernel G of the central inclusion*

$$\alpha : \mathscr{C} \hookrightarrow \mathscr{G} \rtimes C$$

defined by $c \mapsto (c^{-1}, \phi(c))$ is pseudo-reductive.

Informally, G is obtained from $\mathscr{G}$ by replacing $\mathscr{C}$ with C.

PROOF. It is elementary to check that $H := \mathscr{G} \rtimes C$ is pseudo-reductive and that any central k-subgroup of H is contained in $\mathscr{C} \times C$. Hence, to prove that $U := \mathscr{R}_{u,k}(G)$ is trivial it suffices to show that the (visibly solvable) smooth connected normal preimage $N \subset H$ of U is central, as then naturally $U = N/\mathscr{C} \hookrightarrow C$, forcing $U = 1$ since C is commutative and pseudo-reductive.

To show that N is central in H it is enough to prove that the smooth connected normal commutator k-subgroup (H, N) in the pseudo-reductive k-group H is unipotent. But for any smooth connected affine k-group H and *solvable* smooth connected normal k-subgroup N, the commutator subgroup (H, N) is unipotent (as we easily check over $\overline{k}$ by working with the maximal reductive quotient $H_{\overline{k}}^{\mathrm{red}}$ in which the solvable normal image of $N_{\overline{k}}$ must be a normal torus, hence central due to the connectedness of $H_{\overline{k}}^{\mathrm{red}}$). □

The main class of $\mathscr{C}$'s of interest for applying Proposition 2.2.1 is the Cartan k-subgroups of $\mathscr{G}$, and in such cases the k-group $C = (\mathscr{C} \times C)/\alpha(\mathscr{C})$ is a Cartan k-subgroup of G. There are many natural examples in which ϕ is not surjective; these arise in the "standard construction" (see Definition 2.2.6) and in the study of both k_s/k-forms and automorphism schemes of general pseudo-semisimple groups.

To apply Proposition 2.2.1 in the special case $\mathscr{G} = \mathrm{R}_{k'/k}(G')$ for a finite extension of fields k'/k and a connected reductive k'-group G', it is convenient (for motivational purposes) to first review how the behavior of $\mathrm{R}_{k'/k}$ on linear algebraic groups is sensitive to whether or not k'/k is separable. If k'/k is separable and $f' : X' \to Y'$ is a surjection between affine k'-schemes of finite type then it is an immediate consequence of (1.2.1.2) and considerations over k_s that $\mathrm{R}_{k'/k}(f')$ is surjective. If we drop the separability condition on k'/k then $\mathrm{R}_{k'/k}(f')$ is surjective provided that f' is also *smooth* (in which case $\mathrm{R}_{k'/k}(f')$ is smooth too) [**CGP**, Cor. A.5.4(1)], but surjectivity fails to be preserved in the absence of smoothness:

EXAMPLE 2.2.2. Consider the p-power endomorphism $f' : \mathrm{GL}_1 \to \mathrm{GL}_1$ over a degree-p inseparable extension k'/k in characteristic p. The map $\mathrm{R}_{k'/k}(f')$ is the p-power endomorphism of the smooth connected affine k-group $\mathrm{R}_{k'/k}(\mathrm{GL}_1)$ of dimension p. The image of $\mathrm{R}_{k'/k}(f')$ lies inside the canonical k-subgroup GL_1 (and so coincides with this k-subgroup) because the image of the p-power map on $\mathrm{R}_{k'/k}(\mathrm{GL}_1)(k_s) = (k' \otimes_k k_s)^\times$ is contained inside $k_s^\times = \mathrm{GL}_1(k_s)$.

Although Weil restriction through a finite extension of fields k'/k does not preserve many properties when k'/k is not separable (Remark 1.2.2), it does preserve the property "smooth and geometrically connected" [**CGP**, Prop. A.5.9] (such as for smooth connected affine groups), and if X' is a smooth affine k'-scheme with pure dimension n then $X := \mathrm{R}_{k'/k}(X')$ is k-smooth (by the infinitesimal criterion) with pure dimension $n[k' : k]$ (as we may check by computing tangent spaces at the Zariski-dense set of k_s-points in $X_{k_s} = \mathrm{R}_{k'_s/k_s}(X' \otimes_{k'} k'_s)$ for $k'_s := k' \otimes_k k_s$).

Also, if $f' : X' \to Y'$ is a torsor for a *smooth* affine k'-group H' then $\mathrm{R}_{k'/k}(f')$ is an $\mathrm{R}_{k'/k}(H')$-torsor [**CGP**, Cor. A.5.4(3)]. For our purposes, the most important example is that the natural map

$$\mathrm{R}_{k'/k}(G')/\mathrm{R}_{k'/k}(H') \longrightarrow \mathrm{R}_{k'/k}(G'/H')$$

is an isomorphism for any affine k'-group scheme G' of finite type and *smooth* closed k'-subgroup H'. (In particular, if such an H' is normal in G' then $\mathrm{R}_{k'/k}(H')$ is normal in $\mathrm{R}_{k'/k}(G')$ and the associated quotient group is $\mathrm{R}_{k'/k}(G'/H')$.)

The k'-smoothness hypothesis on H' is crucial, since Example 2.2.2 shows that inseparable Weil restriction generally does not carry isogenies to surjections, even when working with smooth connected affine groups. The bad behavior of inseparable Weil restriction with respect to isogenies has interesting consequences in the context of connected semisimple groups, such as a non-perfect inseparable Weil restriction of such a group:

EXAMPLE 2.2.3. Let k be imperfect of characteristic p, and let k'/k be a nontrivial finite extension satisfying $k'^p \subset k$. The smooth connected affine k-group $\mathrm{R}_{k'/k}(\mathrm{PGL}_p)$ is *not* perfect. To understand this, and more generally to analyze the structure of this k-group, the quotient presentation $\mathrm{PGL}_p \simeq \mathrm{SL}_p/\mu_p$ over k' is not useful because $\mathrm{R}_{k'/k}$ is not compatible with the formation of this central quotient by the non-smooth μ_p (as we shall see).

We need a quotient presentation entirely in terms of smooth k'-groups. The central quotient description $\mathrm{GL}_p/\mathrm{GL}_1$ could be used, but for later purposes it is more convenient to consider another central quotient description with smooth k'-groups, as follows.

Let $T' \subset \mathrm{SL}_p$ be a maximal k'-torus (such as the diagonal k'-torus), and define $\overline{T}' := T'/\mu_p$ to be its maximal k'-torus image in PGL_p. The conjugation action of SL_p on itself naturally factors through an action on SL_p by the central quotient PGL_p, so in this way the k'-subgroup $\overline{T}' \subset \mathrm{PGL}_p$ naturally acts on SL_p. This yields a k'-isomorphism

$$\mathrm{PGL}_p \simeq \mathrm{SL}_p/\mu_p \simeq (\mathrm{SL}_p \rtimes \overline{T}')/T',$$

where $T' \hookrightarrow \mathrm{SL}_p \rtimes \overline{T}$ is the central anti-diagonal inclusion $t' \mapsto ({t'}^{-1}, t' \bmod \mu_p)$. The right side involves only smooth k'-groups and so yields a k-isomorphism

$$\mathrm{R}_{k'/k}(\mathrm{PGL}_p) \simeq (\mathrm{R}_{k'/k}(\mathrm{SL}_p) \rtimes \mathrm{R}_{k'/k}(\overline{T}'))/\mathrm{R}_{k'/k}(T') \tag{2.2.3.1}$$

in which $\mathrm{R}_{k'/k}(\overline{T}')$ acts on $\mathrm{R}_{k'/k}(\mathrm{SL}_p)$ by applying the functoriality of $\mathrm{R}_{k'/k}$ to the $\overline{T}'$-action on SL_p. (Beware that $\mathrm{R}_{k'/k}(T') \to \mathrm{R}_{k'/k}(\overline{T}')$ is *not* surjective, as $T'_{k'_s} \to \overline{T}'_{k'_s}$ is the direct product of GL_1^{p-2} against the p-power map $\mathrm{GL}_1 \twoheadrightarrow \mathrm{GL}_1$.)

On the right side of (2.2.3.1) we have an instance of the cokernel construction in Proposition 2.2.1, and the k-group $\mathrm{R}_{k'/k}(\mathrm{SL}_p)$ is perfect since its group of k_s-points $\mathrm{SL}_p(k'_s)$ is perfect (as $\mathrm{SL}_n(F)$ is generated by subgroups of the form $\mathrm{SL}_2(F)$ for any field F). Thus, the commutativity of $\mathrm{R}_{k'/k}(\overline{T}')$ implies that

$$\mathscr{D}(\mathrm{R}_{k'/k}(\mathrm{PGL}_p)) = \mathrm{R}_{k'/k}(\mathrm{SL}_p)/\mathrm{R}_{k'/k}(\mu_p),$$

with $\mathrm{R}_{k'/k}(\mu_p) = \ker([p] : \mathrm{R}_{k'/k}(\mathrm{GL}_1) \twoheadrightarrow \mathrm{GL}_1)$ of dimension $p - 1 > 0$ (as noted in Example 1.2.4). But $\mathrm{R}_{k'/k}(\mathrm{PGL}_p)$ and $\mathrm{R}_{k'/k}(\mathrm{SL}_p)$ have the same dimension, so it follows that $\mathrm{R}_{k'/k}(\mathrm{PGL}_p)$ is *not* perfect.

To put the construction in Example 2.2.3 into a broader framework, as a first step we record an important result that explains the dichotomy between the perfectness of $\mathrm{R}_{k'/k}(\mathrm{SL}_p)$ and the failure of perfectness of $\mathrm{R}_{k'/k}(\mathrm{PGL}_p)$ above:

PROPOSITION 2.2.4. *If k'/k is a finite extenion of fields and G' is a connected semisimple k'-group that is simply connected then $G := \mathrm{R}_{k'/k}(G')$ is perfect (and hence is pseudo-semisimple).*

PROOF. Suppose the commutative quotient $H := G/\mathscr{D}(G)$ is nontrivial, so the Lie algebra $\mathfrak{h}$ of H is a nonzero G-equivariant quotient of $\mathfrak{g} := \mathrm{Lie}(G)$ with trivial G-action. Hence, it suffices to show that the space $\mathfrak{g}_G$ of G-coinvariants of $\mathfrak{g}$ vanishes. By treating the factor fields of $k' \otimes_k k_s$ separately we may assume $k = k_s$, so $\mathfrak{g}_G = \mathfrak{g}_{G(k)}$. Identifying $G(k)$ and $G'(k')$ is compatible with identifying $\mathfrak{g}$ and the underlying k-vector space of $\mathfrak{g}'$ [**CGP**, Cor. A.7.6], so it suffices to prove $\mathfrak{g}'_{G'} = 0$.

The simply connectedness hypothesis implies that a maximal torus T' of G' is the direct product of coroot groups $a^\vee(\mathrm{GL}_1)$ for roots a in a basis Δ of $\Phi(G', T')$, and pairs of opposite root groups (relative to the positive system of roots associated to Δ) generate SL_2's inside G'. The Lie algebras of these SL_2's span $\mathrm{Lie}(G')$ (as one sees via consideration of an open cell), so the vanishing of G'-coinvariants under $\mathrm{Ad}_{G'}$ reduces to the case $G' = \mathrm{SL}_2$ that is verified by direct calculation. □

2.2.5. The following construction of a large class of pseudo-reductive groups will admit a refined formulation via Proposition 2.2.4. Let k be a field, k' a nonzero finite reduced k-algebra, and G' a smooth affine k'-group whose fibers over the factor fields of k' are connected reductive. (The reason we consider such a product k' of fields as a single k-algebra, rather than treat its factor fields k'_i and the corresponding fiber groups G'_i of G' separately, is due to convenience in later Galois descent arguments since scalar extension along $k \to k_s$ generally does not carry fields to fields.)

Let $T' \subset G'$ be a maximal k'-torus, and let $\mathrm{R}_{k'/k}(G'/Z_{G'})$ act on $\mathrm{R}_{k'/k}(G')$ by applying $\mathrm{R}_{k'/k}$ to the natural $G'/Z_{G'}$-action on G'. (If the k'-group $Z_{G'}$ is non-étale over some point of $\mathrm{Spec}(k')$ that is not k-étale then $\mathrm{R}_{k'/k}(G'/Z_{G'})$ is generally larger than $\mathrm{R}_{k'/k}(G')/\mathrm{R}_{k'/k}(Z_{G'})$.) Finally, consider a commutative pseudo-reductive k-group C equipped with a factorization

$$\mathrm{R}_{k'/k}(T') \xrightarrow{\phi} C \longrightarrow \mathrm{R}_{k'/k}(T'/Z_{G'}) \tag{2.2.5.1}$$

of the natural k-homomorphism $\mathrm{R}_{k'/k}(T') \to \mathrm{R}_{k'/k}(T'/Z_{G'})$ (which is generally not surjective when $Z_{G'}$ is not k'-étale over some point where $\mathrm{Spec}(k')$ is not k-étale).

Since T' is a Cartan k'-subgroup of G', $\mathrm{R}_{k'/k}(T')$ is a Cartan k-subgroup of $\mathrm{R}_{k'/k}(G')$ [**CGP**, Prop. A.5.15(3)]. Thus, by Proposition 2.2.1 the central quotient

$$(\mathrm{R}_{k'/k}(G') \rtimes C)/\mathrm{R}_{k'/k}(T') \tag{2.2.5.2}$$

modulo the anti-diagonal inclusion $\mathrm{R}_{k'/k}(T') \hookrightarrow \mathrm{R}_{k'/k}(G') \rtimes C$ is pseudo-reductive.

DEFINITION 2.2.6. A *standard pseudo-reductive k-group* is a k-group that is k-isomorphic to (2.2.5.2) for some 4-tuple $(G', k'/k, T', C)$ as above equipped with a factorization (2.2.5.1).

Note that every commutative pseudo-reductive k-group is standard, by letting $k' = k$ and $G' = 1$. The pseudo-semisimple k-groups in Example 1.2.4 that do not arise as a k-isogenous quotient of the Weil restriction of a connected reductive group over any finite extension of k are nonetheless standard; see Example 2.2.8. In practice, to solve problems for a standard pseudo-reductive group one can often reduce to the study of $\mathrm{R}_{k'/k}(G')$; this makes standardness a useful notion.

Beware that using different 4-tuples (equipped with respective factorizations (2.2.5.1)) as the data in Definition 2.2.6 can yield the same G. For instance, the data specifies a Cartan k-subgroup $C \subset G$, and if there is a proper k-subalgebra $k'_0 \subset k'$ over which k' is étale then we can replace $(G', k'/k)$ with $(\mathrm{R}_{k'/k'_0}(G'), k'_0/k)$. Hence, for non-commutative standard pseudo-reductive k-groups G, two questions arise:

(1) Does G admit a "standard" description relative to *any* Cartan k-subgroup?
(2) Can $(G', k'/k)$ be chosen so that the fibers of G' over the factor fields of k' are absolutely simple (to avoid the artificial presence of separable Weil restriction in G')?

The answers are affirmative, and (provided that G' has absolutely simple and simply connected fibers over the factor fields of k', as may always be arranged in the non-commutative case) this allows us to arrange that the data $(G', k'/k, T', C)$ and (2.2.5.1) are *uniquely determined* up to unique isomorphism by the pair (G, C). Most of the proofs involve root groups and are addressed in a more general setting later (see Corollary 10.2.6 and Proposition 10.2.7). For now we only need the existence aspect in (2), so we address that and then introduce dynamic constructions underlying a robust theory of root groups in the pseudo-reductive case.

PROPOSITION 2.2.7. *Any non-commutative standard pseudo-reductive group G arises from a 4-tuple $(G', k'/k, T', C)$ and factorization* (2.2.5.1) *such that the fibers of $G' \rightarrow \mathrm{Spec}(k')$ are semisimple, absolutely simple, and simply connected.*

PROOF. Choose an initial 4-tuple $(G', k'/k, T', C)$ and diagram (2.2.5.1) giving rise to G. Since $\mathrm{R}_{k'/k}(T') \times C$ is a Cartan k-subgroup of $\mathrm{R}_{k'/k}(G') \rtimes C$ (see [**CGP**, Prop. A.5.15(3)]), it follows that in the quotient G the inclusion $C \hookrightarrow G$ is a Cartan k-subgroup. The derived group $\mathscr{D}(G)$ is perfect (Proposition 2.1.1(iii)), so by the commutativity of C it follows that the image of $\mathscr{D}(\mathrm{R}_{k'/k}(G'))$ in G is $\mathscr{D}(G)$. In particular, G' is non-commutative since $\mathscr{D}(G) \neq 1$ by hypothesis.

Let $\mathscr{G}' = \mathscr{D}(G')$, so $\mathscr{T}' := T' \cap \mathscr{G}'$ is a maximal k'-torus in $\mathscr{G}'$ [**CGP**, Cor. A.2.7]. (Some fibers of G' over $\mathrm{Spec}(k')$ might be commutative, and the corresponding fibers of $\mathscr{G}'$ are trivial.) For the simply connected central cover $\pi : \widetilde{G}' \twoheadrightarrow \mathscr{G}'$, the preimage $\widetilde{T}' = \pi^{-1}(\mathscr{T}')$ is a maximal k'-torus and the image of $\mathrm{R}_{k'/k}(\pi)$ is $\mathscr{D}(\mathrm{R}_{k'/k}(\mathscr{G}'))$ by Proposition 2.2.4 (since the commutator morphism $\mathscr{G}' \times \mathscr{G}' \rightarrow \mathscr{G}'$ factors through π, so likewise after applying $\mathrm{R}_{k'/k}$). The pseudo-reductivity of $\mathrm{R}_{k'/k}(G')$ implies that $\mathscr{D}(\mathrm{R}_{k'/k}(G'))$ is perfect (Proposition 2.1.1(iii)), yet the latter derived group is contained in $\mathrm{R}_{k'/k}(\mathscr{G}')$ since

$$1 \longrightarrow \mathrm{R}_{k'/k}(\mathscr{G}') \longrightarrow \mathrm{R}_{k'/k}(G') \longrightarrow \mathrm{R}_{k'/k}(G'/\mathscr{G}')$$

is exact with commutative final term, so

$$\mathscr{D}(\mathrm{R}_{k'/k}(G')) = \mathrm{image}(\mathrm{R}_{k'/k}(\widetilde{G}') \longrightarrow \mathrm{R}_{k'/k}(\mathscr{G}')).$$

In other words, $\mathscr{D}(G)$ is the image of $\mathrm{R}_{k'/k}(\widetilde{G}') \to G$.

The composite map $\widetilde{G}' \twoheadrightarrow \mathscr{G}' \hookrightarrow G'$ carries $\widetilde{T}'$ into T' and $Z_{\widetilde{G}'}$ into $Z_{G'}$ inducing an isomorphism $\widetilde{G}'/Z_{\widetilde{G}'} \simeq G'/Z_{G'}$ between maximal adjoint semisimple quotients, so likewise we obtain a natural isomorphism $\widetilde{T}'/Z_{\widetilde{G}'} \simeq T'/Z_{G'}$. Using the diagram

$$\mathrm{R}_{k'/k}(\widetilde{T}') \longrightarrow \mathrm{R}_{k'/k}(T') \stackrel{\phi}{\longrightarrow} C \longrightarrow \mathrm{R}_{k'/k}(T'/Z_{G'}) \simeq \mathrm{R}_{k'/k}(\widetilde{T}'/Z_{\widetilde{G}'})$$

whose composition is the natural map, we can make a standard pseudo-reductive k-group

$$H := (\mathrm{R}_{k'/k}(\widetilde{G}') \rtimes C)/\mathrm{R}_{k'/k}(\widetilde{T}')$$

equipped with an evident k-homomorphism

$$f : H \longrightarrow (\mathrm{R}_{k'/k}(G') \rtimes C)/\mathrm{R}_{k'/k}(T') =: G = \mathscr{D}(G) \cdot C$$

that is visibly surjective.

We claim that $\ker f = 1$. It suffices to show that if a point $(\widetilde{g}', c) \in \mathrm{R}_{k'/k}(\widetilde{G}') \rtimes C$ (valued in a k-algebra) maps into the anti-diagonal k-subgroup

$$\mathrm{R}_{k'/k}(T') \hookrightarrow \mathrm{R}_{k'/k}(G') \rtimes C$$

then $\widetilde{g}' \in \mathrm{R}_{k'/k}(\widetilde{T}')$. This follows immediately from the compatibility of $\mathrm{R}_{k'/k}$ with fiber products, as that gives

$$\mathrm{R}_{k'/k}(\widetilde{T}') = \mathrm{R}_{k'/k}(\mathscr{T}') \times_{\mathrm{R}_{k'/k}(\mathscr{G}')} \mathrm{R}_{k'/k}(\widetilde{G}') = \mathrm{R}_{k'/k}(T') \times_{\mathrm{R}_{k'/k}(G')} \mathrm{R}_{k'/k}(\widetilde{G}').$$

By replacing (G', T') with $(\widetilde{G}', \widetilde{T}')$ and working with the factorization diagram

$$\mathrm{R}_{k'/k}(\widetilde{T}') \longrightarrow C \longrightarrow \mathrm{R}_{k'/k}(\widetilde{T}'/Z_{\widetilde{G}'})$$

built above, we reduce to the case that all fibers of G' over factor fields of k' are semisimple and simply connected. Those factor fields k'_i of k' for which the fiber G'_i of G' is trivial may clearly be dropped from consideration, so we may assume that every G'_i is nontrivial.

By working over each factor field of k' separately, it is well-known (see [**CGP**, Prop. A.5.14]) that there exists a finite étale cover $\mathrm{Spec}(K') \to \mathrm{Spec}(k')$ and smooth affine K'-group H' whose fibers are connected semisimple, absolutely simple, and simply connected such that $\mathrm{R}_{K'/k'}(H') \simeq G'$. By [**CGP**, Prop. A.5.15(1),(2)] we have $Z_{G'} = \mathrm{R}_{K'/k'}(Z_{H'})$ and there exists a unique maximal K'-torus S' in H' such that $\mathrm{R}_{K'/k'}(S') = T'$, so $\mathrm{R}_{K'/k}(S'/Z_{H'}) = \mathrm{R}_{k'/k}(T'/Z_{G'})$ and $\mathrm{R}_{K'/k}(S') = \mathrm{R}_{k'/k}(T')$ since K' is k'-étale. Hence, using the 4-tuple $(H', K'/k, S', C)$ and corresponding factorization diagram recovers G in the desired manner. □

EXAMPLE 2.2.8. The nontrivial standard pseudo-semisimple k-groups G are precisely the pseudo-reductive central quotients $\mathrm{R}_{k'/k}(G')/Z$ where k' is a nonzero finite reduced k-algebra and G' is a smooth affine k'-group whose fibers over the factor fields of k' are connected semisimple, absolutely simple, and simply connected.

Indeed, if we describe G using a 4-tuple as in Proposition 2.2.7 then the k-group $G = \mathscr{D}(G)$ is the image of the map $\mathrm{R}_{k'/k}(G') \to G$, and the kernel Z of this latter map is exactly

$$\ker(\phi : \mathrm{R}_{k'/k}(T') \longrightarrow C) \subset \ker(\mathrm{R}_{k'/k}(T') \longrightarrow \mathrm{R}_{k'/k}(T'/Z_{G'})) = \mathrm{R}_{k'/k}(Z_{G'}),$$

so Z is central in $\mathrm{R}_{k'/k}(G')$.

Conversely, for such pairs $(k'/k, G')$ and a closed k-subgroup $Z \subset Z_{\mathrm{R}_{k'/k}(G')} = \mathrm{R}_{k'/k}(Z_{G'})$ (see [**CGP**, Prop. A.5.15(1)] for the equality), we have to show that any central quotient $G := \mathrm{R}_{k'/k}(G')/Z$ that is *pseudo-reductive* is necessarily standard. Let $T' \subset G'$ be a maximal k'-torus, so $C := \mathrm{R}_{k'/k}(T')$ is a Cartan k-subgroup of $\mathrm{R}_{k'/k}(G')$ due to [**CGP**, Prop. A.5.15(3)]. Hence, C/Z is a Cartan k-subgroup of $\mathrm{R}_{k'/k}(G')/Z$, so if $\mathrm{R}_{k'/k}(G')/Z$ is pseudo-reductive then C/Z is pseudo-reductive. The converse holds too: since $\mathrm{R}_{k'/k}(G')/Z \simeq (\mathrm{R}_{k'/k}(G') \rtimes (C/Z))/C$ with C/Z acting through its natural homomorphism into $\mathrm{R}_{k'/k}(T'/Z_{G'})$, if C/Z is pseudo-reductive then $\mathrm{R}_{k'/k}(G')/Z$ is given by the "standard" construction and thus is pseudo-reductive.

2.3. Dynamic techniques and pseudo-parabolic subgroups. The structure of split connected reductive groups over a field k rests on the fact that a connected semisimple k-group with a split maximal k-torus of dimension 1 is k-isomorphic to SL_2 or PGL_2. In particular, the construction of root groups and root data for split connected reductive groups ultimately rests on this rank-1 classification. Nothing similar is available early in the study of pseudo-reductive groups.

It is true that if $\mathrm{char}(k) \neq 2$ then a pseudo-semisimple k-group with a split maximal k-torus of dimension 1 is k-isomorphic to $\mathrm{R}_{k'/k}(\mathrm{SL}_2)$ or $\mathrm{R}_{k'/k}(\mathrm{PGL}_2)$ for a purely inseparable finite extension k'/k, but (i) the proof requires the full force of the techniques to be discussed in this section, and (ii) in characteristic 2 there is no comparable result. Hence, to develop a *characteristic-free* structure theory involving root groups and root systems we need an alternative viewpoint.

A systematic study of limiting behavior of orbits under 1-parameter subgroups provides an adequate substitute for the lack of a uniform rank-1 classification early on (even if we were to avoid characteristic 2). Motivation for this arises from a description of parabolic subgroups, their unipotent radicals, and Levi factors in GL_n entirely in terms of 1-parameter subgroups (without reference to the usual definitions of parabolicity, unipotence, or Levi factors). Consider an increasing flag

$$0 = F_0 \subsetneq F_1 \subsetneq \cdots \subsetneq F_m = V$$

of subspaces of a nonzero finite-dimensional k-vector space V; this corresponds to a parabolic k-subgroup $P \subset G := \mathrm{GL}(V)$ as the stabilizer of the flag $F_\bullet$. Let V_j be a linear complement to F_{j-1} in F_j for $1 \leqslant j \leqslant m$, so the stabilizer L of the ordered m-tuple $(V_1, \dots, V_m)$ is a Levi factor of P (i.e., $P = L \ltimes U$ for the k-descent $U = \mathscr{R}_{u,k}(P)$ of $\mathscr{R}_u(P_{\overline{k}})$). The decomposition $\bigoplus V_j$ of V is encoded in terms of a 1-parameter k-subgroup $\lambda : \mathrm{GL}_1 \to G$ by making GL_1 act on V_j through the character $t \mapsto t^{a_j}$ for integers $a_1 > \cdots > a_m$; the V_j's are the weight spaces for the weights occurring in this GL_1-action on V.

Choose ordered bases for each V_j and use these to make an ordered basis $\{v_1, \dots, v_n\}$ for V by putting the basis vectors from V_j before V_{j+1} for all j. Denote the unique V_j containing v_r as V_{j_r} (so $j_1 \leqslant \ldots \leqslant j_n$). Under the resulting identification $V = k^n$, any point $g = (x_{rs}) \in \mathrm{GL}_n(R) = G(R)$ (valued in a k-algebra R) satisfies

$$\lambda(t) g \lambda(t)^{-1} = (t^{a_{j_r} - a_{j_s}} x_{rs}).$$

Thus, if we make GL_1 act on G via conjugation through λ then the orbit map $(\mathrm{GL}_1)_R \to G_R$ through g defined by $t \mapsto t.g := \lambda(t) g \lambda(t)^{-1}$ extends to an R-scheme map $\mathbf{A}^1_R \to G_R$ (i.e., the map of coordinate rings $R[G] \to R[T, 1/T]$ lands inside $R[T]$) precisely when $x_{rs} = 0$ whenever $j_r > j_s$, which is to say if and only if $g \in P(R)$. In such cases, we express the existence of this extended map on $\mathbf{A}^1_R$ by saying "$\lim_{t\to 0} t.g$ exists", and the image in $G(R)$ of the zero section under this extended map is referred to as $\lim_{t\to 0} t.g$. Hence, the condition that $\lim_{t\to 0} t.g$ exists and is equal to 1 is precisely that $x_{rs} = \delta_{rs}$ whenever $j_r \geqslant j_s$, which is to say if and only if $g \in U(R)$. Finally, a point $g \in G$ centralizes λ, or equivalently $t \mapsto t.g$ is the constant R-morphism to $g \in G(R)$, if and only if g preserves each $(V_j)_R$, which is to say $g \in L(R)$.

The preceding calculations show that P, U, and L can be recovered dynamically in terms of the GL_1-action on G via $(t, g) \mapsto \lambda(t)g\lambda(t)^{-1}$. Observe that not only is P equal to $L \ltimes U$, but the opposite parabolic P^- relative to L is obtained upon replacing λ with the reciprocal 1-parameter subgroup $t \mapsto \lambda(1/t) = \lambda(t)^{-1}$; the traditional additive notation for characters and cocharacters leads us to denote this latter cocharacter as $-\lambda$ rather than as $1/\lambda$.

REMARK 2.3.1. For the k-unipotent radical U^- of P^-, the multiplication map of k-schemes $U^- \times P = U^- \times L \times U \to G = \mathrm{GL}_n$ is an open immersion. Indeed, we may assume $k = \overline{k}$, and it is a general fact in algebraic geometry that a map between smooth k-varieties is an open immersion if it is injective on k-points and bijective on tangent spaces at k-points of the source.

Injectivity on k-points is clear since $U^-(k) \cap P(k) = 1$ by inspection. Using left translation by $U^-(k)$ and right translation by $P(k)$ reduces bijectivity on tangent spaces at k-points to the bijectivity of the addition map $\mathrm{Lie}(U^-) \oplus \mathrm{Lie}(P) \to \mathrm{Lie}(G)$. Under the adjoint action of GL_1 on $\mathrm{Lie}(G)$, $\mathrm{Lie}(U^-)$ is the span of the negative weight spaces and $\mathrm{Lie}(P)$ is the span of the non-negative weight spaces.

The above considerations with GL_n inspire the following generalization to GL_1-actions on arbitrary affine group schemes of finite type over fields. First, we make a definition over rings. For any ring R and map of affine R-schemes $f : (\mathrm{GL}_1)_R \to X$, we say "$\lim_{t\to 0} f(t)$ exists" if f extends to an R-scheme map $\widetilde{f} : \mathbf{A}^1_R \to X$, which is to say that $f^* : R[X] \to R[T, 1/T]$ lands inside $R[T]$. Such an $\widetilde{f}$ is obviously unique if it exists, in which case the R-point $\widetilde{f}(0) \in X(R)$ is referred to as $\lim_{t\to 0} f(t)$.

LEMMA 2.3.2. *Let $(t, x) \mapsto t.x$ be a GL_1-action on an affine scheme X of finite type over a field k. The functor of points $x \in X$ such that $\lim_{t\to 0} t.x$ exists is represented by a closed subscheme of X.*

PROOF. The coordinate ring $k[X]$ is the direct sum of its weight spaces $k[X]_n$ under the GL_1-action (with $n \in \mathbf{Z}$); i.e., GL_1 acts on $k[X]_n$ via $t.f = t^n f$. The ideal generated by the k-subspaces $k[X]_n$ for $n < 0$ defines a closed subscheme of X which does the job. See [**CGP**, Lemma 2.1.4] for details. □

2.3.3. In the special case that GL_1 acts on an affine k-group scheme G of finite type through conjugation against a k-homomorphism $\lambda : \mathrm{GL}_1 \to G$, we denote the closed subscheme of G arising from Lemma 2.3.2 as $P_G(\lambda)$. Since $(t.g)(t.g') = t.(gg')$ for points g, g' of G valued in a common k-algebra, it is clear that $P_G(\lambda)$ is stable under multiplication inside G. Likewise, $P_G(\lambda)$ passes through the identity point and is stable under inversion, so $P_G(\lambda)$ is a k-subgroup scheme of G.

The scheme-theoretic intersection

$$Z_G(\lambda) := P_G(\lambda) \cap P_G(-\lambda)$$

represents the functorial centralizer of λ in G because for any k-algebra R the only R-scheme maps $\mathbf{P}^1_R \to G_R$ into the affine target G_R are constant maps to elements of $G(R)$. Finally, the scheme-theoretic kernel

$$U_G(\lambda) := \ker(P_G(\lambda) \longrightarrow G)$$

of the map $g \mapsto \lim_{t \to 0} t.g$ clearly has trivial schematic intersection with $Z_G(\lambda)$.

For any positive integer m we have

$$P_G(\lambda^m) = P_G(\lambda),\ U_G(\lambda^m) = U_G(\lambda),\ Z_G(\lambda^m) = Z_G(\lambda) \tag{2.3.3.1}$$

since whether or not an element of $R[T, 1/T]$ lies in $R[T]$ is unaffected by replacing T with T^m; in particular, the k-subgroups $P_G(\lambda), U_G(\lambda), Z_G(\lambda)$ only depend on λ through the subset $\mathbf{Q}_{>0} \cdot \lambda \subset \mathrm{X}_*(T)_{\mathbf{Q}}$.

By functorial considerations, if G' is an affine k-group scheme of finite type and $G \subset G'$ is a k-subgroup inclusion (always a closed immersion [**SGA3**, $\mathrm{VI_B}$, 1.4.2]) then obviously

$$G \cap P_{G'}(\lambda) = P_G(\lambda),\ \ G \cap U_{G'}(\lambda) = U_G(\lambda),\ \ G \cap Z_{G'}(\lambda) = Z_G(\lambda).$$

The case $G' = \mathrm{GL}_n$ thereby helps to reduce some problems for general G to the case of GL_n. For example, $U_G(\lambda)$ is always a *unipotent* k-group scheme because upon choosing a k-subgroup inclusion of G into GL_n (as we may always do [**CGP**, Prop. A.2.3]), $U_G(\lambda)$ is a k-subgroup scheme of the k-group $U_{\mathrm{GL}_n}(\lambda)$ that has been seen to be the unipotent radical of a parabolic k-subgroup of GL_n.

REMARK 2.3.4. Unipotence for a k-group scheme is defined without smoothness hypotheses in [**SGA3**, XVII, 1.3]: it means that over $\overline{k}$ there is a finite composition series of closed subgroup schemes such that each successive quotient is isomorphic to a $\overline{k}$-subgroup of $\mathbf{G}_{\mathrm{a}}$. A review of this notion for our purposes is given in [**CGP**, A.1.3–A.1.4].

The main properties of the preceding dynamic group scheme constructions are recorded in the following important result.

THEOREM 2.3.5. *Define* $\mathfrak{g} = \mathrm{Lie}(G)$ *equipped with the* GL_1*-action through the adjoint representation. Let* $\mathfrak{g}_0 = \mathfrak{g}^{\mathrm{GL}_1}$*, define* $\mathfrak{g}_+$ *to be the span of the weight spaces in* $\mathfrak{g}$ *for the positive weights, and define* $\mathfrak{g}_-$ *likewise with negative weights.*

(i) *Inside* $\mathfrak{g}$, $\mathrm{Lie}(Z_G(\lambda)) = \mathfrak{g}_0$ *and* $\mathrm{Lie}(U_G(\pm\lambda)) = \mathfrak{g}_\pm$.

(ii) *The natural multiplication map* $Z_G(\lambda) \ltimes U_G(\lambda) \to P_G(\lambda)$ *is an isomorphism of* k*-schemes, and the natural multiplication map*

$$\Omega_G(\lambda) := U_G(-\lambda) \times P_G(\lambda) \longrightarrow G$$

is an open immersion. In particular, if G *is smooth then* $P_G(\lambda)$, $Z_G(\lambda)$, *and* $U_G(\lambda)$ *are smooth, and if* G *is connected then each of these three* k*-groups is connected.*

(iii) *If* $H \subset G$ *is a closed* k*-subgroup through which* λ *factors then* $H \cap \Omega_G(\lambda) = \Omega_H(\lambda)$.

(iv) *The unipotent* k*-group scheme* $U_G(\lambda)$ *is connected, and if* G *is smooth then the smooth connected unipotent* k*-group* $U_G(\lambda)$ *is* k*-split.*

PROOF. The details are given in [**CGP**, Prop. 2.1.8] (which works more generally over rings) except for the k-split assertion in (iv) that is [**CGP**, Prop. 2.1.10]. Here we limit ourselves to a few remarks.

The proof of (i) is a consequence of the functorial characterizations of $Z_G(\lambda)$ and $U_G(\pm\lambda)$ applied to points valued in the dual numbers over k. The crux of the first isomorphism in (ii) is that for any $g \in P_G(\lambda)(R)$ (with a k-algebra R), the limit $\lim_{t\to 0} t.g \in G(R)$ lies in $Z_G(\lambda)(R)$. The intuition is that for any point t' of GL_1, the limiting behavior of $t'.(tg) = (t't).g$ as $t \to 0$ is independent of t'.

The open immersion assertion in (ii) has been discussed earlier for GL_n, and the general case is reduced to this by applying (iii) to an inclusion $G \hookrightarrow \mathrm{GL}_n$. That is, (iii) has to be proved before (ii). The idea behind the proof of (iii) is to pick a linear representation of G for which H is the scheme-theoretic stabilizer of a line, and to study how that description of H interacts with the dynamically-defined k-subgroups under consideration; this is a non-trivial task.

The connectedness of $U_G(\lambda)$ in (iv) is clear because any geometric point u of $U_G(\lambda)$ is connected to the identity via a rational curve arising from the map $\mathbf{A}^1_{\overline{k}} \to U_G(\lambda)_{\overline{k}}$ extending $t \mapsto t.u$. The k-split property of $U_G(\lambda)$ for smooth G lies much deeper because we cannot deduce the general case from the easy case of GL_n via an inclusion of G into some GL_n; the problem is that k-split unipotent smooth connected k-groups often contain *non-split* smooth connected k-subgroups when k is imperfect! (For example, if $p = \mathrm{char}(k) > 0$ and $c \in k - k^p$ then the smooth connected 1-dimensional k-subgroup $y^p = x - cx^p$ of the k-split $\mathbf{G}_\mathrm{a}^2$ is not k-split [**CGP**, B.2.3].) To overcome this difficulty, one has to use substantial input from Tits' unpublished work [**Ti2**] on the general structure of smooth connected unipotent groups over imperfect fields; see [**CGP**, App. B] for a modern account of that structure theory. □

In the general theory of pseudo-reductive groups G over a field k, the role of parabolic k-subgroups for the reductive case is replaced with the dynamically-defined subgroups $P_G(\lambda)$ for 1-parameter k-subgroups λ. To see why this is done, consider a nontrivial purely inseparable finite extension of fields k'/k and a connected reductive k'-group G' with a proper parabolic k'-subgroup P'. The quotient

$$\mathrm{R}_{k'/k}(G')/\mathrm{R}_{k'/k}(P') \simeq \mathrm{R}_{k'/k}(G'/P')$$

is *never* proper [**CGP**, Ex. A.5.6]. On the other hand, the parabolic k'-subgroups of G' are precisely the k'-subgroups of the form $P_{G'}(\lambda')$ for 1-parameter k'-subgroups $\lambda' : \mathrm{GL}_1 \to G'$ [**CGP**, Prop. 2.2.9], so choosing λ' such that $P_{G'}(\lambda') = P'$ yields the description

$$\mathrm{R}_{k'/k}(P') = \mathrm{R}_{k'/k}(P_{G'}(\lambda')) = P_{\mathrm{R}_{k'/k}(G')}(\lambda)$$

where $\lambda : \mathrm{GL}_1 \to \mathrm{R}_{k'/k}(G')$ corresponds to λ' via the mapping property of $\mathrm{R}_{k'/k}$ [**CGP**, Prop. 2.1.13].

The k-subgroups $Q \subset G$ of the form $P_G(\lambda)$ generally do not admit a characterization in terms of properties of G/Q beyond the reductive case (see Remark 4.2.5). To make a definition for this class of k-subgroups applicable beyond the pseudo-reductive case (as is sometimes convenient in proofs) we incorporate the k-unipotent radical:

DEFINITION 2.3.6. If G is a smooth connected affine group over a field k then a k-subgroup P of G is *pseudo-parabolic* if $P = P_G(\lambda)\mathscr{R}_{u,k}(G)$ for a 1-parameter k-subgroup $\lambda : \mathrm{GL}_1 \to G$.

EXAMPLE 2.3.7. If k'/k is a finite extension of fields and G' is a connected reductive k'-group then the pseudo-parabolic k-subgroups of $\mathrm{R}_{k'/k}(G')$ are *precisely* the k-subgroups $\mathrm{R}_{k'/k}(P')$ for parabolic k'-subgroups P' of G'; see [**CGP**, Prop. 2.2.13] for a generalization.

If G is a pseudo-reductive k-group and P is a pseudo-parabolic k-subgroup of G then $\mathscr{R}_{u,k}(P)$ is k-split since writing $P = P_G(\lambda)$ implies $\mathscr{R}_{u,k}(P) = U_G(\lambda)$ (because the torus centralizer $Z_G(\lambda)$ is pseudo-reductive). In the reductive case this recovers the well-known fact that k-unipotent radicals of parabolic k-subgroups are k-split.

Pseudo-parabolicity behaves well under passage to quotients in the pseudo-reductive case. More generally, we have the extremely useful:

PROPOSITION 2.3.8. *If $f : G \twoheadrightarrow \overline{G}$ is an arbitrary surjective homomorphism between smooth connected affine k-groups and $\lambda : \mathrm{GL}_1 \to G$ is a 1-parameter k-subgroup then for $\overline{\lambda} = f \circ \lambda$ the inclusions*

$$f(P_G(\lambda)) \subset P_{\overline{G}}(\overline{\lambda}), \quad f(U_G(\lambda)) \subset U_{\overline{G}}(\overline{\lambda}), \quad f(Z_G(\lambda)) \subset Z_{\overline{G}}(\overline{\lambda}), \tag{2.3.8.1}$$

are equalities.

See [**CGP**, Cor. 2.1.9] for a more general result without smoothness hypotheses but assuming f to be flat.

PROOF. The inclusions have closed image, so $f(\Omega_G(\lambda))$ is a closed subset of the dense open $\Omega_{\overline{G}}(\overline{\lambda})$. But $f(\Omega_G(\lambda))$ is dense in $\overline{G}$ since f is dominant, so it follows that $f(\Omega_G(\lambda)) = \Omega_{\overline{G}}(\overline{\lambda})$. Hence, all three inclusions above are equalities. □

Applying the preceding to the maximal pseudo-reductive quotient $q : G \twoheadrightarrow G/\mathscr{R}_{u,k}(G)$ of a smooth connected affine k-group G, if P is a pseudo-parabolic k-subgroup of G then $q(P)$ is a pseudo-parabolic k-subgroup of $G/\mathscr{R}_{u,k}(G)$. Moreover, $P \mapsto q(P)$ is a bijection between the sets of pseudo-parabolic k-subgroups of G and $G/\mathscr{R}_{u,k}(G)$ with inverse $\overline{P} \mapsto q^{-1}(\overline{P})$ [**CGP**, Prop. 2.2.10].

We will now prove the following useful result which shows that there is considerable flexibility in the choice of $\lambda : \mathrm{GL}_1 \to G$ in the description of pseudo-parabolic k-subgroup P in Definition 2.3.6.

LEMMA 2.3.9. *Let G be a smooth connected affine group over a field k and $P = P_G(\lambda)\mathscr{R}_{u,k}(G)$ a pseudo-parabolic k-subgroup of G. Let T be a maximal k-torus of P. There exists $g \in \mathscr{R}_{u,k}(P)(k)$ such that the k-homomorphism $\mu : \mathrm{GL}_1 \to G$ given by $t \mapsto g\lambda(t)g^{-1}$ is valued in T and $P = P_G(\mu)\mathscr{R}_{u,k}(G)$.*

PROOF. Let $\pi : P \to \overline{P} := P/\mathscr{R}_{u,k}(P)$ be the quotient map. The k-group $U := \mathscr{R}_{u,k}(P)$ contains $U_G(\lambda)\mathscr{R}_{u,k}(G)$, so $\overline{P}$ is a quotient of $Z_G(\lambda)$. Hence, the image of λ in $\overline{P}$ is a central torus, so it lies in the maximal k-torus $\overline{T} := \pi(T)$ of $\overline{P}$. We conclude that $\lambda(\mathrm{GL}_1)$ is contained in the smooth connected solvable k-subgroup $H := \pi^{-1}(\overline{T}) = T \ltimes U$. All maximal k-tori of H are $U(k)$-conjugate to each other [**Bo2**, 19.2], so there exists $g \in U(k)$ such that the 1-parameter k-subgroup $\mu : t \mapsto g\lambda(t)g^{-1}$ is valued in T. Since $P_G(\mu) = gP_G(\lambda)g^{-1}$, we see that $P_G(\mu)\mathscr{R}_{u,k}(G) = P$. □

REMARK 2.3.10. The flexibility in the choice of λ for describing a given P leads to some subtleties:

(i) It is not obvious if pseudo-parabolicity descends through arbitrary separable extension of the ground field (even in the Galois case). This is a contrast with parabolicity, whose geometric definition clearly descends through any extension of the ground field. It is true that for separable field extensions K/k (especially k_s/k), a k-subgroup P of a pseudo-reductive k-group G is pseudo-parabolic when P_K is pseudo-parabolic in G_K (the converse is obvious), and this is essential for the utility of pseudo-parabolicity. The proof of this descent result requires substantial input from the theory of root groups in pseudo-reductive groups; see Proposition 4.3.4.

(ii) Since there is no "geometric" characterization of pseudo-parabolicity in the spirit of parabolicity (see Remark 4.2.5 for a precise statement), it is not at all evident if pseudo-parabolicity is transitive with respect to subgroup inclusions: for a pseudo-parabolic k-subgroup P of a smooth connected affine k-group G and a smooth connected k-subgroup Q of P, is Q pseudo-parabolic in G if and only if Q is pseudo-parabolic in P?

Neither implication is obvious. For instance, if Q is pseudo-parabolic in G then it isn't clear if $\mathscr{R}_{u,k}(P) \subset Q$, and if $Q = P_P(\lambda)\mathscr{R}_{u,k}(P)$ for a 1-parameter k-subgroup $\lambda : \mathrm{GL}_1 \to P$ then generally $Q \neq P_G(\lambda)\mathscr{R}_{u,k}(G)$ (as one sees even in the split reductive case by considering the positions of closed half-spaces relative to roots). This problem is settled affirmatively by using root systems for pseudo-reductive groups; see Corollary 4.3.5.

(iii) If P is a pseudo-parabolic k-subgroup of a pseudo-reductive k-group G and Q is a smooth closed k-subgroup of G containing P then is Q pseudo-parabolic in G? This is not easy, in contrast with the analogue for parabolicity, and the affirmative proof involves many arguments with root systems; see Proposition 4.3.7.

In the next section we use dynamic methods to develop a theory of root systems and root groups in the pseudo-reductive setting.

3. Roots, root groups, and root systems

3.1. Root groups. For a split connected reductive group H over a field F, a split maximal F-torus $S \subset H$, and $a \in \Phi(H,S)$, the root group U_a admits a dynamic description as follows. Consider the codimension-1 subtorus $S_a := (\ker a)^0_{\mathrm{red}}$ killed by a. The centralizer $Z_H(S_a)$ is a connected reductive F-group containing S whose set of S-roots is $\Phi(H,S) \cap \mathbf{Q} \cdot a = \{\pm a\}$ (as the root system $\Phi(H,S)$ is reduced). The natural map $S_a \times a(\mathrm{GL}_1) \to S$ is an isogeny and $Z_H(S_a)$ is an isogenous quotient of $S_a \times H_a$, where $H_a := \langle U_a, U_{-a}\rangle$ is F-isomorphic to SL_2 or PGL_2 with $a(\mathrm{GL}_1)$ going over to the diagonal F-torus.

Since $U_{\pm a} \simeq \mathbf{G}_{\mathrm{a}}$ and the S-action on $U_{\pm a}$ is thereby identified with multiplication on $\mathbf{G}_{\mathrm{a}}$ through $\pm a$, inspection of the open cell $U_{-a} \times S \times U_a \subset Z_H(S_a)$ shows that $P_{Z_H(S_a)}(\pm a^\vee) = S \cdot U_{\pm a}$. Thus, $U_{\pm a} = U_{Z_H(S_a)}(\pm a^\vee)$. It is useful to modify this to remove the appearance of the coroot, as follows.

For a cocharacter $\lambda : \mathrm{GL}_1 \to S$ such that $\lambda(\mathrm{GL}_1)$ is an isogeny-complement to S_a inside S (equivalently, $\langle a, \lambda\rangle \neq 0$), by replacing λ with $-\lambda$ if necessary to arrange that $\langle a, \lambda\rangle > 0$ we may use the same reasoning to obtain

$$U_a = U_{Z_H(S_a)}(\lambda).$$

This dynamic description of root groups in the reductive case motivates:

DEFINITION 3.1.1. A smooth connected affine group G over a field k is *pseudo-split* if there exists a split maximal k-torus $T \subset G$. We denote by $\Phi(G,T)$ the set of nontrivial T-weights occurring on $\mathrm{Lie}(G)$, and for $a \in \mathrm{X}(T)_{\mathbf{Q}} - \{0\}$ the codimension-1 subtorus of T killed by every $na \in \mathrm{X}(T)$ with integers $n \neq 0$ is denoted T_a. Define $U_{(a)} = U^G_{(a)} := U_{Z_G(T_a)}(\lambda_a)$ for any $\lambda_a \in \mathrm{X}_*(T)$ satisfying $\langle a, \lambda_a \rangle > 0$.

REMARK 3.1.2. The k-split unipotent smooth connected k-subgroup $U_{(a)} \subset G$ is independent of λ_a due to (2.3.3.1) because the isogeny $T_a \times \mathrm{GL}_1 \to T$ via $(t,x) \mapsto t \cdot \lambda_a(x)$ implies that for any other choice λ'_a there exist integers $m, m' > 0$ such that $m'\lambda'_a = m\lambda_a + \mu$ for some $\mu \in \mathrm{X}_*(T)$ valued in the *central* torus T_a of $Z_G(T_a)$.

EXAMPLE 3.1.3. Let k'/k be a finite purely inseparable extension of fields and G' a connected reductive k'-group with a split maximal k'-torus T'. The pseudo-reductive k-group $G := \mathrm{R}_{k'/k}(G')$ is pseudo-split because the split maximal k-torus T in $\mathrm{R}_{k'/k}(T')$ is a maximal k-torus in G (as the natural map $G_{k'} \to G'$ is surjective with smooth connected unipotent kernel [**CGP**, Prop. A.5.11(1),(2)]).

We have $Z_G(T) = \mathrm{R}_{k'/k}(Z_{G'}(T')) = \mathrm{R}_{k'/k}(T')$ [**CGP**, Prop. A.5.15(1)], and under the natural identification $\mathrm{X}(T) \simeq \mathrm{X}(T')$ the set $\Phi(G,T)$ is carried onto $\Phi(G',T')$ [**CGP**, Ex. 2.3.2]. If $a' \in \Phi(G',T')$ corresponds to $a \in \Phi(G,T)$ then inspection of Lie algebras shows that the evident inclusion $\mathrm{R}_{k'/k}(U_{a'}) \subset U^G_{(a)}$ of smooth connected k-subgroups of G is an equality. (Here we use the natural identification of the functor $\mathrm{Lie} \circ \mathrm{R}_{k'/k}$ with "underlying Lie algebra over k" [**CGP**, Cor. A.7.6].)

For any k-torus $S \subset G$ the functorial definition of $Z_G(S)$ implies via consideration of points valued in the dual numbers that $\mathrm{Lie}(Z_G(S)) = \mathrm{Lie}(G)^S$. Thus, for nonzero $a \in \mathrm{X}(T)_{\mathbf{Q}}$, $\mathrm{Lie}(Z_G(T_a))$ is the span of $\mathrm{Lie}(Z_G(T)) = \mathrm{Lie}(G)^T$ and the T-weight spaces for all $b \in \Phi(G,T)$ that are trivial on T_a (equivalently $b \in \mathbf{Q} \cdot a$). Hence, by Theorem 2.3.5(ii) and T-weight space considerations we obtain all but the final assertion in:

PROPOSITION 3.1.4. *The Lie algebra* $\mathrm{Lie}(U_{(a)})$ *is the span of the* T*-weight spaces for all* $b \in \Phi(G,T) \cap \mathbf{Q}_{>0} \cdot a$*; in particular,* $U_{(a)} \neq 1$ *if and only if* $\mathbf{Q}_{>0} \cdot a$ *meets* $\Phi(G,T)$*. The* k*-subgroups* $Z_G(T)$ *and* $\{U_{(a)}\}_{a \in \Phi(G,T)}$ *generate* G*. If* G *is perfect then the* $U_{(a)}$*'s, for* $a \in \Phi(G,T)$*, generate* G*.*

PROOF. We just need to explain why the k-subgroup N generated by the $U_{(a)}$'s coincides with G when G is perfect. Since each $Z_G(T)$ normalizes $U_{(a)}$ (as we may verify using k_s-points), so $Z_G(T)$ normalizes N, and G is generated by N and $Z_G(T)$, it follows that N is normal in G. The quotient G/N has trivial T-action on its Lie algebra since $\mathrm{Lie}(G/N) = \mathrm{Lie}(G)/\mathrm{Lie}(N)$, so for the maximal torus image $\overline{T}$ of T in G/N we see that the inclusion $Z_{G/N}(\overline{T}) \subset G/N$ between *smooth* connected affine k-groups is an equality on Lie algebras and hence is an equality.

In other words, G/N is a perfect smooth connected affine k-group with a central maximal k-torus. But then the quotient by that central maximal torus is unipotent (as any smooth connected affine k-group which does not contain a nontrivial k-torus is unipotent), so G/N is solvable. Perfectness of G/N then forces $G/N = 1$; i.e., $N = G$. □

Now we focus on pseudo-split pseudo-reductive G, for which we shall see that the k-subgroups $U_{(a)}$ have some structural properties reminiscent of root groups in the reductive case. Example 3.1.3 illustrates that such $U_{(a)}$ can be vector groups over k with very large dimension, in contrast with the split reductive case that always has 1-dimensional root groups, though such examples arise from 1-dimensional vector groups over a finite extension k'/k.

A consequence of the subsequent structure theory of pseudo-reductive groups will be that in the pseudo-split case such $U_{(a)}$'s always arise from 1-dimensional vector groups over finite extensions of k *except* when k is imperfect with $\mathrm{char}(k) = 2$ (for which counterexamples are given by constructions in §7.3, §7.2, §8.2–§8.3, and §10.1). The first step towards establishing good properties of $U_{(a)}$'s in the pseudo-reductive case is to relate $\Phi(G,T)$ to $\Phi(G^{\mathrm{red}}_{\overline{k}}, T_{\overline{k}})$ up to rational multipliers:

LEMMA 3.1.5. *Let G be a pseudo-reductive k-group with a split maximal k-torus T. Each $a \in \Phi(G,T)$ admits a unique $\mathbf{Q}_{>0}$-multiple in $\Psi := \Phi(G^{\mathrm{red}}_{\overline{k}}, T_{\overline{k}})$ and every element of Ψ arises in this way from some such a. Moreover, for all $a \in \Phi(G,T)$ the k-group $U_{(a)}$ is commutative and if $\mathrm{char}(k) = p > 0$ then $U_{(a)}$ is p-torsion.*

PROOF. By Proposition 2.3.8 and the compatibility of torus centralizers with quotient maps between smooth affine groups, for any nonzero $a \in \mathrm{X}(T)_{\mathbf{Q}} = \mathrm{X}(T_{\overline{k}})_{\mathbf{Q}}$ the quotient map $\pi : G_{\overline{k}} \twoheadrightarrow G^{\mathrm{red}}_{\overline{k}} =: H$ carries $(U^G_{(a)})_{\overline{k}}$ onto $U^H_{(a)}$ upon identifying $T_{\overline{k}}$ with a maximal torus in H. It is a general fact that a smooth connected k-subgroup U of G satisfying $U_{\overline{k}} \subset \mathscr{R}_u(G_{\overline{k}})$ must be trivial [**CGP**, Lemma 1.2.1], so $U^G_{(a)} \neq 1$ if and only if $U^H_{(a)} \neq 1$. This implies that each $a \in \Phi(G,T)$ admits a $\mathbf{Q}_{>0}$-multiple in $\Psi := \Phi(G^{\mathrm{red}}_{\overline{k}}, T_{\overline{k}})$, with the multiplier being unique since Ψ is reduced. Similarly, Proposition 3.1.4 implies that any element of Ψ admits a $\mathbf{Q}_{>0}$-multiple in $\Phi(G,T)$.

For any $a \in \Phi(G,T)$ we have $U^H_{(a)} = \mathbf{G}_{\mathrm{a}}$, so $\pi(\mathscr{D}(U^G_{(a)})_{\overline{k}}) = 1$, forcing $\mathscr{D}(U^G_{(a)}) = 1$; i.e., $U^G_{(a)}$ is commutative. By the same reasoning, if $p = \mathrm{char}(k) > 0$ then $U^G_{(a)}$ is killed by p since $U^H_{(a)}$ is killed by p. □

A p-torsion commutative smooth connected affine group over a field k of characteristic $p > 0$ need *not* be a vector group (i.e., a direct product of copies of $\mathbf{G}_{\mathrm{a}}$) when k is imperfect; e.g., if $c \in k - k^p$ then the 1-dimensional $y^p = x - cx^p$ is not a vector group [**CGP**, B.2.3]. But the k-groups $U^G_{(a)}$ considered in Lemma 3.1.5 will turn out to always be vector groups because they satisfy an additional property: they are normalized by T, and the resulting T-action on their Lie algebra has only nontrivial T-weights (in fact, only $\mathbf{Q}_{>0}$-multiples of a).

Tits proved the remarkable fact [**CGP**, Thm. B.4.3] that every p-torsion smooth connected commutative unipotent group U in characteristic $p > 0$ equipped with an action by a torus T such that $\mathrm{Lie}(U)^T = 0$ is necessarily a vector group and even admits a *linear structure* (i.e., $\mathbf{G}_{\mathrm{a}}$-module scheme structure) that is T-equivariant. Combining this with the unique (algebraic exponential) isomorphism $U \simeq \mathrm{Lie}(U)$ inducing the identity on Lie algebras for commutative unipotent groups U in characteristic 0 yields the first part of:

PROPOSITION 3.1.6. *Let G be a pseudo-reductive k-group admitting a split maximal k-torus T.*

(i) *For each $a \in \Phi(G,T)$, $U_{(a)}$ is a vector group admitting a T-equivariant linear structure.*
(ii) *If $a, b \in \Phi(G,T)$ and $ra + sb \notin \Phi(G,T)$ for all $r, s \in \mathbf{Q}_{>0}$ then $U_{(a)}$ commutes with $U_{(b)}$.*

PROOF. Let $\pi : G_{\overline{k}} \to G_{\overline{k}}^{\mathrm{red}} =: H$ be the canonical quotient map. Lemma 3.1.5 provides $q, q' \in \mathbf{Q}_{>0}$ such that $qa, q'b \in \Phi(H, T_{\overline{k}})$, and the images $\pi((U_{(a)}^G)_{\overline{k}})$ and $\pi((U_{(b)}^G)_{\overline{k}})$ respectively coincide with the root groups U_{qa}^H and $U_{q'b}^H$ in H for the respective roots qa and $q'b$.

Necessarily $q'b \neq -qa$, as otherwise $2a + (q'/q)b = a \in \Phi(G, T)$, contradicting the hypotheses. Since $qa + q'b$ is not a root of H (again, due to the hypotheses), it follows that the root groups U_{qa}^H and $U_{q'b}^H$ in the reductive group H commute. Hence, the commutator $(U_{(a)}, U_{(b)})$ is killed by π over $\overline{k}$, so this commutator is trivial since G is pseudo-reductive [**CGP**, Lemma 1.2.1]. □

For any pseudo-split pseudo-reductive k-group G and split maximal k-torus T, Lemma 3.1.5 gives that the two subsets $\Phi(G, T)$ and $\Phi(G_{\overline{k}}^{\mathrm{red}}, T_{\overline{k}})$ of $\mathrm{X}(T) = \mathrm{X}(T_{\overline{k}})$ coincide up to $\mathbf{Q}_{>0}$-multipliers on their elements. It is important that these rational multipliers can be very tightly controlled:

THEOREM 3.1.7. *The sets $\Phi(G, T)$ and $\Phi(G_{\overline{k}}^{\mathrm{red}}, T_{\overline{k}})$ coincide except possibly when k is imperfect of characteristic 2 and $G_{\overline{k}}^{\mathrm{red}}$ contains a connected semisimple normal subgroup that is simply connected of type C_n ($n \geqslant 1$). In general $\Phi(G_{\overline{k}}^{\mathrm{red}}, T_{\overline{k}}) \subset \Phi(G, T)$, and if $a \in \Phi(G, T)$ is not a $T_{\overline{k}}$-root for $G_{\overline{k}}^{\mathrm{red}}$ then $2a$ is such a root.*

This result is proved in [**CGP**, Thm. 2.3.10], and the basic idea goes as follows. From the explicit description of irreducible root systems, we see that the only irreducible and reduced semisimple root datum admitting a root that is twice a weight is simply connected type C. Thus, we can replace G with $\mathscr{D}(Z_G(T_a))$ for suitable nonzero $a \in \mathrm{X}(T)_{\mathbf{Q}}$ to reduce to the case $\dim T = 1$ (using [**CGP**, Lemma 1.2.5(iii)] to control $\dim(T \cap \mathscr{D}(Z_G(T_a)))$). In the rank-1 case, $G_{\overline{k}}^{\mathrm{red}}$ is isomorphic to SL_2 or PGL_2. Nontrivial computations using Proposition 3.1.6(i) and the position of roots in the character lattices of SL_2 and PGL_2 eventually yield the result.

REMARK 3.1.8. The exceptional case in Theorem 3.1.7 with a root that is twice another root does occur over any imperfect field of characteristic 2, with $\Phi(G_{\overline{k}}^{\mathrm{red}}, T_{\overline{k}})$ of type C_n for any desired $n \geqslant 1$. The construction of such G is a highly nontrivial matter (as we shall discuss in §8).

DEFINITION 3.1.9. Let G be a pseudo-split pseudo-reductive k-group, and $T \subset G$ a split maximal k-torus. Elements of $\Phi(G, T)$ are called *roots*, and $a \in \Phi(G, T)$ is called *divisible* (resp. *multipliable*) if $a/2 \in \Phi(G, T)$ (resp. $2a \in \Phi(G, T)$).

The preceding terminology is reasonable because $\Phi(G, T)$ is a root system in its $\mathbf{Q}$-span inside $\mathrm{X}(T)_{\mathbf{Q}}$ (though in contrast with the reductive case, it can be non-reduced when k is imperfect with characteristic 2); see Proposition 3.2.7. Note in particular that $\Phi(G_{\overline{k}}^{\mathrm{red}}, T_{\overline{k}})$ is always the set of non-multipliable elements of $\Phi(G, T)$.

COROLLARY 3.1.10. *For $a \in \Phi(G, T)$ there exists a unique smooth connected k-subgroup $U_a \subset G$ normalized by T such that $\mathrm{Lie}(U_a)$ is the a-weight space in $\mathrm{Lie}(G)$ when a is not multipliable and is the span of the weight spaces for a and $2a$ when a is multipliable. The k-group U_a is a vector group admitting a T-equivariant linear structure, and this linear structure is unique when a is not multipliable.*

We call U_a the *root group* associated to a.

PROOF. When a is not divisible then $U_{(a)}$ does the job. If a is divisible then upon choosing a T-equivariant linear structure on $U_{(a)}$, the a-weight space for this linear structure does the job. If a is not divisible then the uniqueness of U_a is a consequence of the good behavior of the dynamic constructions with respect to intersections against equivariant subgroups, but the divisible case requires rather more effort (using centralizers of μ_2-actions in characteristic 2); see [**CGP**, Prop. 2.3.11] for details. □

By uniqueness (or by construction), the formation of U_a commutes with any separable extension on k. Hence, by working over k_s and using uniqueness we see that U_a is normalized by $Z_G(T)$.

REMARK 3.1.11. The pseudo-semisimple derived group $\mathscr{D}(G)$ of a pseudo-split pseudo-reductive k-group G is pseudo-split and generated by the root groups U_a relative to a split maximal k-torus T. To prove this, choose a T-equivariant linear structure on U_a. The absence of the trivial T-weight on $\mathrm{Lie}(U_a)$ implies that $(T, U_a) = U_a$. Hence, each U_a lies inside $\mathscr{D}(G)$.

The maximal k-torus $T' = T \cap \mathscr{D}(G)$ of $\mathscr{D}(G)$ [**CGP**, Cor. A.2.7] is certainly split, and if Z is the maximal central k-torus in G then the natural map $T' \times Z \to T$ is an isogeny [**CGP**, Lemma 1.2.5(iii)]. Thus, under the finite-index inclusion $\mathrm{X}(T) \hookrightarrow \mathrm{X}(T') \oplus \mathrm{X}(Z)$ we see that $\Phi(G, T)$ is identified with $\Phi(\mathscr{D}(G), T') \times \{0\}$. In particular, the T-root groups of G are the *same* as the T'-root groups of $\mathscr{D}(G)$. Since $\mathscr{D}(G)$ is perfect, we may conclude via the final assertion in Proposition 3.1.4 (applied to $\mathscr{D}(G)$).

3.2. Pseudo-simplicity and root systems. The core of the theory of connected semisimple k-groups is the absolutely simple case, characterized by irreducibility of the root system over k_s. One source of the importance of irreducibility is that the Weyl group of an irreducible root system acts transitively on the set of roots with a given length. In the study of pseudo-reductive groups, the case of irreducible root systems over k_s will play a similarly important role.

Before discussing root systems in the pseudo-reductive context, it is convenient to develop the analogue of "absolutely simple" by group-theoretic means (to be informed by the properties of $G_{\overline{k}}^{\mathrm{red}}$ from the existing structure theory of reductive groups). We begin with an elementary lemma (see [**CGP**, Def. 3.1.1, Lemma 3.1.2]):

LEMMA 3.2.1. *Let G be a smooth connected affine group over a field k. The following conditions are equivalent:*

(i) *G is non-commutative and it does not contain a nontrivial smooth connected proper normal k-subgroup;*
(ii) *G is pseudo-semisimple and the connected semisimple maximal reductive quotient $G_{k_{\mathrm{p}}}^{\mathrm{red}}$ over the perfect closure k_{p} of k is k_{p}-simple.*

Any G satisfying the equivalent conditions in Lemma 3.2.1 is called *pseudo-simple* (over k); if G_{k_s} is pseudo-simple then we say that G is *absolutely pseudo-simple.* In other words, an absolutely pseudo-simple k-group is a pseudo-semisimple k-group G such that the connected semisimple group $G_{\overline{k}}^{\mathrm{red}}$ is simple.

Just as every nontrivial connected semisimple k-group is a central isogenous quotient of the direct product of its k-simple smooth connected normal k-subgroups (which pairwise commute), we have an analogue in the pseudo-semisimple case.

This rests on lifting k_{p}-simple connected semisimple "factors" of $G_{k_{\mathrm{p}}}^{\mathrm{red}}$ to pseudo-simple normal k-subgroups of G (the latter built by means of Galois descent and root groups over k_s):

PROPOSITION 3.2.2. *Let G be a pseudo-semisimple k-group. For each pseudo-semisimple normal k-subgroup $N \subset G$, let $\overline{N}$ denote the connected semisimple normal image of $N_{k_{\mathrm{p}}}$ in $\overline{G} := G_{k_{\mathrm{p}}}^{\mathrm{red}}$. Then $N \mapsto \overline{N}$ is a bijection (inclusion-preserving in both directions) between the sets of perfect smooth connected normal k-subgroups of G and smooth connected normal k_{p}-subgroups of $\overline{G}$. Moreover, N is pseudo-simple (over k) if and only if $\overline{N}$ is k_{p}-simple.*

PROOF. The asserted properties of $N \mapsto \overline{N}$ permit us to reduce to the case $k = k_s$ via Galois descent, so all k-tori are split and hence we may build root groups. Let T be a maximal k-torus in G, so $\overline{T} := T_{\overline{k}}$ is a maximal torus in the connected semisimple $\overline{k}$-group $\overline{G} = G_{\overline{k}}^{\mathrm{red}}$. We may assume $G \neq 1$, so $\overline{G} \neq 1$. Let $\Phi = \Phi(\overline{G}, \overline{T})$, and let $\{\Phi_i\}_{i \in I}$ be its (non-empty) set of irreducible components.

The structure theory of connected reductive groups ensures that the minimal nontrivial smooth connected normal subgroups of $\overline{G}$ pairwise commute and that the set of these subgroups is in natural bijective correspondence with I, where to $i \in I$ we associate the k-subgroup $\overline{N}_i$ generated by the root groups $\overline{U}_a$ for $a \in \Phi_i$. Likewise, the set of connected semisimple normal subgroups of $\overline{G}$ is in bijective correspondence with the set of subsets of I, by associating to any $J \subset I$ the k-subgroup $\overline{N}_J$ generated by $\{\overline{N}_j\}_{j \in J}$.

For each $i \in I$ we define N_i to be the k-subgroup of G generated by the groups $U_{(a)}$ for $a \in \Phi_i$. Since every non-multipliable element of $\Phi(G,T)$ lies in Φ, it follows that the N_i's generate the same k-subgroup that is generated by the $U_{(a)}$'s for all $a \in \Phi(G,T)$. The final assertion in Proposition 3.1.4 ensures the N_i's generate G, and Proposition 3.1.6(ii) implies that $N_{i'}$ commutes with N_i for all $i' \neq i$. Hence, each N_i is normal in G, so N_i is pseudo-reductive and $(N_i)_{\overline{k}}$ has image $\overline{N}_i$ in $\overline{G}$.

The intersection $T_i := T \cap N_i$ is a maximal k-torus in N_i [**CGP**, Cor. A.2.7] and $(T_i)_{\overline{k}}$ maps isomorphically onto the maximal torus $\overline{T} \cap \overline{N}_i$ in $\overline{N}_i$. In view of the identification of Φ_i with $\Phi(\overline{N}_i, \overline{T}_i)$ (compatibly with the equality $\mathrm{X}(\overline{T})_{\mathbf{Q}} = \bigoplus_i \mathrm{X}(\overline{T}_i)_{\mathbf{Q}}$), the restriction $a|_{T_i}$ is nontrivial for all $a \in \Phi_i$. Hence, a choice of T-equivariant linear structure on $U_{(a)}$ shows that $(T_i, U_{(a)}) = U_{(a)}$ for all $a \in \Phi_i$, so $U_{(a)} \subset \mathscr{D}(N_i)$ for all $a \in \Phi_i$. In view of how N_i was defined, it follows that $N_i = \mathscr{D}(N_i)$ for all i.

For any $J \subset I$, the smooth connected k-subgroup N_J of G generated by $\{N_j\}_{j \in J}$ is perfect and normal, hence pseudo-semisimple. The image of $(N_J)_{\overline{k}}$ in $\overline{G}$ is $\overline{N}_J$, so for each $\overline{N}$ we have built an N giving rise to $\overline{N}$. The construction of N from $\overline{N}$ rests on the choice of T, but N is independent of that choice since all such T are $G(k)$-conjugate to each other and the subgroups $\overline{N} \subset \overline{G}$ and $N \subset G$ are normal. By construction, it is clear that $\overline{N} \mapsto N$ is inclusion-preserving.

It remains to show that if N' is a perfect smooth connected normal k-subgroup of G such that $N'_{\overline{k}}$ is carried onto $\overline{N}$ then $N' = N$. For the perfect smooth connected normal k-subgroup $N'' := (N, N')$, the natural map $N''_{\overline{k}} \to \overline{N}$ is surjective. The smooth connected affine quotients N/N'' and N'/N'' are therefore unipotent (as we may check over $\overline{k}$, using that $\ker(G_{\overline{k}} \twoheadrightarrow \overline{G})$ is unipotent), yet each quotient is perfect, so these quotients are trivial. Hence, $N' = N'' = N$ as desired. $\square$

REMARK 3.2.3. The perfectness condition on N and $\overline{N}$ in Proposition 3.2.2 can be considerably relaxed: a different method of proof (unrelated to root groups) yields a generalization to arbitrary smooth connected affine k-groups G (see [**CGP**, Prop. 3.1.6]), using smooth connected normal subgroups that are "generated by tori" (these constitute a larger class than smooth connected perfect subgroups). In that generality, perfectness of N is equivalent to the same for $\overline{N}$. The "generated by tori" hypothesis cannot be dropped since smooth connected normal k-subgroups of a pseudo-semisimple k-subgroup need not be perfect (see [**CGP**, Ex. 1.6.4] for counterexamples over every imperfect field).

Part (ii) of the following result is a pseudo-semisimple analogue of the isogeny decomposition of a connected semisimple k-group into k-simple "factors".

PROPOSITION 3.2.4. *Let G be a pseudo-reductive k-group, and $\{N_i\}_{i\in I}$ the set of minimal nontrivial perfect smooth connected normal k-subgroups.*

(i) *The N_i's are pseudo-simple over k and pairwise commute.*
(ii) *The natural map $\pi : \prod_{i\in I} N_i \to \mathscr{D}(G)$ is surjective with central kernel that contains no nontrivial smooth connected k-subgroup.*
(iii) *The set of perfect smooth connected normal k-subgroups of G is in bijective correspondence with the set of subsets J of I, where to each J we associate the k-subgroup N_J generated by $\{N_j\}_{j\in J}$. Moreover, $N_J \subset N_{J'}$ if and only if $J \subset J'$.*

See [**CGP**, Prop. 3.1.8] for a generalization to arbitrary smooth connected affine k-groups. Beware that in (ii), $\ker \pi$ generally has *positive* dimension (unlike the reductive case); see [**CGP**, Ex. 3.1.9] for such examples over any imperfect field.

PROOF. The case of commutative G is trivial (using empty I), so we may assume G is non-commutative; i.e., the pseudo-semisimple derived group $\mathscr{D}(G)$ is nontrivial. By Proposition 2.1.2 in the case of perfect normal k-subgroups, normality is transitive among perfect smooth connected k-subgroups. (The more general assertion in Proposition 2.1.2 without perfectness hypotheses *cannot* be used here since the proof in that generality uses the present proposition to reduce to the settled case of perfect subgroups!) Thus, we may replace G with $\mathscr{D}(G)$ to reduce to the case where G is pseudo-semisimple. The same transitivity implies that the N_i's are pseudo-simple over k.

If the result is settled over k_s in general then by Galois descent and the transitivity of normality in the perfect case it would follow that the pseudo-simple normal k-subgroups of G correspond to the $\mathrm{Gal}(k_s/k)$-orbits of pseudo-simple normal k_s-subgroups of G_{k_s}. Thus, by Galois descent we may and do now assume $k = k_s$. The construction of all N's (in the proof of Proposition 3.2.2) from the smooth connected normal subgroups of $G_{\overline{k}}^{\mathrm{red}}$ yields everything we wish except for the assertions concerning $\ker \pi$ in (ii).

For any $(n_i) \in \ker \pi$ valued in a k-algebra A, we have that $n_i \in N_i(A) \cap N_i'(A)$ where N_i' is the k-subgroup generated by $\{N_{i'}\}_{i'\neq i}$. Clearly N_i commutes with N_i' and these two k-subgroups generate G, so $N_i \cap N_i'$ is contained in the scheme-theoretic center Z_G of G. In other words, $n_i \in Z_G(A)$ for all i, so $\ker \pi$ is central.

To show that the only smooth connected k-subgroup $H \subset \ker \pi$ is the trivial subgroup, observe that by centrality of $\ker \pi$ any such H has central image in each N_i. But N_i is pseudo-simple over k, so H has trivial image in N_i. The inclusion $H \hookrightarrow \prod N_i$ therefore has trivial image, so $H = 1$. □

COROLLARY 3.2.5. *If G is pseudo-semisimple and $Z_G = 1$ then $G \simeq \mathrm{R}_{k'/k}(G')$ for a finite étale k-algebra k' and smooth affine k'-group G' whose fiber over each factor field of k' is absolutely pseudo-simple with trivial center. The pair $(k'/k, G')$ is unique up to unique isomorphism: for another $(k''/k, G'')$, any k-isomorphism $\mathrm{R}_{k'/k}(G') \simeq \mathrm{R}_{k''/k}(G'')$ arises from a unique pair (α, φ) where $\alpha : k' \simeq k''$ is a k-algebra isomorphism and $\varphi : G' \simeq G''$ is a group isomorphism over α.*

PROOF. This is a straightforward exercise in Galois descent (using Proposition 3.2.4(iii) over k_s); see [**CP**, Lemma 6.3.13]. □

COROLLARY 3.2.6. *Pseudo-split pseudo-simple k-groups are absolutely pseudo-simple.*

PROOF. Let T be a split maximal k-torus in a pseudo-split pseudo-simple k-group G, so T_{k_s} is a maximal k_s-torus in G_{k_s}. Let $\{N_i\}$ be as in Proposition 3.2.4(i) applied to G_{k_s}. By construction in the proof of Proposition 3.2.2, the pseudo-simple normal k_s-subgroups N_i of G_{k_s} correspond to the irreducible components Φ_i of $\Phi(G^{\mathrm{red}}_{\overline{k}}, T_{\overline{k}}) \subset \mathrm{X}(T_{\overline{k}}) = \mathrm{X}(T_{k_s})$; explicitly, N_i is generated by the k_s-groups $U^{G_{k_s}}_{(a)}$ for $a \in \Phi_i$. But T is k-split, so $\mathrm{X}(T_{k_s}) = \mathrm{X}(T)$ and hence the k-groups $U^{G}_{(a)}$ make sense inside G for all $a \in \Phi_i$. It follows that every N_i descends to a nontrivial perfect smooth connected normal k-subgroup of G, but G is pseudo-simple over k, so there is only one N_i. This says that G_{k_s} is pseudo-simple, or in other words that G is absolutely pseudo-simple. □

We are finally in a position to construct coroots and thereby define the *root datum* associated to a pseudo-split pseudo-reductive k-group G. Let T be a split maximal k-torus of G, so $\Phi(G^{\mathrm{red}}_{\overline{k}}, T_{\overline{k}})$ is the set of non-multipliable elements of $\Phi(G, T)$. Recall from Remark 3.1.11 that for the split maximal k-torus $T' := T \cap \mathscr{D}(G) \subset \mathscr{D}(G)$ [**CGP**, Cor. A.2.7] that is an isogeny complement in T to the maximal central k-torus Z, we have $\Phi(G, T) = \Phi(\mathscr{D}(G), T')$ via the natural identification of $\mathrm{X}(T')_{\mathbf{Q}}$ as a direct summand of $\mathrm{X}(T)_{\mathbf{Q}}$.

For each non-multipliable $a \in \Phi(G, T)$, define $a^\vee \in \mathrm{X}_*(T) = \mathrm{X}_*(T_{\overline{k}})$ to correspond to the coroot for $a_{\overline{k}} \in \Phi(G^{\mathrm{red}}_{\overline{k}}, T_{\overline{k}})$. If $a \in \Phi(G, T)$ is multipliable then we define $a^\vee = 2(2a)^\vee \in \mathrm{X}_*(T)$. The set of cocharacters $a^\vee$ for $a \in \Phi(G, T)$ is denoted $\Phi(G, T)^\vee$, and its elements are called *coroots* for (G, T). It is clear from the reductive case over $\overline{k}$ that $a \mapsto a^\vee$ is a bijection from $\Phi(G, T)$ onto $\Phi(G, T)^\vee$.

PROPOSITION 3.2.7. *Let G be a pseudo-split pseudo-reductive k-group, and $T \subset G$ a split maximal k-torus.*

(i) *The 4-tuple $R(G, T) := (\mathrm{X}(T), \Phi(G, T), \mathrm{X}_*(T), \Phi(G, T)^\vee)$ is a root datum.*
(ii) *The finite étale k-group $W(G, T) := N_G(T)/Z_G(T)$ is constant and the natural map $W(G, T)(k) \to \mathrm{Aut}(\mathrm{X}(T))$ is injective onto $W(\Phi(G, T))$.*
(iii) *Let Z be the maximal central k-torus in G and $\{G_i\}$ the set of pseudo-simple normal k-subgroups of G. For the associated k-split maximal k-tori $T_i = G_i \cap T \subset G_i$, the multiplication map $Z \times \prod T_i \to T$ is an isogeny identifying $\{\Phi(G_i, T_i)\}$ with the set of irreducible components of $\Phi(G, T)$.*

In particular, for the split maximal k-torus $T' := T \cap \mathscr{D}(G) \subset \mathscr{D}(G)$ that is an isogeny complement to Z in T, $\Phi(G, T)$ is a root system with $\mathbf{Q}$-span $\mathrm{X}(T')_{\mathbf{Q}}$ and if G is pseudo-semisimple then it is (absolutely) pseudo-simple if and only if $\Phi(G, T)$ is irreducible.

A finer analysis shows that the inclusion $N_G(T)(k)/Z_G(T)(k) \hookrightarrow W(G,T)(k) = W(\Phi(G,T))$ is bijective; we will address this in Proposition 4.1.3.

PROOF. Apart from the final assertion relating (absolute) pseudo-simplicity and irreducibility of a root system in the pseudo-semisimple case, the rest is largely an exercise in bootstrapping from the reductive case by using Propositions 3.2.2 and 3.2.4; see [**CGP**, Lemma 1.2.5(ii), 3.2.5–3.2.10] for the details.

Now assume G is pseudo-semisimple, so by the classical link between simple isogeny factors of a split connected semisimple group and the irreducible components of its root system it follows that the connected semisimple $\overline{k}$-group $G_{\overline{k}}^{\mathrm{red}}$ is simple if and only if the root system $\Phi(G_{\overline{k}}^{\mathrm{red}}, T_{\overline{k}})$ is irreducible. But $\Phi(G_{\overline{k}}^{\mathrm{red}}, T_{\overline{k}})$ is the set of non-multipliable elements of $\Phi(G,T)$, and a root system is irreducible if and only if the root system of its non-multipliable elements is irreducible. Thus, it remains to observe that G is absolutely pseudo-simple if and only if the connected semisimple $G_{\overline{k}}^{\mathrm{red}}$ is simple, due to Proposition 3.2.2. □

REMARK 3.2.8. For any $a \in \Phi(G,T)$, the 1-dimensional torus $a^\vee(\mathrm{GL}_1)$ is maximal in an absolutely pseudo-simple k-subgroup of G attached to a similarly to the reductive case, as follows. Defining $G_a := \langle U_a, U_{-a}\rangle$, the equality $a^\vee(\mathrm{GL}_1) = T \cap G_a$ is an easy consequence of the well-known analogue for $G_{\overline{k}}^{\mathrm{red}}$; see [**CGP**, Prop. 3.2.3]. The absolute pseudo-simplicity of G_a shall now be deduced from a description of G_a in terms of derived groups and centralizers of k-subgroup schemes of tori.

If a is not divisible then, as in the reductive case, we have $G_a = \mathscr{D}(Z_G(T_a))$ for the codimension-1 subtorus $T_a = (\ker a)^0_{\mathrm{red}} \subset T$ killed by a; this is easy to prove since (i) the pseudo-split pseudo-reductive $Z_G(T_a)$ has as its roots only $\{\pm a\}$ or $\{\pm a, \pm 2a\}$ due to the non-divisibility of a, and (ii) the pseudo-split pseudo-semisimple group $\mathscr{D}(Z_G(T_a))$ is generated by its root groups (Proposition 3.1.4). If instead a is divisible (so k is imperfect with characteristic 2) then similar arguments show that $G_a = \mathscr{D}(H_a)$ for $H_a = Z_{Z_G(T_a)}(\mu)^0$ with μ denoting the infinitesimal k-group scheme $(T \cap G_a)[2] \simeq \mu_2$, but the pseudo-reductivity of H_a lies much deeper than that of $Z_G(T_a)$; see [**CGP**, Prop. 3.4.1] for further details. The description of G_a in each case implies that $T \cap G_a$ is a maximal k-torus in G_a (because the intersection of a maximal k-torus with a smooth connected normal k-subgroup N is a maximal k-torus in N [**CGP**, Cor. A.2.7]).

3.3. Open cell. For a split connected reductive k-group (G,T) and the Borel k-subgroup B containing T for which $\Phi(B,T)$ coincides with a given positive system of roots Φ^+ in $\Phi(G,T)$, the k-unipotent radical $\mathscr{R}_{u,k}(B)$ is directly spanned in any order by the positive root groups $\{U_a\}_{a\in\Phi^+}$; i.e., if $\{a_1,\dots,a_n\}$ is *any* enumeration of Φ^+ then the multiplication map $\prod U_{a_i} \to \mathscr{R}_{u,k}(B)$ is an isomorphism of k-schemes. The analogous result for k-unipotent radicals of parabolic k-subgroups containing B is classical. Traditional proofs of these results rest crucially on the 1-dimensionality of the root groups and on an inductive procedure to build up Φ^+ from well-chosen subsets.

In the pseudo-reductive case such a "direct spanning" result holds for the k-unipotent radicals of pseudo-parabolic k-subgroups, but the proof is necessarily completely different from the traditional arguments in the reductive case since the root groups are generally not 1-dimensional. Rather generally, one considers a split k-torus S acting on any smooth affine k-group G and seeks to describe certain S-stable smooth connected k-subgroups $H \subset G$ in terms of the subsemigroup of $\mathrm{X}(S)$

generated by the S-weights that occur in $\mathrm{Lie}(H)$. With the aid of dynamic methods (especially the k-groups $U_{H'}(\lambda)$ for S-stable smooth connected k-subgroups $H' \subset G$ and $\lambda \in \mathrm{X}_*(S)$), a study of the S-action on coordinates rings of geometrically integral closed subschemes of G passing through 1 yields a crucial result [**CGP**, Prop. 3.3.6]:

PROPOSITION 3.3.1. *If $A \subset \mathrm{X}(S)$ is a subsemigroup then among all S-stable smooth connected k-subgroups $H \subset G$ such that all S-weights on* $\mathrm{Lie}(H)$ *lie in A, there is one such k-subgroup $H_A(G)$ that contains all others. If $0 \notin A$ then $H_A(G)$ is unipotent.*

EXAMPLE 3.3.2. If G is pseudo-reductive with a split maximal k-torus T and $a \in \Phi(G,T)$ then for the semigroup $A = \langle a \rangle$ of multiples na with positive integers n, the k-subgroup $H_A(G)$ is the a-root group U_a. This special case is easily proved since $\mathrm{Lie}(U_a)$ is the span of all T-weight spaces in $\mathrm{Lie}(G)$ for weights in A.

By investigating the functorial behavior of $H_A(G)$ upon varying A and G (e.g., for an S-stable smooth closed k-subgroup $G' \subset G$, when does $G' \cap H_A(G) = H_A(G')$?), as is carried out in [**CGP**, 3.3.8–3.3.10], one obtains a vast generalization [**CGP**, Thm. 3.3.11] of the direct spanning of $\mathscr{R}_u(B)$ by positive root groups in the connected reductive case:

THEOREM 3.3.3. *Let S be a split k-torus and U a nontrivial smooth connected unipotent k-group equipped with an S-action such that the set Ψ of S-weights occurring on* $\mathrm{Lie}(U)$ *does not contain* 0. *For any decomposition $\Psi = \coprod_{j=1}^{n} \Psi_j$ into disjoint non-empty subsets such that the semigroup $A_j = \langle \Psi_j \rangle$ is disjoint from $\Psi_{j'}$ for all $j' \neq j$, the natural multiplication map*

$$H_{A_1}(U) \times \cdots \times H_{A_n}(U) \longrightarrow U$$

is an isomorphism of k-schemes.

EXAMPLE 3.3.4. Let G be a pseudo-split pseudo-reductive group, with T a split maximal k-torus. Let Φ^+ be a positive system of roots in $\Phi := \Phi(G,T)$, so by general facts in the theory of root systems (see [**CGP**, Prop. 2.2.8(3)]) Φ^+ is the locus where Φ meets an open half-space $\{\lambda > 0\}$ for some $\lambda \in \mathrm{X}_*(T)$ that is non-vanishing on Φ. We apply Theorem 3.3.3 to $U = U_G(\lambda)$ and $\Psi = \Phi^+$ with Ψ_j's taken to be where Ψ meets half-lines in $\mathrm{X}(T)_{\mathbf{Q}}$ (so each Ψ_j is either a singleton consisting of a non-divisible non-multipliable positive root or has the form $\{a, 2a\}$ for a multipliable positive root a). It follows that the root groups U_a for non-divisible $a \in \Phi^+$ directly span *in any order* a smooth connected unipotent k-subgroup $U_{\Phi^+} \subset G$. (This k-subgroup is $U_G(\lambda)$ by another name, but clearly depends only on Φ^+ rather than on the choice of λ.)

If $\{a_1, \ldots, a_n\}$ is an enumeration of the non-divisible elements of Φ^+ then by Theorem 2.3.5(ii) the natural multiplication map

$$\prod U_{-a_i} \times Z_G(T) \times \prod U_{a_i} = U_{-\Phi^+} \times Z_G(T) \times U_{\Phi^+} \longrightarrow G$$

is an open immersion; this is called the *open cell* attached to Φ^+. The k-subgroup $P_{\Phi^+} = Z_G(T) \ltimes U_{\Phi^+} = P_G(\lambda)$ is pseudo-parabolic, and $\Phi^+ \mapsto P_{\Phi^+}$ is a bijection from the set of choices of Φ^+ onto the set of minimal pseudo-parabolic k-subgroups P of G such that $T \subset P$ (see Proposition 3.3.7 below); in the reductive case this recovers the well-known link between Borel subgroups and positive systems of roots (but by an entirely different method of proof!).

Since we will use root systems to control pseudo-parabolic subgroups, we now recall the combinatorial notion that corresponds to pseudo-parabolicity:

DEFINITION 3.3.5. A subset Ψ of a root system Φ is *parabolic* if it is closed (i.e., $a+b \in \Psi$ for any $a, b \in \Psi$ such that $a+b \in \Phi$) and $\Psi \cup -\Psi = \Phi$.

Letting V be the $\mathbf{Q}$-span of Φ, it is a classical fact (see [**CGP**, Prop. 2.2.8]) that the parabolic subsets are precisely the intersections

$$\Phi_{\lambda \geqslant 0} := \Phi \cap \{\lambda \geqslant 0\}$$

for linear forms $\lambda : V \to \mathbf{Q}$; in geometric terms, these are precisely the intersections of Φ with a closed half-space in $V_{\mathbf{R}}$. For λ that is non-vanishing at all points of Φ (the "generic" case), the resulting parabolic subsets $\Phi_{\lambda \geqslant 0} = \Phi_{\lambda > 0}$ are precisely the positive systems of roots. (Moreover, closed subsets of Φ are precisely the intersections $\Phi \cap A$ with a subsemigroup $A \subset V$ [**CGP**, Prop. 2.2.7].)

EXAMPLE 3.3.6. If G is pseudo-reductive and pseudo-split with a split maximal k-torus T, for the pseudo-parabolic k-subgroups $P \subset G$ containing T the subsets $\Phi(P,T) \subset \Phi$ (consisting of nontrivial T-weights occuring in $\mathrm{Lie}(P)$) are precisely the parabolic subsets of Φ. Indeed, we can choose $\lambda : \mathrm{GL}_1 \to T$ such that $P = P_G(\lambda)$ by Lemma 2.3.9, so then $\Phi(P,T) = \Phi_{\lambda \geqslant 0}$ by Theorem 2.3.5(i),(ii).

PROPOSITION 3.3.7. *Let G be a pseudo-split pseudo-reductive k-group, with T a split maximal k-torus and $\Phi = \Phi(G,T)$. Consider pseudo-parabolic k-subgroups P of G that contain T. The set $\Phi(P,T)$ is a positive system of roots in Φ if and only if P is minimal as a pseudo-parabolic k-subgroup of G, and $P \mapsto \Phi(P,T)$ is a bijection from the set of such minimal P onto the set of positive systems of roots in Φ.*

Since $W(G,T)(k) = W(\Phi)$ by Proposition 3.2.7(ii), we have a simply transitive action of $W(G,T)(k)$ on the set of such P since $W(\Phi)$ acts simply transitively on the set of positive systems of roots in Φ for any root system Φ.

PROOF. By Lemma 2.3.9 we can choose $\lambda \in \mathrm{X}_*(T)$ such that $P = P_G(\lambda)$, so Theorem 2.3.5(i),(ii) implies

$$\Phi(P,T) = \Phi_{\lambda \geqslant 0} := \{a \in \Phi \mid \langle a, \lambda \rangle \geqslant 0\}.$$

This is a positive system of roots precisely when λ is non-vanishing on all elements of Φ. Suppose that $\Phi(P,T)$ is *not* a positive system of roots, so the hyperplane $\{\lambda = 0\}$ in $\mathrm{X}(T)_{\mathbf{Q}}$ meets Φ. Choose $\lambda' \in \mathrm{X}_*(T)_{\mathbf{Q}}$ sufficiently near λ so that it is positive on the finite set $\Phi_{\lambda>0}$ and negative on $\Phi_{\lambda<0}$ but non-vanishing on $\Phi_{\lambda=0}$. Hence, for $\mu := n\lambda' \in \mathrm{X}_*(T)$ with an integer $n > 0$ that is sufficiently divisible, the set $\Phi_{\mu \geqslant 0} = \Phi_{\mu > 0}$ is a positive system of roots contained in $\Phi_{\lambda \geqslant 0}$. Thus, for the pseudo-parabolic k-subgroup $Q := P_G(\mu)$ the containment $Q \cap P = P_Q(\lambda) \subset Q$ between smooth connected k-subgroups is an equality on Lie algebras and so is an equality of groups; i.e., $Q \subset P$. But $\mathrm{Lie}(Q)$ is a proper subspace of $\mathrm{Lie}(P)$, so P is not minimal in G. In other words, if P is minimal then $\Phi(P,T)$ is a positive system of roots.

Suppose instead that P is not minimal. We wish to show that $\Phi(P,T)$ is not a positive system of roots. For *this purpose* it is harmless to extend scalars to k_s. It is likewise harmless to replace T with a $P(k)$-conjugate. Letting P' be a pseudo-parabolic k-subgroup of G strictly contained in P, consider a maximal k-torus

$T' \subset P'$. Note that T' is split since we have arranged (for present purpose) that $k = k_s$. It is a well-known result of Grothendieck that maximal tori in a smooth affine group over a separably closed field are rationally conjugate to each other [**CGP**, Prop. A.2.10] (this is much more elementary than the rational conjugacy of maximal split tori in smooth affine – or even just connected reductive! – groups over general fields, which we will address in Theorem 4.2.9). Hence, via suitable $P(k)$-conjugation to carry T onto T' we can assume $T \subset P'$.

The strict containment of P' in P implies a strict containment $\mathrm{Lie}(P') \subsetneq \mathrm{Lie}(P)$. Since $T \subset P'$, so $P' = P_G(\lambda')$ for some $\lambda' \in \mathrm{X}_*(T)$ by Lemma 2.3.9, clearly $P' \supset Z_G(\lambda') \supset Z_G(T)$. Thus, $\mathrm{Lie}(G)^T \subset \mathrm{Lie}(P')$ and each of $\mathrm{Lie}(P')$ and $\mathrm{Lie}(P)$ is spanned by $\mathrm{Lie}(G)^T$ and the root spaces for roots respectively in $\Phi(P', T)$ and $\Phi(P, T)$. Hence, the *parabolic* subset $\Phi(P', T)$ of Φ inside $\Phi(P, T)$ must be a proper subset of $\Phi(P, T)$, so $\Phi(P, T)$ is not a positive system of roots.

Now we return to general k and consider minimal pseudo-parabolic k-subgroups P and Q of G that contain T such that $\Phi(P, T) = \Phi(Q, T)$. We need to show that $P = Q$. But each of P and Q is generated by $Z_G(T)$ and root groups of G for the T-weights that appear in the respective Lie algebras (e.g., if $a \in \Phi(P', T) = \Phi_{\lambda' \geqslant 0}$ then the containment $U_{(a)}^{P'} \subset U_{(a)}^{G}$ is an equality on Lie algebras and thus an equality of k-groups, so $U_{(a)}^{G} \subset P'$), so obviously $P = Q$. □

4. Structure theory

4.1. Bruhat decomposition. For a connected reductive k-group G, the subgroup structure of $G(k)$ is governed by the *Bruhat decomposition* as follows. If S is a maximal split k-torus (with associated relative root system ${}_k\Phi = \Phi(G, S)$ that may be non-reduced) and P is a minimal parabolic k-subgroup of G containing S then the relative Weyl group ${}_kW := N_G(S)(k)/Z_G(S)(k)$ (which maps isomorphically onto $W({}_k\Phi)$) labels the $P(k)$-double cosets in $G(k)$: the natural map

$$ {}_kW \longrightarrow P(k)\backslash G(k)/P(k) $$

is bijective.

Writing $n_w \in N_G(S)(k)$ to denote a representative of $w \in {}_kW$, in the *split* case the locally closed subsets Pn_wP constitute a stratification of G whose closure relations can be expressed entirely in terms of the combinatorics of Coxeter groups and root systems (via the "Bruhat order" on ${}_kW$ defined by a choice of basis of ${}_k\Phi$). If G is not assumed to be split then $P = Z_G(S) \ltimes U$ where $U := \mathscr{R}_{u,k}(P)$ is k-split and directly spanned in any order by the root groups associated to members of the positive system of roots $\Phi(P, S) \subset {}_k\Phi$, and the Bruhat decomposition

$$ G(k) = \coprod_{w \in {}_kW} P(k) n_w P(k) $$

has only group-theoretic rather than geometric meaning.

The preceding Bruhat decomposition is a consequence of general results concerning groups equipped with a Tits system [**Bou**, Ch. IV, §2.3, Thm. 1], so the main work in its proof is to show that the 4-tuple $(G(k), P(k), N_G(S)(k), R)$ is a Tits system (see Definition 4.1.6), where $R = \{r_a\}_{a \in \Delta}$ is the set of reflections in ${}_kW = W({}_k\Phi)$ associated to a basis Δ of ${}_k\Phi$. The equality ${}_kW = W({}_k\Phi)$ is an essential step in relating the structure of $G(k)$ to the theory of Coxeter groups, and it rests on finding $n_a \in N_G(S)(k)$ inducing the reflection $r_a : x \mapsto x - \langle x, a^\vee \rangle a$ in each root $a \in {}_k\Phi$.

An analogous structure is available for pseudo-reductive groups, both in the pseudo-split case (using root systems and root groups as introduced in §3) as well as in the general case. This development involves a Bruhat decomposition for $G(k)$ relative to maximal k-split tori S and minimal *pseudo-parabolic* k-subgroups $P \supset S$, as well as $G(k)$-conjugacy of all such pairs (S, P). The general $G(k)$-conjugacy results will be discussed in §4.2 and §5.1, and we now focus on the pseudo-split case because ultimately the proof of the general k-rational Bruhat decomposition in §5.2 rests on the k_s-rational Bruhat decomposition.

REMARK 4.1.1. We shall see (in the proof of Theorem 4.1.7) that Tits systems are used to prove the Bruhat decomposition in the pseudo-split case. In contrast, the Bruhat decomposition in the general pseudo-reductive case over k [**CGP**, Thm. C.2.8] rests on the settled (pseudo-split) case over k_s, whereas (akin to the general connected reductive case) verifying the Tits system axioms over k [**CGP**, Thm. C.2.20] rests on the Bruhat decomposition over k. This will be discussed more fully in §5.3.

As a first step, for a pseudo-reductive k-group G and split maximal k-torus T we shall construct representatives in $N_G(T)(k)$ for reflections in $W(\Phi(G,T))$ attached to roots in $\Phi(G,T)$. It is instructive to recall motivation from the rank-1 split connected semisimple case:

EXAMPLE 4.1.2. Let G be a split connected semisimple k-group of rank 1, $T \subset G$ a split maximal k-torus, and $a \in \Phi(G,T)$ one of the two roots. Choose a nontrivial element $u \in U_a(k) - \{1\}$. We may pick an isomorphism from G onto SL_2 or PGL_2 carrying T onto the diagonal k-torus such that U_a is carried into the upper-triangular unipotent k-subgroup. In this way, u goes over to an element $\left(\begin{smallmatrix} 1 & x \\ 0 & 1 \end{smallmatrix}\right)$ with $x \in k^\times$.

An elementary calculation with SL_2 and PGL_2 shows that there exist *unique* $u', u'' \in U_{-a}(k)$ such that $m(u) := u'uu'' \in N_G(T)(k)$, and that necessarily $u' = u'' \neq 1$ with $m(u)$ representing the unique nontrivial element in $N_G(T)(k)/T(k)$. Explicitly, $u' = u'' = \left(\begin{smallmatrix} 1 & 0 \\ -1/x & 1 \end{smallmatrix}\right) = m(u)um(u)^{-1}$ and $m(u)^2 = \mathrm{diag}(-1,-1) = a^\vee(-1)$ regardless of the choice of u.

We want to adapt Example 4.1.2 to the rank-1 pseudo-split *pseudo-semisimple* case. Two immediate difficulties are: (i) there is no concrete description of the rank-1 possibilities at this stage of the theory, and (ii) over imperfect fields of characteristic 2 the cases with root system BC_1 are especially difficult to describe even after we have developed a lot more theory.

PROPOSITION 4.1.3. *Let G be a pseudo-split pseudo-reductive k-group and $T \subset G$ a split maximal k-torus. Choose $a \in \Phi := \Phi(G,T)$, $u \in U_a(k) - \{1\}$. There exist unique $u', u'' \in U_{-a}(k)$ such that $m(u) := u'uu'' \in N_G(T)(k)$. Moreover, $u' = u'' = m(u)um(u)^{-1} \neq 1$, $m(u)^2 = a^\vee(-1)$, and the image of $m(u)$ in $W(\Phi)$ is the reflection $r_a : x \mapsto x - \langle x, a^\vee \rangle a$ arising from the root datum $R(G,T)$.*

In particular, the natural inclusion $N_G(T)(k)/Z_G(T)(k) \hookrightarrow W(\Phi)$ is an equality and hence if G is absolutely pseudo-simple (so Φ is irreducible) then $N_G(T)(k)$ acts transitively on the set of roots in Φ with a given length.

See Proposition 5.4.2 (and Proposition 5.3.1 and Theorem 5.3.2(i)) for a version beyond the pseudo-split case.

PROOF. We provide a sketch of the main ideas, referring the reader to [**CGP**, Prop. 3.4.2] for complete details. Let $N = N_G(T)$ and $U^{\pm} = U_{\pm a}$, and let C denote the Cartan k-subgroup $Z_G(T)$. By using Galois descent and centralizer considerations as near the end of Remark 3.2.8, the general case reduces to the rank-1 case with non-divisible a and $k = k_s$ that we now address.

The two possibilities for Φ are $\{\pm a\}$ or $\{\pm a, \pm 2a\}$ (the latter only possible when k is imperfect of characteristic 2), and $N(k)/C(k) = (N/C)(k) = W(\Phi)$ has order 2 (see Proposition 3.2.7(ii)). Thus, C and $N - C$ are the connected components of N, so upon choosing $n \in N(k) - C(k) = (N - C)(k)$ we have $N - C = nC$. Note that $U^- = nU^+n^{-1}$ since the nontrivial n-conjugation on T must act via a nontrivial automorphism of the rank-1 root system Φ and hence negates the roots.

The natural quotient map $\pi : G_{\overline{k}} \to G^{\mathrm{red}}_{\overline{k}}$ need not be injective on $G(k)$ (since C may have nontrivial étale p-torsion when k is imperfect with characteristic p [**CGP**, Ex. 1.6.3]). However, the kernels $\ker(\pi|_{U^{\pm}(k)})$ are trivial. To prove this triviality, first note that $\ker(\pi|_{U^{\pm}(k)}) \subset \mathscr{R}_u(G_{\overline{k}})$. Thus, by pseudo-reductivity of G and [**CGP**, Lemma 1.2.1], the Zariski closure of $\ker(\pi|_{U^{\pm}(k)})$ has trivial identity component. This says that $\ker(\pi|_{U^{\pm}(k)})$ is *finite*. Each restriction $\pi|_{U^{\pm}(k)}$ is equivariant with respect to $T(k) \to T(\overline{k})$, so $\ker(\pi|_{U^{\pm}(k)})$ is stable inside $U^{\pm}(k)$ under conjugation by $T(k)$. But $U^{\pm}$ admits a T-equivariant linear structure with only nontrivial weights, so the finite group $\ker(\pi|_{U^{\pm}(k)})$ must be trivial; i.e., $\pi|_{U^{\pm}(k)}$ is *injective*.

Since π carries the "open cell" $(U^- \times C \times U^+)_{\overline{k}} \subset G_{\overline{k}}$ into the corresponding open cell in $G^{\mathrm{red}}_{\overline{k}}$, the injectivity of $\pi|_{U^{\pm}(k)}$ reduces the proofs of the uniqueness of u', u'' (given their existence!) and the identities $u' = u'' = m(u)um(u)^{-1} \neq 1$ in $U^-(k)$ to the settled reductive case over $\overline{k}$. Hence, the main problem is existence of u' and u''.

REMARK 4.1.4. The other desired identity, $m(u)^2 \stackrel{?}{=} a^{\vee}(-1)$, does not take place inside $U^{\pm}(k)$ and so does not reduce to the reductive case over $\overline{k}$. Its proof involves separate arguments depending on whether or not $\mathrm{char}(k) = 2$. If $\mathrm{char}(k) = 2$ then we use that $U^{\pm}(k)$ is 2-torsion, and if $\mathrm{char}(k) \neq 2$ then we use that there exists $t \in T(k)$ such that $a(t) = -1 \neq 1$ (since $k = k_s$ and a root is at worst divisible by 2 in $\mathrm{X}(T)$).

Continuing with the proof, we shall construct u' and u'' such that $u'uu'' \in N - C = nC$ by studying the multiplication map $\mu : U^- \times U^+ \times U^- \to G$. Working with points valued in k-algebras, an identity of the form $u'uu'' = nc$ for $u', u'' \in U^-$ and $c \in C$ can be rewritten as $n^{-1}u = (n^{-1}u'^{-1}n)cu''^{-1} \in U^+ \times C \times U^-$. Since $n^{-1}u \in G(k)$ and the multiplication map $U^+ \times C \times U^- \to G$ is an open immersion, it suffices to construct u' and u'' as $\overline{k}$-points! We likewise see that the preimage $\mu^{-1}(N - C)$ projects isomorphically onto the open subscheme $\Omega \subset U^+$ of points whose left n^{-1}-translate lies in the open cell $U^+CU^- \subset G$. Thus, it suffices to prove $U^+(k) - \{1\} \subset \Omega$.

Injectivity of $\pi|_{U^+(k)}$ ensures that $U^+(k) - \{1\}$ is disjoint from $R := \mathscr{R}(G_{\overline{k}})$. Thus, it suffices to show that $U^+_{\overline{k}} \cap (G_{\overline{k}} - R) \subset \Omega_{\overline{k}}$ (in fact, equality holds), or equivalently that

$$n(U^+_{\overline{k}} \cap (G_{\overline{k}} - R)) \subset (U^+CU^-)_{\overline{k}}. \tag{4.1.4.1}$$

For this purpose we can replace G with $\langle U^+, U^-\rangle = \mathscr{D}(G)$ since $R \cap N = \mathscr{R}_u(N)$ for any smooth connected subgroup $N \subset G_{\overline{k}}$ [**CGP**, Prop. A.4.8] (applied with $N = \mathscr{D}(G)_{\overline{k}}$). Now G is pseudo-semisimple and $G^{\mathrm{red}}_{\overline{k}}$ is equal to SL_2 or PGL_2. Hence, the analogue of (4.1.4.1) for $G^{\mathrm{red}}_{\overline{k}}$ is a trivial calculation, so to prove (4.1.4.1) it suffices to show $(U^+CU^-)_{\overline{k}}$ is the preimage of its image under π.

In other words, for $\lambda = a^\vee_{\overline{k}}$ we need to show that $U_{G_{\overline{k}}}(\lambda)P_{G_{\overline{k}}}(-\lambda)$ is stable under right multiplication against R. But R is normal in $G_{\overline{k}}$, so it is the same to show that

$$U_{G_{\overline{k}}}(\lambda)P_{G_{\overline{k}}}(-\lambda) = U_{G_{\overline{k}}}(\lambda)RP_{G_{\overline{k}}}(-\lambda).$$

Since R is a *solvable* smooth connected affine group, it coincides with its own "open cell" relative to *any* GL_1-action [**CGP**, Rem. 2.1.11, Prop. 2.1.12(1)]. Hence, making GL_1 act on the normal subgroup $R \subset G_{\overline{k}}$ through λ-conjugation, the open immersion $U_R(\lambda) \times P_R(-\lambda) \to R$ via multiplication is an isomorphism and so we are done. □

REMARK 4.1.5. By Proposition 3.2.7 we have $(N_G(T)/Z_G(T))(k) = W(\Phi)$, so Proposition 4.1.3 implies that the short exact sequence of k-groups

$$1 \longrightarrow Z_G(T) \longrightarrow N_G(T) \longrightarrow N_G(T)/Z_G(T) \longrightarrow 1$$

induces a short exact sequence on k-points. This is remarkable because it admits no cohomological explanation. To explain this point, note that the cohomological obstruction to short-exactness on k-points lies in $\mathrm{H}^1(k, Z_G(T))$. If G is reductive then this cohomology group vanishes by Hilbert's Theorem 90 since $Z_G(T) = T$ is a split k-torus.

The general structure of the commutative pseudo-reductive k-group $Z_G(T)$ is mysterious (as the unipotent $Z_G(T)/T$ *never* contains $\mathbf{G}_{\mathrm{a}}$ as a k-subgroup [**CGP**, Ex. B.2.8]), so it isn't clear if $\mathrm{H}^1(k, Z_G(T))$ vanishes. In fact, over imperfect k with a sufficiently nontrivial Brauer group there exist "standard" pseudo-split absolutely pseudo-simple k-groups G for which $\mathrm{H}^1(k, Z_G(T)) \neq 1$ (this occurs whenever k coincides with the rational function field $\kappa(u, v)$ over an algebraically closed field κ with positive characteristic); see [**CGP**, Ex. 3.4.4] for such examples.

Finally, we can adapt techniques from the Borel–Tits structure theory of arbitrary connected reductive groups to establish the Bruhat decomposition for *pseudo-split* pseudo-reductive groups (to be generalized to smooth connected affine groups in Theorem 5.2.2). The complete result in this case involves the following important notion from [**Bou**, IV, §2]:

DEFINITION 4.1.6. A BN-*pair* for a group G is an ordered pair (B, N) of subgroups such that:

(BN1) $\mathsf{B} \cup \mathsf{N}$ generates G, and $\mathsf{B} \cap \mathsf{N}$ is normal in N,

(BN2) $\mathsf{W} := \mathsf{N}/(\mathsf{B} \cap \mathsf{N})$ is generated by a set R of elements of order 2 that do *not* normalize B,

(BN3) for any $n \in \mathsf{N}$ and representative $s \in \mathsf{N}$ of an element of R, $s\mathsf{B}n \subset \mathsf{B}n\mathsf{B} \cup \mathsf{B}sn\mathsf{B}$.

If there exists a nilpotent normal subgroup $\mathsf{U} \subset \mathsf{B}$ such that $\mathsf{B} = (\mathsf{B} \cap \mathsf{N})\mathsf{U}$ then the BN-pair is *weakly split*, and if $\mathsf{B} \cap \mathsf{N} = \bigcap_{w \in \mathsf{W}} w\mathsf{B}w^{-1}$ then the BN-pair is *saturated.* Any such 4-tuple $(\mathsf{G}, \mathsf{B}, \mathsf{N}, \mathrm{R})$ is called a *Tits system.*

The group W is called the *Weyl group* of the BN-pair, and the set R is uniquely determined by the triple $(\mathsf{G}, \mathsf{B}, \mathrm{N})$ [**Bou**, IV, §2.4, Rem.(1)].

THEOREM 4.1.7. *Let G be a pseudo-split pseudo-reductive k-group with a split maximal k-torus T. Let P be a minimal pseudo-parabolic k-subgroup of G containing T, and define $N = N_G(T)$ and $Z = Z_G(T)$.*

(i) *The pair $(P(k), N(k))$ is a saturated* BN*-pair for $G(k)$ with associated Weyl group $W(\Phi(G,T))$.*

(ii) *(Bruhat decomposition) The natural map*

$$N(k)/Z(k) \longrightarrow P(k)\backslash G(k)/P(k) \tag{4.1.7.1}$$

is bijective.

We have not yet addressed (and do not presently need) the $G(k)$-rational conjugacy of all pairs (T, P); this will be proved in §4.2 and §5.1 (not relying on the present considerations).

PROOF. We shall sketch the proof, and refer the reader to [**CGP**, Thm. 3.4.5] for omitted details. Let $N = N_G(T)$, $Z = Z_G(T)$, and $\Phi = \Phi(G,T)$. Define $\Gamma \subset G(k)$ to be the subgroup generated by $Z(k)$ and $\{U_a(k)\}_{a \in \Phi}$. (Eventually we will see that $\Gamma = G(k)$, but we do not yet know this.)

Since the natural map $N(k) \to W(\Phi)$ is surjective by Proposition 4.1.3, for each $a \in \Phi$ we may define a $Z(k)$-coset $M_a \subset N(k)$ to be the preimage in $N(k)$ of the reflection $r_a \in W(\Phi)$ attached to a (i.e., $r_a : x \mapsto x - \langle x, a^\vee \rangle a$). In the work of Bruhat–Tits on the structure of reductive groups over local fields, the notion of a "generating root datum" (of type Φ) [**BrTi**, (6.1.1)] is defined via 6 axioms and a generating property that we do not state here.

The data $(Z(k), (U_a(k), M_a)_{a \in \Phi})$ satisfies the 6 axioms due to several earlier results: Theorem 2.3.5(ii), the direct spanning of $\mathscr{R}_{u,k}(P)$ by its T-root groups in any order (see Example 3.3.4), and Proposition 4.1.3 (as well as [**CGP**, Cor. 3.3.13(2)]). The remaining ingredient to establish that $(Z(k), (U_a(k), M_a)_{a \in \Phi})$ is a generating root datum (of type Φ) is that $\Gamma = G(k)$.

By Proposition 3.3.7, $\Phi^+ := \Phi(P,T)$ is a positive system of roots in Φ. For the k-groups $U_{\pm\Phi^+}$ as in Example 3.3.4, clearly $U_{\pm\Phi^+}(k)$ are generated by the subgroups $\{U_a(k)\}_{a \in \pm\Phi^+}$. Thus, for the open cell $\Omega := U_{-\Phi^+} \times Z \times U_{\Phi^+} \hookrightarrow G$ (via multiplication), clearly $\Omega(k) \subset \Gamma$. Hence, to prove $\Gamma = G(k)$ it is enough to show that $\Omega(k)$ generates $G(k)$. More specifically, we claim that for every $g \in G(k)$, the dense open $\Omega_g := g\Omega \cap \Omega \subset G$ contains a k-point. This is rather more delicate than in the reductive case since G is generally *not* unirational (see Example 2.1.5). Nonetheless, dynamic arguments establish that $\Omega_g(k)$ is non-empty. (The idea for proving $\Omega_g(k) \neq \emptyset$ when k is infinite is to show that $\Omega_g(k)/P(k) = (\Omega_g/P)(k)$, a useful equality because Ω_g/P is clearly a dense open subscheme of the k-scheme $\Omega/P = U^-$ that is an affine space and hence has Zariski-dense locus of k-points. The case of finite k is part of the standard Borel–Tits structure theory for connected reductive groups.)

Let Δ be the set of simple roots in Φ^+. Since $(Z(k), (U_a(k), M_a)_{a \in \Phi})$ is a generating root datum for $G(k)$, by [**BrTi**, 6.1.11(ii), 6.1.12] it follows that

$(G(k), P(k), N(k), \{r_a\}_{a\in\Delta})$ is a saturated Tits system with Weyl group $N(k)/Z(k)$ (as $N \cap P = Z$, since $W(\Phi)$ acts freely on the set of positive systems of roots in Φ). But $N(k)/Z(k) = W(\Phi)$ by Proposition 4.1.3, so the Bruhat decomposition in (ii) is a consequence of the Bruhat decomposition for groups equipped with a Tits system [**Bou**, Ch. IV, §2.3, Thm. 1]. □

4.2. Pseudo-completeness. A lacuna in our formulation of the Bruhat decomposition for $G(k)$ in the pseudo-split pseudo-reductive case in Theorem 4.1.7 is that we have not yet proved $G(k)$-conjugacy of all pairs (T, P) (with minimal P). A new concept will be required in order to settle this issue.

To motivate where the difficulty lies, recall that in the split connected reductive case such conjugacy results are proved via Borel's fixed point theorem for the action of a k-split solvable smooth connected affine group on a proper k-scheme; the proper k-scheme to which this is applied is G/P. But in the pseudo-split pseudo-reductive case the quotient G/P modulo a proper pseudo-parabolic k-subgroup P is generally *not* proper (see [**CGP**, Ex. A.5.6]), so Borel's fixed point theorem does not apply. Fortunately, G/P satisfies a weaker property that is adequate for establishing an analogue of Borel's theorem:

DEFINITION 4.2.1. A k-scheme X is *pseudo-complete* if it is separated, of finite type, and satisfies the valuative criterion for properness with discrete valuation rings R over k whose residue field is separable over k.

This is only of interest for imperfect k, as otherwise all extensions of k are separable and hence pseudo-completeness over k recovers properness (due to the valuation criterion). By [**CGP**, Prop. C.1.2], pseudo-completeness is insensitive to separable extension of the ground field and to check pseudo-completness we only need to consider those R that are also complete and have separably closed residue field. Arguments with Artin approximation imply that it is even enough to consider only $R = k_s[\![x]\!]$ (see [**CGP**, Rem. C.1.4]); we will never use this, but it recovers the definition considered by Tits.

EXAMPLE 4.2.2. Let k'/k be a finite extension of fields, G' a connected reductive k'-group, and $P' \subset G'$ a proper parabolic k'-subgroup. Let $G = \mathrm{R}_{k'/k}(G')$ and $P = \mathrm{R}_{k'/k}(P')$, so G is pseudo-reductive over k and P is a proper pseudo-parabolic k-subgroup of G (see [**CGP**, Prop. 2.2.13]). The quotient G/P is identified with $\mathrm{R}_{k'/k}(G'/P')$ where G'/P' is smooth and projective with positive dimension. Hence, if k'/k is not separable then G/P is *never* proper (see [**CGP**, Ex. A.5.6]). Nonetheless, G/P is pseudo-complete.

More generally, if X' is a projective k'-scheme then we claim that the separated k-scheme $\mathrm{R}_{k'/k}(X')$ of finite type is pseudo-complete. Since pseudo-completness is insensitive to separable extension on k, we may extend scalars to k_s at the cost of replacing k' with the individual factor fields of $k' \otimes_k k_s$ (and X' with its base change over such fields) to reduce to the case that k'/k is purely inseparable. By definition, we need to show that if A is a discrete valuation ring over k with fraction field K such that the residue field F of A is separable over k then $\mathrm{R}_{k'/k}(X')(A) = \mathrm{R}_{k'/k}(X')(K)$, or equivalently $X'(k' \otimes_k A) = X'(k' \otimes_k K)$. Since X' is pseudo-complete over k', it suffices to show that if k/k is any purely inseparable extension then $\mathsf{A} := \mathsf{k} \otimes_k A$ is a discrete valuation ring with fraction field $\mathsf{k} \otimes_k K$ and residue field $\mathsf{k} \otimes_k F$. For a uniformizer t of A it suffices to prove every nonzero element of A is a unit multiple of t^n for a unique $n \geqslant 0$, so we may assume $[\mathsf{k} : k] < \infty$. But then A is visibly noetherian and local with $1 \otimes t$ a non-nilpotent element generating the maximal ideal (as $\mathsf{k} \otimes_k \kappa$ is a field, since κ/k is separable), so A is a discrete valuation ring by [**Ser**, Prop. 2, §2, Ch. I].

The proof of Borel's fixed point theorem for a k-split solvable smooth connected affine k-group acting on a proper k-scheme with a k-point involves extending to $\mathbf{P}^1$ certain k-scheme maps from $\mathbf{G}_{\mathrm{a}}$ or GL_1. By elementary denominator-chasing, these extension problems only involve the completed local ring $k[\![x]\!]$ at 0 or ∞, so we only need to work with $R = k[\![x]\!]$ to construct the desired extension. This establishes:

PROPOSITION 4.2.3. *If H is a k-split solvable smooth connected affine k-group and X is a pseudo-complete k-scheme equipped with an action by H such that $X(k) \neq \emptyset$ then $X(k)$ contains a point fixed by H.*

Pseudo-completeness underlies a generalization [**CGP**, Prop. C.1.6] of Example 4.2.2:

THEOREM 4.2.4. *If P is a pseudo-parabolic k-subgroup of a smooth connected affine k-group G then G/P is pseudo-complete over k.*

PROOF. We may assume $k = k_s$ and G is pseudo-reductive (as $\mathscr{R}_{u,k}(G) \subset P$ by definition of pseudo-parabolicity, with $P/\mathscr{R}_{u,k}(G)$ pseudo-parabolic in $G/\mathscr{R}_{u,k}(G)$ by Proposition 2.3.8). Let G' denote the maximal geometric reductive quotient $G^{\mathrm{red}}_{\overline{k}}$ of G, and let P' be the image of $P_{\overline{k}}$ in G'. If k is perfect, so $k = \overline{k}$, then there is nothing to do because over k pseudo-completeness coincides with properness and pseudo-parabolicity in G coincides with parabolicity by [**CGP**, Prop. 2.2.9]. Hence, we may assume $p = \mathrm{char}(k) > 0$.

We explain why P' is parabolic in G', and refer the reader to the proof of [**CGP**, Prop. C.1.6] for the rest of the argument. By definition, $P = P_G(\lambda)$ for some $\lambda : \mathrm{GL}_1 \to G$. For the maximal geometric reductive quotient $G' = G^{\mathrm{red}}_{\overline{k}}$ of $G_{\overline{k}}$, the image P' of $P_{\overline{k}}$ in G' is $P_{G'}(\lambda_{\overline{k}})$ by Proposition 2.3.8. Thus, P' is parabolic in G' [**CGP**, Prop. 2.2.9]. □

REMARK 4.2.5. It is natural to ask if the converse to Theorem 4.2.4 holds (providing a "geometric" characterization of pseudo-parabolicity). Unfortunately, the converse essentially always fails away from the reductive case, thereby explaining why pseudo-parabolicity is developed via dynamic rather than geometric means.

To make this failure precise, assume G is pseudo-reductive (a harmless hypothesis since $P \mapsto P/\mathscr{R}_{u,k}(G)$ is a bijection between the sets of pseudo-parabolic k-subgroups of G and $G/\mathscr{R}_{u,k}(G)$ [**CGP**, Prop. 2.2.10]). By [**CGP**, Thm. C.1.9], the following two conditions are equivalent:

(i) the smooth closed k-subgroups Q of G for which G/Q is pseudo-complete are precisely the pseudo-parabolic k-subgroups,
(ii) every Cartan k-subgroup of G is a torus.

Since parabolicity and pseudo-parabolicity are the same in the connected reductive case [**CGP**, Prop. 2.2.9], it follows that for connected reductive G the parabolic k-subgroups are precisely the smooth closed k-subgroups $Q \subset G$ such that G/Q is pseudo-complete. If instead G is pseudo-reductive but *not* reductive (so k is imperfect) then the equivalent conditions (i) and (ii) *always* fail except for precisely the special cases to be described in Theorem 7.3.3 that occur over k if and only if k is imperfect with characteristic 2.

As an application of the analogue of the Borel fixed point theorem in the pseudo-complete setting (Proposition 4.2.3), we can establish rational conjugacy theorems in the smooth connected affine case that generalize well-known results in the connected reductive case. To get started, we require a nontrivial lemma:

LEMMA 4.2.6. *If G is a smooth connected affine k-group and P is a pseudo-parabolic k-subgroup then $G(k) \to (G/P)(k)$ is surjective.*

The connected reductive case is part of [**Bo2**, 20.5].

PROOF. This problem is not easily reduced to the pseudo-reductive case because $\mathscr{R}_{u,k}(G)$ might not be k-split. The general case is treated in [**CGP**, Lemma C.2.1], and here we give a proof when G is pseudo-reductive with $G(k)$ Zariski-dense in G. (This case plays a role in the proof for general G, via an inductive argument to handle the possibility that $G(k)$ may not be Zariski-dense in G, as can happen even for pseudo-reductive G over infinite k; see Example 2.1.5.)

Assume G is pseudo-reductive, so $P = P_G(\lambda)$ for some $\lambda : \mathrm{GL}_1 \to G$, and that $G(k)$ is Zariski-dense in G. For the dense open subscheme $\Omega := U_G(-\lambda) \times P \subset G$ (via multiplication), the translates $\{g\Omega\}_{g \in G(k)}$ constitute an open cover of G (as this can be checked on $\overline{k}$-points, using that $G(k)$ is Zariski-dense in $G_{\overline{k}}$). Passing to the quotient modulo P, for the dense open $\overline{\Omega} := \Omega/P \subset G/P$ the translates $\{g\overline{\Omega}\}_{g \in G(k)}$ constitute an open cover of G/P. But $U_G(-\lambda)(k) \to \overline{\Omega}(k)$ is bijective, so we are done for such G. □

LEMMA 4.2.7. *Let G be a smooth connected affine k-group and P a pseudo-parabolic k-subgroup. Every k-split solvable smooth connected k-subgroup $H \subset G$ admits a $G(k)$-conjugate contained in P. In particular, P contains a $G(k)$-conjugate of any split k-torus $S \subset G$.*

PROOF. For $g \in G(k)$ we have $g^{-1}Hg \subset P$ if and only if $HgP \subset gP$, which is to say that the image of g in $(G/P)(k)$ is fixed under the left H-action. But G/P is pseudo-complete by Theorem 4.2.4, so the fixed point theorem (Proposition 4.2.3) provides a point in $(G/P)(k)$ fixed by the left H-action. By Lemma 4.2.6, this k-point of G/P lifts to $G(k)$, so we get the desired $G(k)$-conjugate of H. □

PROPOSITION 4.2.8. *Let G be a pseudo-split pseudo-reductive k-group. A pseudo-parabolic k-subgroup P of G is minimal if and only if $P/\mathscr{R}_{u,k}(P)$ is commutative.*

PROOF. A split maximal k-torus of G admits a $G(k)$-conjugate contained in P by Lemma 4.2.7, so P contains a split maximal k-torus T of G. Choose $\lambda \in \mathrm{X}_*(T)$ such that $P = P_G(\lambda)$; such λ exists by Lemma 2.3.9. Clearly $P = Z_G(\lambda) \ltimes U_G(\lambda)$ and $\mathscr{R}_{u,k}(P) = U_G(\lambda)$, so $P/\mathscr{R}_{u,k}(P) \simeq Z_G(\lambda)$.

As $T \subset Z_G(T) \subset Z_G(\lambda)$ and the Cartan k-subgroup $Z_G(T)$ is commutative, $Z_G(\lambda)$ is commutative if and only if the inclusion $Z_G(T) \subset Z_G(\lambda)$ is an equality. Such equality of smooth connected groups is equivalent to equality of their Lie algebras. These Lie algebras coincide if and only if $\Phi_{\lambda=0}$ is empty, which is the case if and only if $\Phi_{\lambda>0}$ is a positive system of roots. Applying Proposition 3.3.7 therefore finishes the proof. □

THEOREM 4.2.9 (Borel–Tits). *Any two maximal split k-tori in a smooth connected affine k-group G are conjugate under $G(k)$.*

PROOF. We proceed by induction on $\dim G$, the 0-dimensional case being clear. If G admits a proper pseudo-parabolic k-subgroup P then by Lemma 4.2.7 every split k-torus in G admits a $G(k)$-conjugate contained in P. Thus, we may rename P as G and conclude by induction on dimension. Hence, we may assume that G does not contain a proper pseudo-parabolic k-subgroup, so the pseudo-reductive quotient $\overline{G} := G/\mathscr{R}_{u,k}(G)$ also does not contain a proper pseudo-parabolic k-subgroup [**CGP**, Prop. 2.2.10]. In other words, for every k-homomorphism $\lambda : \mathrm{GL}_1 \to \overline{G}$ we have

$$\overline{G} = P_{\overline{G}}(-\lambda) = Z_{\overline{G}}(-\lambda) \ltimes U_{\overline{G}}(-\lambda),$$

so $U_{\overline{G}}(\lambda) = 1$. Likewise, $U_{\overline{G}}(-\lambda) = 1$, so $\overline{G} = Z_{\overline{G}}(\lambda)$. This says that *every* λ is central in $\overline{G}$, so every k-split torus in $\overline{G}$ is central! In particular, there is a unique maximal k-split torus $\overline{S}$ in $\overline{G}$ and it is central.

Consider the preimage H of $\overline{S}$ under $\pi : G \twoheadrightarrow \overline{G}$. It is clear that every k-split torus in G must be carried by π into $\overline{S}$ and so lies inside H. Thus, the problem for G reduces to the same for H. But H is a smooth connected solvable k-group, and in such a group any two maximal k-tori (and hence any two maximal split k-tori) are conjugate to each other by an element of $H(k)$ [**Bo2**, 19.2]. □

4.3. Properties of pseudo-parabolic subgroups. In addition to torus centralizers, proper parabolic k-subgroups P in connected reductive k-groups are a useful source of inductive arguments since $\mathscr{R}_{u,k}(P)$ is k-split and $P/\mathscr{R}_{u,k}(P)$ is reductive of smaller dimension. The relative root system ${}_k\Phi = \Phi(G,S)$ for a maximal split k-torus S (all choices of which are $G(k)$-conjugate to each other) controls the collection of P's containing S as well as the structure of $\mathscr{R}_{u,k}(P)$ in terms of S-root groups for such P. These root groups can have large dimension and ${}_k\Phi$ can be non-reduced (for k of any characteristic, even $k = \mathbf{R}$). For semisimple G, k-anisotropicity is equivalent to G having no proper parabolic k-subgroup.

The analogous notion of relative roots for pseudo-reductive G will be discussed in §5.3, resting on a robust theory of pseudo-parabolic subgroups in the pseudo-split case (such as over k_s). In this section we will address several basic structural results for pseudo-parabolic k-subgroups of pseudo-reductive groups, sometimes in the pseudo-split case and sometimes more generally. Everything we do in the pseudo-split case here will be extended to the general case in §5.3–§5.4.

PROPOSITION 4.3.1. *Consider a pseudo-reductive k-group G containing a split maximal k-torus T. Let $\Phi = \Phi(G,T)$ and P be a pseudo-parabolic k-subgroup of G containing T.*

(i) *The subspace* $\mathrm{Lie}(P) \subset \mathrm{Lie}(G)$ *is the span of* $\mathrm{Lie}(Z_G(T))$ *and the T-weight spaces for roots in $\Phi(P,T)$, and if P' is a second pseudo-parabolic k-subgroup containing T then $P' = P$ if and only if* $\mathrm{Lie}(P') = \mathrm{Lie}(P)$.
(ii) *If A is the subsemigroup of Φ spanned by the set Ψ of non-divisible roots in $\Phi(P,T)$ outside $-\Phi(P,T)$ then $\mathscr{R}_{u,k}(P) = H_A(G)$ is the k-subgroup directly spanned by the root groups for roots in Ψ.*
(iii) *For pseudo-parabolic k-subgroups P, Q containing T, the following are equivalent: $P \subset Q$,* $\mathrm{Lie}(P) \subset \mathrm{Lie}(Q)$, $\Phi(P,T) \subset \Phi(Q,T)$.

This result (along with Proposition 4.2.8) is [**CGP**, Prop. 3.5.1], for which $\mathrm{Lie}(Z_G(T))$ in (i) is mistakenly written as $\mathrm{Lie}(T)$, a typographical error not affecting the proof there.

PROOF. The essential point is to reconstruct P from the set $\Phi(P,T)$ of non-trivial T-weights on $\mathrm{Lie}(P)$. By Lemma 2.3.9, there exists $\lambda \in \mathrm{X}_*(T)$ such that $P = P_G(\lambda) = Z_G(\lambda) \ltimes U_G(\lambda)$, so $U_G(\lambda) = \mathscr{R}_{u,k}(P)$. By Example 3.3.6 we have $\Phi(P,T) = \Phi_{\lambda \geqslant 0}$. For each $a \in \Phi(P,T)$, the dynamic definition of $U_{(a)}$ and the parabolicity of $\Phi(P,T)$ imply that $U_{(a)}^G = U_{(a)}^P \subset P$. Since $Z_G(T) \subset Z_G(\lambda) \subset P$, so $Z_G(T) = Z_P(T)$, by applying Proposition 3.1.4 to P we conclude that P is generated by $Z_G(T)$ and $\{U_{(a)}^G\}_{a \in \Phi(P,T)}$. Passing to Lie algebras yields (i) and (iii). In the setting of (ii) we have $A \cap \Phi = \Phi_{\lambda > 0}$, so $\mathscr{R}_{u,k}(P) = U_G(\lambda) \subset H_A(G)$ by the maximality property of $H_A(G)$ in Proposition 3.3.1. This containment between smooth connected k-subgroups of G induces an equality on Lie algebras, so it is an equality of k-subgroups, establishing (ii). □

REMARK 4.3.2. Based on experience in the reductive case, it is natural to inquire if the equivalence of "$P \subset Q$" and "$\mathrm{Lie}(P) \subset \mathrm{Lie}(Q)$" in Proposition 4.3.1(iii) is valid more generally for arbitrary pseudo-parabolic k-subgroups P, Q in a pseudo-reductive k-group G *without* assuming P and Q share a common split maximal k-torus. The answer is affirmative; see Proposition 5.1.4(i) (whoose proof uses Proposition 4.3.1(iii) over k_s).

PROPOSITION 4.3.3. *Let G be a pseudo-reductive k-group with a split maximal k-torus T. Let $G' = G_{\overline{k}}^{\mathrm{red}}$, and let $T' \subset G'$ be the (isomorphic) image of $T_{\overline{k}}$. Assigning to each pseudo-parabolic k-subgroup $P \subset G$ containing T the image P' of $P_{\overline{k}} \to G' := G_{\overline{k}}^{\mathrm{red}}$ is an inclusion-preserving bijection in both directions between the set of such P and the set of parabolic subgroups of G' that contain T'. Moreover, $\Phi(P',T') = \Phi(P,T) \cap \Phi(G',T')$ inside* $\mathrm{X}(T) = \mathrm{X}(T')$.

PROOF. By Lemma 2.3.9, the k-groups P are exactly $P_G(\lambda)$ for $\lambda \in \mathrm{X}_*(T)$. In particular, the image P' of $P_{\overline{k}}$ in G' is $P_{G'}(\lambda_{\overline{k}})$ by Proposition 2.3.8 (applied to $G_{\overline{k}} \twoheadrightarrow G'$). The parabolic subgroups of G' containing T' are exactly $P_{G'}(\mu)$ for $\mu \in \mathrm{X}_*(T')$, by [**CGP**, Prop. 2.2.9]. Since $\mathrm{X}_*(T) = \mathrm{X}_*(T_{\overline{k}})$ via $\lambda \mapsto \lambda_{\overline{k}}$, as we vary $P \supset T$ the associated subgroups $P' \subset G'$ vary through precisely the parabolic subgroups of G' containing T'.

Let $\Phi = \Phi(G,T)$ and $\Phi' = \Phi(G',T')$, so Φ' is the set of non-multipliable elements of Φ (Theorem 3.1.7). For any root system spanning a vector space V, the parabolic subsets are precisely those with non-negative pairing against a linear form [**CGP**, Prop. 2.2.8], so $\Psi \mapsto \Psi \cap \Phi'$ is an inclusion-preserving bijection (in both directions) between the sets of parabolic subsets of Φ and of Φ'. By Proposition 4.3.1(iii), if $P, Q \subset G$ are pseudo-parabolic k-subgroups containing T, then $P \subset Q$ if and only if $\Phi(P,T) \subset \Phi(Q,T)$ inside $\Phi(G,T)$. Thus, to complete the proof we just have to establish the formula $\Phi(P,T) \cap \Phi' = \Phi(P',T')$. Writing $P = P_G(\lambda)$ with $\lambda \in \mathrm{X}_*(T)$, we have $P' = P_{G'}(\lambda')$ for $\lambda' = \lambda_{\overline{k}}$. Thus, $\Phi(P,T) = \Phi_{\lambda \geqslant 0}$ and $\Phi(P',T') = \Phi'_{\lambda' \geqslant 0}$. Since $\Phi_{\lambda \geqslant 0} \cap \Phi' = \Phi'_{\lambda' \geqslant 0}$, we are done. □

Now we are finally in a position to address some subtle points that were noted at the end of §2.3: does pseudo-parabolicity descend through separable extension of the ground field, and is it transitive with respect to subgroup inclusions? Fortunately, both answers are affirmative. We begin with separable extension of the ground field, as an application of Proposition 4.3.1.

PROPOSITION 4.3.4. *Let G be a smooth connected affine k-group, P a smooth connected k-subgroup, and K/k a separable extension of fields. Then P is pseudo-parabolic in G if and only if P_K is pseudo-parabolic in G_K, and for a maximal k-torus $T \subset G$ the map $P \mapsto \Phi((P/\mathscr{R}_{u,k}(G))_{k_s}, T_{k_s})$ is a bijection between the set of pseudo-parabolic k-subgroups of G containing T and the set of* $\mathrm{Gal}(k_s/k)$*-stable parabolic sets of roots in* $\Phi((G/\mathscr{R}_{u,k}(G))_{k_s}, T_{k_s})$.

PROOF. By Galois descent and Proposition 4.3.1(i), the bijectivity assertion follows from the equivalence of pseudo-parabolicity for P and P_K. Since it is obvious that P_K is pseudo-parabolic when P is pseudo-parabolic, we assume P_K is pseudo-parabolic and must show that P is pseudo-parabolic. By definition of pseudo-parabolicity and the equality in (1.2.1.1) for separable K/k, we have

$$\mathscr{R}_{u,k}(G)_K = \mathscr{R}_{u,K}(G_K) \subset P_K,$$

so $\mathscr{R}_{u,k}(G) \subset P$. Thus, we may pass to $G/\mathscr{R}_{u,k}(G)$ so that G is pseudo-reductive.

Suppose $k = k_s$, and choose a maximal k-torus $T \subset P$. By Lemma 2.3.9 (applied to the pseudo-parabolic K-subgroup $P_K \subset G_K$ containing the K-split T_K) there exists $\lambda \in \mathrm{X}_*(T_K) = \mathrm{X}_*(T)$ such that $P_K = P_{G_K}(\lambda_K) = P_G(\lambda)_K$, so $P = P_G(\lambda)$ is pseudo-parabolic as desired. Hence, our remaining problem for general k is one of descent from k_s to k.

Let T be a (possibly non-split) maximal k-torus in P, so it is also maximal in G. By Lemma 2.3.9, we may write $P_{k_s} = P_{G_{k_s}}(\mu)$ for some $\mu \in \mathrm{X}_*(T_{k_s})$, and the problem is that μ might not be $\mathrm{Gal}(k_s/k)$-invariant. Indeed, if μ were Galois-invariant then it would descend to a k-homomorphism $\mu_0 : \mathrm{GL}_1 \to T$ and so $P_{k_s} = P_{G_{k_s}}((\mu_0)_{k_s}) = P_G(\mu_0)_{k_s}$, yielding that $P = P_G(\mu_0)$ is pseudo-parabolic as desired. To overcome this problem we shall use Proposition 4.3.1.

Let k'/k be a finite Galois extension splitting T, so a k'-homomorphism $\mu' : \mathrm{GL}_1 \to T_{k'}$ exists that descends μ. For each $\sigma \in \mathrm{Gal}(k'/k)$, the natural identification of $G_{k'}$ with its σ-twist $\sigma^*(G_{k'})$ implies that as k'-subgroups of $G_{k'}$ we have

$$P_{k'} = \sigma^*(P_{k'}) = P_{G_{k'}}(\sigma.\mu').$$

Comparing Lie algebras yields

$$\Phi_{\sigma.\mu' \geqslant 0} = \Phi_{\mu' \geqslant 0},$$

so for each $a \in \Phi$ either $\langle a, \sigma.\mu' \rangle \geqslant 0$ for all σ or $\langle a, \sigma.\mu' \rangle < 0$ for all σ. Hence, for the Galois-invariant $\lambda' = \sum_\sigma \sigma.\mu'$ we have $\langle a, \lambda' \rangle \geqslant 0$ precisely when $\langle a, \mu' \rangle \geqslant 0$, which is to say $\Phi(P_{k'}, T_{k'}) = \Phi_{\lambda' \geqslant 0} = \Phi(P_{G_{k'}}(\lambda'), T_{k'})$. By Proposition 4.3.1(i) it follows that $P_{k'} = P_{G_{k'}}(\lambda')$. As we saw above, this implies the pseudo-parabolicity of the k-subgroup P in G since λ' is $\mathrm{Gal}(k'/k)$-invariant. □

As an application of the two preceding propositions, we can establish the transitivity of pseudo-parabolicity:

COROLLARY 4.3.5. *Let P be a pseudo-parabolic k-subgroup of a smooth connected affine k-group G. A smooth connected k-subgroup Q of P is pseudo-parabolic in P if and only if it is pseudo-parabolic in G.*

The idea behind the proof of Corollary 4.3.5 is as follows. By Proposition 4.3.4 we may assume $k = k_s$ (so all k-tori are split). The argument for pseudo-reductive G involves a detailed study of root groups, building on the description of $\mathscr{R}_{u,k}(P)$

in Proposition 4.3.1(ii). The most difficult part is to show that every pseudo-parabolic k-subgroup of $P/\mathscr{R}_{u,k}(P)$ is the image of a pseudo-parabolic k-subgroup of G contained in P. This rests on a description (given in [**CGP**, Prop. 2.2.8(2)]) of parabolic sets of roots in terms of a basis of a root system (rather than via the construction $\Phi_{\lambda \geqslant 0}$). We refer the reader to [**CGP**, Lemma 3.5.5] for the details.

In the theory of connected reductive groups, it is an important theorem that every parabolic subgroup is its own normalizer. In traditional developments this is proved at the level of geometric points, and the stronger result of equality with its scheme-theoretic normalizer is [**SGA3**, XXII, 5.8.5]. (See [**CGP**, p. 469] for the existence of the scheme-theoretic normalizer of any smooth closed k-subgroup of a smooth k-group.) In the general case the same strengthened normalizer result holds for pseudo-parabolic subgroups:

PROPOSITION 4.3.6. *Every pseudo-parabolic k-subgroup P of a smooth connected affine k-group G coincides with its own scheme-theoretic normalizer.*

PROOF. We may assume $k = k_s$. For a smooth k-subgroup $H \subset G$ and $h \in H(k)$ let $f_h(g) = hgh^{-1}g^{-1}$ (so $f_h(1) = 1$). Since $H(k)$ is Zariski-dense in H, the scheme-theoretic normalizer $N_G(H)$ of H in G is $\bigcap_{h \in H(k)} f_h^{-1}(H)$ by construction. The pointed map f_h induces $\mathrm{Ad}_G(h) - \mathrm{id}$ on the tangent space $\mathrm{Lie}(G)$ at 1, so

$$\mathrm{Lie}(N_G(H)) = \{X \in \mathrm{Lie}(G) \mid \mathrm{Ad}_G(h)(X) - X \in \mathrm{Lie}(H) \text{ for all } h \in H(k)\}.$$

The first step is to show that $\mathrm{Lie}(P) = \mathrm{Lie}(N_G(P))$, which is to say that the smooth closed k-subgroup P of $N_G(P)$ has full Lie algebra and hence coincides with $N_G(P)^0$ as schemes; in particular, $N_G(P)$ would then be k-smooth. Pick a maximal k-torus $T \subset P$, so T is split. If the inclusion $\mathrm{Lie}(P) \subset \mathrm{Lie}(N_G(P))$ of T-stable subspaces of $\mathrm{Lie}(G)$ were strict then (as T is split) we could find an element X in a T-weight space of $\mathrm{Lie}(N_G(P))$ such that X is not in $\mathrm{Lie}(P)$. Let $a \in \mathrm{X}(T)$ be the eigencharacter for X. Writing $P = P_G(\lambda)\mathscr{R}_{u,k}(G)$ for some $\lambda \in \mathrm{X}_*(T)$ (Lemma 2.3.9), we have $(a(t) - 1)X \in \mathrm{Lie}(P)$ for all $t \in T(k)$. This forces $a = 1$, which is to say

$$X \in \mathrm{Lie}(G)^T = \mathrm{Lie}(Z_G(T)) \subset \mathrm{Lie}(Z_G(\lambda)) \subset \mathrm{Lie}(P),$$

a contradiction.

We have proved that $N_G(P)$ is smooth with identity component P, so since $k = k_s$ it remains only to show that any $g \in G(k)$ normalizing P lies in $P(k)$. We may pass to $G/\mathscr{R}_{u,k}(G)$ since $\mathscr{R}_{u,k}(G) \subset P$, so now G is pseudo-reductive. Since $P(k)$ is Zariski-dense in P (as $k = k_s$), it is the same to show that $P(k)$ is its own normalizer in $G(k)$ (i.e., every $g \in G(k)$ satisfying $gP(k)g^{-1} = P(k)$ lies in $P(k)$). Choose a minimal pseudo-parabolic k-subgroup $B \subset P$ of G such that B contains T, as may be done by Propositions 3.3.7, 4.3.1, and 4.3.4. By Theorem 4.1.7, $(B(k), N_G(T)(k))$ is a BN-pair for $G(k)$. For any group G equipped with a BN-pair (B, N), every subgroup $\mathsf{P} \subset \mathsf{G}$ containing B is equal to its own normalizer in G [**Bou**, IV, §2.6, Thm. 4(iv)]. Thus, we are done. □

In Remark 2.3.10(iii) we noted that (in contrast with parabolicity) it is not at all obvious if a smooth closed subgroup Q of a smooth connected affine k-group G is necessarily pseudo-parabolic when it contains a pseudo-parabolic k-subgroup. As in the reductive case, the answer is affirmative:

PROPOSITION 4.3.7. *A smooth closed k-subgroup Q of a smooth connected affine k-group G is pseudo-parabolic if it contains a pseudo-parabolic k-subgroup $P \subset G$.*

PROOF. There is no harm in assuming $k = k_s$, shrinking P to be minimal, and passing to $G/\mathscr{R}_{u,k}(G)$, so G is pseudo-reductive. Choose a maximal k-torus $T \subset P$, so $\Phi(P,T)$ is a positive system of roots in G. Writing $P = P_G(\lambda)$ for some $\lambda \in \mathrm{X}_*(T)$ not vanishing on Φ, the image P' of $P_{\overline{k}}$ in $G_{\overline{k}}^{\mathrm{red}}$ is the (pseudo-)parabolic subgroup $P_{G_{\overline{k}}^{\mathrm{red}}}(\lambda_{\overline{k}})$ that is visibly a Borel subgroup (as $\Phi(G,T)$ and $\Phi(G_{\overline{k}}^{\mathrm{red}}, T_{\overline{k}})$ coincide up to rational multipliers). Hence, the image Q' of $Q_{\overline{k}}$ in $G_{\overline{k}}^{\mathrm{red}}$ contains a Borel subgroup, so it is parabolic. Thus, Q' corresponds to a parabolic set of roots in $\Phi(G_{\overline{k}}^{\mathrm{red}}, T_{\overline{k}})$ containing $\Phi(P', T_{\overline{k}})$; i.e., it is the set of roots with non-negative pairing against some $\mu \in \mathrm{X}_*(T_{\overline{k}}) = \mathrm{X}_*(T)$.

The idea is to use an analysis of root groups to show that $P_G(\mu) \subset Q$, and to show that this inclusion is an equality on Lie algebras, so $P_G(\mu) = Q^0$. It would then follow that Q normalizes the pseudo-parabolic $P_G(\mu)$, so Proposition 4.3.6 would imply that $Q = P_G(\mu)$. The study of root groups to compare $P_G(\mu)$ and Q rests on first showing that the natural map $W(Q,T) \to W(Q', T_{\overline{k}})$ is an isomorphism (and using transtivity of the $W(Q', T_{\overline{k}})$-action on the set of positive systems of roots in $\Phi({Q'}^{\mathrm{red}}, T_{\overline{k}})$); see the proof of [**CGP**, Prop. 3.5.8] for the details. □

5. Refined structure theory

5.1. Further rational conjugacy. As a supplement to Theorem 4.2.9 we wish to establish $G(k)$-conjugacy of all minimal pseudo-parabolic k-subgroups, as well as rational conjugacy for maximal k-split unipotent and maximal k-split solvable smooth connected k-subgroups. We begin with two preliminary results, the first of which is an application of Tits' structure theory for unipotent groups in positive characteristic [**CGP**, App. B]:

THEOREM 5.1.1 (Tits). *For any smooth connected affine k-group H, the formation of its maximal k-split smooth connected unipotent normal k-subgroup $\mathscr{R}_{us,k}(H)$ commutes with separable extension on k and $\mathscr{R}_{u,k}(H)/\mathscr{R}_{us,k}(H)$ does not contain $\mathbf{G}_{\mathrm{a}}$ as a k-group.*

See [**CGP**, Cor. B.3.5] for a proof. In general, a smooth connected unipotent k-group U not containing $\mathbf{G}_{\mathrm{a}}$ as a k-subgroup is called *k-wound* (see [**CGP**, Def. B.2.1, Prop. B.3.2] for alternative characterizations). A very useful property of such k-groups is that they admit *no nontrivial action* by a k-torus [**CGP**, Prop. B.4.4]. This is used in the proof of:

PROPOSITION 5.1.2. *For a smooth connected affine k-group G, pseudo-parabolic k-subgroup P, and a maximal split k-torus S in P, the centralizer $Z_G(S)$ is contained in P (so S is maximal as a k-split torus in G). Moreover, P is minimal in G if and only if $P = Z_G(S)\mathscr{R}_{us,k}(P)$.*

In the reductive case this is a well-known result (essentially part of [**Bo2**, 20.6]).

PROOF. Assuming $Z_G(S) \subset P$, so $Z_G(S)$ normalizes every normal k-subgroup of P, we claim that $Z_G(S)\mathscr{R}_{us,k}(P) = Z_G(S)\mathscr{R}_{u,k}(P)$ or equivalently that the image of S in $P/\mathscr{R}_{us,k}(P)$ centralizes $U := \mathscr{R}_{u,k}(P)/\mathscr{R}_{us,k}(P)$. But U is k-wound, so any action on it by a k-torus must be trivial [**CGP**, Prop. B.4.4].

For the rest of the argument we may and do consider $\mathscr{R}_{u,k}(P)$ instead of $\mathscr{R}_{us,k}(P)$. Since $\mathscr{R}_{u,k}(G) \subset \mathscr{R}_{u,k}(P)$ by definition of pseudo-parabolicity, and the image of S in $P/\mathscr{R}_{u,k}(G)$ is clearly a maximal split k-torus, we can pass to $G/\mathscr{R}_{u,k}(G)$ so that G is pseudo-reductive. Now $P = P_G(\lambda)$ for some $\lambda : \mathrm{GL}_1 \to P$, so $\mathscr{R}_{u,k}(P) = U_G(\lambda)$. Let T be a maximal k-torus of P containing S. We can assume that λ is valued in T (Lemma 2.3.9) and then, as S is the maximal split subtorus of T, λ is actually valued in S. Thus, $Z_G(S) \subset Z_G(\lambda) \subset P_G(\lambda) = P$. In particular, $Z_G(S) = Z_P(S)$.

The pseudo-parabolic k-subgroups of P are precisely the pseudo-parabolic k-subgroups of G that are contained in P (Corollary 4.3.5), so P is minimal in G if and only if the pseudo-reductive quotient $\overline{P} := P/\mathscr{R}_{u,k}(P)$ does not contain any proper pseudo-parabolic k-subgroup. The image $\overline{S}$ of S in $\overline{P}$ is clearly a maximal split k-torus in $\overline{P}$, and $Z_P(S) \to Z_{\overline{P}}(\overline{S})$ is surjective. Thus, we may rename $\overline{P}$ as G to reduce to showing that if G is pseudo-reductive then it has no non-central split k-tori if and only if it has no proper pseudo-parabolic k-subgroup. This equivalence is [**CGP**, Lemma 2.2.3(1)]. □

THEOREM 5.1.3 (Borel–Tits). *The minimal pseudo-parabolic k-subgroups of a smooth connected affine k-group G are pairwise $G(k)$-conjugate.*

PROOF. Let P be a minimal pseudo-parabolic k-subgroup of G and let Q be any pseudo-parabolic k-subgroup of G. We seek a $G(k)$-conjugate of P that is contained in Q. By Proposition 5.1.2, $P = Z_G(S)U$ for a maximal split k-torus $S \subset G$ and $U := \mathscr{R}_{us,k}(P)$. The smooth connected affine k-group $H := SU = S \ltimes U$ is k-split solvable, so by Lemma 4.2.7 applied to H and Q we can replace the triple (P, S, U) by a suitable $G(k)$-conjugate so that $H \subset Q$. But maximality of S in G implies that S is a maximal split k-torus in Q, so Q contains $Z_G(S)$ by Proposition 5.1.2. Hence, Q contains $Z_G(S)U = P$. □

In §5.2–§5.4 we will extend to general smooth connected affine k-groups the Borel–Tits structure theory of arbitrary connected reductive k-groups (replacing parabolic k-subgroups with pseudo-parabolic k-subgroups). This requires the following generalization of a well-known result in the reductive case:

PROPOSITION 5.1.4. *Consider pseudo-parabolic k-subgroups P, Q of a smooth connected affine k-group G.*

(i) *The k-group scheme $P \cap Q$ is smooth and connected, and its maximal k-tori are maximal in G. Moreover, $P \subset Q$ if and only if* $\mathrm{Lie}(P) \subset \mathrm{Lie}(Q)$.
(ii) *The image of $P \cap Q$ in $P/\mathscr{R}_{u,k}(P)$ is pseudo-parabolic.*
(iii) *If $P \cap Q$ is pseudo-parabolic then the k-subgroups P and Q are $G(k)$-conjugate if and only if $P = Q$.*

PROOF. Without loss of generality we may assume $k = k_s$ and G is pseudo-reductive. We shall first use the pseudo-split Bruhat decomposition in Theorem 4.1.7 to find a (split) maximal k-torus T of G contained in both P and Q. For this purpose, there is no harm in first shrinking P and Q to be minimal. Now choose a maximal k-torus S of G contained in P; we shall find a $P(k)$-conjugate of S contained in Q. By Theorem 5.1.3, $Q = gPg^{-1}$ for some $g \in G(k)$ since P and Q are minimal. The pseudo-split Bruhat decomposition provides $p, p' \in P(k)$ and $n \in N_G(S)(k)$ such that $g = pnp'$. Hence, $Q = pnPn^{-1}p^{-1}$, so Q

contains $pnSn^{-1}p^{-1} = pSp^{-1}$. But clearly P also contains pSp^{-1}, so $T := pSp^{-1}$ is contained in both P and Q.

We can describe P and Q in terms of the Cartan k-subgroup $Z_G(T)$ and suitable T-root groups. This provides a mechanism one can use to prove smoothness of $P \cap Q$ in (i) via the study of Lie algebras as T-representation spaces, after which the final assertion in (i) is clear since $\mathrm{Lie}(P \cap Q) = \mathrm{Lie}(P) \cap \mathrm{Lie}(Q)$. A lengthy analysis of root systems and root groups is required to deduce (ii). The reader is referred to the proof of [**CGP**, Prop. 3.5.12] for the details. □

The results on rational conjugacy of maximal split tori (Theorem 4.2.9) and minimal pseudo-parabolic subgroups (Theorem 5.1.3) admit analogues announced by Borel and Tits for maximal split (smooth connected) unipotent subgroups and maximal split (smooth connected) solvable subgroups. The essential step is to prove that the maximal k-split smooth connected unipotent k-subgroups of a pseudo-reductive k-group are precisely $\mathscr{R}_{us,k}(P)$ for minimal pseudo-parabolic k-subgroups P. A proof inspired by ideas of Kempf [**Kem**] is given in [**CGP**, C.3], to which we also refer for a complete discussion of the following consequences:

THEOREM 5.1.5. *Let U be a k-split smooth connected unipotent k-subgroup of a pseudo-reductive k-group G, and let H be a (possibly disconnected) smooth closed k-subgroup of G normalizing U. There exists a pseudo-parabolic k-subgroup P of G containing H such that $U \subset \mathscr{R}_{us,k}(P)$.*

In the special case $H = 1$ this says that there exists a pseudo-parabolic k-subgroup P satisfying $U \subset \mathscr{R}_{us,k}(P)$. But if a k-subgroup $Q \subset P$ is pseudo-parabolic (either in P or in G, equivalent conditions on Q by Corollary 4.3.5) then $\mathscr{R}_{us,k}(P)$ is normal in Q and hence by computing over k_s we see that $\mathscr{R}_{us,k}(P) \subset \mathscr{R}_{us,k}(Q)$.

Thus, a special case of Theorem 5.1.5 is that the maximal k-split smooth connected unipotent k-subgroups U in a pseudo-reductive k-group G are precisely $\mathscr{R}_{us,k}(Q)$ for the minimal pseudo-parabolic k-subgroups Q of G, all of which are $G(k)$-conjugate to each other by Theorem 5.1.3. (In Corollary 5.1.7 we will see that this description of the maximal U remains valid *without* a pseudo-reductivity hypothesis on G.)

It follows that if G is an arbitrary smooth connected affine k-group with *no* proper pseudo-parabolic k-subgroup (equivalently, all k-split tori in $G/\mathscr{R}_{u,k}(G)$ are central [**CGP**, Lemma 2.2.3(1)]) then the image in $G/\mathscr{R}_{u,k}(G)$ of any k-split smooth connected unipotent k-group $U \subset G$ must be trivial and so $U \subset \mathscr{R}_{u,k}(G)$. But $\mathscr{R}_{u,k}(G)/\mathscr{R}_{us,k}(G)$ is k-wound, so it receives no nontrivial k-homomorphism from U, forcing $U \subset \mathscr{R}_{us,k}(G)$. In other words:

THEOREM 5.1.6. *If a smooth connected affine k-group G has no proper pseudo-parabolic k-subgroup then $\mathscr{R}_{us,k}(G)$ contains every k-split unipotent smooth connected k-subgroup of G. In particular, if G is pseudo-reductive then it contains a non-central* GL_1 *if and only if it contains* $\mathbf{G}_{\mathrm{a}}$ *as a k-subgroup.*

COROLLARY 5.1.7. *For a smooth connected affine k-group G, the maximal k-split smooth connected unipotent k-subgroups U of G are precisely $\mathscr{R}_{us,k}(P)$ for the minimal pseudo-parabolic k-subgroups P of G.*

PROOF. Since all such P are $G(k)$-conjugate to each other by Theorem 5.1.3, we may choose one such P and seek a $G(k)$-conjugate of U contained in $\mathscr{R}_{us,k}(P)$.

By Lemma 4.2.7, by passing to such a conjugate we can arrange that $U \subset P$. But $\overline{P} := P/\mathscr{R}_{u,k}(P)$ is a pseudo-reductive k-group with no proper pseudo-parabolic k-subgroup (due to the minimality of P and Corollary 4.3.5), so $\overline{P}$ does not contain any nontrivial k-split smooth connected unipotent k-subgroup. Hence, U has trivial image in $\overline{P}$, so $U \subset \mathscr{R}_{u,k}(P)$. The k-wound quotient $\mathscr{R}_{u,k}(P)/\mathscr{R}_{us,k}(P)$ receives no nontrivial k-homomorphism from the k-split U, so $U \subset \mathscr{R}_{us,k}(P)$. □

Beware that the assertion in Theorem 5.1.6 relating non-central split tori and split unipotent subgroups in pseudo-reductive groups has no analogue without the "split" hypothesis, even in the semisimple case. More specifically, for suitable k there exist k-anisotropic connected semisimple groups that contain (necessarily k-wound!) nontrivial smooth connected unipotent k-subgroups. Examples of adjoint type A over every local function field are given in [**CGP**, Rem. C.3.10], and examples in the simply connected case are given in [**GQ**].

Theorem 5.1.3 and Corollary 5.1.7 yield the unipotent case of:

THEOREM 5.1.8. *For a smooth connected affine k-group G, the maximal k-split unipotent smooth connected k-subgroups of G are pairwise $G(k)$-conjugate and likewise for the maximal k-split solvable smooth connected k-subgroups of G.*

By using Proposition 5.1.2, the conjugacy of maximal k-split solvable smooth connected k-subgroups can be deduced without difficulty from the conjugacy of maximal k-split unipotent smooth connected k-subgroups if G is pseudo-reductive (as pseudo-reductivity of G implies that $\mathscr{R}_{u,k}(P)$ is k-split for any pseudo-parabolic k-subgroup $P \subset G$). However, for general G we cannot pass to the pseudo-reductive case since the quotient map $G \to G/\mathscr{R}_{u,k}(G)$ can fail to be surjective on k-points when $\mathscr{R}_{u,k}(G)$ is not k-split. See [**CGP**, Thm. C.3.12] for the additional arguments to overcome this problem.

PROPOSITION 5.1.9. *Let G be a smooth connected affine k-group that is quasi-reductive (i.e., $\mathscr{R}_{us,k}(G) = 1$). Any maximal proper smooth connected k-subgroup M of G either is quasi-reductive or is pseudo-parabolic in G.*

PROOF. Assume M is not quasi-reductive, so $U := \mathscr{R}_{us,k}(M) \neq 1$. By applying Theorem 5.1.5 to the images of U and M in the maximal pseudo-reductive quotient $G/\mathscr{R}_{u,k}(G)$ of G, we obtain a pseudo-parabolic k-subgroup P of G containing M such that $U \subset \mathscr{R}_{us,k}(P)$. Since $\mathscr{R}_{us,k}(P) \neq 1$ (as $U \neq 1$), P is a proper k-subgroup of G. Thus, maximality of M implies that $M = P$. □

The following result was proved by V. V. Morozov over fields of characteristic 0 and was announced by Borel and Tits in general in [**BoTi3**].

PROPOSITION 5.1.10. *Let G be a smooth connected affine k-group. A smooth closed k-subgroup H of G is pseudo-parabolic if and only if it is the maximal smooth closed k-subgroup of G normalizing $U := \mathscr{R}_{us,k}(H^0)$.*

PROOF. If H is pseudo-parabolic in G then $H(k_s)$ is the normalizer in $G(k_s)$ of $U(k_s)$ by [**CGP**, Cor. 3.5.10], so the desired maximality property for H holds.

Now assume that H is the maximal smooth closed k-subgroup of G normalizing U. Let P be a pseudo-parabolic k-subgroup of G containing H such that $U \subset \mathscr{R}_{us,k}(P)$; the existence of such a P is easily seen by applying Theorem 5.1.5 to the images of U and H in the maximal pseudo-reductive quotient $G/\mathscr{R}_{u,k}(G)$ of G.

We will show that $U = \mathscr{R}_{us,k}(P)$, so P normalizes U. Then the maximality of H forces the inclusion $H \subset P$ to be an equality, establishing the desired converse.

Let $V = \mathscr{R}_{us,k}(P)$, and define V_0 to be the center of V if $\mathrm{char}(k) = 0$ and to be the maximal k-split smooth connected p-torsion central k-subgroup of V when $\mathrm{char}(k) = p > 0$ (so $V_0 \neq 1$ when $V \neq 1$, using [**CGP**, Cor. B.3.3] if $\mathrm{char}(k) > 0$). Iterate this for V/V_0 to obtain k-split smooth connected k-subgroups $V_0 \subset V_1 \subset \cdots \subset V_n = V$ normalized by P such that V_0 is central in V and V_j/V_{j-1} is central in V/V_{j-1} for $0 < j \leqslant n$.

Suppose V_j normalizes U, as happens for $j = 0$ (by centrality of V_0 in $V \supset U$). The smooth connected k-subgroup $\langle H, V_j \rangle \subset G$ containing H normalizes U, so the maximality hypothesis on H forces $\langle H, V_j \rangle = H$; i.e., $V_j \subset H$. But V_j is then normal in H (since $H \subset P$ and V_j is normal in P), so the k-split smooth connected unipotent V_j is contained in U. Now U/V_j makes sense and is a k-subgroup of V/V_j. If $j < n$ then the central $V_{j+1}/V_j \subset V/V_j$ certainly normalizes U/V_j, so V_{j+1} normalizes U. We may induct on j to eventually obtain that $V = V_n \subset U$, so $V = U$ as desired. □

5.2. General Bruhat decomposition. We will give a proof of a general Bruhat decomposition (announced by Borel and Tits) that removes the pseudo-split and pseudo-reductivity hypothesis in Theorem 4.1.7. This requires an important preliminary result:

PROPOSITION 5.2.1. *The intersection of two pseudo-parabolic k-subgroups in a smooth connected affine k-group G contains $Z_G(S)$ for some maximal split k-torus $S \subset G$.*

PROOF. Let P and P' be two pseudo-parabolic k-subgroups of G. By Proposition 5.1.4(i) we can find a maximal k-torus T of G contained in $P \cap P'$. Let T_0 be the maximal split k-torus in T. By Lemma 2.3.9, there exist $\lambda, \lambda' \in \mathrm{X}_*(T)$ such that $P = P_G(\lambda)\mathscr{R}_{u,k}(G)$ and $P' = P_G(\lambda')\mathscr{R}_{u,k}(G)$. The k-homomorphisms $\lambda, \lambda' : \mathrm{GL}_1 \rightrightarrows T$ are valued in T_0, so clearly $Z_G(T_0) \subset Z_G(\lambda) \cap Z_G(\lambda') \subset P \cap P'$. Let S be a maximal k-split torus of G containing T_0. Then $Z_G(S) \subset Z_G(T_0) \subset P \cap P'$. □

THEOREM 5.2.2 (General Bruhat decomposition). *Let G be a smooth connected affine k-group. For any maximal split k-torus S and minimal pseudo-parabolic k-subgroup P containing S, $G(k) = \coprod_{w \in W} P(k) n_w P(k)$ where $W := N(k)/Z(k)$ for $N = N_G(S)$ and $Z = Z_G(S)$ with $n_w \in N(k)$ a representative of $w \in W$.*

PROOF. First we show that every $P(k)$-double coset in $G(k)$ meets $N(k)$. For $g \in G(k)$, by Proposition 5.2.1 we may choose a maximal split k-torus S' of G contained in $P \cap gPg^{-1}$, so the tori S, S', and $g^{-1}S'g$ are maximal split k-tori in P. They are $P(k)$-conjugate by Theorem 4.2.9, so we obtain $p, p' \in P(k)$ such that $pS'p^{-1} = S = p'g^{-1}S'gp'^{-1}$. Hence, p^{-1} and gp'^{-1} each conjugate S into S', so $pgp'^{-1} \in N(k)$; i.e., $g \in p^{-1}N(k)p'$. (Note that the proof of Proposition 5.2.1 rests on Proposition 5.1.4(i), whose proof uses the *pseudo-split* Bruhat decomposition in Theorem 4.1.7 over k_s.)

By Proposition 5.1.2, in such cases $P = ZU$ for $U := \mathscr{R}_{u,k}(P)$. Group-theoretic manipulations resting on Theorem 4.2.9 (given at the end of the proof of [**CGP**, Thm. C.2.8]) allow one to reduce the pairwise disjointness of the double cosets to the disjointness of $P(k)nP(k)$ from $P(k)$ for any $n \in N(k) - Z(k)$. An even stronger statement is true for such n: the locally closed subset $PnP \subset G$ is disjoint from P. Equivalently, we claim that $P(\overline{k}) \cap N(\overline{k}) = Z(\overline{k})$.

That is, if an element $g \in P(\overline{k})$ normalizes $S_{\overline{k}}$ then we claim that its conjugation action on $S_{\overline{k}}$ is trivial. The natural map $S \to P/U$ is a k-subgroup inclusion since S is a torus and U is unipotent, so it suffices to check that g-conjugation on $(P/U)_{\overline{k}}$ is trivial on $S_{\overline{k}}$. But $P = ZU$ with $Z = Z_G(S)$, so this is clear. □

REMARK 5.2.3. Consider minimal P in the setting of Theorem 5.2.2. In view of the disjointness of PnP and P inside G for $n \in N(k) - Z(k)$ as established in the proof above, it is natural to ask more generally if the locally closed subsets PnP and $Pn'P$ are disjoint for $n, n' \in N(k)$ that lie in distinct $Z(k)$-cosets. This is equivalent to disjointness of sets $P(\overline{k})nP(\overline{k})$ and $P(\overline{k})n'P(\overline{k})$.

The elementary group-theoretic argument which reduces the disjointness of $P(k)nP(k)$ and $P(k)n'P(k)$ to the disjointness of $P(k)$ and $P(k)n^{-1}n'P(k)$ rests on Theorem 4.2.9 (which is sensitive to extension of the ground field) and so does not carry over to the level of geometric points. The disjointness does hold on geometric points, but its proof requires an entirely different approach, making use of dynamic considerations (especially that the open immersion in Theorem 2.3.5(ii) is an equality in the solvable case) after preliminary reduction to the pseudo-reductive case. See [**CGP**, Rem. C.2.9] for further details.

5.3. Relative roots. The structure of general connected reductive groups is controlled by relative root systems (which treat the anisotropic case as a black box), and in this section we sketch how it can be extended to arbitrary smooth connected affine groups.

As an application of rational conjugacy theorems for maximal k-split tori and minimal pseudo-parabolic k-subgroups, as well as the disjointness of P and PnP (rather than just of $P(k)$ and $P(k)nP(k)$) for minimal P and $n \in N(k) - Z(k)$ shown in the proof of Theorem 5.2.2, there is a good notion of relative Weyl group beyond the pseudo-split pseudo-reductive case (in Proposition 4.1.3):

PROPOSITION 5.3.1. *Let S be a maximal split k-torus in a smooth connected affine k-group G. The finite étale quotient $W(G,S) := N_G(S)/Z_G(S)$ is constant, and the inclusion $N_G(S)(k)/Z_G(S)(k) \hookrightarrow W(G,S)(k)$ is an equality.*

PROOF. Let $N = N_G(S)$, $Z = Z_G(S)$, and $W = W(G,S)$. For any $n \in N(k_s)$ and $\gamma \in \mathrm{Gal}(k_s/k)$, the conjugation actions of n and $\gamma(n)$ on S_{k_s} are related through γ-twisting (using the canonical k_s-isomorphism between S_{k_s} and its γ-twist), but all k_s-automorphisms of S_{k_s} descend to k-automorphisms of S because S is k-split. Hence, these two conjugations on S_{k_s} coincide, which is to say $\gamma(n)n^{-1} \in Z(k_s)$. This says exactly that $W(k_s)$ has trivial Galois action, or in other words that W is constant.

There is a natural action of $W(k) = W(k_s) = N(k_s)/Z(k_s)$ on the set $\mathscr{P}$ of minimal pseudo-parabolic k-subgroups of G containing S: if P is such a k-subgroup and $n \in N(k_s)$ then $nP_{k_s}n^{-1}$ only depends on n through its $Z(k_s)$-coset $w \in W(k)$ since $Z \subset P$ (by Proposition 5.1.2). But $\gamma(n)$ is in the same coset for all $\gamma \in \mathrm{Gal}(k_s/k)$, so $nP_{k_s}n^{-1}$ is $\mathrm{Gal}(k_s/k)$-stable inside G_{k_s} and thus descends to a pseudo-parabolic k-subgroup of G containing S. This descent is minimal in G for dimension reasons, due to Theorem 5.1.3. In this way, $W(k)$ acts on $\mathscr{P}$.

We saw in the proof of Theorem 5.2.2 that for every $P \in \mathscr{P}$ the inclusion $Z \subset P \cap N$ is an equality on $\overline{k}$-points, so the $W(k)$-action on $\mathscr{P}$ is free. Hence, to finish the proof it suffices to show that $N(k)$ acts transitively on $\mathscr{P}$, which in turn is immediate from Theorem 5.1.3 and Theorem 4.2.9 (the latter applied to an element of $\mathscr{P}$). □

We want to upgrade Proposition 5.3.1 by showing that in the pseudo-reductive case $\Phi(G,S)$ is a root system in its $\mathbf{Q}$-span and that its Weyl group is naturally identified with $W(G,S)(k)$. For later purposes with Tits systems, it is convenient to avoid a pseudo-reductivity hypothesis on G, though one cannot expect $\Phi(G,S)$ to be a root system without any hypotheses on $\mathscr{R}_{u,k}(G)$ (e.g., consider the case where G is a vector group equipped with a linear S-action). If $\mathscr{R}_{u,k}(G)$ is k-wound then it admits no non-trivial action by a k-torus, so in such cases $\mathscr{R}_{u,k}(G)$ centralizes S and hence does not contribute to $\Phi(G,S)$.

THEOREM 5.3.2. *Let G be a smooth connected affine k-group, S a maximal split k-torus, P a minimal pseudo-parabolic k-subgroup containing S.*

(i) *The set ${}_k\overline{\Phi} = \Phi(G/\mathscr{R}_{u,k}(G),S)$ is a root system in its $\mathbf{Q}$-span in $\mathrm{X}(S)_{\mathbf{Q}}$, its subset $\Phi(P/\mathscr{R}_{u,k}(G),S)$ is a positive system of roots, and the natural map $N_G(S)(k)/Z_G(S)(k) \to W({}_k\overline{\Phi})$ is an isomorphism.*

(ii) *The set of pseudo-parabolic k-subgroups of G containing S is in bijection with the set of parabolic sets of roots in ${}_k\overline{\Phi}$ via $P' \mapsto {}_k\overline{\Phi}_{P'} := \Phi(P'/\mathscr{R}_{u,k}(G),S)$, and $P' \subset P''$ if and only if ${}_k\overline{\Phi}_{P'} \subset {}_k\overline{\Phi}_{P''}$.*

(iii) *There is a root datum $({}_k\overline{\Phi}, \mathrm{X}(S), {}_k\overline{\Phi}^{\vee}, \mathrm{X}_*(S))$ using a canonically associated subset ${}_k\overline{\Phi}^{\vee} \subset \mathrm{X}_*(S) - \{0\}$.*

(iv) *Assume $\mathscr{R}_{u,k}(G)$ is k-wound. The root system ${}_k\overline{\Phi}$ consists of the nontrivial S-weights on* Lie(G) *and its $\mathbf{Q}$-span coincides with $\mathrm{X}(S')_{\mathbf{Q}}$, where S' is the subtorus $(S \cap \mathscr{D}(G))^0_{\mathrm{red}}$ in S that is an isogeny complement to the maximal split central k-torus $S_0 \subset G$. Moreover, ${}_k\overline{\Phi}^{\vee} \subset \mathrm{X}_*(S')$.*

The proof of this theorem is rather long; we refer to [**CGP**, Thm. C.2.15] for the details and explain here just two points: why $N_G(S)(k)/Z_G(S)(k)$ is unaffected by passing to $\overline{G} := G/\mathscr{R}_{u,k}(G)$ (even though $G(k) \to \overline{G}(k)$ is generally not surjective) and how coroots are built (since the method has nothing to do with a rank-1 classification as in the reductive case).

We have $N_G(S)(k)/Z_G(S)(k) = W(G,S)(k)$ by Proposition 5.3.1, and the map of finite étale (even constant) k-groups $W(G,S) \to W(\overline{G},S)$ is an isomorphism due to [**CGP**, Lemma 3.2.1] (using that $\ker(G \twoheadrightarrow \overline{G})$ is unipotent), so passing to k-points gives the desired invariance under passage to $\overline{G}$ (and so reduces the problem of relating $W(G,S)(k)$ and $W({}_k\overline{\Phi})$ to the case where G is pseudo-reductive).

For each $a \in {}_k\overline{\Phi}$, we shall define the associated cocharacter $a^{\vee} \in \mathrm{X}_*(S)$ using the scheme-theoretic kernel $\ker a$ that is of multiplicative type (contained in S) and so has smooth scheme-theoretic centralizer $Z_G(\ker a)$ [**CGP**, Prop. A.8.10(1),(2)]. For $G_a := Z_G(\ker a)^0$ containing S, by [**CGP**, Prop. A.8.14] its maximal pseudo-reductive quotient maps isomorphically onto the analogue $\overline{G}_a$ for $\overline{G}$. The centrality of $\ker a$ in G_a implies that the finite group $W(G_a,S)(k)$ has order at most 2; its order is actually 2 because $N_{G_a}(S)(k)$ acts transitively on the set of minimal pseudo-parabolic k-subgroups of G_a (Theorem 5.1.3) and there are two such subgroups [**CGP**, Lemma C.2.14] (proved by dynamic considerations with the maximal pseudo-reductive quotient $G_a/\mathscr{R}_{u,k}(G_a) \simeq \overline{G}_a$ in which S is non-central due to nontriviality of its adjoint representation).

Since $W(G_a, S)(k)$ is naturally a subgroup of $W(G, S)(k)$, we may define $r_a \in W(G, S)(k)$ to be the unique nontrivial element in $W(G_a, S)(k)$. The endomorphism of S defined by $s \mapsto s/r_a(s)$ kills $\ker a$ because $\ker a$ is central in G_a, so there exists a unique $a^\vee \in X_*(S)$ such that $s/r_a(s) = a^\vee(a(s))$ since the map $S/(\ker a) \to \mathrm{GL}_1$ defined via a is an isomorphism. This yields the habitual formula $r_a(x) = x - \langle x, a^\vee\rangle a$ on $X(S)$.

Building on Theorem 5.3.2, we now associate a BN-pair (in the sense of Definition 4.1.6) to the triple (G, S, P), allowing us to analyze the structure of $G(k)$ (especially when G is pseudo-reductive, or more generally when $\mathscr{R}_{u,k}(G)$ is k-wound):

THEOREM 5.3.3. *In the setting of Theorem* 5.3.2, $(P(k), N_G(S)(k))$ *is a* BN-*pair for* $G(k)$ *with associated Weyl group* $W({}_k\overline{\Phi})$ *and distinguished set of involutions* $R := \{r_a\}_{a\in\Delta}$ *for the basis* Δ *of the positive system of roots* ${}_k\overline{\Phi}_P \subset {}_k\overline{\Phi}$.

In the pseudo-split pseudo-reductive case this recovers Theorem 4.1.7.

PROOF. We sketch a few main points, referring to [**CGP**, Thm. C.2.20] for full details. That $P(k)$ and $N_G(S)(k)$ generate $G(k)$ is immediate from the Bruhat decomposition in Theorem 5.2.2, the proof of which showed $P(k) \cap N_G(S)(k) = Z_G(S)(k)$ (so (BN1) in Definition 4.1.6 holds). Hence, the final assertion in Theorem 5.3.2(i) identifies the associated Weyl group with $W({}_k\overline{\Phi})$, and by the theory of root systems the latter is generated by R.

To verify (BN2), it remains to show that elements $r \in R$ (or rather, their representatives in $N_G(S)(k)$) do not normalize $P(k)$. Such elements certainly do not normalize P, since $P = N_G(P)$ by Proposition 4.3.6 and $P \cap N_G(S)(k) = Z_G(S)(k)$, but working with just $P(k)$ rather than P will require a finer technique with "root groups" for G (not assumed to be pseudo-reductive) since $P(k)$ is generally not Zariski-dense in P (see Example 2.1.5).

For any $b \in X(S) - \{0\}$, define the smooth connected *root group* $U_b := H_{\langle b\rangle}(G)$ via the construction in Proposition 3.3.1. This is unipotent, and in the pseudo-split pseudo-reductive case it recovers the notion of root group considered already in such cases in Corollary 3.1.10 when $U_b \neq 1$ (since in any root system, such as ${}_k\overline{\Phi}$, the only possible root that is a nontrivial $\mathbf{Q}_{>0}$-multiple of a given root c is either $2c$ or $c/2$ and not both). We do not make any claims yet concerning the commutativity of U_b for non-multipliable $b \in {}_k\overline{\Phi}$, even assuming G is pseudo-reductive but possibly not pseudo-split (we will address this later, in Proposition 5.4.2).

Since the basis Δ lies inside the positive system of roots ${}_k\overline{\Phi}_P$, $U_a \subset P$ for any $a \in \Delta$ due to the dynamic description of $\overline{P} := P/\mathscr{R}_{u,k}(G)$ inside $\overline{G} := G/\mathscr{R}_{u,k}(G)$. (Indeed, $\overline{P} = P_{\overline{G}}(\lambda)$ for some $\lambda \in X_*(S)$ satisfying $\langle a, \lambda\rangle \geqslant 0$, so the inclusion $\overline{U}_a \cap \overline{P} = P_{\overline{U}_a}(\lambda) \subset \overline{U}_a$ between smooth connected k-groups is an equality on Lie algebras and hence an equality of k-groups. Thus, $\overline{U}_a \subset \overline{P}$; by the dynamic construction of $U_a := H_{\langle a\rangle}(G)$ in the proof of Proposition 3.3.1, the quotient map $G \to \overline{G}$ carries U_a into $\overline{U}_a \subset \overline{P}$ and so $U_a \subset P$ as desired.) Hence, the r_a-conjugate of P contains $U_{r_a(a)} = U_{-a}$, so to verify (BN2) it is sufficient to show that $U_{-a}(k) \not\subset P(k)$ for $a \in \Delta$. This is established using properties of the "$H_A(G)$"-construction (for subsemigroups A of $X(S)$).

The verification of (BN3) is essentially the same as in the connected reductive case, using calculations with the Bruhat decomposition (which can be applied here, due to Theorem 5.2.2). □

REMARK 5.3.4. The BN-pair (B, N) arising in Theorem 5.3.3 is well-defined up to $G(k)$-conjugation, and is called *standard* for $G(k)$ (relative to the specification of the k-group G). This BN-pair satisfies some additional properties, as follows. Firstly, the associated Weyl group W is obviously finite; this is the *spherical* condition. Moreover, the BN-pair is saturated and weakly split in the sense of Definition 4.1.6 (using $\mathsf{U} := \mathscr{R}_{us,k}(P)(k)$ for the nilpotent normal subgroup of B in the weakly-split property); the verification of these two properties is given in [**CGP**, Rem. C.2.22]. Finally, by root group considerations, if k is infinite then $\mathsf{B} \cap r\mathsf{B}r^{-1}$ is of infinite index in B for any $r \in \mathsf{R}$.

Remarkably, there is a converse result when k is infinite and $\mathscr{R}_{u,k}(G)$ is k-wound (e.g., G is pseudo-reductive): any weakly-split saturated spherical BN-pair (B, N) for $G(k)$ (with associated set of involutions in its Weyl group denoted as R) such that $\mathsf{B} \cap r\mathsf{B}r^{-1}$ is of infinite index in B for all $r \in R$ *must* arise from a pair (S, P) in G provided that the BN-pair satisfies a further mild group-theoretic hypothesis related to the k-isotropic minimal normal k-simple pseudo-semisimple k-subgroups of G; see [**P**, Thm. B, Rem. 1] for a precise statement.

5.4. Applications of refined structure. The formalism of BN-pairs provides a unified approach to properties of the subgroup structure of $G(k)$ for connected semisimple k-groups G (of interest with finite k for finite group theory, and $k = \mathbf{R}$ for Lie theory); uniform simplicity proofs for $G(k)/Z_G(k)$ with simply connected G are an especially useful application of this perspective. For any group G equipped with a BN-pair (B, N) and the associated set R of involutions in the Weyl group W, there are $2^{\#\mathsf{R}}$ subgroups of G containing B; these are parameterized by the subsets I of R [**Bou**, IV, §2.5, Thm. 3(b)].

Relative to the Bruhat decomposition $\mathsf{G} = \coprod_{w \in \mathsf{W}} \mathsf{B}w\mathsf{B}$, the subgroup G_I associated to I is uniquely determined by the conditions that it contains B and meets N in the preimage of the subgroup $\mathsf{W}_I \subset \mathsf{W}$ generated by I. Equivalently, the B-double cosets in G_I are precisely the ones labelled by W_I via the Bruhat decomposition. When this result for BN-pairs is applied to $\mathsf{G} = G(k)$ equipped with its standard BN-pair, one gets precise group-theoretic control over the pseudo-parabolic k-subgroups of G (even though $P(k)$ need not be Zariski-dense in P!):

THEOREM 5.4.1. *Let G be a smooth connected affine k-group, and choose a minimal pseudo-parabolic k-subgroup P and a maximal split k-torus $S \subset P$. The map $Q \mapsto Q(k)$ is a bijection from the set of smooth closed k-subgroups of G containing P onto the set of subgroups of $G(k)$ containing $P(k)$. Moreover, for any two such Q and Q', we have $Q \subset Q'$ if and only if $Q(k) \subset Q'(k)$.*

PROOF. We saw above that the set of subgroups of $G(k)$ containing $P(k)$ is naturally labeled by the set of subsets of the basis Δ of ${}_k\overline{\Phi}_P$. By construction, this labeling is inclusion-preserving in both directions (i.e., $G(k)_I \subset G(k)_{I'}$ if and only if $I \subset I'$). Since the sets of pseudo-parabolic k-subgroups of G containing S and of $\overline{G} = G/\mathscr{R}_{u,k}(G)$ containing (the naturally isomorphic image of) S are in bijective correspondence via reduction modulo $\mathscr{R}_{u,k}(G)$, it follows that the possibilities for Q can be labelled after passing to the pseudo-reductive case (but beware that $Q(k) \to (Q/\mathscr{R}_{u,k}(G))(k)$ can fail to be surjective when $\mathscr{R}_{u,k}(G)$ is not k-split).

Recall from Theorem 4.3.7 that any smooth closed k-subgroup of G containing P is pseudo-parabolic. For pseudo-reductive G, the map $Q \mapsto \Phi(Q, S)$ is a bijection, inclusion-preserving in both directions, from the set of Q's containing S onto the set of parabolic subsets of $\Phi(G, S)$ (see Theorem 5.3.2(ii),(iv)). Hence, if G is pseudo-reductive then the set of those Q containing P is labeled by the set of parabolic subsets of $\Phi(G, S)$ containing the positive system of roots $\Phi(P, S)$. But for any root system Ψ, it is well-known that the set of parabolic subsets containing a given positive system of roots Ψ^+ is naturally labeled by the set of subsets of the basis of simple roots in Ψ^+ [**Bou**, VI, §1.7, Lemma 3]. Thus, the possibilities for $Q \supset P$ are parameterized by the set of subsets of Δ.

In the proof of Theorem 5.3.2(ii) (Step 5 in the proof of [**CGP**, Thm. C.2.15]) it is shown that for pseudo-reductive G these two bijections onto the set of subsets of Δ are compatible with the map $Q \mapsto Q(k)$, and both bijections just considered are inclusion-preserving in each direction, so the pseudo-reductive case is settled. For more general G additional arguments are required; see [**CGP**, Thm. C.2.23] for a complete treatment. □

As a further application of the general Bruhat decomposition in Theorem 5.2.2, we can prove an important extension of Proposition 4.1.3 to the general pseudo-reductive case (i.e., no pseudo-split hypothesis), as follows. Consider a pseudo-reductive k-group G, and a maximal split k-torus $S \subset G$, so $\Phi = \Phi(G, S)$ is a root system (Theorem 5.3.2(i)). For each $a \in \Phi$ we defined the unipotent smooth connected *root group* $U_a := H_{\langle a \rangle}(G) \subset G$ in the proof of Theorem 5.3.3. Inspired by the case of relative root groups in connected reductive groups, we now prove that U_a is a vector group when a is not multipliable, and much more:

PROPOSITION 5.4.2. *Using notation and hypotheses as above, if a is not multipliable then U_a is a vector group whereas if a is multipliable then U_{2a} is central in U_a and U_a/U_{2a} is commutative. For any nontrivial $u \in U_a(k)$, the following hold:*

(i) *There exist unique $u', u'' \in U_{-a}(k)$ such that $m(u) := u'uu'' \in N_G(S)(k)$. The effect of $m(u)$-conjugation on $\mathrm{X}(S)$ is r_a, and $u', u'' \neq 1$.*
(ii) *If a is not multipliable then $u' = u'' = m(u)^{-1}um(u)$ and $m(u)^2 \in S(k)$.*
(iii) *The formation of $m(u)$ is $Z_G(S)$-equivariant in the sense that for any extension field k'/k and $z \in Z_G(S)(k')$ satisfying $zuz^{-1} \in U_a(k)$, necessarily $zu'z^{-1}, zu''z^{-1} \in U_{-a}(k)$ and $m(zuz^{-1}) = zm(u)z^{-1}$.*

The method of proof for the pseudo-split case in Proposition 4.1.3 involves bootstrapping from calculations with $G^{\mathrm{red}}_{\overline{k}}$ after passing to the rank-1 case. The proof of Proposition 5.4.2 is entirely different, involving no use of $\overline{k}$-groups; this allows the result to be extended (with appropriate formulation) to a wider class of smooth connected affine k-groups (including those whose k-unipotent radical is k-wound); see [**CGP**, Prop. C.2.24] for this additional generality.

PROOF. We sketch the proof of (i) (and the proofs of (ii) and (iii) amount to group-theoretic computations, aided by the dynamic relation $U_G(-\lambda) \cap P_G(\lambda) = 1$); for further details we refer to [**CGP**, Prop. C.2.24]. As a first step, by replacing G with $Z_G(\ker a)^0$ (pseudo-reductive due to [**CGP**, Prop. A.8.14(2)]) we can pass to the case of a rank-1 root system in which a is non-divisible. Now $\Phi = \{\pm a\}$ or $\Phi = \{\pm a, \pm 2a\}$. By [**CGP**, Lemma 3.3.8] (and its proof), U_{na} is a normal k-subgroup of U_a with $(U_a, U_{na}) \subset U_{(n+1)a}$ and each $U_{na}/U_{(n+1)a}$ is a vector group admitting an S-equivariant linear structure with na as the unique S-weight on the Lie algebra when this quotient is nontrivial. That settles the assertions concerning U_a and (in the multipliable case) U_{2a}.

Define $N = N_G(S)$ and $Z = Z_G(S)$, so $N(k) - Z(k) = Z(k)r_a$ since the Weyl group $N(k)/Z(k)$ has order 2. (We adopt the standard abuse of notation by writing r_a where we really intend a representative of r_a in $N(k)$; for our calculations this will be harmless, but note that the representative may not be an involution.) Fix a nontrivial element $u \in U_a(k)$, and consider the Bruhat decomposition relative to the minimal pseudo-parabolic k-subgroup $P = Z \ltimes U_{-a}$. Dynamic considerations via Theorem 2.3.5(ii) imply that $U_a \cap P = 1$, so the nontrivial u lies in the complement

$$G(k)-P(k) = P(k)r_aP(k) = U_{-a}(k)Z(k)r_aU_{-a}(k) = U_{-a}(k)(N(k)-Z(k))U_{-a}(k).$$

This provides $u', u'' \in U_{-a}(k)$ such that $u = u'^{-1}nu''^{-1}$ for some $n \in N(k) - Z(k)$, so $u'uu'' = n$ acts on S through r_a. This proves existence of u' and u'' in (i).

For uniqueness in (i), observe that if elements $u', u'' \in U_{-a}(k)$ satisfy $u'uu'' \in N(k)$ then necessarily $u'uu'' \in N(k) - Z(k) = Z(k)r_a$ because otherwise $u \in u'^{-1}Z(k)u''^{-1} \subset U_{-a}(k)Z(k) = P(k)$, contradicting that $P \cap U_a = 1$. Hence, to prove uniqueness we are reduced to showing that if elements $v', v'' \in U_{-a}(k)$ and $n \in N(k)$ satisfy $v'nv'' = nz$ for some $z \in Z(k)$ then $v' = 1 = v''$. But $v'nv''n^{-1} = nzn^{-1} \in Z(k)$ and $nv''n^{-1} \in U_a(k)$, so it is enough to prove that $(U_{-a}(k)U_a(k)) \cap Z(k) = 1$. This triviality is immediate since $P \cap U_a = 1$.

Finally, to prove that $u', u'' \neq 1$ it suffices to show that $U_{-a}(k)(U_a(k) - \{1\})$ is disjoint from $N(k)$. Suppose an element $n \in N(k)$ has the form $n = v'v$ for $v' \in U_{-a}(k)$ and nontrivial $v \in U_a(k)$. Clearly $v' = nv^{-1} = (nv^{-1}n^{-1})n \in U_{-a}(k)n$, so $n \in U_{-a}(k)$ and hence $v = v'^{-1}n \in U_{-a}(k)$. This is an absurdity since $P \cap U_a = 1$ and we assumed $v \neq 1$. □

In [**St**, Thm. 5.4], Steinberg gave a new proof of the Isomorphism Theorem for split connected reductive groups over a field k. Given two such groups G and G' equipped with respective split maximal k-tori $T \subset G$ and $T' \subset G'$, we assume that an isomorphism of root data $\phi : R(G,T) \simeq R(G',T')$ is given and we wish to construct a k-isomorphism of pairs $f : (G',T') \simeq (G,T)$ giving rise to ϕ (and to show that f is unique up to the action of $(T/Z_G)(k)$).

The idea of Steinberg's proof is to construct f by constructing its graph Γ_f as a k-subgroup of $G' \times G$ satisfying specified conditions (e.g., the graph of the isomorphism $T' \simeq T$ arising from ϕ is a split maximal k-torus in this graph). Briefly, ϕ determines which root group of (G',T') is to be carried to a given root group of (G,T), the isomorphism between such root groups is specified for roots from compatible bases, and then the isomorphism is extended to all matching pairs of root groups via Weyl-group actions. In effect, we try to build Γ_f as the smooth connected k-subgroup of $G' \times G$ generated by the graphs of specific isomorphisms between certain root groups, and the work is to show that such a k-subgroup has desired properties (e.g., it is reductive and projects isomorphically onto G').

Steinberg's method for constructing a k-subgroup of $G' \times G$ with specified properties is generalized in Theorem 5.4.3 below to prove a much more general result concerning the existence and uniqueness inside a given smooth connected affine k-group of pseudo-reductive k-subgroups for which we have specified its maximal split k-torus S, its S-centralizer, a basis Δ for its relative root system, and the root groups for the roots in Δ. This is extremely powerful: it provides a unified approach to the construction of certain "exotic" non-standard pseudo-reductive groups (built as k-subgroups of a Weil restriction), it implies the existence of Levi k-subgroups of pseudo-split pseudo-reductive groups (see Theorem 5.4.4, a very useful result), and it leads to an "Isomorphism Theorem" (Theorem 9.1.7) for pseudo-split pseudo-reductive groups.

THEOREM 5.4.3. *Let G be a smooth connected affine k-group, and let $S \subset G$ be a nontrivial split k-torus. Fix a smooth connected k-subgroup $C \subset Z_G(S)$ in which S is a maximal split k-torus, and a non-empty linearly independent subset $\Delta \subset \mathrm{X}(S)$.*

For each $a \in \Delta$ let F_a be a pseudo-reductive k-subgroup of G containing S such that $Z_{F_a}(S) = C$ and $\{\pm a\} \subset \Phi(F_a, S) \subset \mathbf{Z}a$. Let $U_{\pm a}$ be the $\pm a$-root groups of F_a, and assume U_a commutes with U_{-b} for all distinct $a, b \in \Delta$. Let $F \subset G$ be the smooth connected k-subgroup generated by $\{F_a\}_{a\in\Delta}$.

(i) *The k-group F is pseudo-reductive with S as a maximal split k-torus, $Z_F(S) = C$, Δ is a basis of $\Phi(F, S)$, and the $\pm a$-root groups of F are $U_{\pm a}$ for all $a \in \Delta$.*
(ii) *If each F_a is reductive then so is F.*
(iii) *The k-group F is functorial with respect to isomorphisms in the 5-tuple $(G, S, C, \Delta, \{F_a\}_{a\in\Delta})$.*

Note that C is generally not commutative (when S is not a maximal k-torus in the F_a's). As a special case, a criterion in [**PR**, Thm. 2.2] for a pair of quasi-split connected semisimple subgroups of a connected semisimple group to generate a quasi-split semisimple subgroup is an immediate consequence of Theorem 5.4.3.

PROOF. We refer the reader to [**CGP**, Thm. C.2.29] for a complete proof (as well as for a more general result in which the pseudo-reductivity hypotheses and conclusion are relaxed). Here we just sketch some ideas in the proof.

Since Δ is linearly independent, we may choose a cocharacter $\lambda \in \mathrm{X}_*(S)$ satisfying $\langle a, \lambda\rangle > 0$ for all $a \in \Delta$ (this corresponds to λ lying in a specific connected component of the complement in $\mathrm{X}_*(S)_{\mathbf{R}}$ of the union of the hyperplanes killed by the elements of Δ). Such λ can also be chosen to not annihilate any of the finitely many nontrivial S-weights that occur on $\mathrm{Lie}(G)$, so $Z_G(\lambda) = Z_G(S)$ (hence $Z_F(\lambda) = Z_F(S)$). Although we know very little about the structure of F, we may nonetheless apply Theorem 2.3.5(ii) to get an open immersion

$$U_F(-\lambda) \times Z_F(S) \times U_F(\lambda) \longrightarrow F$$

via multiplication. Since $C \subset F$, clearly $C \subset Z_F(S)$.

For a fixed sign, let $U_\pm \subset U_F(\pm\lambda)$ be the smooth connected k-subgroup generated by $\{U_{\pm a}\}_{a\in\Delta}$, so the multiplication map

$$U_- \times C \times U_+ \longrightarrow F$$

is a locally closed immersion. An inductive argument now shows that the Zariski closure of this locally closed subset is stable under left multiplication by $U_{\pm a}$ for all $a \in \Delta$, as well as obviously stable under left multiplication by C, so it is stable under left multiplication by F and therefore coincides with F.

We conclude that $U_- \times C \times U_+$ is *open* in F, so the closed immersion

$$U_- \times C \times U_+ \hookrightarrow U_F(-\lambda) \times Z_F(S) \times U_F(\lambda)$$

is an equality. In other words, $Z_F(S) = C$ and $U_\pm = U_F(\pm\lambda)$. This implies that S is a maximal split k-torus in F (by the maximality hypothesis on S in C) and that (for a fixed sign) the S-weights occurring in $\mathrm{Lie}(U_F(\pm\lambda))$ lie in the subsemigroup

$A^{\pm} \subset \mathrm{X}(S) - \{0\}$ generated by $\pm\Delta$. In other words, the set $\Phi(F, S)$ of nontrivial S-weights occuring in $\mathrm{Lie}(F)$ is contained in $A^{+} \cup A^{-}$.

Systematic application of properties of the unipotent $H_{\Sigma}(\cdot)$-construction from Proposition 3.3.1 with varying subsemigroups $\Sigma \subset \mathrm{X}(S)$ not containing 0 (especially the direct spanning property in Theorem 3.3.3) enables one to prove that $U_{\pm a} = H_{\pm a}(F)$ for all $a \in \Delta$ (so $U_{\pm a}$ is the $\pm a$-root group of F once F is shown to be pseudo-reductive) and that no positive integral multiple of any $a \in \Delta$ is a weight on $\mathrm{Lie}(\mathscr{R}_{u,k}(F))$. In particular, Δ is contained in the root system ${}_k\overline{\Phi} := \Phi(F/\mathscr{R}_{u,k}(F), S)$ that in turn lies inside the set of S-weights $\Phi(F, S) \subset A^{+} \cup A^{-}$. Thus, Δ satisfies the condition that uniquely characterizes a basis of ${}_k\overline{\Phi}$, so by Theorem 5.3.2(i) applied to F the Weyl group ${}_kW$ of this root system is generated by reflections $\{r_a\}_{a\in\Delta}$ represented by elements of $N_F(S)(k)$.

The k-group C is pseudo-reductive since it is a torus-centralizer in the pseudo-reductive k-group F_a for any $a \in \Delta$, so the smooth connected unipotent normal k-subgroup $U := \mathscr{R}_{u,k}(F)$ of F cannot be contained in $C = Z_F(S)$ if it is nontrivial. Assuming $U \neq 1$, we seek a contradiction. The S-action on U must be nontrivial, so there exists a nontrivial S-weight b occurring in $\mathrm{Lie}(U)$. This S-weight lies in A^{+} or A^{-}, and we have shown that Δ is a basis of ${}_k\overline{\Phi}$, so by normality of U in F we may use the action of ${}_kW$ to arrange that $b \in A^{+}$. The long Weyl element w in ${}_kW$ relative to Δ carries b into A^{-}. An inductive argument using an expression for w in reflections r_a ($a \in \Delta$) eventually produces an element of Δ that occurs as an S-weight in $\mathrm{Lie}(U)$, a contradiction. Thus, $U = 1$, which is to say F is pseudo-reductive. This finishes our sketch of the proof of (i). By applying (i) over $\overline{k}$ we immediately get (ii).

The proof of (iii) amounts to a generalization of Steinberg's graph method for proving the Isomorphism Theorem in the split connected reductive case. More specifically, the graph of the isomorphism we seek to build must be a pseudo-reductive k-subgroup of $G' \times G$, where $(G', S', C', \Delta', \{F'_{a'}\}_{a'\in\Delta'})$ is the other 5-tuple under consideration. This pseudo-reductive k-subgroup of $G' \times G$ can be constructed as an application of (i) in exactly the same way that Steinberg proved the Isomorphism Theorem. □

An important application of Theorem 5.4.3 is the existence of Levi k-subgroups of pseudo-split pseudo-reductive groups. Recall that a *Levi k-subgroup* of a smooth connected affine k-group G is a smooth connected k-subgroup L such that the natural map $L_{\overline{k}} \to G^{\mathrm{red}}_{\overline{k}}$ is an isomorphism.

Such subgroups need not exist in positive characteristic, even over an algebraically closed ground field. For example, if F is algebraically closed of positive characteristic then for any $n \geqslant 2$ the F-group corresponding to $\mathrm{SL}_n(W_2(F))$ (with W_2 denoting the functor of length-2 Witt vectors) has no Levi F-subgroup; see [**CGP**, A.6] for a proof.

The existence of Levi k-subgroups in the pseudo-split pseudo-reductive case is proved in [**CGP**, Thm. 3.4.6] by an indirect process. We now give a completely different and conceptually simpler proof by using Proposition 5.4.2 and Theorem 5.4.3 (adapting the proof of a more general existence result in [**CGP**, Thm. C.2.30] for split connected reductive k-subgroups of smooth connected affine k-groups):

THEOREM 5.4.4. *Let G be a pseudo-split pseudo-reductive k-group, and $T \subset G$ a split maximal k-torus. Let Δ be a basis of the reduced root system Φ' of non-multipliable roots in $\Phi(G,T)$, and U_a^G the a-root group of G for $a \in \Phi(G,T)$.*

For each set $\{E_a\}_{a\in\Delta}$ of 1-dimensional smooth connected k-subgroups $E_a \subset U_a^G$ normalized by T, there exists a unique Levi k-subgroup L of G containing T such that $U_a^L = E_a$ for all $a \in \Delta$.

The hypothesis that G is pseudo-split cannot be dropped: for any local or global function field k over a finite field, [**CGP**, Ex. 7.2.2] provides an absolutely pseudo-simple k-group with no Levi k-subgroup.

PROOF. Define the k-subgroup scheme $M := \bigcap_{a\in\Phi'} \ker a \subset T$ of multiplicative type. Any Levi k-subgroup L of G containing T is generated by its maximal central k-torus (necessarily contained in T) and its T-root groups. But $\Phi(L,T) = \Phi'$ for any such L by Theorem 3.1.7, so all root groups of L are centralized by M and hence L is centralized by M. Thus, all Levi k-subgroups L containing T are contained in $Z_G(M)^0$. This identity component inherits pseudo-reductivity from G since M is a k-subgroup of T (see [**CGP**, Prop. A.8.14(2)]), and since $\mathrm{X}(T/M) = \sum_{a\in\Phi'} \mathbf{Z}a = \bigoplus_{c\in\Delta} \mathbf{Z}c$ we have $\Phi(Z_G(M)^0, T) = \Phi'$ by [**CGP**, Prop. A.8.14(3)]. Thus, we may replace G with $Z_G(M)^0$ so that $\Phi := \Phi(G,T)$ is reduced. In particular, $\Phi(G_{\overline{k}}^{\mathrm{red}}, T_{\overline{k}}) = \Phi$.

Next we prove that any $L \supset T$ is determined inside G by its T-root groups for the roots in Δ. Since L is generated by T and its root groups for roots in $\pm\Delta$ (as Δ is a basis for $\Phi = \Phi(L,T)$), it suffices to show that for each $a \in \Phi$ (or even just $a \in \Delta$), U_{-a}^L is uniquely determined inside G by U_a^L and T. Pick a nontrivial $u_a \in U_a^L(k)$ (this exists since $U_a^L \simeq \mathbf{G}_{\mathrm{a}}$). By Proposition 5.4.2(i),(ii), there exists a (necessarily nontrivial) unique $u_{-a} \in U_{-a}^G(k)$ such that

$$n_a := u_{-a} u_a u_{-a} \in N_G(T)(k),$$

and n_a-conjugation on T induces the reflection r_a on $\mathrm{X}(T)_{\mathbf{Q}}$. By Proposition 5.4.2(i) applied to L, we see that $u_{-a} \in U_{-a}^L(k)$. Since u_{-a} is nontrivial and there exists a k-isomorphism $U_{-a}^L \simeq \mathbf{G}_{\mathrm{a}}$ carrying T-conjugation over to scaling against the nontrivial character $-a$, the Zariski-closure in G of the T-orbit of u_{-a} coincides with U_{-a}^L. Thus, the uniqueness for L is established.

For any $a \in \Phi$, reducedness of Φ implies that U_a^G is a vector group admitting a T-equivariant linear structure (see Corollary 3.1.10), so there exists a T-equivariant isomorphism between U_a^G and a direct sum of copies of the 1-dimensional representation of T through a. Hence, any 1-dimensional smooth connected k-subgroup $E \subset U_a^G$ that is normalized by T is a line in the T-representation space U_a^G. In particular, $E \simeq \mathbf{G}_{\mathrm{a}}$, so $E(k) \neq 1$. Upon choosing such $E_a \subset U_a^G$ for all $a \in \Delta$, we seek a Levi k-subgroup $L \subset G$ containing T such that $U_a^L = E_a$ for all $a \in \Delta$.

STEP 1. The first task is to define a candidate $E_{-a} \subset U_{-a}^G$ for U_{-a}^L. Motivated by the preceding calculations, choose $u_a \in E_a(k) - \{1\}$ and define $u'_a \in U_{-a}^G(k)$ in terms of u_a via Proposition 5.4.2(i): it is the unique element of $U_{-a}^G(k)$ such that $n_a := u'_a u_a u'_a \in N_G(T)(k)$. Define E_{-a} to be the smooth connected Zariski-closure in U_{-a}^G of the T-orbit of u'_a under conjugation; this is a line in the T-representation space U_{-a}^G. To see that E_{-a} is unaffected if we replace u_a with another nontrivial $u \in E_a(k)$, note that $u = t u_a t^{-1}$ for some $t \in T(\overline{k})$ since $a : T(\overline{k}) \to \overline{k}^{\times}$ is surjective, so the associated u' satisfies $u' = t u_{-a} t^{-1}$ by Proposition 5.4.2(iii). Thus, replacing

u_a with u replaces $(E_{-a})_{\overline{k}}$ with $t(E_{-a})_{\overline{k}}t^{-1} = (E_{-a})_{\overline{k}}$. This proves that E_{-a} is independent of the choice of u_a.

Consider the smooth connected k-subgroup $L \subset G$ generated by $\{E_{\pm a}\}_{a\in\Delta}$ and T. It remains to prove: L is reductive with T as a (split) maximal k-torus, the natural map $L_{\overline{k}} \to G^{\mathrm{red}}_{\overline{k}}$ is an isomorphism, and $U_a^L = E_a$ for all $a \in \Delta$. There is nothing to do in the reductive case (as $L = G$ in such cases), so we may assume k is infinite. We will first treat the case where Φ has rank 1, and then we will reduce the general case to that case.

STEP 2. Now assume that the reduced root system Φ has rank 1, so $\Phi = \{\pm a\}$ for some nontrivial $a \in \mathrm{X}(T)$. For ease of notation, let $E_\pm := E_{\pm a}$ and $n := u'uu'$ for a choice of nontrivial $u := u_a \in E(k)$, so $n^2 \in T(k)$ and $nE_+n^{-1} = E_-$ by Proposition 5.4.2(ii). Define the subset

$$\Gamma = E_+(k)\{1, n\}T(k)E_+(k) = E_+(k)T(k) \coprod E_+(k)nT(k)E_+(k)$$

(disjoint by the Bruhat decomposition for $G(k)$ in Theorem 5.2.2), so Γ generates a Zariski-dense subgroup of L due to the Zariski-density of $T(k)$ in T and of $E_\pm(k)$ in $E_\pm$ (recall T is k-split, $E_\pm \simeq \mathbf{G}_{\mathrm{a}}$, and k is infinite).

LEMMA 5.4.5. *The subset $\Gamma \subset G(k)$ is a subgroup. In particular, Γ is Zariski-dense in L.*

PROOF. It is clear that Γ is stable under left and right multiplication against $E_+(k)$ and $T(k)$ (using that $T(k)$ normalizes $E_+(k)$), so we just have to check that $n\Gamma \subset \Gamma$ or more specifically that $nE_+(k)n^{-1} \subset \Gamma$ (as $n^2 \in T(k)$ and Γ is stable under left and right multiplication against $T(k)E_+(k)$). Since $n = u'uu'$ by definition and $u' = n^{-1}un$ by Proposition 5.4.2(ii), so

$$n = n(u'uu')n^{-1} = (nu'n^{-1})(nun^{-1})(nu'n^{-1}) = u(nun^{-1})u,$$

we see that $nun^{-1} = u^{-1}nu \in E_+(k)nE_+(k) \subset \Gamma$.

A nontrivial $v \in E_+(k)$ has the form tut^{-1} for some $t \in T(\overline{k})$, so

$$nvn^{-1} = (ntn^{-1})(nun^{-1})(ntn^{-1})^{-1} \in (ntn^{-1})(E_+(k)nE_+(k))(ntn^{-1})^{-1}.$$

Hence, it suffices to show that ntn^{-1}-conjugation preserves $E_+(k)$ and carries n to a k-point in $nT(k)$. The effect of ntn^{-1}-conjugation on $(E_+)_{\overline{k}} \simeq \mathbf{G}_{\mathrm{a}}$ is scaling against $(n^{-1}.a)(t) = a(t)^{-1}$ since n acts on $\mathrm{X}(T)$ through the reflection r_a, and $a(t) \in k^\times$ since $u, v \in E_+(\overline{k})$ are nontrivial k-points related through scaling against $a(t)$. Likewise, the ntn^{-1}-conjugate of n is $ntnt^{-1}n^{-1} \in nT(\overline{k})$ (since n normalizes T) yet is a k-point since $tnt^{-1} = m(v)$ by Proposition 5.4.2(iii). $\square$

Clearly $\Gamma = n\Gamma = E_-(k)T(k)E_+(k) \coprod nE_+(k)T(k)$ yet Γ is Zariski-dense in L and $nE_+(k)T(k)$ has Zariski-closure $nE_+ \rtimes T \subset nU_a^G \rtimes T$ that is a proper closed subset of L, so $E_-(k)T(k)E_+(k)$ is Zariski-dense in L. For $\lambda : \mathrm{GL}_1 \to T$ satisfying $\langle a, \lambda\rangle > 0$ we have $E_\pm \subset U_L(\pm\lambda)$, so we have a closed immersion

$$j : E_- \times T \times E_+ \hookrightarrow U_L(-\lambda) \times Z_L(T) \times U_L(\lambda) =: \Omega.$$

But the multiplication map $\Omega \to L$ is an open immersion by Theorem 2.3.5(ii), so the density of $E_-(k)T(k)E_+(k)$ in L implies that j is an equality. Equivalently, $Z_L(T) = T$ and $E_\pm = U_L(\pm\lambda)$. In particular, the solvable smooth connected k-subgroups $B_\pm := T \ltimes E_\pm \subset L$ (via multiplication) have codimension 1.

Since n-conjugation on T swaps the two roots $\pm a \in \Phi(G, T)$, there exist $t \in T(\overline{k})$ such that the commutator $(ntn^{-1})t^{-1} \in T(\overline{k})$ is nontrivial. Hence, $\mathscr{D}(L)$ is not unipotent, so the smooth connected affine k-group L is not solvable. It follows that the k-subgroups $B_{\pm}$ are Borel subgroups, so $\mathscr{R}_u(L_{\overline{k}}) \subset (B_{\pm})_{\overline{k}}$. But working inside the open subscheme $\Omega \subset L$ shows that $B_+ \cap B_- = T$, so $\mathscr{R}_u(L_{\overline{k}}) \subset T_{\overline{k}}$, forcing $\mathscr{R}_u(L_{\overline{k}}) = 1$; i.e., L is reductive.

The preceding calculations show $E_{\pm}$ are the T-root groups of L, so $\Phi(L,T) = \{\pm a\} = \Phi$. Thus, the natural map $f : L_{\overline{k}} \to G_{\overline{k}}^{\mathrm{red}}$ induces an isomorphism between maximal tori and between the root systems. The induced map f_b between root groups for each common root b is $T_{\overline{k}}$-equivariant and hence is *linear* between 1-dimensional root groups, so $\ker(f_b) = 1$ (as otherwise $\ker(f_b)$ is the b-root group of L, forcing the unipotent *normal* subgroup scheme $\ker f = L_{\overline{k}} \cap \mathscr{R}_u(G_{\overline{k}})$ in $L_{\overline{k}}$ to contain a nontrivial smooth connected subgroup, contradicting that $\mathscr{R}_u(L_{\overline{k}}) = 1$). Thus, f induces an isomorphism between open cells, so it is an isomorphism. The case of Φ with rank 1 is done.

STEP 3. If Φ is empty then $G_{\overline{k}}^{\mathrm{red}}$ is commutative and hence G is solvable. In such cases $L = T$, and this is obviously a Levi k-subgroup of G. Thus, we may assume Φ has positive rank. The settled rank-1 case applies to $G_a := Z_G(T_a)$ for the codimension-1 torus $T_a := (\ker a)_{\mathrm{red}}^0 \subset T$ with any $a \in \Delta$, so for all such a the smooth connected k-subgroup $L_a \subset G_a$ generated by T, E_a, and E_{-a} is reductive with maximal k-torus T and $(L_a)_{\overline{k}} \to (G_a)_{\overline{k}}^{\mathrm{red}}$ is an isomorphism. In particular, for dimension reasons $E_{\pm a}$ are the T-root groups of L_a.

By [**CGP**, Prop. A.4.8], the natural map $(G_a)_{\overline{k}}^{\mathrm{red}} \to G_{\overline{k}}^{\mathrm{red}}$ is an isomorphism onto the $(T_a)_{\overline{k}}$-centralizer in $G_{\overline{k}}^{\mathrm{red}}$; i.e., $(G_a)_{\overline{k}}^{\mathrm{red}} = (G_{\overline{k}}^{\mathrm{red}})_a$. These latter groups generate $G_{\overline{k}}^{\mathrm{red}}$, so the natural map $f : L_{\overline{k}} \to G_{\overline{k}}^{\mathrm{red}}$ satisfies three properties: it is surjective, it carries $T_{\overline{k}}$ isomorphically onto a maximal torus of the target, and it carries $(E_a)_{\overline{k}}$ isomorphically onto the a-root group of the target for every $a \in \Delta$.

Note that L is generated by $\{L_a\}_{a\in\Delta}$, $Z_{L_a}(T) = T$ for all $a \in \Delta$, and E_a commutes with E_{-b} for all distinct $a, b \in \Delta$ because U_a^G commutes with U_{-b}^G for such a and b (as $ma + n(-b)$ is not a root for any integers $m, n \geqslant 1$; see [**CGP**, Cor. 3.3.13(2)]). Hence, by Theorem 5.4.3 (applied to the collection of k-subgroups $L_a \subset G$ for $a \in \Delta$) the k-group L is reductive with T as a maximal k-torus, and by design $\Delta \subset \Phi(L,T) \subset \Phi(G,T) =: \Phi$ with Φ a reduced root system having basis Δ. Moreover, by Theorem 5.4.3(i), Δ is a basis of $\Phi(L,T)$ and the $\pm a$-root groups of L are $E_{\pm a}$ for all $a \in \Delta$.

The natural subgroup inclusion $W(L,T) \subset W(\Phi)$ is an equality because the element $u_a' u_a u_a' \in N_L(T)(k)$ is carried to the reflection r_a for every $a \in \Delta$ (and such reflections generate $W(\Phi)$), so $W(\Phi)\cdot\Delta \subset \Phi(L,T)$. Reducedness of Φ implies that every $W(\Phi)$-orbit in Φ meets Δ, so $\Phi(L,T) = \Phi$ as required. Each root in $\Phi(L,T)$ is $N_L(T)(k)$-conjugate to a root in Δ, so for each $b \in \Phi(L,T) = \Phi$ the map f carries the b-root group of $L_{\overline{k}}$ isomorphically onto the b-root group of $G_{\overline{k}}^{\mathrm{red}}$ since the special case $b \in \Delta$ has been verified. It follows that f restricts to an isomorphism between open cells, so it is a birational homomorphism between smooth connected groups and thus is an isomorphism. □

There are conditions that provide a complete characterization of semisimplicity, such as in the following result (proved using Theorem 5.4.4 and additional ideas):

PROPOSITION 5.4.6. *Let G be a pseudo-semisimple k-group. If all root groups of G_{k_s} are 1-dimensional then G is reductive. Likewise, if there exists a pseudo-parabolic k-subgroup $P \subset G$ such that G/P is proper and P does not contain any k-simple pseudo-semisimple normal k-subgroup of G then G is reductive.*

See [**CGP**, Thm. 3.4.9] for a proof. (The assertion concerning G/P is harder, and the role of properness is to ensure that $P_{\overline{k}}$ contains $\mathscr{R}_u(G_{\overline{k}})$, due to the Borel fixed point theorem applied to the translation of $\mathscr{R}_u(G_{\overline{k}})$ on $G_{\overline{k}}/P_{\overline{k}} = (G/P)_{\overline{k}}$.)

We finish this section by recording an interesting result going beyond the more widely-known reductive case:

THEOREM 5.4.7. *Assume k is infinite, and let G be a k-isotropic pseudo-simple k-group. Let $G(k)^+$ denote the subgroup of $G(k)$ generated by the maximal k-split unipotent smooth connected k-subgroups of G. Then $G(k)^+$ is Zariski-dense in G and is also perfect. Moreover, any non-central subgroup of $G(k)$ that is normalized by $G(k)^+$ must contain $G(k)^+$. In particular, the quotient of $G(k)^+$ modulo its center is a simple group.*

See [**CGP**, Thm. C.2.34] for a proof of Theorem 5.4.7 (in a more general formulation); this rests on Proposition 5.4.2 and a generalization of Proposition 4.3.1(ii) that drops pseudo-split hypotheses (see [**CGP**, Prop. C.2.26]).

6. Central extensions and standardness

6.1. Central quotients. An important feature of connected reductive k-groups G is that the formation of the scheme-theoretic center Z_G is compatible with the formation of central quotients $\overline{G} := G/Z$ (for a closed k-subgroup scheme $Z \subset Z_G$); i.e., $Z_{\overline{G}} = Z_G/Z$. This property ultimately rests on the structure of open cells over k_s, and is specific to the reductive case; e.g., any smooth connected unipotent k-group U is nilpotent, so $Z_U \neq 1$ if $U \neq 1$ and hence $Z_{U/Z_U} \neq 1$ if U is non-commutative. As a special case, if G is connected reductive then G/Z_G has trivial center (and it is even perfect, or equivalently semisimple).

For a pseudo-reductive k-group G, two new phenomena occur:

- a central quotient G/Z need not be pseudo-reductive (see Example 2.1.3),
- the central quotient G/Z_G may not be perfect; e.g., $\mathrm{R}_{k'/k}(\mathrm{PGL}_p)$ is not perfect (Example 1.2.4) but it has trivial center [**CGP**, Prop. A.5.15(1)].

Despite this behavior that deviates from the reductive case, central quotients remain a useful tool in the pseudo-reductive case due to:

PROPOSITION 6.1.1. *If G is a pseudo-reductive k-group then a central quotient $\overline{G} := G/Z$ is pseudo-reductive if and only if Z_G/Z does not contain a nontrivial smooth connected unipotent k-subgroup, in which case $Z_{\overline{G}} = Z_G/Z$. In particular, G/Z_G is pseudo-reductive and has trivial center, and a composition of central homomorphisms between pseudo-reductive groups is central.*

The proof that $Z_{\overline{G}} = Z_G/Z$ when $\overline{G}$ is pseudo-reductive rests on an analysis of root groups and open cells over k_s; see [**CP**, Lemma 4.1.1]. The necessity of the condition that Z_G/Z does not contain a nontrivial smooth connected unipotent k-subgroup is immediate from the observations that $Z_{\overline{G}}$ lies inside any Cartan k-subgroup $\overline{C} \subset \overline{G}$ and that $\overline{C}$ is commutative pseudo-reductive if $\overline{G}$ is pseudo-reductive (so $\overline{C}$ cannot contain a nontrivial smooth connected unipotent k-subgroup in such cases).

The sufficiency of this condition for pseudo-reductivity of $\overline{G}$ clearly reduces to the assertion that G/Z_G is pseudo-reductive, but such pseudo-reductivity is rather nontrivial, as it rests on a study of automorphism schemes. Since such automorphism schemes pervade proofs of the deeper structure of pseudo-reductive groups, we now review the relevant existence results and properties for these schemes (and then illustrate their use to prove the pseudo-reductivity of G/Z_G).

PROPOSITION 6.1.2. *Let G be a smooth connected affine k-group, and C a Cartan k-subgroup. The functor $\underline{\mathrm{Aut}}_{G,C}$ assigning to any k-algebra R the group of R-automorphisms of G_R restricting to the identity on C_R is represented by an affine k-group scheme $\mathrm{Aut}_{G,C}$ of finite type. If G is pseudo-reductive then the maximal smooth closed k-subgroup $Z_{G,C}$ of $\mathrm{Aut}_{G,C}$ is commutative and $Z^0_{G,C}$ is pseudo-reductive.*

PROOF. Let T be the maximal k-torus in C. After extending scalars to k_s to split T, by Proposition 3.1.4 we see that G is generated by C and the (generally non-commutative!) subgroups $U_{(a)}$ for the nontrivial T-weights a that occur in $\mathrm{Lie}(G)$. Thus, there is a finite sequence $\{a_i\}_{i\in I}$ of such weights (possibly with repetitions!) such that the multiplication map $C \times \prod_{i\in I} U_{(a_i)} \to G$ is dominant.

The representability of $\underline{\mathrm{Aut}}_{G,C}$ rests on a detailed study of T-equivariant filtrations of the coordinate rings $k[U_{(a)}]$ and the fact that any R-automorphism f of G_R restricting to the identity on C_R must act T_R-equivariantly on each $(U_{(a)})_R$. This allows us to realize $\underline{\mathrm{Aut}}_{G,C}$ as a subfunctor of the direct product $F = \prod_{i\in I} \underline{\mathrm{Aut}}_{U_{(a_i)},T}$ of T-equivariant automorphism functors of the $U_{(a_i)}$'s.

Although the automorphism functor of each $U_{(a)}$ is *not* representable when $\mathrm{char}(k) > 0$, the subfunctor $\underline{\mathrm{Aut}}_{U_{(a)},T}$ is representable [**CGP**, Lemma 2.4.2]. This allows one (after more work) to identify $\underline{\mathrm{Aut}}_{G,C}$ with a *closed* subfunctor of F. We refer the reader to [**CGP**, Thm. 2.4.1] for the details (and see [**CGP**, Cor. 2.4.4] for a variant in which C is replaced with a maximal k-torus of G). □

REMARK 6.1.3. Assume G is reductive. The representability and structure of $\mathrm{Aut}_{G,C}$ can be understood in another way (at the cost of invoking much deeper input): the automorphism functor of G (without reference to C) is represented by a smooth k-group $\mathrm{Aut}_{G/k}$ whose identity component is G/Z_G and whose étale component group over k_s injects into the automorphism group of the based root datum. Since C coincides with its own maximal k-torus T due to reductivity of G, we conclude that $\mathrm{Aut}_{G,C} = T/Z_G$; in particular, $Z_{G,C} = T/Z_G$ is *connected.* This alternative approach through $\mathrm{Aut}_{G/k}$ is much more sophisticated than the proof of Proposition 6.1.2 since the existence and structure of $\mathrm{Aut}_{G/k}$ requires the Isomorphism Theorem for split reductive groups over k-algebras, not just over fields.

In contrast with the reductive case, the automorphism functor of a general pseudo-reductive k-group G is *not* representable (see [**CGP**, Ex. 6.2.1] for commutative counterexamples). However, representability holds in the pseudo-semisimple case since the deformation theory of fiberwise maximal tori in smooth affine group schemes [**SGA3**, XI] allows one to build an affine representing object as a non-commutative pushout analogous to the standard construction; see [**CP**, Prop. 6.2.2] (which does not depend on earlier results in [**CP**]).

Since $G = C \cdot \mathscr{D}(G)$ and $C' := C \cap \mathscr{D}(G)$ is a Cartan k-subgroup of the pseudo-semisimple $\mathscr{D}(G)$, clearly $\underline{\mathrm{Aut}}_{G,C} = \underline{\mathrm{Aut}}_{\mathscr{D}(G),C'}$. Hence, again at the cost of deeper input, one can alternatively deduce the existence of $\mathrm{Aut}_{G,C}$ from the representability of $\mathrm{Aut}_{\mathscr{D}(G)/k}$. However, this does not illuminate the structure of $Z_{G,C} = Z_{\mathscr{D}(G),C'}$ since $\mathrm{Aut}_{\mathscr{D}(G)/k}$ is generally *not* smooth [**CP**, Ex. 6.2.3] and its identity component is generally *larger* than $\mathscr{D}(G)/Z_{\mathscr{D}(G)}$ [**CP**, Rem. 6.2.5].

It is true that $Z_{G,C}$ is always connected [**CP**, Prop. 6.1.4], but this rests on a comprehensive understanding of the structure of rank-1 pseudo-split absolutely pseudo-simple groups with trivial center, an especially delicate task in characteristic 2; we will address the connectedness of $Z_{G,C}$ in §9.1.

The merit of $Z^0_{G,C}$ being a commutative pseudo-reductive group is that it allows us to define a *pseudo-reductive* k-group

$$(G \rtimes Z^0_{G,C})/C$$

as in Proposition 2.2.1 (using the evident k-homomorphism $C \to Z^0_{G,C}$ respecting the natural actions on each on G). The normal image of G under $g \mapsto (g,1) \bmod C$ clearly coincides with G/Z_G, establishing the pseudo-reductivity of G/Z_G!

Root systems and root groups behave as nicely with respect to central quotients as in the reductive case:

PROPOSITION 6.1.4. *Let $f : G \twoheadrightarrow \overline{G}$ be a central quotient map between pseudo-reductive k-groups, and assume G admits a split maximal k-torus T. Let $\overline{T} = f(T)$. Then $\Phi(\overline{G}, \overline{T}) \to \Phi(G,T)$ is bijective and for corresponding roots the map f induces an isomorphism between the associated root groups.*

In particular, if C is a standard pseudo-reductive k-group then the root system for G_{k_s} is reduced.

The invariance under central quotients in Proposition 6.1.4 is a simple application of the dynamic definition of root groups and properties of dynamic constructions; see [**CGP**, Prop. 2.3.15] (which establishes such a result for central quotients of any smooth connected affine k-group). In the standard case, G is a central quotient of $\mathrm{R}_{k'/k}(G') \rtimes C$ with a commutative k-group C, so the compatibility of the formation of root systems with respect to Weil restrictions (see Example 3.1.3) implies the asserted reducedness of the root system of G_{k_s} in the standard case.

6.2. Central extensions. To prove that a large class of pseudo-reductive groups G is standard, the essential case is when G is absolutely pseudo-simple. Our aim in this section is to explain how the task of proving standardness of a given absolutely pseudo-simple k-group G is (under suitable hypotheses) closely related to the study of certain central extensions. We first seek a mechanism to construct intrinsically from a central quotient G of $\mathrm{R}_{k'/k}(G')$, for a purely inseparable finite extension k' of k and a semisimple k'-group G', the pair $(k'/k, G')$ and the (central) quotient map from $\mathrm{R}_{k'/k}(G')$ onto G.

As an initial step we show how k'/k can be recovered from the group $\mathrm{R}_{k'/k}(G')$. The key notion for this purpose is "minimal field of definition" for a closed subscheme after a ground field extension, as follows. If X is any scheme over a field k and Z is a closed subscheme of X_K for an extension field K/k then among all subfields $L \subset K$ over k for which Z descends to a closed subscheme of X_L there is one such L that is contained in all others [**EGA**, IV$_2$, §4.8ff.]; we call L/k the *minimal field of definition over k for Z inside X_K*.

EXAMPLE 6.2.1. Consider a purely inseparable finite extension of fields k'/k and a nontrivial connected reductive k'-group G', and define $G := \mathrm{R}_{k'/k}(G')$. The

$\overline{k}$-subgroup $\mathscr{R}_u(G_{\overline{k}}) \subset G_{\overline{k}}$ descends to a k'-subgroup of $G_{k'}$ since the smoothness of G' and structure of $k' \otimes_k k'$ imply immediately that the natural map $q : G_{k'} \to G'$ defined functorially on points valued in any k'-algebra A' via

$$G_{k'}(A') = G(A') = G'(k' \otimes_k A') = G'(k' \otimes_k k' \otimes_{k'} A') \longrightarrow G'(A')$$

(using the natural quotient map $k' \otimes_k k' \twoheadrightarrow k'$) is a surjection with smooth connected unipotent kernel; i.e., $\ker q$ is a k'-descent of $\mathscr{R}_u(G_{\overline{k}})$. It is a much deeper fact that k'/k is *minimal* as a field of definition for the geometric unipotent radical; see [**CGP**, Prop. A.7.8(2)] (whose proof establishes that k'/k is even minimal as a field of definition for $\mathrm{Lie}(\mathscr{R}_u(G_{\overline{k}}))$ as a $\overline{k}$-subspace of $\mathrm{Lie}(G_{\overline{k}}) = \mathrm{Lie}(G)_{\overline{k}}$, a rather surprising fact in positive characteristic).

Ultimately we are interested in central quotients of pseudo-semisimple groups of the form $\mathrm{R}_{k'/k}(G')$, so it is essential to know that k'/k is characterized by the same minimality property for any such (pseudo-reductive) central quotient. This is a consequence of:

PROPOSITION 6.2.2. *Let H be a perfect smooth connected affine k-group and $\overline{H} := H/Z$ for a central closed k-subgroup $Z \subset H$. Then the minimal fields of definition over k for the geometric unipotent radicals of H and $\overline{H}$ coincide.*

PROOF. The induced map $H_{\overline{k}}^{\mathrm{red}} \to \overline{H}_{\overline{k}}^{\mathrm{red}}$ between connected semisimple groups has central kernel, so it induces an isomorphism between maximal adjoint semisimple quotients. Hence, to prove the proposition it suffices to show that the field of interest for each of H and $\overline{H}$ is unaffected if we work with the maximal adjoint semisimple quotient over $\overline{k}$ rather than with the maximal reductive (equivalently, semisimple) quotient over $\overline{k}$. This is a nontrivial problem due to the possibility that the scheme-theoretic center of a connected semisimple $\overline{k}$-group may not be étale. We refer the reader to (the self-contained proof of) [**CP**, Prop. 3.2.6] for this step, based on the fact that any perfect smooth connected affine k_s-group is generated by its maximal k_s-tori [**CGP**, Cor. A.2.11]. □

REMARK 6.2.3. The perfectness of H in Proposition 6.2.2 cannot be dropped. Indeed, over every imperfect field k there exists a non-reductive pseudo-reductive k-group $H \neq \mathscr{D}(H)$ such that H/Z_H is semisimple (of adjoint type) [**CGP**, Ex. 4.2.6]! The same example shows that for a smooth connected affine k-group H, the finite purely inseparable minimal field of definition K/k for the kernel

$$\mathscr{R}_u(H_{\overline{k}}) = \ker(H_{\overline{k}} \twoheadrightarrow H_{\overline{k}}^{\mathrm{red}})$$

of projection onto the maximal geometric reductive quotient can be strictly larger than the analogous subextension K'/k associated to the kernel of the projection $H_{\overline{k}} \twoheadrightarrow H_{\overline{k}}^{\mathrm{red}}/Z_{H_{\overline{k}}^{\mathrm{red}}}$ onto the maximal geometric adjoint semisimple quotient (whereas the equality $K = K'$ holds for *perfect* H; this is the key step in the proof of Proposition 6.2.2).

Now we can relate standardness to the splitting of central extensions. This rests on the following construction:

DEFINITION 6.2.4. Let G be a smooth connected affine k-group, and let K/k be the minimal field of definition over k for $\mathscr{R}_u(G_{\overline{k}}) \subset G_{\overline{k}}$ (so K/k is purely inseparable of finite degree). For $G' := G_K^{\mathrm{red}} = G_K/\mathscr{R}_{u,K}(G_K)$, define

$$i_G : G \longrightarrow \mathrm{R}_{K/k}(G')$$

to be the natural map corresponding to the quotient map $G_K \twoheadrightarrow G'$ via the universal property of $\mathrm{R}_{K/k}$. For perfect G we further define ξ_G to be the unique map

$$\xi_G : G \longrightarrow \mathscr{D}(\mathrm{R}_{K/k}(G'))$$

through which i_G factors.

EXAMPLE 6.2.5. Consider $(k'/k, G')$ as in Example 6.2.1 and define $G := \mathrm{R}_{k'/k}(G')$, so $K = k'$ and i_G corresponds to the natural quotient map $q : G_{k'} \to G'$ described in Example 6.2.1. By [**CGP**, Thm. 1.6.2(2)], under the universal property of Weil restriction q corresponds to the identity map $G \to \mathrm{R}_{k'/k}(G')$. Thus, i_G is an isomorphism (and is even identified with the identity map of G).

EXAMPLE 6.2.6. For an imperfect field k with characteristic p and $a \in k - k^p$, let $k' = k(a^{1/p^{2n}})$ and $\mathsf{k} = k(a^{1/p^n})$ for an integer $n > 0$. By [**CGP**, Ex. 5.3.7],

$$G := \mathrm{R}_{k'/k}(\mathrm{SL}_{p^n})/\mathrm{R}_{\mathsf{k}/k}(\mu_{p^n})$$

is a standard pseudo-reductive k-group that is absolutely pseudo-simple and the kernel $\ker i_G = \ker \xi_G = \mathrm{R}_{k'/k}(\mu_{p^n})/\mathrm{R}_{\mathsf{k}/k}(\mu_{p^n})$ has dimension $(p^n - 1)^2$.

We shall be interested in ξ_G primarily for absolutely pseudo-simple G since the extension field K/k is a poor invariant for other groups. Nonetheless, this map has an interesting property in the general pseudo-semisimple case:

PROPOSITION 6.2.7. *Let H be a smooth connected affine k-group and $\overline{H} := H/Z$ for a central closed k-subgroup $Z \subset H$. If H and $\overline{H}$ are both pseudo-semisimple then the surjectivity of ξ_H is equivalent to that of $\xi_{\overline{H}}$.*

The proof of this proposition, given in [**CP**, Prop. 4.1.5], rests on a study of open cells and the insensitivity of root groups and root systems with respect to central quotient maps between pseudo-reductive groups.

6.2.8. Let G, K/k, and G' be as in Definition 6.2.4. If G is perfect then the K-group G' is semisimple, so for such G it makes sense to consider the simply connected central cover $\pi : \widetilde{G}' \twoheadrightarrow G'$. Letting $\mu' := \ker \pi \subset Z_{\widetilde{G}'}$ and $\mathscr{Z} := \mathrm{R}_{K/k}(\mu')$, we see from [**CGP**, Prop. 1.3.4] that $\mathscr{D}(\mathrm{R}_{K/k}(G')) = \mathrm{R}_{K/k}(\widetilde{G}')/\mathscr{Z}$. Thus, we can rewrite ξ_G as a map

$$\xi_G : G \longrightarrow \mathrm{R}_{K/k}(\widetilde{G}')/\mathscr{Z}.$$

Assume G is absolutely pseudo-simple and standard. Then by Example 2.2.8 it is the central quotient $\mathrm{R}_{k'/k}(H')/Z$ for a purely inseparable finite extension k'/k and a connected semisimple k'-group H' that is absolutely simple and simply connected. As k' is the minimal field of definition over k for the geometric unipotent radical of $\mathrm{R}_{k'/k}(H')$ (see [**CGP**, Prop. A.7.8(2)]), $k' = K$ as a purely inseparable extension of k by Proposition 6.2.2. Let $q : \mathrm{R}_{K/k}(H')_K \to H'$ be as in Example 6.2.1 (with H' in place of G') and define $\mu = q(Z_K) \subset Z_{H'}$, so $Z \subset \mathrm{R}_{K/k}(\mu)$. The maximal reductive quotient of $\mathrm{R}_{K/k}(H')_K$ is H', so the maximal reductive quotient G' of G_K is H'/μ. Thus, H' is the simply connected central cover of G' and we identify it with $\widetilde{G}'$ over G', so $\mu = \mu'$.

Hence, if G is absolutely pseudo-simple and standard then $G = \mathrm{R}_{K/k}(\widetilde{G}')/Z$, where $\pi : \widetilde{G}' \to G'$ is the simply connected central cover of G', $Z \subset \mathscr{Z}$, and the map ξ_G is the natural quotient map $G = \mathrm{R}_{K/k}(\widetilde{G}')/Z \to \mathrm{R}_{K/k}(\widetilde{G}')/\mathscr{Z}$ between central quotients of $\mathrm{R}_{K/k}(\widetilde{G}')$. Thus, ξ_G is *surjective* with kernel that is *central* in G. Moreover, we claim that if the order of μ' is not divisible by char(k) (such as when $\mu' = 1$; i.e., when G' is simply connected) then $\xi_G\,(= i_G)$ is an isomorphism. To see this we may assume $k = k_s$, so the finite étale K-group $\mu = \mu'$ is constant and hence the only k-subgroup $Z \subset \mathscr{Z} := \mathrm{R}_{K/k}(\mu')$ for which $q : Z_K \to \mu' = \mu$ is surjective is $Z = \mathscr{Z}$.

Near the end of [**Ti3**, Cours 1992-93, II], Tits raised the question of characterizing those non-reductive absolutely pseudo-simple k-groups G for which i_G is an isomorphism. He settled most cases for which the root lattice and weight lattice in the root system *coincide*: E_8, F_4 away from characteristic 2, and G_2 away from characteristic 3. We can now give a criterion for i_G to be an isomorphism; we will revisit the topic in Remark 10.2.12 (after we have a good understanding of the non-standard case).

PROPOSITION 6.2.9. *Let G be a non-reductive absolutely pseudo-simple group over a field k of characteristic $p > 0$. Then i_G is an isomorphism if and only if G is standard and the order of the fundamental group of $G^{\mathrm{ss}}_{\overline{k}}$ is not divisible by p.*

The k-group $G = \mathscr{D}(\mathrm{R}_{K/k}(\mathrm{PGL}_p))$ with K/k purely inseparable of degree p explicitly exhibits the failure of i_G to be surjective when the order of the fundamental group of $G^{\mathrm{ss}}_{\overline{k}}$ is divisible by p.

PROOF. Let K be the minimal field of definition over k for the geometric unipotent radical of G, and define $G' = G^{\mathrm{ss}}_{K}$. Let us assume first that i_G is an isomorphism. Then, by definition, G is standard. To prove that p doesn't divide the order of the fundamental group of $G^{\mathrm{ss}}_{\overline{k}}$, we may and do assume $k = k_s$. Let $q : \widetilde{G}' \to G'$ be the simply connected central cover of G'. We need to prove that p doesn't divide the order of $\mu' := \ker q$.

Surjectivity of i_G implies that $\mathrm{R}_{K/k}(G')$ is perfect, and (as we saw in the proof of Proposition 2.2.7) the derived group of $\mathrm{R}_{K/k}(G')$ coincides with $\mathrm{R}_{K/k}(\widetilde{G}')/\mathrm{R}_{K/k}(\mu')$ due to the perfectness of $\mathrm{R}_{K/k}(\widetilde{G}')$. But $\mathrm{R}_{K/k}(\widetilde{G}')$ and $\mathrm{R}_{K/k}(G')$ have the same dimension, namely $[K : k]d$ for the common dimension d of G' and $\widetilde{G}'$, so $\mathrm{R}_{K/k}(\mu')$ must be 0-dimensional. If p divides $\#\mu'$ then $\dim \mathrm{R}_{K/k}(\mu_p) = 0$ since $\mu_p \subset \mu'$ (as K is separably closed and μ' is finite of multiplicative type). The purely inseparable finite extension K/k is nontrivial since G is assumed to be non-reductive. Letting k_0/k be a degree-p subextension, we obtain that $\dim \mathrm{R}_{k_0/k}(\mu_p) = 0$. But this latter dimension is $p - 1 > 0$ (see [**CGP**, Ex. 1.3.2]), giving a contradiction.

Conversely, if G is standard and the order of the fundamental group of $G^{\mathrm{ss}}_{\overline{k}}$ is not divisible by p, then as we saw in 6.2.8 the map i_G is an isomorphism. □

To go further, we require a framework in which certain central quotient maps

$$\mathrm{R}_{k'/k}(G') \to \mathrm{R}_{k'/k}(G')/Z$$

(with simply connected semisimple G') satisfy properties reminiscent of the "simply connected central cover" of a connected semisimple k-group. This begins with:

DEFINITION 6.2.10. Let Z be a commutative affine k-group scheme of finite type. We say that Z is *k-tame* if it does not contain a nontrivial unipotent k-subgroup scheme. A central extension

$$1 \longrightarrow Z \longrightarrow E \longrightarrow G \longrightarrow 1$$

of an affine k-group scheme G of finite type is *k-tame* if Z is k-tame.

EXAMPLE 6.2.11. If K/k is any finite extension of fields and M is a K-group scheme of multiplicative type then $\mathrm{R}_{K/k}(M)$ is k-tame (use the universal property of $\mathrm{R}_{K/k}$); the same holds more generally if M is a K-tame commutative affine K-group of finite type, and clearly a closed k-subgroup of a k-tame group is k-tame.

For example, if $E = \mathrm{R}_{k'/k}(G')$ for a nonzero finite reduced k-algebra k' and smooth affine k'-group G' with connected reductive fibers over the factor fields of k' then the center $Z_E = \mathrm{R}_{k'/k}(Z_{G'})$ is k-tame. Hence, every central closed k-subgroup of such E is k-tame.

Arguments with specialization and relative Verschiebung morphisms in positive characteristic ensure that if k'/k is a separable extension of fields then Z is k-tame if and only if $Z_{k'}$ is k'-tame [**CP**, Prop. 5.1.2]. This is used frequently without comment, especially for $k' = k_s$.

The interest in k-tameness is that the perfect smooth connected affine k-tame central extensions E of an arbitrary perfect smooth connected affine k-group G behave similarly to connected semisimple central extensions of connected semisimple groups. To make this precise, if E_1 and E_2 are two such k-tame central extensions of such a G then a *morphism* $f : E_1 \to E_2$ is a k-homomorphism over G. It is easy to verify that such an f is unique if it exists (as E_1 is perfect) and is automatically surjective; this defines a partial ordering ("$E_1 \geqslant E_2$") among such k-tame central extensions of G.

THEOREM 6.2.12. *Fix a perfect smooth connected affine k-group G, and let K/k be the minimal field of definition for $\mathscr{R}_u(G_{\overline{k}}) \subset G_{\overline{k}}$. The functor $E \rightsquigarrow E_K^{\mathrm{red}} := E_K/\mathscr{R}_{u,K}(E_K)$ is an equivalence between the category of perfect smooth connected k-tame central extensions of G and the category of connected semisimple central extensions of the connected semisimple K-group G_K^{red}.*

The perfect smooth connected k-tame central extension $\widetilde{G}$ of G corresponding to the simply connected central cover of G_K^{red} satisfies the following properties:

(i) *$\widetilde{G}$ is initial among all smooth k-tame central extensions of G,*
(ii) *if G is pseudo-semisimple then so is $\widetilde{G}$,*
(iii) *the formation of $\widetilde{G}$ is compatible with separable extension on k.*

Proposition 6.2.2 ensures that E_K^{red} is reductive (so also semisimple, as it is perfect).

PROOF. To establish the equivalence of categories, by Galois descent we may and do assume $k = k_s$ (so all k-tori are split). The essential step is to make a natural construction in the opposite direction: given a connected semisimple central extension E' of $G' := G_K^{\mathrm{red}}$, we seek a perfect smooth connected k-tame central extension E of G such that $E_K^{\mathrm{red}} \simeq E'$ over G'.

The simply connected central cover $\widetilde{G}'$ of G' is identified with that of E', so

$$\mu_{E'} := \ker(\widetilde{G}' \twoheadrightarrow E'),\ \ \mu := \ker(\widetilde{G}' \twoheadrightarrow G')$$

satisfy $\mu_{E'} \subset \mu$. Hence, it makes sense to form the fiber product

$$\mathscr{G}(E') = G \times_{\mathrm{R}_{K/k}(\widetilde{G}')/\mathrm{R}_{K/k}(\mu)} (\mathrm{R}_{K/k}(\widetilde{G}')/\mathrm{R}_{K/k}(\mu_{E'})).$$

The evident projection $\mathscr{G}(E') \to G$ is clearly surjective with central kernel that is k-tame since

$$\mathrm{R}_{K/k}(\mu_{E'})/\mathrm{R}_{K/k}(\mu) \subset \mathrm{R}_{K/k}(\mu_{E'}/\mu)$$

with $\mu_{E'}/\mu$ of multiplicative type.

The k-group scheme $\mathscr{G}(E')$ is generally not smooth (see [**CP**, Ex. 5.1.6] for non-smooth examples over any imperfect field), so this does not provide the desired reverse construction. Arguments with an "open cell" $\Omega_G(\lambda)$ as in Theorem 2.3.5(iii) yield that the maximal smooth closed k-subgroup

$$E := \mathscr{D}((\mathscr{G}(E')^{\mathrm{sm}})^0)$$

inside $\mathscr{G}(E')$ is a perfect connected k-group for which the projection $E \to G$ is central and surjective; see the proof of [**CP**, Thm. 5.1.3] for the details (showing that $E' \rightsquigarrow E$ is the sought-after reverse construction).

To verify (i), (ii), and (iii) we return to considering general k. Properties (i) and (ii) are rather formal, and explained near the beginning of [**CP**, §5.2]. For (iii) we recall that if Z is k-tame then $Z_{k'}$ is k'-tame for separable k'/k. □

In view of Theorem 6.2.12(i), we call $\widetilde{G}$ the *universal smooth k-tame central extension* of G.

EXAMPLE 6.2.13. If $G = \mathrm{R}_{k'/k}(\widetilde{G}')/Z$ as in 6.2.8 for k-tame Z then $\widetilde{G}$ is equal to $\mathrm{R}_{k'/k}(\widetilde{G}')$ equipped with the evident central quotient map onto G.

Here is an interesting application of the existence of $\widetilde{G}$:

PROPOSITION 6.2.14. *An absolutely pseudo-semisimple k-group G is standard if and only if G_{k_s} is standard.*

This result holds for any pseudo-reductive k-group, but we do not need that generality until much later and so postpone it to Corollary 10.2.8.

PROOF. It is immediate from the definition of standardness that if G is standard then so is G_{k_s}. Conversely, assume G_{k_s} is standard. Let K/k be the minimal field of definition for the geometric unipotent radical of G, so $K_s = K \otimes_k k_s$ has the same property for G_{k_s} over k_s by Galois descent. Let $G' = G_K/\mathscr{R}_{u,K}(G_K)$ and $\widetilde{G}'$ be the simply connected cover of G'. Then, by 6.2.8 the k_s-group G_{k_s} is a central quotient $\mathrm{R}_{K_s/k_s}(\widetilde{G}'_{K_s})/Z$. Since Z is k_s-tame by Example 6.2.11, $\mathrm{R}_{K_s/k_s}(\widetilde{G}'_{K_s})$ is the universal smooth k_s-tame central extension of G_{k_s}.

By Theorem 6.2.12, the universal smooth k-tame central extension $\widetilde{G}$ of G is pseudo-semisimple and $\widetilde{G}_{k_s}$ is the universal smooth k_s-tame central extension $\mathrm{R}_{K_s/k_s}(\widetilde{G}'_{K_s})$ of G_{k_s} (so $\widetilde{G}$ is absolutely pseudo-simple). Since every pseudo-reductive central quotient of a standard pseudo-semisimple group is standard (by Example 2.2.8 and the preservation of centrality under composition of quotient maps in Proposition 6.1.1), we may replace G with $\widetilde{G}$ to reduce to the case that $G_{k_s} \simeq \mathrm{R}_{K_s/k_s}(\widetilde{G}'_{K_s})$. Thus, $i_{G_{k_s}} (= (i_G)_{k_s})$ is an isomorphism (see 6.2.8). But then i_G is an isomorphism, so G is certainly standard. □

Recall that if G is standard then the root system of G_{k_s} is reduced (Proposition 6.1.4). We saw in 6.2.8 that $\ker \xi_G$ is central in the standard absolutely pseudo-simple case. An important ingredient in standardness proofs is a partial converse:

PROPOSITION 6.2.15. *Let G be a pseudo-reductive k-group such that G_{k_s} has a reduced root system. Then* $\ker i_G$ $(= \ker \xi_G)$ *is central in* G.

PROOF. A detailed proof of this important result is given in [**CP**, Prop. 2.3.4], and here we sketch the main ideas. We may replace k with k_s to arrange that k is separably closed. Let K and $G' = G_K/\mathscr{R}_{u,K}(G_K)$ be as in Definition 6.2.4. Let T be a maximal k-torus of G and $\Phi = \Phi(G,T)$. View $T' := T_K$ as a maximal K-torus of G', so $\Phi(G',T') = \Phi$ since the root system Φ has been assumed to be reduced. For $a \in \Phi$, let U_a be the corresponding root group of G, and U'_a that of G'.

Using the natural actions of T on U_a and of T' on U'_a, these commutative smooth connected unipotent groups over k and K respectively admit unique linear structures equivariant for the respective actions (by Corollary 3.1.10). Using the resulting linear structure on $\mathrm{R}_{K/k}(U'_a)$, the map $i_G|_{U_a} : U_a \to \mathrm{R}_{K/k}(U'_a)$ is equivariant with respect to the inclusion $T \hookrightarrow \mathrm{R}_{K/k}(T')$ and thus is linear, so $\ker(i_G|_{U_a})$ is a vector group. This kernel is therefore a smooth connected k-subgroup, but its geometric fiber is contained in $\mathscr{R}_u(G_{\overline{k}})$. Pseudo-reductivity of G implies that $\ker(i_G|_{U_a}) = 1$, so the a-weight space of $\mathrm{Lie}(\ker(i_G))$ is trivial for all $a \in \Phi$.

As $\ker(i_G)_{\overline{k}} \subset \mathscr{R}_u(G_{\overline{k}})$, we conclude using [**CGP**, Prop. 2.1.12(2)] that $\ker i_G \subset C := Z_G(T)$. Since C is commutative, and T was an arbitrary maximal k-torus, it follows that $\ker i_G$ commutes with every Cartan k-subgroup. Any smooth connected affine k-group is generated by its Cartan k-subgroups, so $\ker i_G$ is central. □

REMARK 6.2.16. The reducedness hypothesis in Proposition 6.2.15 is essential: for every $n \geqslant 1$ and imperfect field k of characteristic 2, by [**CGP**, Thm. 9.8.1(2),(4)] there exist pseudo-split absolutely pseudo-simple k-groups G with root system BC_n such that $\ker \xi_G$ is connected, commutative and directly spanned by nontrivial closed k-subgroups of root groups for multipliable roots (relative to a fixed split maximal k-torus). Since Z_G is contained in every Cartan k-subgroup of G, consideration of an open cell shows that $\ker \xi_G$ is necessarily non-central.

PROPOSITION 6.2.17. *If G is an absolutely pseudo-simple k-group such that G_{k_s} has a reduced root system then G is standard if and only if ξ_G is surjective. In such cases, i_G is an isomorphism when the connected semisimple $\overline{k}$-group $G^{\mathrm{red}}_{\overline{k}}$ is simply connected.*

PROOF. Since the root system is reduced, $\ker \xi_G$ is central by Proposition 6.2.15. If G is standard, then surjectivity of ξ_G was shown in 6.2.8. For the converse, assume ξ_G is surjective. Let $\widetilde{G} \twoheadrightarrow G$ be the universal smooth k-tame central extension of G, so $\xi_{\widetilde{G}}$ is surjective by Proposition 6.2.7. If $\widetilde{G}$ is standard then so is its pseudo-reductive central quotient G by Example 2.2.8 (and 6.2.8) Thus, we may assume $G^{\mathrm{ss}}_{\overline{k}}$ is simply connected. In such cases, the isomorphism property for i_G was proved in 6.2.8 for standard G.

It remains to show that standardness must hold under our current hypotheses (with $G^{\mathrm{ss}}_{\overline{k}}$ simply connected). The k-group G fits into a central extension

$$1 \to Z \to G \to \mathrm{R}_{K/k}(G') \to 1 \tag{6.2.17.1}$$

where K/k is a purely inseparable extension field and G' is a connected semisimple K-group that is absolutely simple and simply connected. Moreover, the finite type affine commutative k-group scheme $Z = \ker \xi_G = \ker i_G$ contains no nontrivial smooth connected k-subgroup, as any such k-subgroup would have to be central and unipotent (since $Z_{\overline{k}} \subset \mathscr{R}_u(G_{\overline{k}})$ by definition of i_G) yet G is pseudo-reductive.

To prove $Z = 1$ (so ξ_G is an isomorphism) it suffices to prove (6.2.17.1) splits, as then Z is a direct factor of the smooth connected G (so Z is smooth and connected).

The splitting of (6.2.17.1) is shown in [**CGP**, Prop. 5.1.3, Ex. 5.1.4], based on a general criterion for splitting a central extension of a nontrivial pseudo-semisimple group H by such Z. The criterion is that a Cartan k_s-subgroup C of H_{k_s} is "rationally generated" by root groups relative to the maximal k_s-torus in C (in a sense made precise in the statement of [**CGP**, Prop. 5.1.3]). The applicability of this criterion to $H = \mathrm{R}_{K/k}(G')$ rests on two properties of simply connected groups (such as G'): a basis of coroots for a maximal k_s-torus T' is a $\mathbf{Z}$-basis of $\mathrm{X}_*(T')$, and the root groups for a pair of opposite roots $\{a, -a\}$ generate SL_2 in which the (diagonal) torus $a^\vee(\mathrm{GL}_1)$ is "rationally generated" by the associated root groups U_a and U_{-a} due to the well-known identity

$$\begin{pmatrix} t & 0 \\ 0 & 1/t \end{pmatrix} = u_+(t)u_-(-1/t)u_+(t-1)u_-(1)u_+(-1)$$

for $u_+(x) := \left(\begin{smallmatrix} 1 & x \\ 0 & 1 \end{smallmatrix}\right)$ and $u_-(x) := \left(\begin{smallmatrix} 1 & 0 \\ x & 1 \end{smallmatrix}\right)$. □

REMARK 6.2.18. The splitting criterion used in the preceding proof is not applicable to generalizations of the standard construction in characteristics 2 and 3 that we will encounter later. For those purposes, the universal smooth k-tame central extension will provide a substitute sufficient for our needs.

7. Non-standard constructions

7.1. Groups of minimal type. For a pseudo-reductive k-group G with minimal field of definition K/k for its geometric unipotent radical and the associated maximal reductive quotient $G_K^{\mathrm{red}} := G_K/\mathscr{R}_{u,K}(G_K)$ over K, consider the map

$$i_G : G \longrightarrow \mathrm{R}_{K/k}(G_K^{\mathrm{red}})$$

introduced in Definition 6.2.4. In §6.2 we analyzed the kernel and image of i_G in the absolutely pseudo-simple case. Note that $\ker i_G$ is not sensitive to the minimality condition on K/k, and it is unipotent since $(\ker i_G)_K \subset \mathscr{R}_{u,K}(G_K)$ (see Remark 2.3.4 for the notion of unipotence without smoothness hypotheses). Example 6.2.6 provides *standard* absolutely pseudo-simple G satisfying $\dim \ker i_G > 0$ over any imperfect field k. One of the main difficulties in any attempt to classify pseudo-reductive groups is the structure of the unipotent group scheme $\ker i_G$.

EXAMPLE 7.1.1. Let k be an imperfect field with characteristic p. There exist commutative pseudo-reductive k-groups C such that $\ker i_C = \mathbf{Z}/p\mathbf{Z}$ [**CGP**, Ex. 1.6.3]; this is interesting because the center of a connected reductive k-group never has nontrivial étale p-torsion (as it is a k-group scheme of multiplicative type). It is natural to ask if there exists an *absolutely pseudo-simple* k-group G such that the unipotent normal k-subgroup scheme $\ker i_G$ is nontrivial and étale (forcing it to be central, by connectedness of G).

If $p > 2$ then no such G exists. The proof is quite nontrivial, and goes as follows. For an absolutely pseudo-simple k-group G that is standard, the k-group $\ker i_G = \ker \xi_G$ is connected by [**CGP**, Thm. 5.3.8] (this applies for all p). According to the classification results in Theorem 7.4.8 and Proposition 7.5.10, an absolutely pseudo-simple k-group is standard except possibly when $p \leqslant 3$, and the only other possibilities for G when $p = 3$ are certain "exotic" constructions for type G_2 that have trivial center.

In contrast, if $p = 2$ and $[k : k^2] \geqslant 16$ then there exist pseudo-split absolutely pseudo-simple k-groups G with root system A_1 and $G_{\overline{k}}^{\mathrm{red}} \simeq \mathrm{SL}_2$ such that $\ker i_G = \mathbf{Z}/2\mathbf{Z}$ [**CGP**, Rem. 9.1.11]. (No such G exists if $[k : k^2] \leqslant 8$ [**CP**, Prop. B.3.1].)

If the root system of G_{k_s} is reduced then the closed normal k-subgroup $\ker i_G$ of G is central by Proposition 6.2.15, so root groups over k_s contribute nontrivially to $\ker i_G$ only when G_{k_s} has a non-reduced root system. In general, for a Cartan k-subgroup $C \subset G$,

$$\mathscr{C}_G := C \cap \ker i_G$$

is the maximal central unipotent k-subgroup scheme of G [**CP**, Prop. 2.3.7]; thus, $\mathscr{C}_G$ is independent of C. Whether or not $\mathscr{C}_G$ is trivial can be detected after an arbitrary separable extension on k because $(\mathscr{C}_G)_{k'} = \mathscr{C}_{G_{k'}}$ inside $G_{k'}$ for separable extension fields k'/k [**CP**, Lemma 2.3.6].

Since $\mathscr{C}_{G/\mathscr{C}_G} = 1$ and the k-groups G and $G/\mathscr{C}_G$ share the same minimal field of definition over k for their geometric unipotent radicals and share the same root data over k_s (see [**CGP**, Prop. 9.4.2, Cor. 9.4.3] for proofs), it is natural to focus attention on the cases for which $\mathscr{C}_G = 1$. We give these a special name:

DEFINITION 7.1.2. A pseudo-reductive k-group G is of *minimal type* if $\mathscr{C}_G = 1$.

If k'/k is a separable extension then G is of minimal type if and only if $G_{k'}$ is of minimal type. The minimal-type property is also inherited by smooth connected normal k-subgroups (such as derived groups) and torus centralizers [**CP**, Lemma 2.3.10]. Likewise, if k'/k is a finite extension of fields and G' is a pseudo-reductive k'-group of minimal type then the Weil restriction $G := \mathrm{R}_{k'/k}(G')$ is of minimal type. (Indeed, we may assume $k = k_s$ upon passing to factor fields of $k' \otimes_k k_s$, so now k'/k is purely inseparable. Thus, $\pi : G_{k'} \to G'$ is surjective by [**CGP**, Prop. A.5.11], so the unipotent k'-group scheme $\pi((\mathscr{C}_G)_{k'})$ is central in G'. Hence, $\mathscr{C}_G \subset \mathrm{R}_{k'/k}(\mathscr{C}_{G'}) = 1$.)

Examples of groups not of minimal type are given by the standard absolutely pseudo-simple k-groups G in Example 6.2.6 (which satisfy $\mathscr{C}_G = \ker i_G \neq 1$). Here are two interesting sources of pseudo-reductive groups H for which $\ker i_H = 1$ (so H is of minimal type):

PROPOSITION 7.1.3. *Let K/k be a purely inseparable finite extension of fields.*

(i) *Let L be a connected reductive k-group. Every smooth connected intermediate k-group $L \subset H \subset \mathrm{R}_{K/k}(L_K)$ is pseudo-reductive with L as a Levi k-subgroup (so the natural map $H_K \to L_K$ is a K-descent of $H_{\overline{k}} \twoheadrightarrow H_{\overline{k}}^{\mathrm{red}}$, the minimal field of definition k'/k for $\mathscr{R}_u(H_{\overline{k}}) \subset H_{\overline{k}}$ is a subextension of K/k, $H \subset \mathrm{R}_{k'/k}(L_{k'})$, and the latter inclusion is i_H).*

(ii) *For any pseudo-reductive k-group G with minimal field of definition K/k for its geometric unipotent radical and $G' := G_K/\mathscr{R}_{u,K}(G_K)$, the k-group $H := i_G(G)$ is pseudo-reductive with minimal field of definition K/k for its geometric unipotent radical and its inclusion into $\mathrm{R}_{K/k}(G')$ is identified with i_H. In particular, H is of minimal type.*

In (i), the inclusion $H \subset \mathrm{R}_{k'/k}(L_{k'})$ and its identification with i_H rest on the fact that via $L \hookrightarrow H$ the map $L_{k'} \to H_{k'}/\mathscr{R}_{u,k'}(H_{k'})$ is a k'-descent of the analogue over K that is a K-isomorphism whose inverse arises from $H_K \twoheadrightarrow L_K$. Note also that in (ii) we permit G_{k_s} to have a non-reduced root system, in which case (as H_{k_s} has a reduced root system) $\ker i_G$ is non-central in G (by Proposition 6.1.4).

PROOF. By Example 6.2.5 the natural map $\mathrm{R}_{K/k}(L_K)_K \to L_K$ (which restricts to the identity on the K-subgroup $L_K \subset \mathrm{R}_{K/k}(L_K)_K$) has smooth connected unipotent kernel, so L is a Levi k-subgroup of $\mathrm{R}_{K/k}(L_K)$. Letting $U = \mathscr{R}_u(\mathrm{R}_{K/k}(L_K)_{\overline{k}})$, the equality $L_{\overline{k}} \ltimes U = \mathrm{R}_{K/k}(L_K)_{\overline{k}}$ implies

$$L_{\overline{k}} \ltimes (U \cap H_{\overline{k}}) = H_{\overline{k}}.$$

But $H_{\overline{k}}$ is smooth and connected, so its unipotent normal subgroup scheme $U \cap H_{\overline{k}}$ is forced to be smooth and connected. We conclude that $U \cap H_{\overline{k}} = \mathscr{R}_u(H_{\overline{k}})$, so L is a Levi k-subgroup of H.

Since the k-subgroup $\mathscr{R}_{u,k}(H)$ of the pseudo-reductive $\mathrm{R}_{K/k}(L_K)$ satisfies

$$\mathscr{R}_{u,k}(H)_{\overline{k}} \subset \mathscr{R}_u(H_{\overline{k}}) \subset U := \mathscr{R}_u(\mathrm{R}_{K/k}(L_K)_{\overline{k}}),$$

it follows from [**CGP**, Lemma 1.2.1] that $\mathscr{R}_{u,k}(H) = 1$. This establishes (i).

Now consider (ii), for which we may assume $k = k_s$. Hence, by Theorem 5.4.4 we can choose a Levi k-subgroup $L \subset G$, so the natural map $L_K \to G'$ is an isomorphism. Using this K-isomorphism, $i_G(G)$ is identified with an intermediate group between $\mathrm{R}_{K/k}(L_K)$ and L. Hence, by (i) we get everything in (ii) except that the minimal field of definition over k for $\mathscr{R}_u(H_{\overline{k}}) \subset H_{\overline{k}}$ is merely a subfield $k' \subset K$ over k and correspondingly $H \subset \mathrm{R}_{k'/k}(L_{k'})$ inside $\mathrm{R}_{K/k}(G') = \mathrm{R}_{K/k}(L_K)$ with this inclusion equal to i_H. Hence, it just has to be shown that $K = k'$.

The composite map

$$G \twoheadrightarrow i_G(G) \hookrightarrow \mathrm{R}_{k'/k}(L_{k'})$$

corresponds to a k'-homomorphism $q : G_{k'} \to L_{k'}$. Clearly q_K corresponds to the map $i_G : G \to \mathrm{R}_{K/k}(L_K) = \mathrm{R}_{K/k}(G')$, so q_K is surjective with a smooth connected unipotent kernel. Hence, q is also surjective with a smooth connected unipotent kernel, so q is a k'-descent of the maximal geometric reductive quotient of G. By minimality of K/k, this forces the inclusion $k' \subset K$ over k to be an equality. □

Pseudo-split pseudo-reductive groups of minimal type with a *reduced* root system always arise as in Proposition 7.1.3(i):

EXAMPLE 7.1.4. Consider pseudo-reductive k-groups G such that the root system of G_{k_s} is reduced (as is automatic except when k is imperfect with characteristic 2). The kernel $\ker i_G$ is central (Proposition 6.2.15), so G is of minimal type if and only if $\ker i_G = 1$. If G is of minimal type and pseudo-split and K/k is the minimal field of definition for the geometric unipotent radical then Theorem 5.4.4 provides a split Levi k-subgroup L of G, so the natural map $L_K \to G_K^{\mathrm{red}}$ is an isomorphism and hence i_G identifies G with a k-subgroup of $\mathrm{R}_{K/k}(L_K)$ containing L.

As an illustration, consider a pseudo-split *pseudo-semisimple* k-group H of minimal type with root system A_1 and denote by K/k the minimal field of definition for its geometric unipotent radical. Fix a split maximal k-torus $T \subset H$, so by Theorem 5.4.4 there exists a Levi k-subgroup $L \subset H$ containing T. We may and do identify L with SL_2 or PGL_2 carrying T onto the diagonal k-torus. Since $\ker i_H = 1$, this identifies H with a k-subgroup of $\mathrm{R}_{K/k}(L_K)$ containing L. For the T-root groups $U^{\pm} \subset \mathrm{R}_{K/k}(\mathbf{G}_{\mathrm{a}})$ of H containing the canonical $\mathbf{G}_{\mathrm{a}} \subset \mathrm{R}_{K/k}(\mathbf{G}_{\mathrm{a}})$, the stability of $U^{\pm}$ under T-conjugation implies that the subsets $V^{\pm} := U^{\pm}(k)$ of K containing k are k-subspaces, and that conjugation on H by the standard Weyl element $w = \left(\begin{smallmatrix} 0 & 1 \\ -1 & 0 \end{smallmatrix}\right) \in L(k)$ swaps the root groups via negation relative to the standard parameterizations of the root groups. Hence, $V^+ = V^-$ inside K.

Denote the common k-subspace $V^+ = V^-$ of K containing 1 as V, and let $\underline{V}$ be the corresponding k-subgroup of $\mathrm{R}_{K/k}(\mathbf{G}_\mathrm{a})$. By Proposition 3.1.4, pseudo-semisimplicity of H implies that H is the k-subgroup of $\mathrm{R}_{K/k}(\mathrm{SL}_2)$ or $\mathrm{R}_{K/k}(\mathrm{PGL}_2)$ generated by $\underline{V}$ inside both root groups relative to the diagonal k-torus. For each $L \in \{\mathrm{SL}_2, \mathrm{PGL}_2\}$, this describes all possibilities for such H up to k-isomorphism in terms of possibilities for V, but the relationship between V and K/k needs to be described, as does the characterization of when two such permissible V's give rise to isomorphic k-groups. These matters will be addressed in §7.2.

The notion of "minimal type" is useful when proving classification results and general structure theorems for pseudo-reductive k-groups G because the central pseudo-reductive quotient $G/\mathscr{C}_G$ is of minimal type and has the same associated extension K/k (so passage to $G/\mathscr{C}_G$ is compatible with the formation of i_G). It is convenient in some proofs to first treat the minimal-type case and then to infer the general case. The proof of standardness of all pseudo-reductive groups away from characteristics 2 and 3 (shown in Theorem 7.4.8 for absolutely pseudo-simple groups, and deduced in general in Corollary 10.2.14) uses such a technique.

There is good behavior of the "minimal type" property with respect to the useful operations of passage to normal k-subgroups and centralizers of subgroup schemes of multiplicative type:

PROPOSITION 7.1.5. *For a pseudo-reductive k-group G of minimal type, every smooth connected normal k-subgroup is of minimal type and $Z_G(M)^0$ is of minimal type for every closed k-subgroup scheme M of a k-torus in G.*

To make sense of the statement of Proposition 7.1.5 it is necessary to first show that the k-subgroups of G being considered are pseudo-reductive; the pseudo-reductivity of smooth connected normal k-subgroups of G is elementary, and for k-subgroups of the form $Z_G(M)^0$ it is rather nontrivial when M is non-smooth; see [**CGP**, Prop. A.8.14(2)]. The idea of the proof of Proposition 7.1.5 is to show for any k-subgroup H of G that is of either of the two types under consideration, the following two properties hold: $H \cap \ker i_G = \ker i_H$, and $H \cap C$ is a Cartan k-subgroup of H for any Cartan k-subgroup C of G when H is normal in G as well as for the specific Cartan k-subgroup $C = Z_G(T)$ when $H = Z_G(M)^0$ for a maximal k-torus T of G and closed k-subgroup scheme $M \subset T$. It is then immediate that $\mathscr{C}_H = H \cap \mathscr{C}_G$ by definition of $\mathscr{C}_H$ (so $\mathscr{C}_H = 1$ when $\mathscr{C}_G = 1$). The equality $H \cap \ker i_G = \ker i_H$ amounts to showing $\ker i_H \subset \ker i_G$ and that $\mathscr{R}_u(H_{\overline{k}}) = H_{\overline{k}} \cap \mathscr{R}_u(G_{\overline{k}})$, which is nontrivial when $H = Z_G(M)^0$ for non-smooth M as above; see [**CGP**, Prop. 9.4.5] for the details.

EXAMPLE 7.1.6. As an illustration of the technical advantages of the "minimal type" case, consider a smooth connected affine k-group G and smooth connected normal k-subgroup N. For the maximal pseudo-reductive quotients

$$G^{\mathrm{pred}} := G/\mathscr{R}_{u,k}(G), \quad N^{\mathrm{pred}} := N/\mathscr{R}_{u,k}(N)$$

there is a natural map $N^{\mathrm{pred}} \to G^{\mathrm{pred}}$ because the smooth connected image of N in the pseudo-reductive k-group G^{pred} is normal and hence pseudo-reductive, and over any imperfect field k we give examples in [**CP**, Rem. 2.3.14] of such G and N for which $\ker(N^{\mathrm{pred}} \to G^{\mathrm{pred}})$ has positive dimension.

The situation is better if we consider the maximal quotients

$$G^{\mathrm{prmt}} := G^{\mathrm{pred}}/\mathscr{C}_{G^{\mathrm{pred}}}, \quad N^{\mathrm{prmt}} := N^{\mathrm{pred}}/\mathscr{C}_{N^{\mathrm{pred}}}$$

that are pseudo-reductive of minimal type. Indeed, there is a natural map $N^{\mathrm{prmt}} \to G^{\mathrm{prmt}}$ since the smooth connected image of N in G^{prmt} is normal and hence pseudo-reductive of minimal type, and we claim that this map always has trivial kernel.

The key to the proof of such triviality, given in detail in [**CP**, Prop. 2.3.13], is that after reducing to the case $k = k_s$ with N pseudo-reductive of minimal type, the unipotent subgroup scheme $N \cap \mathscr{R}_{u,k}(G)$ is *central* in N. To establish this centrality, observe that the smooth connected normal commutator k-subgroup $(N, \mathscr{R}_{u,k}(G)) \subset N$ is unipotent since it is contained in $\mathscr{R}_{u,k}(G)$. Thus, this commutator subgroup is contained in $\mathscr{R}_{u,k}(N)$, and $\mathscr{R}_{u,k}(N) = 1$ since N is pseudo-reductive, so the centrality follows. But we arranged that $\mathscr{C}_N = 1$, so N has no nontrivial unipotent central k-subgroup. Hence, $N \cap \mathscr{R}_{u,k}(G) = 1$, so we may replace G with $G/\mathscr{R}_{u,k}(G)$ to make G pseudo-reductive. We need to show that $N \cap \mathscr{C}_G = 1$. The proof of the normal case in Proposition 7.1.5 yields that $N \cap \mathscr{C}_G = \mathscr{C}_N$, yet we arranged that N is of minimal type, so $\mathscr{C}_N = 1$.

EXAMPLE 7.1.7. Let G be a pseudo-split pseudo-reductive k-group, and T a split maximal k-torus in G. For $a \in \Phi(G,T)$, let $G_a := \langle U_a, U_{-a} \rangle$ be the smooth connected k-subgroup generated by the $\pm a$-root groups.

We saw in Remark 3.2.8 that G_a is pseudo-split and absolutely pseudo-simple with 1-dimensional maximal k-torus $a^\vee(\mathrm{GL}_1) = T \cap G_a$, and that G_a is described in terms of passage to successive k-subgroups considered in Proposition 7.1.5 (depending on whether or not a is divisible). Hence, G_a is of minimal type whenever G is of minimal type (and its root system is $\{\pm a\}$ when a is not a multipliable root of G, and is $\{\pm a, \pm 2a\}$ otherwise). This often permits passage to the absolutely pseudo-simple rank-1 case when proving general theorems for pseudo-reductive groups of minimal type.

A very important feature of passage to such k-subgroups G_a is that it interacts well with the maps i_G and i_{G_a}. To explain this, the key point is that the explicit description of G_a, depending on whether or not a is divisible, yields the equality of group schemes

$$\mathscr{R}_u((G_a)_{\overline{k}}) = (G_a)_{\overline{k}} \cap \mathscr{R}_u(G_{\overline{k}})$$

(use [**CGP**, Prop. A.4.8] for non-divisible a, and [**CGP**, Prop. A.8.14(2)] for divisible a). It follows that $(G_a)_K \cap \mathscr{R}_{u,K}(G_K)$ is a K-descent of $\mathscr{R}_u((G_a)_{\overline{k}})$, so the minimal field of definition K_a/k for the geometric unipotent radical of G_a is a subextension of K/k and the restriction $i_G|_{G_a}$ is the composition of i_{G_a} with the *inclusion* of k-group schemes

$$\mathrm{R}_{K_a/k}((G_a)') \hookrightarrow \mathrm{R}_{K/k}((G_a)'_K) \hookrightarrow \mathrm{R}_{K/k}(G'),$$

where $G' := G_K/\mathscr{R}_{u,K}(G_K)$ and $(G_a)' := (G_a)_{K_a}/\mathscr{R}_{u,K_a}((G_a)_{K_a})$. In particular, naturally $i_G(G_a) \simeq i_{G_a}(G_a)$, so Proposition 7.1.3(ii) applied to G_a implies that $i_G(G_a)$ is pseudo-split and absolutely pseudo-simple of minimal type with minimal field of definition K_a/k for its geometric unipotent radical and root system $\{\pm a'\}$, where $a' = a$ in the non-multipliable case and $a' = 2a$ in the multipliable case.

7.2. Rank-1 groups and applications. The structure theory of split connected reductive groups rests on the fact that SL_2 and PGL_2 are the only split connected semisimple groups of rank 1. (For example, this result is the reason that root groups for split connected reductive groups are 1-dimensional.) Likewise, our classification of pseudo-reductive groups will require a description of all pseudo-split pseudo-semisimple groups of minimal type with root system A_1 or BC_1 (the latter only relevant over imperfect fields of characteristic 2).

In this section we describe the A_1-cases, and in §7.4 use that to define a useful invariant called the *root field*. The proof of exhaustiveness of our list of groups will use the (pseudo-split) Bruhat decomposition, in contrast with the reductive case.

Recall that in Example 7.1.4, for any purely inseparable finite extension K/k we described all pseudo-split pseudo-simple k-subgroups of minimal type with root system A_1 and minimal field of definition K/k for the geometric unipotent radical. This description was given in terms of certain k-subspaces $V \subset K$ containing 1. However, we did not characterize exactly which V can occur, and for any two such V we did not determine when the associated k-groups are k-isomorphic. The most interesting case is when k is imperfect of characteristic 2 and $[k : k^2] > 2$, since in all other cases it will turn out that necessarily $V = K$. Thus, we shall begin by describing the rank-1 pseudo-split pseudo-simple construction in Example 7.1.4 from a broader point of view over imperfect fields of characteristic 2.

Let k be imperfect with $\mathrm{char}(k) = 2$, K/k a nontrivial purely inseparable finite extension, and $V \subset K$ a nonzero kK^2-subspace such that the k-subalgebra $k\langle V\rangle$ generated by the ratios v/v' for $v, v' \in V - \{0\}$ coincides with K. (If $[k : k^2] = 2$ then $V = K$.) Identify the root groups of $\mathrm{R}_{K/k}(\mathrm{SL}_2)$ and $\mathrm{R}_{K/k}(\mathrm{PGL}_2)$ relative to their diagonal k-tori with $\mathrm{R}_{K/k}(\mathbf{G}_a)$ in the standard manner, and let $\underline{V}^+$ and $\underline{V}^-$ be the k-subgroups of these root groups corresponding to $V \subset K$ (with $\underline{V}^+$ inside the upper-triangular root group, and $\underline{V}^-$ inside the lower-triangular root group).

DEFINITION 7.2.1. Let $H_{V,K/k} \subset \mathrm{R}_{K/k}(\mathrm{SL}_2)$ be the k-subgroup generated by $\underline{V}^{\pm}$, and let $\mathrm{P}H_{V,K/k} \subset \mathrm{R}_{K/k}(\mathrm{PGL}_2)$ be defined similarly (so there is a natural surjection $H_{V,K/k} \to \mathrm{P}H_{V,K/k}$).

REMARK 7.2.2. It is generally difficult to describe the kernel of $H_{V,K/k} \twoheadrightarrow \mathrm{P}H_{V,K/k}$ as a k-subgroup of the center $\mathrm{R}_{K/k}(\mu_2)$ of $\mathrm{R}_{K/k}(\mathrm{SL}_2)$. Examples of pairs $(K/k, V)$ for which this kernel is a proper k-subgroup of $\mathrm{R}_{K/k}(\mu_2)$ can be built over any k satisfying $[k : k^2] \geqslant 16$; see [**CP**, Rem. 3.1.5].

PROPOSITION 7.2.3. *Let L be SL_2 or PGL_2, and let $D \subset L$ be the diagonal k-torus. Let H denote the corresponding k-subgroup $H_{V,K/k}$ or $\mathrm{P}H_{V,K/k}$ of $\mathrm{R}_{K/k}(L_K)$, with $K = k\langle V\rangle$.*

(i) *The k-group H is absolutely pseudo-simple of minimal type with root system A_1 and the minimal field of definition for its geometric unipotent radical is K/k. It contains $D = \mathrm{GL}_1$, and $Z_H(D) \subset \mathrm{R}_{K/k}(D_K) = \mathrm{R}_{K/k}(\mathrm{GL}_1)$ coincides with the k-subgroup $V^*_{K/k}$ generated by the ratios $v/v' \in K^\times = \mathrm{R}_{K/k}(\mathrm{GL}_1)(k)$ for nonzero $v, v' \in V$, and the D-root groups of H are equal to $\underline{V}^{\pm}$.*

(ii) *If V' is another nonzero kK^2-subspace of K such that $k\langle V'\rangle = K$ and H' denotes the associated k-subgroup of $\mathrm{R}_{K/k}(L_K)$ then $H' \simeq H$ if and only if $V' = cV$ for some $c \in K^\times$.*

The notation $V^*_{K/k}$ does not keep track of which of the two possibilities for L is under consideration, but the context will always make the intended meaning clear.

PROOF. The action of $\mathrm{diag}(c, 1) \in \mathrm{PGL}_2(K)$ on $\mathrm{R}_{K/k}(L_K)$ carries $\underline{V}^+$ onto $\underline{cV}^+$ and carries $\underline{V}^-$ onto $\underline{c^{-1}V}^- = \underline{cV}^-$ (equality since V is a kK^2-subspace of

K). By choosing $c = 1/v_0$ for a nonzero $v_0 \in V$, we reduce the verification of the properties of H to the cases for which $1 \in V$. By construction H contains the k-subgroups $\underline{V}^{\pm} \subset \mathrm{R}_{K/k}(\mathbf{G}_{\mathrm{a}})$ that now contain $\mathbf{G}_{\mathrm{a}}$; these $\mathbf{G}_{\mathrm{a}}$'s are the D-root groups of L. Since L is generated by such root groups, now $L \subset H$. Hence, by Proposition 7.1.3(i), H is pseudo-reductive of minimal type with L as a Levi k-subgroup and K/k is a (not necessarily minimal) field of definition for $\mathscr{R}_u(H_{\overline{k}}) \subset H_{\overline{k}}$.

It is not at all clear that K/k is minimal as a field of definition for $\mathscr{R}_u(H_{\overline{k}}) \subset H_{\overline{k}}$, nor that H is perfect with D-root groups $\underline{V}^{\pm}$ and $Z_H(D)$ generated by the ratios v/v'. The verification of these properties rests on explicit calculations in $L(K)$ and dynamic considerations with the open cell of H relative to a 1-parameter subgroup $\mathrm{GL}_1 \simeq D \subset H$ (Theorem 2.3.5(ii)); see [**CP**, Prop. 3.1.4] for the details.

Consider V' and H' as in (ii) such that there exists a k-isomorphism $f : H' \simeq H$. We want to show that V' is a $K^{\times}$-multiple of V. By Theorem 4.2.9, we may compose f with an $H(k)$-conjugation so that $f(D) = D$. The effect of f on $D = \mathrm{GL}_1$ is either the identity or inversion. Proposition 4.1.3 provides an element in $N_H(D)(k)$ whose effect on $D = \mathrm{GL}_1$ is inversion, so composing f with conjugation by such an element if necessary allows us to arrange that f restricts to the identity on D. Hence, the associated K-isomorphism $f_K^{\mathrm{red}} : {H'}_K^{\mathrm{red}} \simeq H_K^{\mathrm{red}}$ is identified with a K-automorphism of L_K restricting to the identity on D_K. But $\mathrm{Aut}_K(L_K)$ is identified with $\mathrm{PGL}_2(K)$, so f_K^{red} is induced by the action of a unique diagonal matrix $\mathrm{diag}(c, 1)$ with $c \in K^{\times}$. By canonicity of i_H and $i_{H'}$, f is induced by $\mathrm{R}_{K/k}(f_K^{\mathrm{red}})$ on $\mathrm{R}_{K/k}(L_K)$. Hence, inspection of D-root groups implies that $cV = V'$. □

REMARK 7.2.4. The construction of H in Proposition 7.2.3 can be carried out if the combined hypotheses that the nonzero k-subspace $V \subset K$ is a kK^2-subspace and that $k\langle V\rangle = K$ are relaxed to the single weaker hypothesis that V is a nonzero $k\langle V^2\rangle$-subspace of K (where $k\langle V^2\rangle$ denotes the k-subalgebra of K generated by the ratios $v^2/{v'}^2$ for nonzero $v, v' \in V$). The proof of Proposition 7.2.3 carries over in this generality essentially without change, except that the minimal field of definition over k for the geometric unipotent radical of H is $k\langle V\rangle$ (a subextension of K/k that might not contain V in general, though certainly can be arranged to contain V after replacing V with $(1/v_0)V$ for a nonzero $v_0 \in V$).

An advantage of this more general context for the construction of H is that it then makes sense to consider how the formation of H interacts with Weil restriction. This is a subtle problem because the structure (and even merely the dimension!) of the Cartan k-subgroup $Z_H(D)$ is generally intractable. We refer the reader to [**CP**, Ex. 3.1.6] for a discussion of these matters.

Now we can complete the analysis of the possibilities for V in Example 7.1.4 by combining the pseudo-split Bruhat decomposition over general fields and Proposition 7.2.3 over imperfect fields of characteristic 2:

THEOREM 7.2.5. *Let G be an absolutely pseudo-simple group over a field k, and assume G_{k_s} has root system of rank 1. Let K/k be the minimal field of definition for the geometric unipotent radical, and $G' = G_K/\mathscr{R}_{u,K}(G_K)$.*

(i) *If $\mathrm{char}(k) \neq 2$ or k is perfect then $i_G : G \to \mathrm{R}_{K/k}(G')$ is an isomorphism.*
(ii) *Assume k is imperfect with characteristic 2 and G is pseudo-split. Let $L \subset G$ be a Levi k-subgroup containing a split maximal k-torus $T \subset G$, so $i_G(G)$ is a k-subgroup of $\mathrm{R}_{K/k}(L_K)$ containing L. Fix a k-isomorphism of L onto SL_2 or PGL_2 such that T is carried onto the diagonal k-torus*

D. There exists a nonzero kK^2-subspace $V \subset K$ satisfying $k\langle V\rangle = K$ such that $i_G(G)$ is equal to $H_{V,K/k}$ or $\mathrm{P}H_{V,K/k}$ respectively.

Part (i) is [**CGP**, Thm. 6.1.1], and the proof we give (via arguments in [**CP**, §3.1]) is a significant simplification in technique due to using Levi k-subgroups and the Bruhat decomposition (in the pseudo-split case).

PROOF. The case of perfect k in (i) is trivial, so when we consider (i) below we will always assume $\mathrm{char}(k) \neq 2$ (so the root system is A_1 since it is reduced in such cases). Since V in (ii) is clearly unique if it exists (by inspection of D-root groups in $i_G(G)$), for the entire proof we may assume $k = k_s$. Hence, we are in the pseudo-split situation as considered in Example 7.1.4; we let T be a split maximal k-torus of G and $L \subset G$ a Levi k-subgroup containing T (provided by Theorem 5.4.4). By Proposition 7.1.3, the quotient $i_G(G)$ is pseudo-reductive of minimal type with the same minimal field of definition K/k for its geometric unipotent radical, and as a k-subgroup of $\mathrm{R}_{K/k}(L_K)$ containing L its inclusion into $\mathrm{R}_{K/k}(L_K)$ is identified with $i_{i_G(G)}$. Hence, to prove (ii) we may replace G with $i_G(G)$ so that G is of minimal type with root system A_1 (rather than BC_1).

Let us show that for the proof of (i) it is also harmless to replace G with $i_G(G)$. By Proposition 6.2.17, if (i) is settled for $i_G(G)$ then in general any such G is at least standard. Being absolutely pseudo-simple, it would follow from 6.2.8 that $G \simeq \mathrm{R}_{K/k}(\mathrm{SL}_2)/Z$ for a closed k-subgroup $Z \subset \mathrm{R}_{K/k}(\mu_2)$. But μ_2 is finite étale since $\mathrm{char}(k) \neq 2$ in (i), so since K/k is purely inseparable it follows that $\mathrm{R}_{K/k}(\mu_2)$ is finite étale. Thus, the natural map $\mu_2 \to \mathrm{R}_{K/k}(\mu_2)$ is an isomorphism by comparison of k-points (recall $k = k_s$). It follows that the only possibilities for Z would be $Z = 1$ and $Z = \mathrm{R}_{K/k}(\mu_2)$, in which case $G \simeq \mathrm{R}_{K/k}(\mathrm{SL}_2)$ and $G \simeq \mathrm{R}_{K/k}(\mathrm{PGL}_2)$ respectively (as $\mathrm{R}_{K/k}$ is compatible with the formation of quotients modulo smooth closed subgroups [**CGP**, Cor. A.5.4(3)]). The isomorphism property for i_G in such cases is then part of Example 6.2.5.

For the rest of the argument, now we may assume G is also of minimal type, so the analysis in Example 7.1.4 is applicable, giving that G is identified with the k-subgroup of $\mathrm{R}_{K/k}(L_K)$ generated by $\underline{V}^{\pm}$ for a k-subspace $V \subset K$ containing 1. It has to be shown that if $\mathrm{char}(k) \neq 2$ then $V = K$ and that if $\mathrm{char}(k) = 2$ then $k\langle V\rangle = K$ and V is a kK^2-subspace of K.

In fact, it suffices in all characteristics to prove that $v^2 \cdot V \subset V$ for all $v \in V$. Indeed, granting this we see that $(1+v)^2 \cdot v', v^2 \cdot v' \in V$ for all $v, v' \in V$, so $vv' \in V$ when $\mathrm{char}(k) \neq 2$. Thus, now assuming $\mathrm{char}(k) \neq 2$, V is a k-subalgebra of K; equivalently, V is a field F between k and K. But then G coincides with the k-subgroup $\mathrm{R}_{F/k}(L_F)$ inside $\mathrm{R}_{K/k}(L_K)$ because $\mathrm{R}_{F/k}(L_F)$ is generated by its root groups relative to the diagonal k-torus D (due to perfectness of $\mathrm{R}_{F/k}(L_F)$ for both possibilities for L when $\mathrm{char}(k) \neq 2$). The equality $G = \mathrm{R}_{F/k}(L_F)$ forces the inclusion $F \subset K$ over k to be an equality by Example 6.2.1.

Suppose instead that $\mathrm{char}(k) = 2$, and continue to assume that $v^2 \cdot V \subset V$ for all $v \in V$. Hence, V is a submodule of K over the k-algebra $k[V^2]$ generated by squares of elements in V. Since $k[V^2]$ must be a field and $1 \in V$, clearly $k[V^2] = k\langle V^2\rangle$. We are therefore in precisely the situation addressed in Remark 7.2.4, so the k-groups $H_{V,K/k}$ and $\mathrm{P}H_{V,K/k}$ make sense and respectively coincide with G depending on whether L is equal to SL_2 or PGL_2, and $k\langle V\rangle$ is the minimal field of definition over

k for its geometric unipotent radical. This forces $k\langle V\rangle = K$, so V is a kK^2-subspace of K.

Finally, it remains to show in every characteristic that $v^2 \cdot v' \in V$ for all $v, v' \in V$ with $v \neq 0$. Since the D-root groups of G coincide with $\underline{V}^{\pm}$, so $G(k)$ meets the strictly upper-triangular subgroup of $L(K)$ in precisely the points of $\underline{V}^{+}(k) = V \subset K$, it suffices to show that $\mathrm{diag}(v, 1/v) \in G(k)$ (understood to mean that this element of $\mathrm{SL}_2(K)$ maps to an element of $G(k) \subset \mathrm{PGL}_2(K)$ when $L = \mathrm{PGL}_2$) since

$$\begin{pmatrix} v & 0 \\ 0 & 1/v \end{pmatrix}\begin{pmatrix} 1 & v' \\ 0 & 1 \end{pmatrix}\begin{pmatrix} 1/v & 0 \\ 0 & v \end{pmatrix} = \begin{pmatrix} 1 & v^2v' \\ 0 & 1 \end{pmatrix}.$$

It remains to find a mechanism to discover $\mathrm{diag}(v, 1/v)$ inside $G(k)$.

Define the elements $u^{+}(x) = \left(\begin{smallmatrix} 1 & x \\ 0 & 1 \end{smallmatrix}\right)$ and $u^{-}(x) = \left(\begin{smallmatrix} 1 & 0 \\ x & 1 \end{smallmatrix}\right)$ inside $L(K)$ for $x \in K$. The key idea is to determine the Bruhat decomposition of $u^{+}(v)$ in

$$G(k) = (\underline{V}^{-}(k)nP(k)) \coprod P(k) \tag{7.2.5.1}$$

relative to the minimal pseudo-parabolic k-subgroup $P := Z_G(D) \ltimes \underline{V}^{-}$ for any $v \in V - \{0\}$, where $n = \left(\begin{smallmatrix} 0 & 1 \\ -1 & 0 \end{smallmatrix}\right) \in N_L(D)(k) - D(k) \subset N_G(D)(k) - Z_G(D)(k)$. Since $v \neq 0$, clearly $u^{+}(v) \notin P(k)$. Hence, $u^{+}(v)$ lies in the first constituent of the decomposition (7.2.5.1). In other words, there exist unique $v', v'' \in V$ and $z \in Z_G(D)(k) \subset D(K)$ such that

$$u^{+}(v) = u^{-}(v')znu^{-}(v''). \tag{7.2.5.2}$$

All terms in (7.2.5.2) aside from z naturally arise from $\mathrm{SL}_2(K)$. The diagonal subgroup of $\mathrm{SL}_2(K)$ is the full preimage of the diagonal subgroup of $\mathrm{PGL}_2(K)$, so if $L = \mathrm{PGL}_2$ then there is a unique $t \in K^{\times}$ such that replacing z with $\mathrm{diag}(t, 1/t)$ in (7.2.5.2) yields an identity in $\mathrm{SL}_2(K)$. Likewise, if $L = \mathrm{SL}_2$ then $z = \mathrm{diag}(t, 1/t)$ for a unique $t \in K^{\times}$. Elementary calculations in $\mathrm{SL}_2(K)$ now imply that $v = t$, so if $L = \mathrm{SL}_2$ then $\mathrm{diag}(v, 1/v) = z \in G(k)$ whereas if $L = \mathrm{PGL}_2$ then $\mathrm{diag}(v, 1/v)$ represents $z \in G(k) \subset L(K)$. $\square$

The good properties of $i_G(G)$ in Proposition 7.1.3(ii) and of G_a's in Example 7.1.7 now yield a consequence of Theorem 7.2.5 that will be crucial in the proofs of later classification results:

PROPOSITION 7.2.6. *Let G be a pseudo-split pseudo-reductive k-group with a split maximal k-torus T and minimal field of definition K/k for its geometric unipotent radical. For each $a \in \Phi(G,T)$, let K_a/k be the analogous subextension defined similarly for the pseudo-split absolutely pseudo-simple k-group $G_a := \langle U_a, U_{-a}\rangle$.*

(i) *The k-group $i_G(G_a)$ is isomorphic to $\mathrm{R}_{K_a/k}(\mathrm{SL}_2)$ or $\mathrm{R}_{K_a/k}(\mathrm{PGL}_2)$ if k is perfect or* $\mathrm{char}(k) \neq 2$, *and if k is imperfect of characteristic 2 then $i_G(G_a)$ is isomorphic to $H_{V_a,K_a/k}$ or $\mathrm{P}H_{V_a,K_a/k}$ for a nonzero kK_a^2-subspace $V_a \subset K_a$ satisfying $k\langle V_a\rangle = K_a$; such V_a is unique up to $K_a^{\times}$-scaling.*
(ii) *If G is perfect then K is generated over k by its subfields K_a for non-divisible $a \in \Phi(G,T)$.*

PROOF. We just need to explain (ii), for which we first introduce some notation. Let $\Phi = \Phi(G,T)$, and for each $a \in \Phi$ define $a' = a$ when a is non-multipliable and $a' = 2a$ when a is multipliable. Let Φ_{nd} denote the set of non-divisible roots.

By Proposition 3.1.4, G is generated by its k-subgroups $\{G_a\}_{a\in\Phi_{\rm nd}}$ since $U_a \subset U_{a/2}$ when a is divisible, so $i_G(G)$ is generated by its k-subgroups $i_G(G_a)$ for non-divisible a. Letting $L \subset G$ be a Levi k-subgroup containing T (Theorem 5.4.4), the k-group $i_G(G)$ lies between $\mathrm{R}_{K/k}(L_K)$ and L. Since $\Phi(L,T)$ is identified with the set of non-multipliable roots in Φ, $a \mapsto a'$ is a bijection from $\Phi_{\rm nd}$ onto $\Phi(L,T)$.

Consider the subfield $K' \subset K$ generated by the K_a's for non-divisible a, so inside $\mathrm{R}_{K/k}(L_K)$ we have

$$i_G(G_a) = i_{G_a}(G_a) \subset \mathrm{R}_{K_a/k}((L_{a'})_{K_a}) \subset \mathrm{R}_{K'/k}((L_{a'})_{K'}) \subset \mathrm{R}_{K'/k}(L_{K'})$$

for all such a. Hence, $i_G(G)$ lies between $\mathrm{R}_{K'/k}(L_{K'})$ and L. By Proposition 7.1.3(i) it follows that K'/k is a field of definition for the geometric unipotent radical of $i_G(G)$, yet by Proposition 7.1.3(ii) the minimal such extension is K/k! Hence, the inclusion $K' \subset K$ over k is an equality. □

COROLLARY 7.2.7. *If the Cartan subgroups of a pseudo-reductive group G over a field k are tori and k is not imperfect of characteristic 2 then G is reductive.*

Theorem 7.3.3 lists all non-reductive pseudo-reductive groups over an imperfect field of characteristic 2 whose Cartan subgroups are tori.

PROOF. Without loss of generality we may and do assume $k = k_s$. Let T be a maximal k-torus in G, so T is split. By Proposition 2.1.1(ii) we have $G = Z_G(T)\cdot\mathscr{D}(G)$, and by [**CGP**, Lemma 1.2.5(ii)] we have $T = Z\cdot T'$ for the maximal central k-torus $Z \subset G$ and maximal k-torus $T' := T\cap\mathscr{D}(G)$ in $\mathscr{D}(G)$. Hence, we may replace G with $\mathscr{D}(G)$ so that G is perfect. Letting K/k and K_a/k respectively denote the minimal fields of definition for the geometric unipotent radicals of G and G_a for any $a \in \Phi(G,T)$, our task is to show that $K = k$. By Proposition 7.2.6, it suffices to show that $K_a = k$ for each a.

Since $G_a = \mathscr{D}(Z_G(T_a))$ for the codimension-1 subtorus $T_a \subset T$ contained in the kernel of a, and T_a is an isogeny complement to $a^\vee(\mathrm{GL}_1) \subset G_a$, it is clear that the Cartan k-subgroups of G_a are tori. Thus, we may assume G has rank 1. By Theorem 7.2.5(i), G is isomorphic to $\mathrm{R}_{K/k}(\mathrm{SL}_2)$ or $\mathrm{R}_{K/k}(\mathrm{PGL}_2)$. These each admit $\mathrm{R}_{K/k}(\mathrm{GL}_1)$ as a Cartan k-subgroup, so $\mathrm{R}_{K/k}(\mathrm{GL}_1)$ is a torus. But K/k is purely inseparable, so $K = k$. □

The following consequence of Proposition 7.2.6 will permit some classification proofs to be reduced to the rank-1 case.

PROPOSITION 7.2.8. *A pseudo-split pseudo-semisimple k-group G of minimal type with a reduced root system and split maximal k-torus T is determined up to k-isomorphism by the isomorphism classes of $G^{\rm red}_{\overline{k}}$ and of the k-groups $G_a := \langle U_a, U_{-a}\rangle$ for all $a \in \Phi := \Phi(G,T)$.*

PROOF. By Theorem 5.4.4, we may choose a Levi k-subgroup $L \subset G$ containing T; the k-group L is uniquely determined up to k-isomorphism as a split k-descent of $G^{\rm red}_{\overline{k}}$ (due to the Existence and Isomorphism Theorems for split connected semisimple k-groups). Let K/k be the minimal field of definition for $\mathscr{R}_u(G_{\overline{k}}) \subset G_{\overline{k}}$. Since G is of minimal type and has a *reduced* root system, so $\ker i_G = 1$ by Proposition 6.2.15, we may and do identify G with a k-subgroup of $\mathrm{R}_{K/k}(L_K)$ containing L.

For $a \in \Phi$, let K_a/k be the minimal field of definition over k for the geometric unipotent radical of G_a. If k is not imperfect of characteristic 2 then for each $a \in \Phi$ we have $G_a = \mathrm{R}_{K_a/k}((L_a)_{K_a})$ inside $\mathrm{R}_{K/k}((L_a)_K)$ by Proposition 7.2.6(i), so G is uniquely determined inside $\mathrm{R}_{K/k}(L_K)$ in such cases because the G_a's generate G (by Proposition 3.1.4, since G is perfect).

Now suppose k is imperfect of characteristic 2. By Proposition 7.2.6(i), for each $a \in \Phi(G,T)$ there is a nonzero kK_a^2-subspace $V_a \subset K_a$, unique up to $K_a^\times$-scaling, such that $k\langle V_a\rangle = K_a$ and G_a is equal to $H_{V_a,K_a/k}$ or $\mathrm{P}H_{V_a,K_a/k}$ inside $\mathrm{R}_{K_a/k}((L_a)_{K_a})$ (depending on whether L_a is equal to SL_2 or PGL_2 respectively). Since V_a is only unique up to $K_a^\times$-scaling and not generally unique as a k-subspace of K_a, we require further arguments to justify that G is uniquely determined up to k-isomorphism by L and the k-isomorphism class of each G_a (equivalently, the $K_a^\times$-homothety class of V_a for each $a \in \Phi$). For a basis Δ of Φ, the equality $N_L(T)(k)/T(k) = W(\Phi)$ implies that G is generated by L and $\{G_a\}_{a\in\Delta}$. In fact, G is generated by $\{G_a\}_{a\in\Delta}$ since L is generated by its k-subgroups $L_a \subset G_a$ for $a \in \Delta$.

Suppose $\mathscr{G}$ is another such k-group so that its root system is identified with Φ in such a manner that $\mathscr{G}_a \simeq G_a$ for all $a \in \Phi$ and $\mathscr{G}_{\overline{k}}^{\mathrm{red}} \simeq G_{\overline{k}}^{\mathrm{red}} = L_{\overline{k}}$. Hence, $\mathscr{G}$ admits L as a Levi k-subgroup and the minimal field of definition for its geometric unipotent radical is also K/k due to Proposition 7.2.6(ii). Thus, $\mathscr{G}$ is a k-subgroup of $\mathrm{R}_{K/k}(L_K)$ containing L, and $\mathscr{G}$ is generated by $\{\mathscr{G}_a\}_{a\in\Delta}$. For each $a \in \Phi$ we have $\mathscr{G}_a \simeq G_a$ by design, so the k-subgroup $\mathscr{G}_a \subset \mathrm{R}_{K_a/k}((L_a)_{K_a})$ arises from the kK_a^2-subspace $\lambda_a V_a \subset K_a$ for some $\lambda_a \in K_a^\times$. It is therefore sufficient to find

$$\overline{t}_0 \in (T/Z_L)(K) = \mathrm{R}_{K/k}((T/Z_L)_K)(k)$$

whose action on $\mathrm{R}_{K/k}(L_K)$ carries G_a onto $\mathscr{G}_a$ for each $a \in \Delta$. But $T/Z_L \simeq \mathrm{GL}_1^\Delta$ via $\overline{t} \mapsto (a(\overline{t}))_{a\in\Delta}$, so the unique $\overline{t}_0$ corresponding to $(\lambda_a) \in (K^\times)^\Delta = (T/Z_L)(K)$ does the job. □

7.3. A non-standard construction. Among non-standard pseudo-reductive groups, it is the absolutely pseudo-simple groups that are the most interesting. Since ultimately it turns out that standardness can only fail in characteristics 2 and 3, any construction of a non-standard pseudo-reductive group must use features specific to these small positive characteristics.

Recall from Proposition 6.1.4 that the formation of the root system and root groups of a pseudo-split pseudo-reductive group is unaffected by passage to a central pseudo-reductive quotient. Thus, by Example 3.1.3 and the reducedness of root systems for connected reductive groups, for any standard pseudo-reductive k-group the root system over k_s is *reduced* and the root groups for roots in a common irreducible component of the root system have the same dimension. In particular, an absolutely pseudo-simple k-group whose root groups over k_s do not all have the same dimension cannot be standard.

Reducedness of the root system can only fail over imperfect fields of characteristic 2 (Theorem 3.1.7), so non-reducedness of the root system over k_s can be an obstruction to standardness only over such fields. In fact, over every imperfect field k of characteristic 2 there *do* exist pseudo-split absolutely pseudo-simple k-groups with root system BC_n for any desired $n \geqslant 1$. The construction of such groups is quite hard; we will discuss it in §8.

To give the reader a flavor of non-standardness at the present stage we now give a different construction of non-standard absolutely pseudo-simple groups that is specific to characteristic 2, realizing variation in dimension of the root groups as an obstruction to standardness. This construction has the virtue that it also solves a natural problem over any field, having nothing to do with standardness: find all non-reductive pseudo-reductive groups whose Cartan subgroups are tori.

Let (V, q) be a quadratic space with dimension $d > 0$ over a field k of characteristic 2. The symmetric bilinear form $B_q(v, v') = q(v + v') - q(v) - q(v')$ is alternating since $\mathrm{char}(k) = 2$. Assume $q \neq 0$. The *defect space* $V^\perp$ is the set of $v \in V$ that satisfy $B_q(v, \cdot) = 0$, so B_q induces a (non-degenerate) symplectic form $\overline{B}_q$ on $V/V^\perp$. In particular, $\dim(V/V^\perp)$ is even. Note that $q : V^\perp \to k$ is additive.

Let Q be the projective quadric hypersurface $(q = 0)$. As is explained at the beginning of [**CP**, §7.1], the quadric Q is regular (equivalently smooth) at its k-points if and only if $q|_{V^\perp}$ is injective, and this property is preserved under separable extension on k; we say q is *regular* in such cases. The smoothness of Q for even d is exactly the condition that $V^\perp = 0$ whereas smoothness of Q for odd d is exactly the condition that $V^\perp$ is a line. If Q is smooth and $d \geqslant 3$ then when $d = 2m$ is even the group scheme $\mathrm{O}(q)$ is an extension of $\mathbf{Z}/2\mathbf{Z}$ by a connected semisimple group $\mathrm{SO}(q)$ of type D_m whereas when $d = 2m + 1$ is odd the group scheme $\mathrm{O}(q)$ is the direct product of μ_2 and a connected absolutely simple group of adjoint type B_m.

Now assume $0 < \dim V^\perp < \dim V =: d$, so $\dim(V/V^\perp) = 2n$ for some integer $n > 0$ and hence $d \geqslant 2n + 1 \geqslant 3$. In these cases Q is smooth precisely when $\dim V^\perp = 1$. We make the weaker hypothesis that q is regular. In concrete terms, this says

$$q - c_1 x_1^2 + \cdots + c_{d-2n} x_{d-2n}^2 + q_0(x_{d-2n+1}, \ldots, x_d)$$

where q_0 is non-degenerate in $2n$ variables and $\{c_1, \ldots, c_{d-2n}\}$ is k^2-linearly independent. In particular, the case $\dim V^\perp > 1$ occurs over k if and only if $[k : k^2] \geqslant 2$.

If $\dim V^\perp = 1$ then it is well-known that $\mathrm{SO}(q)$ concides with the maximal smooth closed k-subgroup $\mathrm{O}(q)^{\mathrm{sm}}$ of $\mathrm{O}(q)$. Hence, when $\dim V^\perp > 1$ we are motivated to make the *definition*

$$\mathrm{SO}(q) := \mathrm{O}(q)^{\mathrm{sm}}.$$

(See §1.3 for references on the existence and uniqueness of a maximal smooth closed subgroup scheme H^{sm} of any group scheme H of finite type over a field and its relationship with the underlying reduced scheme H_{red}.) Note that $\mathrm{SO}(q) = \mathrm{SO}(cq)$ for any $c \in k^\times$, so there is generally no harm in assuming that $1 \in q(V^\perp)$ (as is sometimes convenient in calculations).

Arguments in [**CP**, 7.1.2–7.1.3] establish several results that we now review (recovering well-known properties of SO_{2m+1} when $\dim V^\perp = 1$). Regularity of q_{k_s} ensures that the kernel of $\pi_q : \mathrm{SO}(q) = \mathrm{O}(q)^{\mathrm{sm}} \to \mathrm{Sp}(\overline{B}_q)$ has no nontrivial k_s-points, and inspection of q in suitable coordinates shows that π_q is surjective, so it follows that $\mathrm{SO}(q)$ has no nontrivial smooth connected unipotent normal k-subgroup. This doesn't immediately imply that $\mathrm{SO}(q)$ is pseudo-reductive since it isn't evident if $\mathrm{SO}(q)$ is connected! (Recall that we are now assuming $\dim V^\perp > 0$.)

PROPOSITION 7.3.1. *Consider regular (V, q) satisfying $0 < \dim V^\perp < \dim V$, and write $2n = \dim(V/V^\perp)$. The k-group $\mathrm{SO}(q)$ is absolutely pseudo-simple with trivial center and root system over k_s of type B_n, and its Cartan k-subgroups are tori. Over k_s the long root groups have dimension 1 whereas the short root groups have dimension $\dim V^\perp$ (with both roots understood to be short when $n = 1$).*

The minimal field of definition over k for the geometric unipotent radical of $\mathrm{SO}(q)$ is the subfield $K \subset k^{1/2}$ generated over k by $\{\sqrt{q(v)/q(v')}\}_{v, v' \in V^\perp - \{0\}}$.

The universal smooth k-tame central extension of $\mathrm{SO}(q)$ is denoted $\mathrm{Spin}(q)$; it coincides with the usual spin group when $\dim V^{\perp} = 1$. This k-group inherits absolute pseudo-simplicity from $\mathrm{SO}(q)$ (use Proposition 3.2.2).

Assume $\dim V^{\perp} > 1$, so k is imperfect and $\mathrm{SO}(q)$ is not reductive (as its short root groups over k_s have dimension $\dim V^{\perp}$). If $n > 1$ then the presence of root groups over k_s with unequal dimensions (1 for long roots, $\dim V^{\perp}$ for short roots) implies that the non-reductive absolutely pseudo-simple k-group $\mathrm{SO}(q)$ is *not standard* (so $\mathrm{Spin}(q)$ is not standard, by the characterization in 6.2.8).

REMARK 7.3.2. Assume $n = 1$ (and $\dim V^{\perp} > 1$). By varying such (V, q), when $[k : k^2] > 2$ the non-reductive absolutely pseudo-simple k-group $\mathrm{SO}(q)$ can be arranged to be either standard or not standard (depending on q) whereas if $[k : k^2] = 2$ then $\mathrm{SO}(q)$ is always standard. We now establish an interesting *characterization* of standardness when $n = 1$: it is equivalent that the k-subspace $q(V^{\perp})^{1/2} \subset K$ with dimension $\dim V^{\perp} \geqslant 2$ is a line over a subfield of K strictly containing k (a property that we can arrange to either hold or fail via suitable choice of (V, q) if $[k : k^2] > 2$, whereas if $[k : k^2] = 2$ then it cannot ever fail since $K = k^{1/2}$ is 2-dimensional over k in such cases).

To prove this characterization we may extend scalars to k_s, and we may replace q with a $k^{\times}$-multiple so that $1 \in q(V^{\perp})$. Use $q^{1/2}|_{V^{\perp}}$ to identify $V^{\perp}$ with a k-subspace of K strictly containing k, and identify $\mathbf{G}_{\mathrm{a}}$ with each root group $U^{\pm}$ of $\mathrm{SO}(q)_K^{\mathrm{red}} = \mathrm{PGL}_2$ relative to the diagonal K-torus via the usual parameterization. Since $k \subset q(V^{\perp})^{1/2}$, the k-subspace $q(V^{\perp})^{1/2}$ of K is a line over a subfield of K containing k if and only if $q(V^{\perp})^{1/2}$ is itself a subfield of K (larger than k).

It is shown in the proof of [**CP**, Prop. 7.2.5] (using [**CP**, 3.1.3–3.1.4]) that $\mathrm{SO}(q)$ is the k-subgroup of $\mathrm{R}_{K/k}(\mathrm{PGL}_2)$ between PGL_2 and $\mathrm{R}_{K/k}(\mathrm{PGL}_2)$ generated by the k-subgroups of $\mathrm{R}_{K/k}(U_K^{\pm}) = \mathrm{R}_{K/k}(\mathbf{G}_{\mathrm{a}})$ corresponding to the k-subspace $q(V^{\perp})^{1/2} \subset K$ strictly containing k. By [**CP**, Prop. 3.1.8(ii)], this k-subgroup of $\mathrm{R}_{K/k}(\mathrm{PGL}_2)$ is standard if and only if $q(V^{\perp})^{1/2}$ is a field $F \subset K$ (in which case $q|_{V^{\perp}}$ is identified with the squaring inclusion $F \to k$), as desired.

The non-reductive centerless k-group $\mathrm{SO}(q)$ has k-automorphisms not arising from $\mathrm{SO}(q)(k)$-conjugation, in contrast with the well-known adjoint absolutely simple case for type B when $\dim V^{\perp} = 1$. Nonetheless, all elements of $\mathrm{Aut}_k(\mathrm{SO}(q))$ arise from a suitable notion of "conformal isometry"; see [**CP**, Prop. 7.2.2(i)].

Remarkably, the $\mathrm{SO}(q)$-construction with regular degenerate quadratic spaces (V, q) satisfying $V^{\perp} \neq V$ underlies *all* non-reductive pseudo-reductive groups whose Cartan subgroups are tori. This is made precise in [**CP**, Prop. 7.3.7]:

THEOREM 7.3.3. *Let k be a field. There exist non-reductive pseudo-reductive k-groups whose Cartan subgroups are tori if and only if k is imperfect of characteristic 2. For such k, these k-groups are precisely $H \times \mathrm{R}_{k'/k}(G')$ where H is a connected reductive k-group, k' is a nonzero finite étale k-algebra, and G' is a smooth affine k'-group whose fiber G'_i over each factor field k'_i of k' is a descent of $\mathrm{SO}(q'_i)$ for a regular degenerate quadratic space (V'_i, q'_i) over $k'_{i,s}$ satisfying ${V'}_i^{\perp} \neq V'_i$.*

The proof of Theorem 7.3.3 requires a detailed understanding of the structure of pseudo-split pseudo-reductive groups with a non-reduced root system; see §8 (such groups can only exist over imperfect fields with characteristic 2, by Theorem 3.1.7). This hard input ensures that the root system over k_s of every non-reductive pseudo-reductive k-group whose Cartan subgroups are tori is *reduced*.

7.4. Root fields and standardness. A further application of the pseudo-split rank-1 classification in Theorem 7.2.5 concerns an auxiliary field that arises in the description of automorphisms. For pseudo-semisimple k-groups G, these auxiliary fields underlie the possible failure of G/Z_G to exhaust the identity component of the maximal smooth closed k-subgroup of the automorphism scheme $\mathrm{Aut}_{G/k}$. This is best understood with some examples:

EXAMPLE 7.4.1. Let K/k be a nontrivial purely inseparable finite extension in characteristic $p > 0$, and let G' be a non-trivial connected semisimple K-group that is simply connected, with $T' \subset G'$ a maximal K-torus. Let $G = \mathrm{R}_{K/k}(G')$. As a special case of our discussion of automorphism functors of pseudo-semisimple groups to be given in §9.1, the automorphism functor $\underline{\mathrm{Aut}}_{G/k}$ is represented by an affine k-group scheme $\mathrm{Aut}_{G/k}$ of finite type [**CP**, Prop. 6.2.2].

In contrast with the semisimple case, $\mathrm{Aut}_{G/k}$ is *never* smooth [**CP**, Ex. 6.2.3]. For the study of Galois-twisted forms and local-global principles, it is the maximal smooth closed k-subgroup $\mathrm{Aut}^{\mathrm{sm}}_{G/k}$ that matters. Using the action of $G'/Z_{G'}$ on G', we can form the standard pseudo-reductive k-group

$$\mathscr{G} := (G \rtimes \mathrm{R}_{K/k}(T'/Z_{G'}))/\mathrm{R}_{K/k}(T'),$$

and the evident k-subgroup inclusion $\mathscr{G} \hookrightarrow \mathrm{Aut}_{G/k}$ has image $(\mathrm{Aut}^{\mathrm{sm}}_{G/k})^0$ due to [**CP**, Prop. 6.3.4(i), Prop. 6.2.4, Lemma 6.1.3].

Clearly $\mathscr{D}(\mathscr{G})$ is the image G/Z_G of G, and the commutative quotient

$$\mathscr{G}/(G/Z_G) = \mathrm{coker}(\mathrm{R}_{K/k}(T') \longrightarrow \mathrm{R}_{K/k}(T'/Z_{G'}))$$

is a smooth connected unipotent k-group with dimension $\dim \mathrm{R}_{K/k}(Z_{G'})$. In particular, $(\mathrm{Aut}^{\mathrm{sm}}_{G/k})^0$ is larger than G/Z_G precisely when $Z_{G'}$ is not K-étale.

EXAMPLE 7.4.2. Let K/k be a nontrivial finite extension in characteristic 2, and let V be a nonzero kK^2-subspace of K satisfying $k\langle V\rangle = K$. Let

$$F = \{\lambda \in K \mid \lambda V \subset V\};$$

this is a kK^2-subalgebra of K, hence a field, and it is the largest subfield of K over which V is a subspace. We will be most interested in the cases with F strictly between kK^2 and K, so since $k\langle V\rangle = K$ we see that in such cases $[K : F] \geqslant 4$ and $[F : kK^2] \geqslant 2$.

Consider the absolutely pseudo-simple k-group $H := H_{V,K/k} \subset \mathrm{R}_{K/k}(\mathrm{SL}_2)$. Using the natural action of PGL_2 on SL_2, for the diagonal k-torus $\mathrm{GL}_1 = \overline{D} \subset \mathrm{PGL}_2$ (via $t \mapsto \mathrm{diag}(t, 1)$) we see the $\mathrm{R}_{K/k}(\overline{D}_K)$-action on $\mathrm{R}_{K/k}(\mathrm{SL}_2)$ preserves the standard root groups $U^{\pm} = \mathrm{R}_{K/k}(\mathbf{G}_{\mathrm{a}}) \subset \mathrm{R}_{K/k}(\mathrm{SL}_2)$ via the scaling $t.x = t^{\pm 1}x$. Hence, by definition of F, the action of $\mathrm{R}_{F/k}(\overline{D}_F)$ on $\mathrm{R}_{K/k}(\mathrm{SL}_2)$ preserves the k-groups $\underline{V}^{\pm} \subset U^{\pm}$ (corresponding to $V \subset K$) that generate H by definition.

For the diagonal k-torus $D \subset \mathrm{SL}_2$, the natural map $D \to \overline{D}$ corresponds to squaring on GL_1. Thus, the Cartan k-subgroup $V^*_{K/k} = Z_H(D)$ (generated by ratios v/v' for nonzero $v, v' \in V$, due to Proposition 7.2.3(i)) is carried into the k-subgroup $R_{F/k}(\overline{D}_F) \subset \mathrm{R}_{K/k}(\overline{D}_K)$ since $K^2 \subset F$ by hypothesis. Consider the resulting central quotient

$$\mathscr{H} := (H \rtimes \mathrm{R}_{F/k}(\overline{D}_F))/V^*_{K/k}$$

(using the anti-diagonal inclusion). This is pseudo-reductive by Proposition 2.2.1, and it is naturally a subgroup of the automorphism functor $\underline{\mathrm{Aut}}_{H/k}$.

By the same results from [**CP**, Ch. 6] as used in Example 7.4.1, $\underline{\mathrm{Aut}}_{H/k}$ is represented by an affine k-group scheme $\mathrm{Aut}_{H/k}$ of finite type and $\mathscr{H} = \mathrm{Aut}^{\mathrm{sm}}_{H/k}$. Hence, $\mathrm{Aut}^{\mathrm{sm}}_{H/k}$ is pseudo-reductive with derived group equal to the image H/Z_H of H, and its quotient by this derived group is the unipotent k-group $\mathrm{R}_{F/k}(\mathrm{GL}_1)/\mathrm{GL}_1$; this is nontrivial precisely when $F \neq k$.

Consider an absolutely pseudo-simple k-group G such that G_{k_s} has a *reduced* root system, and let K/k be the minimal field of definition for its geometric unipotent radical, so for $G' := G_K/\mathscr{R}_{u,K}(G_K)$ the natural map $i_G : G \to \mathrm{R}_{K/k}(G')$ has kernel that is central (Proposition 6.2.15). Hence, for any maximal k-torus $T \subset G$ we have

$$\ker(\mathrm{Lie}(i_G)) = \mathrm{Lie}(\ker i_G) \subset \mathrm{Lie}(Z_G(T)) = \mathfrak{g}^T$$

for $\mathfrak{g} := \mathrm{Lie}(G)$. The complete reducibility of finite-dimensional linear representations of T provides a unique T-equivariant linear complement $\mathfrak{g}(T)$ to $\mathfrak{g}^T$ in $\mathfrak{g}$, and the map

$$\mathrm{Lie}(i_G) : \mathfrak{g}(T) \longrightarrow \mathrm{Lie}(\mathrm{R}_{K/k}(G')) = \mathrm{Lie}(G')$$

is *injective*. Hence, the following definition makes sense:

DEFINITION 7.4.3. The *root field* $F \subset K$ is the subextension of K/k consisting of those $\lambda \in K$ such that multiplication by λ on $\mathrm{Lie}(G')$ carries $\mathfrak{g}(T)$ into itself.

If we use the more precise notation F_T for the root field, indicating the possible dependence on T, then it is clear that the formation of F_T is compatible with separable extension on k. Hence, we can use the $G(k_s)$-conjugacy of maximal k_s-tori in G_{k_s} to deduce that F_T is independent of T.

EXAMPLE 7.4.4. Assume G as above is pseudo-split of minimal type with root system A_1, and let L be a Levi k-subgroup of G. By Theorem 7.2.5(i), if k is not imperfect of characteristic 2 then $G = \mathrm{R}_{K/k}(L_K)$, so clearly $F = K$ in such cases. If k is imperfect of characteristic 2 then (by Theorem 7.2.5(ii) and Proposition 6.2.15) G is equal to $H_{V,K/k}$ or $\mathrm{P}H_{V,K/k}$ for some nonzero kK^2-subspace $V \subset K$ satisfying $k\langle V\rangle = K$. In such cases F is as described in terms of V in Example 7.4.2.

If G as above with arbitrary rank has a split maximal k-torus T, so $\Phi := \Phi(G,T)$ is *reduced*, then the root field $F \subset K$ of G is determined by the root fields $F_a \subset K_a \subset K$ of the k-groups $G_a := \langle U_a, U_{-a}\rangle$ for all $a \in \Phi$ via the formula $F = \bigcap_{a\in\Phi} F_a$ that is immediate from the direct product structure of an open cell and the compatibility of $i_G|_{G_a}$ and i_{G_a} (as discussed in Example 7.1.7). The computation of root fields may always be reduced to the case of the *central* quotient $i_G(G)$ of minimal type because the formation of root fields is unaffected by passage to pseudo-reductive central quotients when the root system is reduced [**CP**, Rem. 3.3.3].

One reason for interest in root fields is that they detect the presence of a Weil restriction in the description of G, at least when G is of minimal type. This is made precise by the following result involving maximal pseudo-reductive quotients of minimal type (as in Example 7.1.6):

PROPOSITION 7.4.5. *If G is an absolutely pseudo-simple k-group of minimal type and G_{k_s} has a reduced root system then for the root field F/k the natural map $G \to \mathscr{D}(\mathrm{R}_{F/k}(G_F^{\mathrm{prmt}}))$ is an isomorphism.*

This is a special case of [**CP**, Prop. 3.3.6], and is proved by computations with a Levi k_s-subgroup and the type-A_1 descriptions in Example 7.4.4. Proposition

7.4.5 allows one to reduce some general problems for G of minimal type over k to the study of $G_F^{\rm prmt}$ over F, and the latter *also has root field* F (as may be deduced from the rank-1 cases in Example 7.4.4). In other words, for some purposes we can arrange that the root field is equal to the ground field.

For a maximal k-torus $T \subset G$ and the irreducible and reduced root system $\Phi = \Phi(G_{k_s}, T_{k_s})$, the root field $F_a \subset K_s$ of $(G_{k_s})_a$ for a root $a \in \Phi$ only depends on a through its $W(\Phi)$-orbit since $W(\Phi) = N_G(T)(k_s)/Z_G(T)(k_s)$. But $W(\Phi)$ acts transitively on the set of roots with a given length (since Φ is irreducible), so in the simply laced case the subfields $F_a \subset K_s$ all coincide and hence they are all equal to the root field $\bigcap_{a \in \Phi} F_a = F_s$ of G_{k_s}.

Likewise, if Φ has two distinct root lengths then F_a depends on a only through its length, so by Galois descent the subfields $F_a \subset K_s = K \otimes_k k_s$ over k_s arise from corresponding subfields $F_>, F_< \subset K$ over k for long and short roots respectively. We call $F_>$ the *long root field* and call $F_<$ the *short root field*, so $F_> \cap F_< = F$. In the simply laced case it is convenient to use the definitions $F_< := F$ and $F_> := F$.

EXAMPLE 7.4.6. Let k be an imperfect field of characteristic 2 and let (V, q) be a finite-dimensional regular quadratic space over k such that $0 < \dim V^{\perp} < \dim V$ (so q is degenerate precisely when $\dim V^{\perp} > 1$). Let $K \subset k^{1/2}$ be the finite extension of k generated by the ratios $\sqrt{q(v)/q(v')}$ for nonzero $v, v' \in V^{\perp}$. In §7.3 we introduced the class of absolutely pseudo-simple k-groups $\mathrm{SO}(q)$ with trivial center and root system B_n over k_s, where $\dim(V/V^{\perp}) = 2n$ for an integer $n \geqslant 1$.

By [**CP**, Ex. 7.1.8], the short root field $F_<$ of $\mathrm{SO}(q)$ consists of precisely those $\lambda \in K$ such that λ-scaling preserves the k-subspace

$$\{\sqrt{q(v)/q(v_0)} \mid v \in V^{\perp}\} \subset K$$

for a fixed $v_0 \in V^{\perp} - \{0\}$ (the choice of which does not matter, as replacing v_0 with $v_0' \in V^{\perp} - \{0\}$ simply multiplies this k-subspace by $\sqrt{q(v_0)/q(v_0')} \in K^{\times}$). If q is non-degenerate then clearly $F_< = K = k$.

Nontriviality of the extension $F_</k$ has concrete meaning in terms of (V, q), as follows. Let $\mathrm{CO}(q)$ denote the maximal smooth k-subgroup of the group scheme of *conformal isometries* of (V, q) (the functor of pairs (L, μ) consisting of a linear automorphism L of V and a unit μ such that $q \circ L = \mu \cdot q$). Since $Z_{\mathrm{SO}(q)} = 1$, there is an evident inclusion of k-groups $j : \mathrm{GL}_1 \times \mathrm{SO}(q) \hookrightarrow \mathrm{CO}(q)$ (well-known to be an equality when q is non-degenerate). In general there is a canonical isomorphism

$$\mathrm{CO}(q)/\mathrm{SO}(q) \simeq \mathrm{R}_{F_</k}(\mathrm{GL}_1)$$

by [**CP**, (7.2.1.2)ff., Prop. 7.2.2(ii)], so $\mathrm{CO}(q)$ is *connected* and by dimension reasons we see that $F_<$ is larger than k precisely when j is not an equality.

For absolutely pseudo-simple k-groups G with a reduced root system over k_s, it is always the case that $F = F_>$. More generally, beyond the rank-1 case (Example 7.4.4) applied to the central quotient $i_G(G)$ of minimal type described in Theorem 7.2.5, the relationships among root fields and minimal fields of definition of geometric unipotent radicals are as follows:

THEOREM 7.4.7. *Consider a pseudo-split absolutely pseudo-simple group G over a field k that is imperfect with characteristic $p > 0$, and let $T \subset G$ be a split maximal k-torus. Assume $n := \dim T \geqslant 2$ and that the rank-n irreducible root system $\Phi := \Phi(G, T)$ is reduced. Let F, K respectively denote the root field for G*

and the minimal field of definition over k for $\mathscr{R}_u(G_{\overline{k}}) \subset G_{\overline{k}}$, and define F_a, K_a similarly for $G_a := \langle U_a, U_{-a}\rangle$ for each $a \in \Phi$.

(i) *If Φ has no edge of multiplicity p in its Dynkin diagram then $F_a = F = K = K_a$ for all $a \in \Phi$.*

(ii) *Assume Φ has an edge of multiplicity p in its diagram, so $p \in \{2,3\}$ and Φ has two root lengths; denote by $K_<$ (resp. $K_>$) the subfield $K_a \subset K$ for $a \in \Phi$ that is short (resp. long). Then*

$$kK^p \subset K_> \subset K_< = K,$$

and if $p = 3$ then $F_> = K_>$ and $F_< = K_<$.

(iii) *Assume Φ has an edge of multiplicity $p = 2$. Then*

$$kK^2 \subset F = F_> \subset K_> \subset F_< \subset K_< = K,$$

and for types F_4 or B_n with $n \geqslant 3$ we have $F_> = K_>$ whereas for types F_4 or C_n with $n \geqslant 3$ we have $F_< = K$.

In (iii) no assertion is made for type $\mathrm{B}_2 = \mathrm{C}_2$. It can happen in such cases that $F_a \neq K_a$ for all roots a; i.e., the nonzero kK_a^2-subspace $V_a \subset K_a$ classifying G_a can be a proper subspace for all roots a (this is discussed in Remark 10.1.8).

We now sketch a few points in the proof of Theorem 7.4.7, referring to [**CP**, Thm. 3.3.8] for the details. The formation of K/k is unaffected by passage to a central pseudo-reductive quotient (Proposition 6.2.2), and likewise for root fields [**CP**, Rem. 3.3.3], so by replacing G with its universal smooth k-tame central extension we may assume the connected semisimple group $G_{\overline{k}}^{\mathrm{red}}$ is simply connected. Since $i_G(G)$ is a central quotient of G (as Φ is reduced) and is of minimal type with the same maximal geometric reductive quotient as G, and moreover $i_G|_{G_a}$ is compatible with i_{G_a} in the sense of Example 7.1.7, we may replace G with $i_G(G)$ to reduce to the case that G is also of minimal type.

A Levi k-subgroup $L \subset G$ containing T exists by Theorem 5.4.4, and it is simply connected since $L_{\overline{k}} \simeq G_{\overline{k}}^{\mathrm{red}}$. (In [**CP**, §3.3] the passage to simply connected L is done via a more explicit procedure involving root groups because the universal smooth k-tame central extension built and studied in §6.2 is not provided until later in [**CP**]. However, its development can be carried out earlier, as we have done in this survey.)

The k-group G lies between L and $\mathrm{R}_{K/k}(L_K)$ because G is of minimal type. For $a \in \Phi$, the k-group $G_a = \langle U_{-a}, U_a\rangle$ lies between L_a and $\mathrm{R}_{K_a/k}((L_a)_{K_a})$. Upon choosing a basis Δ for Φ, the effect of conjugation by $Z_{G_a}(T \cap G_a) = G_a \cap Z_G(T)$ on the b-root group $U_b \subset G$ for adjacent $a, b \in \Delta$ can be described by computing inside $\mathrm{R}_{K/k}(L_K)$. Combining this description with the list of possibilities for G_a given in Theorem 7.2.5 yields all of the asserted relations among fields (since the group $N_L(T)(k)/T(k) = W(\Phi)$ acts transitively on the set of roots with a fixed length due to the irreducibility of Φ). This completes our sketch of the proof of Theorem 7.4.7.

The control of root fields in Theorem 7.4.7 underlies the proof of our first major classification result:

THEOREM 7.4.8. *An absolutely pseudo-simple k-group G is standard except possibly when k is imperfect with $p := \mathrm{char}(k) \in \{2,3\}$ and the root system Φ of G_{k_s} satisfies one of the following conditions:* (i) *its Dynkin diagram has an edge of* *multiplicity* p, (ii) *it is non-reduced* (*as can only happen when* $p = 2$), *or* (iii) *it is of type* A_1 *with* $p = 2$.

Before we prove Theorem 7.4.8, we note that this result is the absolutely pseudo-simple case of [**CGP**, Cor. 6.3.5, Prop. 6.3.6], as well as of [**CP**, Thm. 3.4.2], and the proof we give below is simpler. In Corollary 10.2.14 we remove the absolute pseudo-simplicity hypothesis.

PROOF. By Proposition 6.2.14 we may assume $k = k_s$, so G is pseudo-split. We may also certainly assume Φ is reduced. The rank-1 case away from imperfect fields of characteristic 2 is settled by Theorem 7.2.5, so we may also assume that Φ has rank $n \geqslant 2$. Finally, we can assume k is imperfect (as otherwise our task is trivial) and that the diagram of Φ does not have an edge of multiplicity $p := \mathrm{char}(k) > 0$.

Since Φ is reduced, by Proposition 6.2.17 it suffices to prove that ξ_G is surjective. We may choose a Levi k-subgroup $L \subset G$ since $k = k_s$, so the target of ξ_G is naturally identified with $\mathscr{D}(\mathrm{R}_{K/k}(L_K))$. This derived group is generated by its root groups relative to a split maximal k-torus of L (Proposition 3.1.4), and these root groups coincide with the root groups of $\mathrm{R}_{K/k}(L_K)$.

Since $K_a = K$ for all $a \in \Phi$ by Theorem 7.4.7(i), by the compatibility of $i_G|_{G_a}$ and i_{G_a} for $a \in \Phi$ it suffices to check that the inclusion $i_{G_a}(G_a) \subset \mathscr{D}(\mathrm{R}_{K/k}((L_a)_K))$ is an equality for each $a \in \Phi$. By Proposition 7.2.6 we are done if $p \neq 2$ by comparing dimensions of root groups relative to $a^\vee(\mathrm{GL}_1)$. The case $p = 2$ is settled since $V_a = K$ due to the equality $F_a = K$ (again see Theorem 7.4.7(i)). $\square$

7.5. Basic exotic constructions. We have encountered two classes of non-standard pseudo-reductive groups, both over imperfect fields k of characteristic 2: the $\mathrm{SO}(q)$-construction for regular degenerate quadratic spaces (V, q) over k in §7.3 (to be discussed more fully in §10.1), and the groups $H_{V,K/k}$ and $\mathrm{P}H_{V,K/k}$ introduced in Definition 7.2.1 for a purely inseparable finite extension K/k and nonzero proper kK^2-subspace $V \subset K$ such that $k\langle V\rangle = K$. Motivated by Theorem 7.4.8, to construct non-standard absolutely pseudo-simple groups G over an imperfect field k of characteristic p we focus on $p \in \{2, 3\}$ and three cases depending on the root system Φ over k_s:

(i) Φ of type F_4, B_n ($n \geqslant 1$), or C_n ($n \geqslant 1$) with $p = 2$ (B_1, C_1 mean A_1),
(ii) Φ of type G_2 with $p = 3$,
(iii) Φ of type BC_n ($n \geqslant 1$) with $p = 2$; i.e., the non-reduced case.

In this section we shall describe constructions in the first two cases, though not the $\mathrm{SO}(q)$-construction for regular degenerate (V, q) (which is an instance of (i) and will be placed into a broader framework in §10.1).

Let k be an imperfect field of characteristic $p \in \{2, 3\}$. As motivation for the non-standard groups to be built over k, first consider a pseudo-split absolutely pseudo-simple k-group G of minimal type with root system of type G_2 if $p = 3$ and of type F_4 if $p = 2$. (In Corollary 7.5.11 we will see that the "minimal type" property is automatic for these root systems.) Let K/k be the minimal field of definition for the geometric unipotent radical of G. By Proposition 7.2.8 and Theorem 7.4.7(ii),(iii), the possibilities for G are determined up to isomorphism by K/k and a subfield $K_> \subset K$ that contains kK^p.

More specifically, let L be a split connected absolutely simple k-group with the chosen root system Φ (G_2 in characteristic 3, F_4 in characteristic 2). The only possibility for G is the smooth connected k-subgroup $\mathscr{G}$ of $\mathrm{R}_{K/k}(L_K)$ generated by the k-subgroups $\mathrm{R}_{K/k}((L_a)_K)$ for short $a \in \Phi$ and $\mathrm{R}_{K_>/k}((L_a)_{K_>})$ for long $a \in \Phi$, where $K_>/k$ is a purely inseparable finite extension contained in K and containing kK^p. Note that $\mathscr{G}$ contains L and hence is pseudo-reductive of minimal type with L as a Levi k-subgroup and Φ as its root system (by Proposition 7.1.3 (i)). Moreover, $\mathscr{G}$ is perfect because each group $\mathrm{R}_{K/k}((L_a)_K) = \mathrm{R}_{K/k}(\mathrm{SL}_2)$ and $\mathrm{R}_{K_>/k}((L_a)_{K_>}) = \mathrm{R}_{K_>/k}(\mathrm{SL}_2)$ is perfect. Further arguments are needed to show that the long root groups of $\mathscr{G}$ have dimension $[K_> : k]$ (rather than larger than $[K_> : k]$). These considerations motivate analyzing a construction permitting types B_n and C_n for $n \geqslant 2$ when $p = 2$ as well:

PROPOSITION 7.5.1. *Let k be an imperfect field of characteristic $p \in \{2, 3\}$, and let L be a split connected absolutely simple k-group that is simply connected with a split maximal k-torus T and root system $\Phi = \Phi(L, T)$ that is irreducible with an edge of multiplicity p. Let K/k be a nontrivial purely inseparable finite extension and $K_> \subset K$ a proper subfield containing kK^p.*

The k-subgroup $\mathscr{G} \subset \mathrm{R}_{K/k}(L_K)$ generated by the k-subgroups $\mathrm{R}_{K/k}((L_a)_K)$ for short $a \in \Phi$ and $\mathrm{R}_{K_>/k}((L_a)_{K_>})$ for long $a \in \Phi$ is absolutely pseudo-simple of minimal type with root system Φ, Levi k-subgroup L, and long root groups with dimension $[K_> : k]$. For a basis Δ of Φ (so $\mathrm{GL}_1^{\Delta} \simeq T$ via $(t_a)_{a\in\Delta} \mapsto \prod_a a^\vee(t_a)$ since L is simply connected), we have

$$Z_{\mathscr{G}}(T) = \prod_{a\in\Delta_<} \mathrm{R}_{K/k}(\mathrm{GL}_1) \times \prod_{a\in\Delta_>} \mathrm{R}_{K_>/k}(\mathrm{GL}_1) \tag{7.5.1.1}$$

inside $\mathrm{R}_{K/k}(T_K) = \mathrm{R}_{K/k}(\mathrm{GL}_1)^{\Delta}$ for the subsets $\Delta_<$ of short roots in Δ and $\Delta_>$ of long roots in Δ. Moreover, $\mathscr{G} = \mathrm{R}_{K_>/k}(\mathscr{G}')$ where $\mathscr{G}'$ is the analogous $K_>$-subgroup of $\mathrm{R}_{K/K_>}(L_K)$.

Informally, inside $\mathrm{R}_{K/K_>}(L_K)$ the $K_>$-subgroup $\mathscr{G}'$ is built by shrinking the long T-root groups to be the ones arising from the $K_>$-subgroup $L_{K_>}$.

PROOF. The arguments in the preceding discussion show that $\mathscr{G}$ is pseudo-semisimple with L as a Levi k-subgroup and root system Φ, so in particular $\mathscr{G}$ is absolutely pseudo-simple. We also clearly have $\mathscr{G} \subset \mathrm{R}_{K_>/k}(\mathscr{G}')$, so by open cell considerations this inclusion is an equality once (7.5.1.1) is established. To prove that the long root groups coincide with those of $\mathrm{R}_{K_>/k}(L_{K_>})$ and that (7.5.1.1) holds, we shall use Theorem 5.4.3.

Define the commutative pseudo-reductive k-subgroup

$$C \subset \mathrm{R}_{K/k}(T_K) = \prod_{a\in\Delta} \mathrm{R}_{K/k}(\mathrm{GL}_1)$$

using the right side of (7.5.1.1), so obviously C normalizes $\mathrm{R}_{K/k}((L_a)_K)$ for all $a \in \Delta$. For $a \in \Delta_<$ we define $F_a := C \cdot \mathrm{R}_{K/k}((L_a)_K)$, so clearly $Z_{F_a}(T) = C$. For $a \in \Delta_>$ the k-group C normalizes $\mathrm{R}_{K_>/k}((L_a)_{K_>})$ because for $b \in \Delta_<$ the action of $t \in \mathrm{R}_{K/k}(\mathrm{GL}_1) = \mathrm{R}_{K/k}(b_K^\vee(\mathrm{GL}_1))$ on $\mathrm{R}_{K/k}((U_{\pm a})_K) = \mathrm{R}_{K/k}(\mathbf{G}_\mathrm{a})$ is via scaling through $t^{\langle a, b^\vee\rangle} \in t^{p\mathbf{Z}} \subset \mathrm{R}_{K_>/k}(\mathrm{GL}_1)$ (recall that $kK^p \subset K_>$). Hence, for $a \in \Delta_>$ the k-group $F_a = C \cdot \mathrm{R}_{K_>/k}((L_a)_{K_>})$ satisfies $Z_{F_a}(T) = C$.

Since $C \cap \mathrm{R}_{K/k}((L_a)_K)$ coincides with $\mathrm{R}_{K/k}(a_K^\vee(\mathrm{GL}_1))$ for all $a \in \Delta_<$ and with $\mathrm{R}_{K_>/k}(a_{K_>}^\vee(\mathrm{GL}_1))$ for $a \in \Delta_>$, clearly the k-groups F_a for $a \in \Delta$ are given by the construction in Proposition 2.2.1 and hence are pseudo-reductive. Now we may apply Theorem 5.4.3 to conclude that the k-group F generated by $\{F_a\}_{a\in\Delta}$ is pseudo-reductive with C as a Cartan k-subgroup and its $\pm a$-root groups coincide with those of F_a for each $a \in \Delta$. In particular, F contains the k-group L generated by the root groups $\{U_{\pm a}\}_{a\in\Delta}$. But it is clear that $F = \mathscr{G}$, so the long T-root groups of $\mathscr{G}$ have dimension $[K_> : k]$ and C is a Cartan k-subgroup of $\mathscr{G}$. This completes the proof. □

Observe that the pseudo-split absolutely pseudo-simple k-groups $\mathscr{G}$ in Proposition 7.5.1 are necessarily *non-standard* since the root groups with distinct lengths have different dimensions. To extend this construction beyond the pseudo-split case, we focus on the essential case where $K_> = k$ and shall use a fiber product construction resting on an exceptional class of isogenies that only exist in characteristics 2 and 3. These isogenies arise from the following result (for which we refer the reader to [**CGP**, Lemma 7.1.2] for a proof based on an analysis of root groups):

LEMMA 7.5.2. *Let k be a field of characteristic $p \in \{2,3\}$, and let G be a connected semisimple k-group that is absolutely simple and simply connected with root system over k_s having an edge of multiplicity p.*

Among all nonzero G-submodules of $\mathrm{Lie}(G)$ *distinct from* $\mathrm{Lie}(Z_G)$*, there is a unique such $\mathfrak{n}$ contained in all others, and it is a p-Lie subalgebra of* $\mathrm{Lie}(G)$*. If G contains a split maximal k-torus T then $\mathfrak{n}$ is spanned by the T-weight spaces for the short roots and the coroot lines* $\mathrm{Lie}(a^\vee(\mathrm{GL}_1))$ *for short $a \in \Phi(G,T)$.*

By [**CGP**, Prop. A.7.14, Ex. A.7.16], if H is an affine k-group scheme of finite type and $\mathfrak{n}$ is a p-Lie subalgebra of $\mathrm{Lie}(H)$ then there is a unique k-subgroup scheme $N \subset H$ with vanishing Frobenius morphism and Lie algebra $\mathfrak{n} \subset \mathrm{Lie}(H)$, and N is normal in H if and only if $\mathfrak{n}$ is stable under the adjoint action of H.

Thus, in the setting of Lemma 7.5.2 we obtain a unique normal k-subgroup scheme $N \subset G$ with vanishing Frobenius such that $\mathrm{Lie}(N) = \mathfrak{n}$ inside $\mathrm{Lie}(G)$. Consequently, we obtain a factorization of the Frobenius isogeny $F_{G/k}$ as

$$F_{G/k} : G \xrightarrow{\pi} G/N \xrightarrow{\overline{\pi}} G^{(p)}.$$

The isogeny $\pi : G \to G/N$ is called a *very special isogeny* (and G/N is called the *very special quotient* of G). Note that if G contains a split maximal k-torus T and $a \in \Phi(G,T)$ is long then π carries G_a *isomorphically* onto its image in G/N because the infinitesimal k-group scheme $G_a \cap N$ is trivial (as we can check on Lie algebras using the description of $\mathrm{Lie}(N)$ in Lemma 7.5.2 via long coroots and short root spaces relative to T).

The relationship between G and its very special quotient $\overline{G} := G/N$ is symmetric in the following sense:

PROPOSITION 7.5.3. *The very special quotient $\overline{G}$ of G is simply connected with root system over k_s dual to that of G_{k_s}, and the isogeny $\overline{\pi} : \overline{G} \to G^{(p)}$ arising in the factorization of $F_{G/k}$ through $\overline{G}$ is the very special isogeny for $\overline{G}$.*

If $T \subset G$ is a split maximal k-torus then for $\overline{T} := \pi(T)$ the map $\pi : G \to \overline{G}$ carries long T-root groups isomorphically onto short $\overline{T}$-root groups and carries short T-root groups U_a onto long $\overline{T}$-root groups $\overline{U}_{pa}$ via the Frobenius morphism $F_{\mathbf{G}_a/k}$.

The proof of this result rests on a direct study of the restriction of π between root groups by analyzing the weight spaces for T that occur in $\mathfrak{n}$; see [**CGP**, Prop. 7.1.5] for the details.

EXAMPLE 7.5.4. The best-known very special isogenies are from type B to type C in characteristic 2. This has a linear algebra intepretation as follows. By the classification of connected absolutely simple groups, for $n \geqslant 1$ the connected absolutely simple groups of type B_n with trivial center over a field are the special orthogonal groups $\mathrm{SO}(q)$ of non-degenerate quadratic spaces (V, q) of dimension $2n+1$. (Note however that $\mathrm{SO}(q)$ only determines (V, q) up to a conformal isometry.) By non-degeneracy, the subspace $V^{\perp}$ consisting of the vectors orthogonal to everything in V relative to the associated symmetric bilinear form $B_q(v, v') = q(v+v') - q(v) - q(v')$ on V is a line. The bilinear form B_q is alternating since $\mathrm{char}(k) = 2$, so it induces a symplectic form $\overline{B}_q$ on $\overline{V} := V/V^{\perp}$. There is an evident k-homomorphism $\mathrm{SO}(q) \to \mathrm{Sp}(\overline{B}_q)$, and the composite map

$$\pi : \mathrm{Spin}(q) \longrightarrow \mathrm{SO}(q) \longrightarrow \mathrm{Sp}(\overline{B}_q)$$

is the very special isogeny for $G := \mathrm{Spin}(q)$ when $n \geqslant 2$. Note that in this case $\overline{G}$ is always split even though G may not be split.

EXAMPLE 7.5.5. In the opposite direction, consider a connected semisimple k-group G that is absolutely simple and simply connected of type C_n $(n \geqslant 2)$. Under the adjoint representation of G on its Lie algebra $\mathfrak{g}$, there is a unique minimal non-central G-submodule $\mathfrak{n} \subset \mathfrak{g}$ [**CGP**, Lemma 7.1.2]. Hence, G naturally acts on $V = \mathfrak{g}/\mathfrak{n}$. The dimension of V is $n(2n+1) - (2n^2 - n - 1) = 2n + 1$, and since $\mathfrak{n}$ is a Lie ideal we see that the resulting representation $\rho : G \to \mathrm{GL}(V)$ kills $\mathfrak{n}$ on Lie algebras, so ρ factors through the simply connected very special quotient $\overline{G}$ of G.

We will show below that there is a *canonical* non-degenerate $\overline{G}$-invariant quadratic map $q : V \to L$ valued in a line L. The map q becomes a quadratic form upon choosing a basis of L, but the resulting k-subgroup $\mathrm{SO}(q) \subset \mathrm{GL}(V)$ does not depend on such a choice. In this manner we get a canonical homomorphism $f : \overline{G} \to \mathrm{SO}(q)$. Since $\overline{G}$ is simply connected, f uniquely factors through a homomorphism $\widetilde{f} : \overline{G} \to \mathrm{Spin}(q)$ that we will show is an isomorphism, so the unique homomorphism $G \to \mathrm{Spin}(q)$ through which $\rho : G \to \mathrm{SO}(q)$ factors is the very special isogeny for G.

A very special isogeny $H \to \overline{H}$ intertwines long roots and the associated coroots for H with short roots and the associated coroots for $\overline{H}$ [**CGP**, Prop. 7.1.5(1)], so the construction of $\mathfrak{n}$ implies that V is identified with the corresponding $\overline{G}$-submodule $\overline{\mathfrak{n}}$ of $\overline{\mathfrak{g}} := \mathrm{Lie}(\overline{G})$ that is also a p-Lie subalgebra of $\overline{\mathfrak{g}}$. The p-operation on a Lie algebra is functorial in the group scheme (see [**CGP**, Lemma A.7.13]), so the p-operation

$$q : V \longrightarrow V$$

on $V = \overline{\mathfrak{n}}$ that is induced by the ones on $\mathfrak{g}$ and $\overline{\mathfrak{g}}$ is equivariant for the natural $\overline{G}$-action on V. The p-operation on a Lie algebra is $D \mapsto D^p$ on global left-invariant derivations of the structure sheaf, and $(D + D')^p = D^p + [D, D'] + D'^p$ since $p = 2$. Hence, on $\mathrm{Lie}(\overline{G})$ we have $(X + X')^{[2]} = X^{[2]} + [X, X'] + X'^{[2]}$ where $[X, X']$ is bilinear in the pair (X, X'), so $X \mapsto X^{[2]}$ is a $\overline{G}$-equivariant *quadratic map* from $\mathrm{Lie}(\overline{G})$ into itself. Thus, q is a $\overline{G}$-equivariant quadratic map whose associated bilinear map B_q is the restriction to V of the Lie bracket.

We claim that $L := [V, V]$ is a line containing $q(V)$ and that the resulting quadratic map $q : V \to L$ is non-degenerate. Once we know that L is a line, the $\overline{G}$-action on L must be trivial (as $\overline{G}$ has no nontrivial characters), so q would be

$\overline{G}$-invariant, giving a canonical homomorphism $\widetilde{f}:\overline{G}\to \mathrm{Spin}(q)$ as explained above (which we will show is an isomorphism). To establish these properties we may and do assume $k=k_s$, so $\overline{G}$ admits a split maximal k-torus $\overline{T}$.

Let Δ be a basis of the root system $\Phi := \Phi(\overline{G},\overline{T})$ of type B_n ($n\geqslant 2$); in particular, Δ contains a unique short root b_0. By design, the subspace $V=\overline{\mathfrak{n}}$ is spanned by the coroot line $L := k\cdot \mathrm{Lie}(b_0^\vee)(\partial_t)=\mathrm{Lie}(\mu_p)$ and the lines $\overline{\mathfrak{g}}_b=\mathrm{Lie}(\alpha_p)$ for b in the set $\Phi_<$ of short roots. For each such b, the coroot line $k\cdot\mathrm{Lie}(b^\vee)(\partial_t)$ coincides with L since the difference of coroots associated to any two short roots for type B_n is twice an element of the coroot lattice (so it induces 0 on Lie algebras in characteristic 2); this is the familiar assertion that for $n\geqslant 2$ any two long roots for type C_n differ by twice an element of the root lattice.

For $b\in\Phi_<$, the lines $\overline{\mathfrak{g}}_b$ and $\overline{\mathfrak{g}}_{-b}$ generate an $\mathfrak{sl}_2$ since $\langle U_b,U_{-b}\rangle=\mathrm{SL}_2\subset\overline{G}$ (as $\overline{G}$ is simply connected). Thus, by functoriality of the p-operation and calculating in $\mathfrak{sl}_2$ we see that for each $b\in\Phi_<$ and nonzero $X_{\pm b}\in\overline{\mathfrak{g}}_{\pm b}$ the vector $B_q(X_b,X_{-b})$ is a nonzero element of L and $B_q(\overline{\mathfrak{g}}_b,L)=0$. The set $\Phi_<$ of short roots for type B_n is the root system A_1^n, so for linearly independent short $b,b'\in\Phi$ the root groups U_b and $U_{b'}$ commute with each other. Hence, $B_q(\overline{\mathfrak{g}}_b,\overline{\mathfrak{g}}_{b'})=[\overline{\mathfrak{g}}_b,\overline{\mathfrak{g}}_{b'}]=0$. Since q kills each line $\overline{\mathfrak{g}}_b$ and has nonzero restriction to L (as the p-operation for α_p vanishes and for μ_p is nonzero), we conclude that $q(V)\subset L=[V,V]$ and that the pairs of root lines for opposite short roots span pairwise B_q-orthogonal hyperbolic planes. In particular, $q:V\to L$ is non-degenerate.

It remains to show that the resulting map $\widetilde{f}:\overline{G}\to\mathrm{Spin}(q)$ is an isomorphism. Although the entire preceding construction makes sense as written only when the rank n is at least 2 (as then there are both short roots and long roots), we shall formulate a rank-1 analogue and reduce our higher-rank problem to the rank-1 analogue that is more amenable to direct calculation. Consider a pair $\{\pm b\}$ of opposite short roots in Φ and the associated k-subgroup $\overline{G}_b=\langle U_b,U_{-b}\rangle\subset\overline{G}$ that meets $\overline{T}$ in $b^\vee(\mathrm{GL}_1)$. Since $\overline{G}_b\simeq\mathrm{SL}_2$, the preceding calculations show that the p-operation $\mathrm{Lie}(\overline{G}_b)\to\mathrm{Lie}(\overline{G}_b)$ is a quadratic map whose image spans L as above and thereby defines a non-degenerate quadratic form $q_b:\mathrm{Lie}(\overline{G}_b)\to L$ that is the restriction of q. We thereby get an analogous homomorphism $\widetilde{f_b}:\overline{G}_b\to\mathrm{Spin}(q_b)$.

Naturally $\mathrm{SO}(q_b)\subset\mathrm{SO}(q)$ since V is the direct sum of $\mathrm{Lie}(\overline{G}_b)$ and the space of vectors in V orthogonal to $\mathrm{Lie}(\overline{G}_b)$, and the natural map $\mathrm{Spin}(q)_b\to\mathrm{SO}(q_b)$ is the quotient by the central μ_2 because the center of $\mathrm{Spin}(q)$ is the μ_2 whose Lie algebra is the common coroot line for every short root. This identifies $\mathrm{Spin}(q)_b$ with $\mathrm{Spin}(q_b)$, and via this inclusion of $\mathrm{Spin}(q_b)$ into $\mathrm{Spin}(q)$ it is clear that $\widetilde{f}|_{\overline{G}_b}=\widetilde{f_b}$.

We will prove that each $\widetilde{f_b}$ is an isomorphism. Granting this, the $\overline{G}$-submodule $\ker\mathrm{Lie}(\widetilde{f})\subset\mathrm{Lie}(\overline{G})$ is contained in the span of the root lines for the long roots and their associated coroots. In particular, this kernel does not contain the Lie algebra of the center, nor does it contain $\overline{\mathfrak{n}}$. But every nonzero $\overline{G}$-submodule of $\mathrm{Lie}(\overline{G})$ must contain one of those two Lie subalgebras by [**CGP**, Lemma 7.1.2], so this forces the kernel to vanish. Thus, $\widetilde{f}$ has étale kernel, so it is an étale isogeny for dimension reasons. Hence, $\widetilde{f}$ is an isomorphism since $\mathrm{Spin}(q)$ is simply connected.

It remains to prove that $\widetilde{f_b}$ is an isomorphism for each $b\in\Phi_<$. This amounts to a concrete assertion in characteristic 2: if $L=\mathrm{Lie}(Z_{\mathrm{SL}_2})\subset\mathfrak{sl}_2$ is the diagonal subspace and $q:\mathfrak{sl}_2\to L$ is the non-degenerate quadratic map induced by $X\mapsto X^{[2]}$ then the representation $\rho:\mathrm{SL}_2\to\mathrm{SO}(q)$ is the quotient by the center (as then the unique factorization through Spin(q) is an isomorphism). But composing ρ with the inclusion $\mathrm{SO}(q) \hookrightarrow \mathrm{GL}(\mathfrak{sl}_2)$ gives $\mathrm{Ad}_{\mathrm{SL}_2}$, and the (scheme-theoretic) kernel of the adjoint representation of any connected reductive group is the center, so comparison of dimensions of SL_2 and $\mathrm{SO}(q)$ implies that ρ is the quotient by the center.

Here is an interesting construction using very special isogenies:

EXAMPLE 7.5.6. Let K/k is a nontrivial finite extension satisfying $K^p \subset k$, and let $\pi : L \to \overline{L}$ be a very special isogeny over k for L as in Proposition 7.5.1 with $K_> = k$ (and $T \subset L$ a split maximal k-torus). Then for $f := \mathrm{R}_{K/k}(\pi_K)$ and the Levi k-subgroup $\overline{L} \subset \mathrm{R}_{K/k}(\overline{L}_K)$ we claim that the k-group $f^{-1}(\overline{L})$ coincides with $\mathscr{G}$ as in Proposition 7.5.1 with $K_> = k$; in particular, $f^{-1}(\overline{L})$ is smooth (and even absolutely pseudo-simple with L as a Levi k-subgroup).

To verify that $\mathscr{G} = f^{-1}(\overline{L})$ inside $\mathrm{R}_{K/k}(L_K)$, we first note that the long T-root groups of L and $\mathscr{G}$ coincide due to 1-dimensionality of each. Thus, $\mathscr{G} \subset f^{-1}(\overline{L})$ since f carries each T-root group of $\mathscr{G}$ into $\overline{L}$ (by applying Proposition 7.5.3 to π_K). This containment is an equality on open cells since π carries L_a isomorphically onto $\overline{L}_a$ for long $a \in \Phi(L, T)$, so to prove it is an equality it is enough to show that $f^{-1}(\overline{L})$ is connected, or more specifically that $\ker f$ is connected. But $\ker f = \mathrm{R}_{K/k}(\ker \pi_K)$ and as a K-scheme (not K-group scheme) $\ker \pi_K$ is isomorphic to a direct product of copies of $\mathrm{Spec}(K[x]/(x^p))$. Since $K^p \subset k$, it is clear that $\mathrm{R}_{K/k}(\mathrm{Spec}(K[x]/(x^p)))$ is geometrically connected and hence $\ker f$ is connected.

The preimage construction in Example 7.5.6 underlies the following remarkable equivalence whose proof rests on arguments with non-smooth group schemes and Theorem 5.4.4:

THEOREM 7.5.7. *Let K/k be a nontrivial purely inseparable finite extension satisfying $K^p \subset k$, and let $\pi' : G' \to \overline{G}'$ be a very special isogeny over K and define $f = \mathrm{R}_{K/k}(\pi')$. For a Levi k-subgroup $\overline{G} \subset \mathrm{R}_{K/k}(\overline{G}')$ (if one exists), the following conditions are equivalent:*

(i) *The k-group scheme $\mathscr{G} := f^{-1}(\overline{G})$ is smooth.*
(ii) *The k-group $\overline{G}$ is contained in the image of f.*
(iii) *The group $\mathscr{G}_{k_s}$ is smooth and contains a Levi k_s-subgroup of $\mathrm{R}_{K/k}(G')_{k_s}$.*

When these conditions hold, $\mathscr{G}$ is absolutely pseudo-simple of minimal type with minimal field of definition K/k for its geometric unipotent radical, $i_{\mathscr{G}}$ is identified with the inclusion of $\mathscr{G}$ into $\mathrm{R}_{K/k}(G')$, and $f(\mathscr{G}) = \overline{G}$. In particular, the quotient map $f : \mathscr{G} \twoheadrightarrow \overline{G}$ is determined by $i_{\mathscr{G}}$ and the very special isogeny $\pi' : G' \twoheadrightarrow \overline{G}'$.

The proof of Theorem 7.5.7 apart from the assertions at the end when (i), (ii), and (iii) hold is given in [**CGP**, Thm. 7.3.1]. Under these conditions it is obvious that $\overline{G} = f(\mathscr{G})$, and to prove the rest we may assume $k = k_s$. Hence, $\mathscr{G}$ contains a Levi k-subgroup G of $\mathrm{R}_{K/k}(G')$, so G is a k-descent of G' (see [**CGP**, Lemma 7.2.1]); this identifies $\pi : G \to \overline{G}$ with a k-descent of π'. By identifying G' with G_K in this manner, $\mathscr{G}$ is an instance of the preimage construction in Example 7.5.6. Thus, the asserted properties for $\mathscr{G}$ are now clear.

REMARK 7.5.8. The existence of the Levi k-subgroup $\overline{G}$ in Theorem 7.5.7 is a nontrivial condition when $\overline{G}'$ is not split (see [**CGP**, Ex. 7.2.2]).

We have finally arrived at a general class of non-standard absolutely pseudo-simple groups:

DEFINITION 7.5.9. A *basic exotic* pseudo-reductive k-group is a k-group that arises as $\mathscr{G}$ in Theorem 7.5.7.

Since split Levi k-subgroups always exist in the pseudo-split case (Theorem 5.4.4), it follows that the pseudo-split basic exotic pseudo-reductive k-groups are precisely the k-groups $\mathscr{G}$ that arise in Proposition 7.5.1 with $K_> = k$. In particular, it is immediate from Proposition 7.5.1 that if $K_>/k$ is a purely inseparable finite extension of fields and $\mathscr{G}'$ is a basic exotic pseudo-reductive $K_>$-group with root system Φ over $(K_>)_s$ then $\mathrm{R}_{K_>/k}(\mathscr{G}')$ is absolutely pseudo-simple of minimal type over k with root system Φ over k_s. (In particular, $\mathrm{R}_{K_>/k}(\mathscr{G}')$ is perfect.) The extension $K_>/k$ is intrinsically determined by such a k-group: it is the *long root field* (as we may check over k_s, via the description provided by Proposition 7.5.1 in the pseudo-split case). The center of a basic exotic k-group admits an explicit description in the presence of a Levi k-subgroup; see [**CGP**, Cor. 7.2.5].

PROPOSITION 7.5.10. *Let k be imperfect with $p := \mathrm{char}(k) \in \{2,3\}$, and let Φ be the root system F_4 when $p = 2$ and G_2 when $p = 3$. Let G be a non-standard absolutely pseudo-simple k-group of minimal type with long root field $K_>$ and root system Φ over k_s. Then $G \simeq \mathrm{R}_{K_>/k}(\mathscr{G}')$ for a basic exotic $K_>$-group $\mathscr{G}'$.*

PROOF. If k'/k and k''/k are purely inseparable finite extensions and $\mathscr{G}'$ and $\mathscr{G}''$ are basic exotic groups over k' and k'' respectively such that $\mathrm{R}_{k'/k}(\mathscr{G}') \simeq \mathrm{R}_{k''/k}(\mathscr{G}'')$ then comparison of long root fields implies $k' = k''$ as purely inseparable extensions of k. In such a situation, any k-isomorphism $f : \mathrm{R}_{k'/k}(\mathscr{G}') \simeq \mathrm{R}_{k'/k}(\mathscr{G}'')$ has the form $\mathrm{R}_{k'/k}(f')$ for a unique k'-isomorphism $f' : \mathscr{G}' \simeq \mathscr{G}''$. Indeed, the natural map

$$\mathrm{R}_{k'/k}(\mathscr{G}')_{k'} \to \mathscr{G}'$$

is the quotient by the k'-unipotent radical (as it is a smooth surjection with connected unipotent kernel [**CGP**, Prop. A.5.11(1),(2)]), and likewise for $\mathscr{G}''$, so $f_{k'}$ dominates a unique isomorphism φ between maximal pseudo-reductive quotients over k' and hence $f = \mathrm{R}_{k'/k}(\varphi)$ (see [**CGP**, Prop. 1.2.2]).

By Galois descent and the preceding canonical description of all possible k-isomorphisms f it follows that we may assume $k = k_s$. In particular, G contains a Levi k-subgroup L (with maximal k-torus T). Moreover, $\ker i_G$ is central in G since Φ is reduced. Thus, since G is minimal type it follows that $\ker i_G = 1$.

By Proposition 7.2.6(i) and Theorem 7.4.7, $K_>$ contains kK^p (over k) and $G_a \simeq \mathrm{R}_{K_>/k}((L_a)_{K_>})$ for long $a \in \Phi(L,T)$ whereas $G_a \simeq \mathrm{R}_{K/k}((L_a)_K)$ for short $a \in \Phi(L,T)$. This implies (by considerations with dimension and minimal fields of definition for geometric unipotent radicals) that the image $G \simeq i_G(G) \subset \mathrm{R}_{K/k}(L_K)$ coincides with the k-group $\mathscr{G} = \mathrm{R}_{K_>/k}(\mathscr{G}')$ as in Proposition 7.5.1 applied to $(L, K/K_>/k)$. □

The following refinement of Proposition 7.5.10 removes the "minimal type" hypothesis.

COROLLARY 7.5.11. *For k and Φ as in Proposition 7.5.10, every non-standard absolutely pseudo-simple k-group G with root system Φ is of minimal type.*

In particular, a pseudo-split absolutely pseudo-simple k-group with root system Φ is uniquely determined up to isomorphism by the minimal field of definition K/k for its geometric unipotent radical and the long root field $K_> \supset kK^p$.

PROOF. The final part follows from the rest by Theorem 7.2.5, Proposition 7.2.8, and Theorem 7.4.7.

In general, the maximal quotient $\mathscr{G} := i_G(G)$ that is pseudo-reductive of minimal type is a central quotient of G since Φ is reduced; i.e., $\mathscr{G} = G/\mathscr{C}_G$. By Proposition 7.5.10, we have $\mathscr{G} \simeq \mathrm{R}_{K_>/k}(\mathscr{G}')$ for the long root field $K_>$ of $\mathscr{G}$ and a basic exotic $K_>$-group $\mathscr{G}'$. Hence, G is a central extension of $\mathscr{G}$ by the unipotent k-group scheme $\mathscr{C}_G$. It is equivalent to show that this is a split extension, as that would force $\mathscr{C}_G = 1$, so we may and do assume $k = k_s$. In particular, G contains a split maximal k-torus T.

Let Δ be a basis for the common root system $\Phi(G,T) = \Phi(\mathscr{G},T) = \Phi(\mathscr{G}',T_{K_>})$, so $Z_{\mathscr{G}}(T) = \prod_{a\in\Delta} \mathrm{R}_{K_a/k}(\mathrm{GL}_1)$ where $K_a = K$ for short a and $K_a = K_>$ for long a. Since $\mathscr{G}_a = \mathrm{R}_{K_a/k}((L_a)_{K_a}) = \mathrm{R}_{K_a/k}(\mathrm{SL}_2)$, the classical formula universally expressing diagonal points in SL_2 as a product of points in the standard root groups allows us to express all points in $Z_{\mathscr{G}}(T)$ universally as a product of points in T-root groups for roots in $\pm\Delta$. It follows from a general splitting criterion for central extensions of pseudo-split pseudo-semisimple groups in [**CGP**, Prop. 5.1.3] that *every* central extension of $\mathscr{G}$ by a commutative affine k-group scheme Z of finite type containing no nontrivial smooth connected k-subgroup is split. We may use $\mathscr{C}_G$ as such a Z to conclude. □

REMARK 7.5.12. In view of Theorem 7.4.8, it follows from Corollary 7.5.11 that away from types B_n and C_n (with $n \geqslant 1$) the basic exotic construction accounts for *all* deviations from standardness with a reduced and irreducible root system.

REMARK 7.5.13. If k is imperfect of characteristic $p = 2$ and $[k : k^2] = 2$, it follows from Theorem 7.2.5(ii) that for any pseudo-split absolutely pseudo-simple k-group G with root system A_1 such that $G^{\mathrm{ss}}_{\overline{k}} \simeq \mathrm{SL}_2$, we have $i_G(G) = \mathrm{R}_{K/k}(\mathrm{SL}_2)$ for a purely inseparable finite extension K/k. But $\ker i_G$ is central since the root system is reduced, so G is a central extension of $\mathrm{R}_{K/k}(\mathrm{SL}_2)$ by the unipotent k-group scheme $\ker i_G$. The same splitting criterion used in the proof of Corollary 7.5.11 then implies that $\ker i_G = 1$. It follows similarly that Corollary 7.5.11 is valid over such k using the root system Φ equal to either of B_n or C_n with any $n \geqslant 2$ when $G^{\mathrm{ss}}_{\overline{k}}$ is simply connected.

The only remaining difficulties in classifying the absolutely pseudo-simple case over k with a *reduced* root system over k_s are for types B and C over imperfect fields k of characteristic 2 (including type A_1, which we have completely described in the pseudo-split minimal type case in Theorem 7.2.5(ii)).

Consider a pseudo-split absolutely pseudo-simple group G with rank $n \geqslant 2$ over a field k of arbitrary characteristic. Assume $G^{\mathrm{ss}}_{\overline{k}}$ is simply connected and let K/k be the minimal field of definition for $\mathscr{R}_u(G_{\overline{k}}) \subset G_{\overline{k}}$. Observe that by Proposition 7.2.8 and Theorem 7.4.7, if L denotes the split K-descent of $G^{\mathrm{ss}}_{\overline{k}}$ then $G \simeq \mathscr{D}(\mathrm{R}_{K/k}(L_K))$ except possibly when the root system Φ has an edge of multiplicity $p = \mathrm{char}(k) > 0$. In the latter cases, Corollary 7.5.11 gives a classification via $K/K_>/k$ for type F_4 with $p = 2$ and type G_2 with $p = 3$. We now record a variant for types B_n and C_n ($n \geqslant 2$) in characteristic 2.

THEOREM 7.5.14. *Let K/k be a purely inseparable finite extension in characteristic 2, $K_> \subset K$ a subfield containing kK^2, and Φ the root system B_n or C_n with $n \geqslant 2$. Choose a nonzero $K_>$-subspace $V \subset K$ satisfying $k\langle V\rangle = K$ and a* *nonzero* kK^2*-subspace* $V_> \subset K_>$ *satisfying* $k\langle V_>\rangle = K_>$. *If* $\Phi = \mathrm{C}_n$ *with* $n \geqslant 3$ *then assume* $V = K$, *and if* $\Phi = \mathrm{B}_n$ *with* $n \geqslant 3$ *then assume* $V_> = K_>$.

There exists a unique pseudo-split absolutely pseudo-simple k*-group* G *of minimal type with root system* Φ *such that:* $G^{\mathrm{ss}}_{\overline{k}}$ *is simply connected,* K/k *is the minimal field of definition for the geometric unipotent radical of* G, $G_a \simeq H_{V_>,K_>/k}$ *for long* $a \in \Phi$, *and* $G_a \simeq H_{V,K/k}$ *for short* $a \in \Phi$.

The appearance of the minimal-type hypothesis and of vector spaces rather than merely fields in Theorem 7.5.14 are a significant constrast with Corollary 7.5.11 (which concerns the root systems F_4 and G_2 in characteristics 2 and 3 respectively). The necessity of the conditions on $K/K_>/k$ and the vector spaces V and $V_>$ in Theorem 7.5.14 is immediate from Theorem 7.4.7. The sufficiency is deeper, and requires constructing a k-subgroup of $\mathrm{R}_{K/k}(L_K)$ containing L and satisfying prescribed properties. Theorem 5.4.3 provides the main technique in this construction; see the proof of [**CP**, Thm. 3.4.1(iii)] for further details.

To conclude our general discussion of basic exotic groups, we record some notable features in the special case $[k : k^p] = p$ (such as for global and local function fields over finite fields of characteristic p), referring to [**CGP**, Prop. 7.3.3, Prop. 7.3.5] for proofs.

PROPOSITION 7.5.15. *Let* $\mathscr{G}$ *be a basic exotic pseudo-reductive* k*-group, where* $\mathrm{char}(k) = p \in \{2, 3\}$ *and* $[k : k^p] = p$. *Let* $f : \mathscr{G} \twoheadrightarrow \overline{G}$ *be the associated surjection as at the end of Theorem* 7.5.7, *with* $\overline{G}$ *a connected semisimple group that is simply connected with root system over* k_s *dual to that of* $\mathscr{G}_{k_s}$.

(i) *The map* f *is bijective on* k*-points, as well as a homeomorphism on adelic points when* k *is global and on* k*-points when* k *is local. Moreover, the natural map* $\mathrm{H}^1(k, \mathscr{G}) \to \mathrm{H}^1(k, \overline{G})$ *induced by* f *is bijective.*

(ii) *If* $\mathscr{G}_1$ *and* $\mathscr{G}_2$ *are basic exotic* k*-groups and* $\overline{G}_j$ *is the associated quotient of* $\mathscr{G}_j$ *then the natural map* $\mathrm{Isom}_k(\mathscr{G}_1, \mathscr{G}_2) \to \mathrm{Isom}_k(\overline{G}_1, \overline{G}_2)$ *is bijective.*

(iii) *The set of isomorphism classes of* k_s/k*-forms of* $\mathscr{G}$ *is in natural bijection with the set of isomorphism classes of* k_s/k*-forms of* $\overline{G}$ *via the analogous construction* $\mathscr{H} \rightsquigarrow \overline{H}$ *for such* k*-forms.*

The significance of this result is that for many arithmetic calculations the intervention of a basic exotic k-group $\mathscr{G}$ *can be replaced* with that of the associated connected semisimple k-group $\overline{G}$. This is crucial in many proofs in [**C2**] to reduce arithmetic problems in the pseudo-reductive case to the *standard* pseudo-reductive case. (One likewise needs analogous results for pseudo-reductive groups with a non-reduced root system in characteristic 2; see Proposition 8.3.10).

The key point in the proof of Proposition 7.5.15, after passing to the pseudo-split case and inspecting open cells, is that if k'/k is a finite extension then the only nontrivial purely inseparable finite extension K/k' satisfying $kK^p \subset k'$ is $K = k'^{1/p}$ and the maps

$$\mathrm{R}_{K/k'}(\mathrm{GL}_1) \to \mathrm{GL}_1, \ \ \mathrm{R}_{K/k'}(\mathbf{G}_\mathrm{a}) \to \mathbf{G}_\mathrm{a}$$

induced by the relative Frobenius endomorphisms of GL_1 and $\mathbf{G}_\mathrm{a}$ over K are bijections between sets of k'-points (but of course are not isomorphisms between k'-groups).

8. Groups with a non-reduced root system

8.1. Preparations for birational constructions. Let k be a field. The construction of pseudo-split absolutely pseudo-simple k-groups G with a non-reduced root system (i.e., BC_n for some $n \geqslant 1$) requires an entirely different approach than the methods based on fiber products and very special isogenies used to build non-standard absolutely pseudo-simple groups of types B_n ($n \geqslant 1$), C_n ($n \geqslant 1$), F_4, and G_2. Letting K/k be the minimal field of definition for the geometric unipotent radical of G, necessarily k is imperfect of characteristic 2 and the quotient $G' = G_K/\mathscr{R}_{u,K}(G_K)$ must be simply connected of type C_n (Theorem 3.1.7).

Let $T \subset G$ be a split maximal k-torus, so for a multipliable root $c \in \Phi(G,T)$ the natural map

$$i_G : G \longrightarrow \mathrm{R}_{K/k}(G') \simeq \mathrm{R}_{K/k}(\mathrm{Sp}_{2n})$$

carries the root group U_c into $\mathrm{R}_{K/k}(U'_{2c})$. By Proposition 3.1.6, U_c is a vector group admitting a T-equivariant linear structure. Upon choosing such a linear structure, the T-action admits as its weights exactly c and $2c$, with U_{2c} precisely the $2c$-weight space. In particular, the T-equivariant linear structure on U_{2c} is *unique* since $2c : T \to \mathrm{GL}_1$ is surjective; uniqueness implies (by working with k_s-points) that this linear structure on U_{2c} is equivariant for the action of $Z_G(T)$.

The T-equivariant linear structure on U_c is generally *not* unique, so it is not evident if it can be chosen to be $Z_G(T)$-equivariant. It is an important fact (see Corollary 8.1.4) that the T-equivariant linear structure on U_c can indeed be chosen to be $Z_G(T)$-equivariant. This enhanced equivariance is a nontrivial condition, insofar as there can exist T-equivariant linear structures on U_c that are not $Z_G(T)$-equivariant.

We wish to illustrate this phenomenon with a "toy example" that will later be seen to account for all possibilities for U_c equipped with its $Z_G(T)$-action. This requires the following useful terminology. For any commutative k-algebra A with $\dim_k A < \infty$, an *$\underline{A}$-module scheme* is a smooth connected commutative affine k-group equipped with a module scheme structure over the ring scheme $\underline{A}$ representing the functor $B \rightsquigarrow A \otimes_k B$ on k-algebras. The functor $\mathscr{M} \rightsquigarrow \mathscr{M}(k)$ defines an equivalence of categories between $\underline{A}$-module schemes and finitely generated A-modules [**CGP**, Lemma 9.3.5].

EXAMPLE 8.1.1. Let K/k be a nontrivial purely inseparable finite extension in characteristic 2 (so $[K : kK^2] \geqslant 2$). Let $V' \subset K$ be a nonzero kK^2-subspace, and $V \subset K^{1/2}$ a nonzero finite-dimensional K-subspace such that for the injective squaring map $q : V \to K$ the nonzero K^2-subspace $q(V) \subset K$ has trivial intersection with V' (so $V' \neq K$, and such pairs (V', V) exist for any K/k). These hypotheses are preserved under scalar extension along $k \to k_s$ (with $k_s \otimes_k K = K_s$).

The associated vector groups $\underline{V'}$ and $\underline{V}$ over k are module schemes over the ring schemes $\underline{kK^2} := \mathrm{R}_{kK^2/k}(\mathbf{G}_\mathrm{a})$ and $\underline{K} := \mathrm{R}_{K/k}(\mathbf{G}_\mathrm{a})$ respectively. Let $\underline{q} : \underline{V} \to \underline{K}$ be the 2-linear map of $\underline{K}$-modules arising from q.

On the k-group $U := \underline{V'} \times \underline{V}$ we define an action of $C := \mathrm{R}_{K/k}(\mathrm{GL}_1)$ via scalar multiplication on $\underline{V}$ (using the K-linear structure on V) and via scalar multiplication on $\underline{V'}$ through squaring on C (using the kK^2-linear structure on V'). Observe that the k-homomorphism $U \to \mathrm{R}_{K/k}(\mathbf{G}_\mathrm{a})$ defined by $(v', v) \mapsto v' + \underline{q}(v)$ is injective on k_s-points and is C-equivariant with $c \in C$ acting on $\mathrm{R}_{K/k}(\mathbf{G}_\mathrm{a})$ through multiplication against $c^2 \in \mathrm{R}_{kK^2/k}(\mathrm{GL}_1) \subset C$.

Let $T = \mathrm{GL}_1$ be the maximal k-torus in C, so the evident linear structure on U (arising from the k-linear structures on V and V') is T-equivariant. This linear structure is also C-equivariant, but there exist other C-equivariant linear structures on U and (when $K^2 \not\subset k$) there exist T-equivariant linear structures on U that are *not* C-equivariant. To build examples of the former, let $k \cdot q(V)$ denote the k-span of $q(V)$ inside K and let $L : k \cdot q(V) \to V'$ be a nonzero kK^2-linear map. Then the map $(v', v) \mapsto (v' + \underline{L}(\underline{q}(v)), v)$ is a C-equivariant k-automorphism of U not respecting the given linear structure. Transporting the given linear structure on the target through this automorphism back onto the source gives a new C-equivariant linear structure

$$\lambda.(v', v) = (\lambda v' + (\lambda - \lambda^2)\underline{L}(\underline{q}(v)), \lambda v)$$

on U. On the other hand, if we choose such an L to be k-linear but not kK^2-linear (as we can always do when $K^2 \not\subset k$) then the same construction using this L is T-equivariant but not C-equivariant (as $\underline{q}(\underline{V}) = \underline{k \cdot q(V)}$ due to the Zariski-density of k^2 inside $\underline{k} = \mathbf{G}_{\mathrm{a}}$).

To motivate how to classify (and construct!) pseudo-split absolutely pseudo-simple G with root system BC_n, we need to describe the possibilities for U_c equipped with its $Z_G(T)$-action and its k-subgroup U_{2c}. We first relate the k-group U_{2c} and the K-group U'_{2c}. For a split k-torus S and nontrivial character $\chi \in \mathrm{X}(S)$, a vector group U over k equipped with an S-action for which $\mathrm{Lie}(U)$ is χ-isotypic admits a unique S-equivariant linear structure [**CGP**, Lemma 2.3.8]. Hence, $U \rightsquigarrow U(k)$ is an equivalence from the category of such U (using S-equivariant k-homomorphisms) onto the category of χ-isotypic finite-dimensional linear representations of S. In particular, the kernels of such k-homomorphisms arise from kernel of k-linear maps and so are smooth and connected. The T-equivariant map $U_{2c} \to \mathrm{R}_{K/k}(U'_{2c})$ induced by i_G therefore has smooth connected kernel.

But $\ker i_G$ contains no nontrivial smooth connected k-subgroup, so i_G carries U_{2c} isomorphically onto a k-subgroup $V'_c \subset \mathrm{R}_{K/k}(U'_{2c})$. This k-subgroup inclusion is also equivariant with respect to the respective actions of $Z_G(T)$ and $\mathrm{R}_{K/k}(\mathrm{GL}_1)$ via the squaring of the composite map

$$\chi_c : Z_G(T) \xrightarrow{i_G} \mathrm{R}_{K/k}(T_K) \xrightarrow{\mathrm{R}_{K/k}(c_K)} \mathrm{R}_{K/k}(\mathrm{GL}_1)$$

Hence, the k_s-subspace $V'_c(k_s) \subset \mathrm{R}_{K/k}(U'_{2c})(k_s) = U'_{2c}(K_s)$ is a subspace over the subfield of K_s generated over k_s by the squares of elements of $\chi_c(Z_G(T)(k_s)) \subset K_s^{\times}$. By Galois descent, the subfield $k_s[\chi_c(Z_G(T)(k_s))] \subset K_s$ arises from a unique subfield $K'_c \subset K$ over k, so V'_c arises from a $kK_c'^2$-subspace of the K-line $U'_{2c}(K)$.

REMARK 8.1.2. In concrete terms, K'_c is the unique minimal field among those subfields $F \subset K$ over k such that χ_c factors through $\mathrm{R}_{F/k}(\mathrm{GL}_1)$. Note that the subfield $K'_c \subset K$ containing k involves the entirety of $Z_G(T)$. For the rank-1 subgroup $G_c = \langle U_c, U_{-c}\rangle$ with split maximal k-torus $T_c = c^\vee(\mathrm{GL}_1) = T \cap G_c$, the Cartan k-subgroup $Z_{G_c}(T_c)$ is contained in $Z_G(T)$. Hence, the subextension of K_c/k analogous to K'_c/k but defined using G_c instead of G is a subextension of K'_c/k that might not equal K'_c.

Consider the natural map

$$q_c : U_c/U_{2c} \longrightarrow \mathrm{R}_{K/k}(U'_{2c})/V'_c$$

induced by i_G. The T-equivariant linear structure on U_c/U_{2c} is unique since the action on the Lie algebra is through the nontrivial character c (so this linear structure must be $Z_G(T)$-equivariant), and this makes q_c a 2-linear map relative to the unique T-equivariant linear structures on its source and target. The 2-linear map over k_s induced by q_c on k_s-points is injective since i_G carries U_{2c} isomorphically onto V'_c and $(U_c \cap \ker i_G)(k_s) = 1$ (as $(\ker i_G)(k_s)$ is finite yet only nontrivial weights – namely c and $2c$ – occur for a choice of T-equivariant linear structure on the vector group U_c). Thus, we may and do view $U_c(k_s)/U_{2c}(k_s)$ as a subgroup of $U'_{2c}(K_s)/V'_c(k_s)$, with 2-linear inclusion between these k_s-vector spaces.

By $Z_G(T)(k_s)$-equivariance, the k_s^2-subspace $U_c(k_s)/U_{2c}(k_s) \subset U'_{2c}(K_s)/V'_c(k_s)$ is stable under the action of $\chi_c(Z_G(T)(k_s))^2$, so $U_c(k_s)/U_{2c}(k_s)$ is a $(K'_c)^2$-subspace of $U'_{2c}(K_s)/V'_c(k_s)$. In view of the 2-linearity over k_s for q_c on k_s-points, it is now reasonable to ask if the unique $Z_G(T)$-equivariant linear structure on U_c/U_{2c} can be enhanced to a $\underline{K'_c}$-linear structure making q_c a linear map over the squaring map of ring schemes $\underline{K'_c} \to \underline{kK'^2_c}$ over k. The answer is affirmative, and this is a crucial first step towards understanding the possibilities for G:

PROPOSITION 8.1.3. *There is a unique $\underline{K'_c}$-module structure on U_c/U_{2c} that is $Z_G(T)$-equivariant and identifies the $Z_G(T)$-action with the composition of χ_c and the $\mathrm{R}_{K'_c/k}(\mathrm{GL}_1)$-action arising from the $\underline{K'_c}$-module structure. Moreover, the natural map*

$$q_c : U_c/U_{2c} \longrightarrow \mathrm{R}_{K/k}(U'_{2c})/V'_c$$

induced by i_G is linear over the squaring map $\underline{K'_c} \to \underline{kK'^2_c}$.

The proof of this result amounts to a delicate analysis of linear structures on vector groups; see [**CGP**, Prop. 9.3.6]. An important consequence of the $Z_G(T)$-equivariant module scheme structure provided by Proposition 8.1.3 is:

COROLLARY 8.1.4. *There exists a $Z_G(T)$-equivariant splitting of U_c as an extension of U_c/U_{2c} by U_{2c}, and the section s to $U_c \twoheadrightarrow U_c/U_{2c}$ can be chosen to make the composite map $U_c/U_{2c} \overset{s}{\hookrightarrow} U_c \to \mathrm{R}_{K/k}(U'_{2c})$ linear over the squaring map of ring schemes $\underline{K'_c} \to \underline{K}$ over k.*

PROOF. Consider the commutative diagram of short exact sequences

$$\begin{array}{ccccccccc} 0 & \longrightarrow & U_{2c} & \longrightarrow & U_c & \longrightarrow & U_c/U_{2c} & \longrightarrow & 0 \\ & & \downarrow{\simeq} & & \downarrow{i_G} & & \downarrow{q_c} & & \\ 0 & \longrightarrow & V'_c & \longrightarrow & \mathrm{R}_{K/k}(U'_{2c}) & \longrightarrow & \mathrm{R}_{K/k}(U'_{2c})/V'_c & \longrightarrow & 0 \end{array}$$

This is equivariant for the action of $Z_G(T)$ on the top row and for the action of $\mathrm{R}_{kK'^2_c/k}(\mathrm{GL}_1)$ on the bottom row by using the square $\chi_c^2 : Z_G(T) \to \mathrm{R}_{kK'^2_c/k}(\mathrm{GL}_1)$ of the map $\chi_c : Z_G(T) \to \mathrm{R}_{K'_c/k}(\mathrm{GL}_1)$, and q_c is *linear* over the squaring map $\underline{K'_c} \to \underline{kK'^2_c}$ by Proposition 8.1.3.

The key observation is that since the left vertical map is an isomorphism, this diagram expresses the top as the pullback of the bottom along q_c. But the bottom is a $\underline{kK'^2_c}$-linear exact sequence of $\underline{kK'^2_c}$-modules, so it is *split* as such due to the equivalence between the categories of smooth connected affine $\underline{kK'^2_c}$-module schemes and finite-dimensional kK'^2_c-vector spaces. Hence, since the $Z_G(T)$-action on U_c/U_{2c} is given by composing χ_c with the natural $\mathrm{R}_{K'_c/k}(\mathrm{GL}_1)$-action through the $\underline{K'_c}$-module structure (Proposition 8.1.3), the q_c-pullback of a $\underline{kK'^2_c}$-linear splitting of the bottom is a $Z_G(T)$-*equivariant* splitting of the top. □

The group $(\ker i_G)(k_s)$ is always finite, hence central in G_{k_s} due to normality of $\ker i_G$ in G, so if G is *of minimal type* then this finite group is trivial. In other words, when G is of minimal type we can establish formulas and identities inside $G(k_s)$ by applying i_G to reduce to computations inside $\mathrm{R}_{K/k}(G')(k_s) = G'(K_s) = \mathrm{Sp}_{2n}(K_s)$. This is rather powerful. For example, suppose G is of minimal type with root system BC_n. It follows from Corollary 8.1.4 that for any multipliable root $c \in \Phi(G,T)$ the k-group U_c equipped with its $Z_G(T)$-action and k-subgroup U_{2c} is given *exactly* by composing the construction in Example 8.1.1 relative to K'_c/k with $\chi_c : Z_G(T) \to \mathrm{R}_{K'_c/k}(\mathrm{GL}_1)$. (The condition in Example 8.1.1 that $V' \cap q(V) = \{0\}$ arises here from the fact that $(\ker i_G)(k_s)$ is trivial.)

This use of Example 8.1.1 to describe U_c equipped with its additional structures in the minimal type case, coupled with verifying formulas in $G(k_s)$ by working inside $G'(K_s) = \mathrm{Sp}_{2n}(K_s)$, has some striking consequences. Here is one:

PROPOSITION 8.1.5. *If G is an absolutely pseudo-simple k-group of minimal type and G_{k_s} has a non-reduced root system then $Z_G = 1$.*

We refer the reader to [**CGP**, Prop. 9.4.9] for the proof of this result; the conclusion is obviously false whenever G is not of minimal type (as $\mathscr{C}_G$ is then a nontrivial central k-subgroup scheme). If $[k : k^2] = 2$ then an absolutely pseudo-simple k-group G for which G_{k_s} has root system BC_n must be of minimal type (as we will prove in Proposition 8.3.9), but if $[k : k^2] > 2$ then for every $n \geqslant 1$ there exist pseudo-split absolutely pseudo-simple k-groups G with root system BC_n such that G is *not* of minimal type; see [**CP**, B.4] for the construction of such G.

Deeper applications of Example 8.1.1 and the triviality of $(\ker i_G)(k_s)$ for G of minimal type require a determination of the possibilities for K'_c as a subfield of K over k, as we shall do without a "minimal type" hypothesis. In the rank-1 case it will be given now, and we shall address the higher-rank case in Proposition 8.1.9.

PROPOSITION 8.1.6. *Assume $\Phi(G,T) = \mathrm{BC}_1$. For each multipliable root c we have $K'_c = K$.*

PROOF. We sketch the main idea of the proof, referring to [**CGP**, Prop. 9.4.6] for complete details. Without loss of generality we may assume $k = k_s$, and we reduce to the case where G is of minimal type by replacing G with its maximal quotient $G/\mathscr{C}_G$ of minimal type (this has no effect on K/k [**CGP**, Cor. 9.4.3], and it has no effect on K'_c due to the characterization of K'_c/k at the start of Remark 8.1.2). Choose a Levi k-subgroup $L \subset G$ containing T, so we may identify L with SL_2 carrying T over to the diagonal k-torus D and the c-root group of L over to the upper-triangular unipotent subgroup of SL_2.

The image $H := i_G(G) \subset \mathrm{R}_{K/k}(G') = \mathrm{R}_{K/k}(\mathrm{SL}_2)$ clearly contains SL_2, and by Proposition 7.1.3(ii) it is absolutely pseudo-simple of minimal type and K/k is the minimal field of definition for its geometric unipotent radical. Thus, by Theorem 7.2.5(ii), $H = H_{V,K/k}$ for some nonzero kK^2-subspace $V \subset K$ such that the ratios among elements of $V - \{0\}$ generate K as a k-algebra.

Since K'_c is generated over k by $\chi_c(Z_G(T)(k))$, it suffices to prove that $v'/v \in \chi_c(Z_G(T)(k))$ for all nonzero $v, v' \in V$. Now we finally use that G is of minimal type: this ensures that the surjective homomorphism $i_G : Z_G(T) \twoheadrightarrow Z_H(D)$ is an *isomorphism*. Hence, since $c : D \to \mathrm{GL}_1$ is inverse to the isomorphism $t \mapsto \mathrm{diag}(t, 1/t)$, the definition of χ_c in terms of $i_G|_{Z_G(T)}$ implies that $\chi_c(Z_G(T)(k)) = Z_H(D)(k)$ inside $\mathrm{R}_{K/k}(D_K)(k) = D(K) = K^\times$. But $Z_H(D)$ coincides with the Zariski closure inside $\mathrm{R}_{K/k}(\mathrm{GL}_1)$ of the subgroup generated by the ratios among nonzero elements of V (Proposition 7.2.3(i)), so we are done. $\square$

A further interesting consequence of the ubiquity of Example 8.1.1 in the minimal type case is that it allows us to explicitly describe the commutator of points in U_c and U_{-c} for multipliable c when G is of minimal type. The resulting explicit formula, given in [**CGP**, Lemma 9.4.8], is a crucial ingredient in the proof of the following important result (whose proof also rests on Corollary 8.1.4 and dynamic methods); we refer the reader to [**CGP**, Thm. 9.4.7] for the details.

THEOREM 8.1.7. *Let G be a pseudo-split absolutely pseudo-simple k-group of minimal type with a split maximal k-torus T, and assume $\Phi(G,T)$ is non-reduced.*

The k-group scheme $\ker i_G$ *is commutative, connected, and non-central, and is directly spanned by its intersections with the root groups for the multipliable roots. Moreover, the T-weights that occur in* $\mathrm{Lie}(\ker i_G)$ *are precisely the multipliable roots.*

REMARK 8.1.8. The Weyl group $W(G,T)$ acts transitively on the set of multipliable roots, so Theorem 8.1.7 implies that $\ker i_G$ is a direct product of copies of the kernel of $i_G : U_c \to \mathrm{R}_{K/k}(U'_{2c})$ for a single multipliable root c. This map on U_c is identified with $\underline{V'} \times \underline{V} \to \mathrm{R}_{K/k}(\mathbf{G}_\mathrm{a})$ defined by $(v', v) \mapsto v' + \underline{q}(v)$ for $(K/k, V', V, q)$ as in Example 8.1.1, so $\ker i_G$ is a direct product of copies of $\underline{q}^{-1}(\underline{V'} \cap \underline{q}(\underline{V}))$. Since $\underline{q}(\underline{V})$ arises from the k-span $k \cdot q(V)$, it follows that $\ker i_G$ is a direct product of copies of $\underline{q}^{-1}(\underline{W})$ for the kK^2-subspace $W := V' \cap (k \cdot q(V)) \subset K$. This is nontrivial even if $W = 0$ since describing the 2-linear nonzero q in coordinates shows that $\ker \underline{q}$ is a nontrivial k-group scheme.

In Proposition 8.1.6 we saw that $K'_c = K$ for any pseudo-split absolutely pseudo-simple k-group G with root system BC_1. For higher-rank G the same equality of fields holds, but the proof is much more difficult because we cannot pass to the rank-1 group G_c in place of G (as this can cause the field K'_c to shrink, and K_c can be a proper subfield of K; examples satisfying $K_c \neq K$ arise with root system BC_n for any $n \geqslant 2$ when k is a rational function field in at least 2 variables over any field of characteristic 2 [**CGP**, Ex. 9.8.18]). Here is the precise result:

PROPOSITION 8.1.9. *Let $T \subset G$ be a split maximal k-torus. Assume $\Phi(G,T) = \mathrm{BC}_n$ with $n \geqslant 2$. Choose a basis Δ of $\Phi(G,T)$, with $c \in \Delta$ the unique multipliable root, and let b be the unique root in Δ adjacent to $2c$ in the Dynkin diagram for the basis $\Delta' = (\Delta - \{c\}) \cup \{2c\}$ of $\Phi(G_K^{\mathrm{ss}}, T_K)$.*

Then $K'_c = K_b = K$, $kK^2 \subset K_c \subset K$, and the map $i_G : U_b \to \mathrm{R}_{K/k}(U'_b) = \mathrm{R}_{K/k}(\mathbf{G}_\mathrm{a})$ is an isomorphism onto a $\underline{K_c}$-submodule.

For the proof of Proposition 8.1.9, after reducing to the case where G is of minimal type, one establishes the asserted relationships among fields by studying the action of $Z_{G_a}(T \cap G_a)$ on $U_{a'}$ for roots $a, a' \in \Delta \cup \{2c\}$. This action is analyzed by combining our explicit knowledge of the possibilities for $i_G(G_a) = i_{G_a}(G_a)$ and $i_G(G_{a'})$ (especially when one of a or a' is multipliable) with calculations similar in spirit to those that arise in the proof of Theorem 7.4.7(iii). See [**CGP**, Prop. 9.5.2] for the details.

The preceding considerations provide precise information on the possibilities for root groups and the $Z_G(T)$-action on them for any pseudo-split absolutely pseudo-simple k-group *of minimal type* with a non-reduced root system. This provides a basic picture for what an open cell in any such G can possibly look like at the level of k_s-points via the injection $i_G : G(k_s) \hookrightarrow \mathrm{R}_{K/k}(G')(k_s) = G'(K_s) = \mathrm{Sp}_{2n}(K_s)$. (Recall that the injectivity of i_G on k_s-points rests on G being of minimal type, but the group scheme $\ker i_G$ is nontrivial; see Remark 8.1.8.)

8.2. Construction via birational group laws. We have not yet *constructed* a pseudo-split absolutely pseudo-simple group with a non-reduced root system. In [**Ti3**, Cours 1991-92, 6.4], Tits constructed some examples of such groups via birational group laws. To give a general construction, we need the pseudo-split rank-1 classification provided by Proposition 7.2.3 and Theorem 7.2.5, as well as the results obtained in §8.1 concerning the structure of the root group U_c for multipliable c. An elegant general discussion of birational group laws and theorems of Weil and Artin on promoting such structures into actual group schemes is given in [**BLR**, Ch. 5]; a summary of some relevant highlights from this theory (tailored to our needs) is provided near the beginning of [**CGP**, §9.6].

The construction of birational group laws and analysis of properties of the associated algebraic groups is always a substantial undertaking. The overview that follows is aimed at conveying the main ideas and difficulties that arise and the motivation for certain parts of the construction of groups with a non-reduced root system. The reader is referred to [**CGP**, §9.6–§9.8] for complete details.

Since the constraints in §8.1 are most definitive in the minimal type case (as it is difficult to work with a Cartan k-subgroup otherwise), below we will give a general construction in the pseudo-split minimal type case over imperfect fields k with characteristic 2. When $[k : k^2] = 2$, this turns out to yield *all* absolutely pseudo-simple k-groups with a non-reduced root system over k_s. For any k satisfying $[k : k^2] > 2$, an alternative method in [**CP**, B.4] builds some rank-n pseudo-split absolutely pseudo-simple k-groups G *not* of minimal type (with any $n \geqslant 1$), but we do not know a general technique for such constructions.

Inspired by our description (via Example 8.1.1 and Corollary 8.1.4) of the possibilities for the root group attached to a multipliable root c, and the fact that $K'_c = K$ (Propositions 8.1.6 and 8.1.9), we begin by choosing the following field-theoretic and linear-algebraic data:

- a nontrivial purely inseparable finite extension K/k,
- a nonzero kK^2-subspace $V' \subset K$,
- a nonzero finite-dimensional K-vector space V equipped with an injective additive map $q : V \to K$ that is 2-linear over K (i.e., $q(\lambda v) = \lambda^2 q(v)$ for $\lambda \in K$ and $v \in V$) such that $V' \cap q(V) = \{0\}$.

Since the composition of q with the square root isomorphism $K \simeq K^{1/2}$ is an injective K-linear map, the map q can be viewed in a rather concrete manner: V is a K-subspace of $K^{1/2}$ and q is the squaring map into K.

REMARK 8.2.1. We could replace the pair (V, q) with the K^2-subspace $q(V) \subset K$ that is identified with the Frobenius twist $V^{(2)}$, and reconstruct V as the K-vector space $V^{(2)} \otimes_{K^2,\iota} K$ where $\iota : K^2 \simeq K$ is the square root isomorphism. It is entirely a matter of taste whether one works with (V, q) or $V^{(2)} \subset K$. The development in [**CGP**, §9.6–§9.8] focuses on the perspective of $V^{(2)}$, but we have chosen to emphasize (V, q) here since this is what emerges more directly from the groups that we aim to construct.

For the pseudo-split k-groups G of minimal type that we seek to construct (with root system BC_n and minimal field of definition K/k for the geometric unipotent radical), we know that necessarily $G' \simeq \mathrm{Sp}_{2n}$ as K-groups and in §8.1 we saw that the k-homomorphism $i_G : G \to \mathrm{R}_{K/k}(G')$ must be injective on k_s-points (because G is to be of minimal type). Hence, we want to describe the possibilities for $G(k_s)$ as a subgroup of $G'(K_s)$, and then use an open cell of $\mathrm{R}_{K/k}(\mathrm{Sp}_{2n})$ as a guide for how to build an open cell for G with a birational group law from which we hope to reconstruct the group. (Strictly speaking, we will work with a left-translate of an open cell by a representative for a long Weyl element, for reasons to be explained later.)

To organize the calculations, it is convenient to begin by specifying a pinning on the K-group Sp_{2n} as follows. Let $D_n \subset \mathrm{GL}_n$ be the diagonal K-torus, $U_n \subset \mathrm{GL}_n$ the upper-triangular unipotent K-subgroup, and $B_n = D_n \ltimes U_n$ the upper-triangular Borel K-subgroup; denote transpose on $n \times n$ matrices as $m \mapsto {}^{\mathrm{t}}m$. We define the maximal K-torus $D := \{(\begin{smallmatrix} d^{-1} & 0 \\ 0 & d \end{smallmatrix}) \,|\, d \in D_n\} \subset \mathrm{Sp}_{2n}$ that normalizes the smooth connected unipotent K-subgroup

$$U = \left\{ \begin{pmatrix} {}^{\mathrm{t}}u^{-1} & mu \\ 0 & u \end{pmatrix} \,|\, u \in U_n,\ m \in \mathrm{Sym}_n \right\}$$

in Sp_{2n}, where Sym_n denotes the affine space of symmetric $n \times n$ matrices over K, and define the Borel K-subgroup

$$B = D \ltimes U = \left\{ \begin{pmatrix} {}^{\mathrm{t}}b^{-1} & mb \\ 0 & b \end{pmatrix} \,|\, b \in B_n,\ m \in \mathrm{Sym}_n \right\}$$

in Sp_{2n}. The maximal k-torus inside $\mathrm{R}_{K/k}(D)$ will be denoted D_0.

The positive system of roots $\Phi^+ := \Phi(B, D) \subset \Phi(\mathrm{Sp}_{2n}, D) =: \Phi$ consists of the following characters: for $1 \leqslant i < j \leqslant n$ the character

$$t = \mathrm{diag}(t_1^{-1}, \dots, t_n^{-1}, t_1, \dots, t_n) \mapsto t_i/t_j$$

corresponds to the root group inside U given by the ij-entry in $u \in U_n$, and for $1 \leqslant i \leqslant j \leqslant n$ the character $t \mapsto 1/(t_i t_j)$ corresponds to the root group given by ii-entry of $m \in \mathrm{Sym}_n$ when $i = j$ and the common ij-entry and ji-entry of m when $i < j$. Letting Δ be the basis of Φ^+, we have $\mathrm{GL}_1^{\Delta} \simeq D$ via $(\lambda_a)_{a \in \Delta} \mapsto \prod_{a \in \Delta} a^\vee(\lambda_a)$ since Sp_{2n} is simply connected.

For $n > 1$, the subset $\Phi_>^+ \subset \Phi^+$ of long positive roots consists of the characters $1/t_i^2$ whose root groups are the diagonal entries of m (so they are 2-divisible in $\mathrm{X}(D)$); the set of short positive roots is denoted $\Phi_<^+$; in the special case $n = 1$ we define $\Phi_>^+ = \Phi^+$ and $\Phi_<^+ = \emptyset$ since the roots for SL_2 are 2-divisible in the character lattice of the diagonal torus. Each positive root group is identified with $\mathbf{G}_\mathrm{a}$ via the matrix-entry coordinatization, and the root groups for long positive roots *pairwise commute* since a sum of distinct long positive roots in type C_n is not a root.

The longest element in $W(\mathrm{Sp}_{2n}, D)$ is represented by the matrix $w = (\begin{smallmatrix} 0 & 1_n \\ -1_n & 0 \end{smallmatrix}) \in \mathrm{Sp}_{2n}(K)$ that has order 2 (as $\mathrm{char}(K) = 2$), with w-conjugation on D equal to inversion and w-conjugation carrying B to the opposite Borel subgroup relative to D via

$$w \begin{pmatrix} {}^{\mathrm{t}}u^{-1} & mu \\ 0 & u \end{pmatrix} w^{-1} = \begin{pmatrix} u & 0 \\ -mu & {}^{\mathrm{t}}u^{-1} \end{pmatrix}.$$

Letting B^- denote the Borel K-subgroup of Sp_{2n} opposite B relative to D, and U^- its unipotent radical, we get an open cell $U^-B = w^{-1}UwB \subset \mathrm{Sp}_{2n}$. Its left w-translate UwB is easier to work with for computations since it involves just points of U and B up to the presence of the 2-torsion point w. Thus, the strategy to build G is not to build a birational group law on a candidate for an open cell, but rather on the left $\mathbf{w}$-translate, where $\mathbf{w}$ is a 2-torsion element lying over $w \in \mathrm{Sp}_{2n}(K)$. We will be guided by the desired homomorphism $i_G : G \to \mathrm{R}_{K/k}(\mathrm{Sp}_{2n})$ that must restrict to an inclusion $C \hookrightarrow \mathrm{R}_{K/k}(D)$ for a Cartan k-subgroup of G that has to be determined *a priori* as a k-subgroup of $\mathrm{R}_{K/k}(D) = \prod_{a\in\Delta} \mathrm{R}_{K/k}(\mathrm{GL}_1)$.

Since U is directly spanned in any order by its positive root groups, upon *choosing* an enumeration of $\Phi^+_>$ and an enumeration of $\Phi^+_<$ we get an isomorphism via multiplication

$$\prod_{a\in\Phi^+_>} U_a \times \prod_{b\in\Phi^+_<} U_b \simeq U.$$

Using standard matrix-entry coordinatizations, each U_c is identified with $\mathbf{G}_\mathrm{a}$ as a K-group. We have the same upon applying $\mathrm{R}_{K/k}$ throughout. The idea now is to replace each long root group $\mathrm{R}_{K/k}(U_a) = \mathrm{R}_{K/k}(\mathbf{G}_\mathrm{a}) \subset \mathrm{R}_{K/k}(U)$ with $\underline{V}' \times \underline{V}$. To make this precise, we define the pointed k-scheme

$$\mathscr{U} := \prod_{c\in(1/2)\Phi^+_>} \mathscr{U}_c \times \prod_{b\in\Phi^+_<} \mathscr{U}_b$$

where $\mathscr{U}_c = \underline{V}' \times \underline{V}$ for $c \in (1/2)\Phi^+_>$ and $\mathscr{U}_b = \mathrm{R}_{K/k}(U_b) = \mathrm{R}_{K/k}(\mathbf{G}_\mathrm{a})$ for $b \in \Phi^+_<$.

The pointed k-scheme $\mathscr{U}$ will turn out to be the k-unipotent radical of a minimal pseudo-parabolic k-subgroup of the k-group that we shall build, and so our first step is to construct a k-group law on $\mathscr{U}$. For this purpose, we will use the map $f_c : \mathscr{U}_c \to \mathrm{R}_{K/k}(U_c)$ defined by $(v', v) \mapsto v' + \underline{q}(v)$ for $c \in (1/2)\Phi^+_>$ and the identity map $f_b : \mathscr{U}_b \to \mathrm{R}_{K/k}(U_b)$ for $b \in \Phi^+_<$. Define $f : \mathscr{U} \to \mathrm{R}_{K/k}(U)$ via multiplication in $\mathrm{R}_{K/k}(U)$ of these componentwise maps. Note that f is injective on k_s-points since U is directly spanned by the positive root groups and f_c is injective on k_s-points for each $c \in (1/2)\Phi^+_>$ (due to the hypotheses on $(K/k, V', V, q)$).

THEOREM 8.2.2. *There is a unique k-group structure μ on $\mathscr{U}$ relative to which $f : \mathscr{U} \to \mathrm{R}_{K/k}(U)$ is a k-homomorphism. The identity $e \in \mathscr{U}(k)$ is the evident base point, and relative to (μ, e) each inclusion $\mathscr{U}_c \hookrightarrow \mathscr{U}$ for $c \in (1/2)\Phi^+_>$ and $\mathscr{U}_b \hookrightarrow \mathscr{U}$ for $b \in \Phi^+_<$ is a k-homomorphism. Moreover, the natural $\mathrm{R}_{K/k}(D)$-action on $\mathrm{R}_{K/k}(U)$ uniquely lifts through f to an action on $\mathscr{U}$.*

We refer the reader to [**CGP**, Thm. 9.6.14] for the details of the long nested induction proof based on the height of positive roots. The success of the induction rests on putting the pairwise-commuting long root groups to the left of the short root groups in the definition of f, together with a fact that is specific to characteristic 2: the short positive root groups of U directly span (in any order) a smooth connected k-subgroup [**CGP**, Thm. 9.6.7]. Ultimately the initial choice of enumerations of the roots does not matter (and for later parts of the construction this is important): using the k-group structure on $\mathscr{U}$ built above, $\mathscr{U}$ is directly spanned *in any order* by the k-groups $\mathscr{U}_a$ for $a \in (1/2)\Phi^+_> \cup \Phi^+_<$, due to Theorem 3.3.3 (using the action just built on $\mathscr{U}$ by $\mathrm{R}_{K/k}(D) \supset D_0$).

Define the k-group $\mathscr{B} = \mathrm{R}_{K/k}(D) \ltimes \mathscr{U}$. The evident k-homomorphism $\mathscr{B} \to \mathrm{R}_{K/k}(B)$ is also denoted as f. Some deeper algebraic geometry (e.g., Zariski's Main

Theorem) and further results specific to characteristic 2 (e.g., the ii-entry of the inverse of an invertible symmetric matrix $(r_{ij}) \in \mathrm{GL}_n(R)$ for an $\mathbf{F}_2$-algebra R has the form $\sum_j f_{ij}^2 r_{jj}$ for some $f_{ij} \in R$) are used to establish the following result, whose proof occupies most of [**CGP**, §9.7]:

THEOREM 8.2.3. *Let $\Omega \subset \mathrm{R}_{K/k}(UwB) \times \mathrm{R}_{K/k}(UwB)$ be the open domain of definition of the birational group law on $\mathrm{R}_{K/k}(UwB)$. For the k-scheme $\mathscr{U}\mathbf{w} \times \mathscr{B}$, where $\mathbf{w}$ is a k-point symbol, define the map $\mathscr{U}\mathbf{w} \times \mathscr{B} \to \mathrm{R}_{K/k}(UwB)$ by $(u\mathbf{w}, b) \mapsto f(u)wf(b)$; denote this map as f too.*

There is a unique birational group law m on $\mathscr{U}\mathbf{w} \times \mathscr{B}$ such that its open domain of definition

$$\mathrm{dom}(m) \subset (\mathscr{U}\mathbf{w} \times \mathscr{B}) \times (\mathscr{U}\mathbf{w} \times \mathscr{B})$$

meets $(f \times f)^{-1}(\Omega)$ and makes f compatible with the birational group laws. Moreover, m is strict (i.e., $\mathrm{dom}(m)$ meets each fiber of the projections $(\mathscr{U}\mathbf{w} \times \mathscr{B})^2 \rightrightarrows \mathscr{U}\mathbf{w} \times \mathscr{B}$).

Since f is generally *not* dominant, the requirement $\mathrm{dom}(m) \cap (f \times f)^{-1}(\Omega) \neq \emptyset$ is not automatic and is required to make sense of $m \circ (f \times f)$ as a rational map. The significance of strictness of m is that for a smooth separated k-scheme X of finite type equipped with a general birational group law μ, only an unknown dense open subset X' of X appears inside the uniquely associated smooth connected k-group provided by Weil's theorem on birational group laws. To perform computations it is very helpful when one can take X' to be X, and for that to happen it is necessary and sufficient to assume μ is strict. See [**CGP**, Thm. 9.6.4] (and references within its proof) for further information on promoting birational group laws to groups, including a functoriality property for the k-group H associated to a *strict* birational group law (X, μ) relative to rational homomorphisms from X into a smooth k-group. Due to this latter functoriality, it follows that if

$$G_{K/k,V',V,q,n}$$

denotes the unique k-group containing $\mathscr{U}\mathbf{w} \times \mathscr{B}$ as a dense open subscheme compatibly with birational group laws, then there is a unique k-homomorphism

$$\phi : G_{K/k,V',V,q,n} \longrightarrow \mathrm{R}_{K/k}(\mathrm{Sp}_{2n})$$

extending f. We will generally denote $G_{K/k,V',V,q,n}$ as G when the context makes the meaning clear.

REMARK 8.2.4. Beware that $\mathscr{U}\mathbf{w} \times \mathscr{B}$ does *not* contain the identity of G, since it is carried by ϕ into the open subset $\mathrm{R}_{K/k}(UwB) \subset \mathrm{R}_{K/k}(\mathrm{Sp}_{2n})$ that does not contain the identity (as $w \notin B$). Hence, *very little* can be easily seen about the structure of G from inspection of $\mathscr{U}\mathbf{w} \times \mathscr{B}$; e.g., it is not obvious yet if $(\mathbf{w}, 1)$ is 2-torsion (all we can detect at the moment is that $(\mathbf{w}, 1)^2 \in (\ker \phi)(k)$). The failure of the identity point to lie in the most tangible open subset $\mathscr{U}\mathbf{w} \times \mathscr{B}$ of G is a source of many headaches when initially trying to analyze G.

Now there are many non-trivial problems to be overcome. To start with the most basic of all: is G affine? As usual with birational group laws, the associated group of interest is built via a gluing process that *a priori* might leave the affine setting. The problem of how to establish affineness *a posteriori* is a serious one when using birational group laws. For example, this difficulty arises in the uniform construction of simply connected Chevalley groups over $\mathbf{Z}$ for all root systems in

[**SGA3**, XXV]. There, the mechanism to prove affineness is *Chevalley's structure theorem* applied on geometric fibers:

THEOREM 8.2.5 (Chevalley). *Every smooth connected affine group over a perfect field is uniquely an extension of an abelian variety by a smooth connected affine group. In particular, if H is a smooth connected group over an arbitrary field F and $H(\overline{F})$ coincides with its own commutator subgroup then H is affine.*

A proof of the first part of Theorem 8.2.5 is given in [**Chev**] (and in [**C1**] using modern terminology); the second part is immediate since abelian varieties are commutative and affineness of an F-scheme can be checked after scalar extension to the perfect field $\overline{F}$. The perfectness hypothesis on the ground field in the first part of Theorem 8.2.5 is unavoidable, since over every imperfect field there are smooth connected affine groups that are not proper yet do not contain a nontrivial smooth connected affine subgroup (see [**CGP**, Ex. A.3.8]).

For our needs, the affineness criterion in Theorem 8.2.5 is not convenient. We will use a different affineness criterion that rests on a little-known but powerful substitute in positive characteristic for Chevalley's structure theorem:

THEOREM 8.2.6. *If H is a smooth connected group over a field F of positive characteristic then H is uniquely a central extension of a smooth connected affine group by a semi-abelian variety with no non-constant global functions.*

In particular, if H contains no nontrivial central F-torus and no nontrivial abelian subvariety then H is affine.

See [**CGP**, Thm. A.3.9] for arguments and references relevant to a proof of Theorem 8.2.6; the proof *uses* the first part of Theorem 8.2.5 (applied over $\overline{F}$). Note that the extension structure in Theorem 8.2.6 is "better" than the one in Theorem 8.2.5 because (i) it is valid without perfectness hypotheses on the ground field (and hence is very useful over local and global function fields), and (ii) it provides a *central* extension.

REMARK 8.2.7. The universal vector extension of any elliptic curve provides counterexamples to the conclusions in Theorem 8.2.6 in characteristic 0, ultimately because (in contrast with positive characteristic) a nonzero endomorphism of $\mathbf{G}_{\mathrm{a}}$ in characteristic 0 is an isomorphism.

Before we use the affineness criterion in Theorem 8.2.6 to prove the affineness of $G := G_{K/k,V',V,q,n}$ we establish a simple but very useful preliminary result:

LEMMA 8.2.8. *The group* $(\ker\phi)(k_s)$ *is trivial.*

This lemma allows us to deduce properties of G by working inside $\mathrm{R}_{K/k}(\mathrm{Sp}_{2n})$, "as if" ϕ were an inclusion. The group scheme $\ker\phi$ turns out to always have positive dimension.

PROOF. By design, on the dense open $\Omega := \mathscr{U}\mathbf{w} \times \mathscr{B} \subset G$ the restriction f of ϕ is injective on k_s-points. If $g \in (\ker\phi)(k_s)$ then for a choice of k_s-point g' in the dense open $\Omega \cap g^{-1}\Omega$ we have $f(gg') = \phi(gg') = \phi(g)\phi(g') = f(g')$. Hence, $gg' = g'$, so $g = 1$ as desired. $\square$

The triviality of $(\ker\phi)(k_s)$ yields many useful further consequences (despite the nontriviality of the group scheme $\ker\phi$). For example, in addition to implying that $(\mathbf{w}, 1)$ is 2-torsion, we note that ϕ carries the smooth closed subscheme

$$C := (\{\mathbf{w}\} \times \mathrm{R}_{K/k}(D)) \cdot (\mathbf{w}, 1)$$

isomorphically onto the Cartan k-subgroup $w\mathrm{R}_{K/k}(D)w = \mathrm{R}_{K/k}(D)$ of $\mathrm{R}_{K/k}(\mathrm{Sp}_{2n})$. Consequently, C must be a k-subgroup of G, and the maximal k-torus $D_0 \subset \mathrm{R}_{K/k}(D)$ is thereby identified with a k-torus of G that is contained in C and is not contained in any strictly larger k-torus of G (as otherwise $\ker\phi$ would kill a nontrivial k-torus for dimension reasons, contradicting Lemma 8.2.8).

By the same reasoning, for any long positive root $a \in \Phi^+ \subset \Phi(\mathrm{Sp}_{2n}, D)$, if we let $\underline{V}'_a$ denote the copy of $\underline{V}'$ inside $\mathscr{U}_{a/2}$ then the smooth closed subscheme

$$\mathscr{U}_a := (\underline{V}'_a\mathbf{w} \times \{1\})(\mathbf{w}, 1) \tag{8.2.8.1}$$

is carried *isomorphically* onto the k-subgroup $\underline{V}'$ of the a-root group $\mathrm{R}_{K/k}(U_a) = \mathrm{R}_{K/k}(\mathbf{G}_\mathrm{a})$ of $\mathrm{R}_{K/k}(\mathrm{Sp}_{2n})$, so it is a k-subgroup of G. Likewise, for any $g \in C(k_s)$ the effect of g-conjugation on G_{k_s} must carry $(\mathscr{U}_a)_{k_s}$ onto itself, so C normalizes $\mathscr{U}_a$ and the k-torus $D_0 \subset C$ thereby acts on $\mathrm{Lie}(\mathscr{U}_a)$ through a. By the same method, the natural maps of smooth k-schemes

$$\mathscr{U} \simeq (\mathscr{U}\mathbf{w} \times \{1\})(\mathbf{w}, 1) \subset G, \;\; \mathscr{B} \longrightarrow C \cdot (\mathscr{U}\mathbf{w} \times \{1\})(\mathbf{w}, 1) \subset G$$

are isomorphisms onto k-subgroups since composing each with the k-homomorphism ϕ (that is injective on k_s-points) respectively gives the k-homomorphism f from Theorem 8.2.2 and its analogue using $B = \mathrm{R}_{K/k}(D) \ltimes U \subset \mathrm{R}_{K/k}(\mathrm{Sp}_{2n})$ (with $\phi : C \to \mathrm{R}_{K/k}(D)$ a k-isomorphism). The C-action on the k-groups $\mathscr{U}_a$ for long roots $a \in \Phi^+$ enables us to prove:

PROPOSITION 8.2.9. *The k-group G is affine.*

PROOF. By Theorem 8.2.6 it suffices to prove that G contains no nontrivial central k-torus or abelian variety as a k-subgroup. Any abelian variety A that is a k-subgroup of G is killed by ϕ since $\mathrm{R}_{K/k}(\mathrm{Sp}_{2n})$ is affine, so $A = 1$ since $(\ker\phi)(k_s) = 1$. But ϕ will turn out to generally not be surjective when $[k : k^2] > 2$ (due to later considerations with root groups), so it is unclear that the image under ϕ of a central torus in G should be central in $\mathrm{R}_{K/k}(\mathrm{Sp}_{2n})$. Thus, to prove the triviality of a central k-torus $Z \subset G$ we proceed in another way.

We have built a copy of D_0 as a k-subgroup of G (contained in C) and showed that D_0 is not contained in any strictly larger k-torus. The multiplication map $D_0 \times Z \to G$ is a k-homomorphism whose image must be a k-torus, so this image is equal to D_0. Hence, $Z \subset D_0$. The centrality of Z in G implies that for each $\mathscr{U}_a$ as in (8.2.8.1), the Z-action on $\mathscr{U}_a$ via D_0-conjugation is trivial. But the associated action of D_0 on $\mathrm{Lie}(\mathscr{U}_a)$ is through the character a, so $Z \subset \ker a$ for every long root $a \in \Phi^+$. The long roots in a type-C_n root system constitute a rank-n root system (of type A_1^n), so such a's span $\mathrm{X}(D_0)_{\mathbf{Q}}$. This forces $Z = 1$. □

It has been shown that D_0 is a maximal k-torus in the smooth connected affine k-group G and that $\phi(C)$ equals the Cartan k-subgroup $\mathrm{R}_{K/k}(D)$ in $\mathrm{R}_{K/k}(\mathrm{Sp}_{2n})$. Hence, the smooth connected k-group $Z_G(D_0) \supset C$ must equal C for dimension reasons since $(\ker\phi)(k_s) = 1$. In other words, C is a Cartan k-subgroup of G.

To prove the pseudo-reductivity of the smooth connected affine k-group G, it suffices to prove that $\phi(\mathscr{R}_{u,k}(G)) = 1$. Here we encounter a more serious manifestation of the problem that arose in the proof of Proposition 8.2.9: the map ϕ is generally not surjective, so there is no evident reason why the smooth connected unipotent k-group $\phi(\mathscr{R}_{u,k}(G))$ should be normal in $\mathrm{R}_{K/k}(\mathrm{Sp}_{2n})$. Hence, there is no obvious way to harness the pseudo-reductivity of $\mathrm{R}_{K/k}(\mathrm{Sp}_{2n})$ to deduce the same for G, so we do not proceed along such lines.

8.3. Properties of birational construction. The smooth connected affine k-group $G = G_{K/k,V',V,q,n}$ built in §8.2 is equipped with a homomorphism $\phi : G \to \mathrm{R}_{K/k}(\mathrm{Sp}_{2n})$ that is injective on k_s-points, and ϕ carries a Cartan k-subgroup C isomorphically onto $\mathrm{R}_{K/k}(D)$. To prove that G is pseudo-reductive, we shall construct a *pseudo-reductive* k-subgroup $H \subset \mathrm{R}_{K/k}(\mathrm{Sp}_{2n})$ via Theorem 5.4.3 and then (after some hard work) prove $\phi(G) = H$ via a comparison of open cells. This implies that $\phi(G)$ is pseudo-reductive (so G is certainly pseudo-reductive). Building on this approach, much of [**CGP**, §9.8] is devoted to proving the following main properties of G (via calculations with root groups and conjugation by points of Cartan subgroups):

THEOREM 8.3.1. *Let $V_0 = V' + k\cdot q(V)$ and $K_0 = k\langle V_0\rangle$. If $n = 1$ then assume $K_0 = K$. Define the k-subgroup $C_0 = (V_0)^*_{K_0/k} \times \prod_b \mathrm{R}_{K/k}(b^\vee(\mathrm{GL}_1)) \subset \mathrm{R}_{K/k}(D)$ where b varies through the short simple roots in $\Phi^+ \subset \Phi := \Phi(\mathrm{Sp}_{2n}, D)$.*

(i) *The k-group G is pseudo-reductive, K/k is the minimal field of definition for the geometric unipotent radicals of G and $\mathscr{D}(G)$, and the maps ϕ and $\phi|_{\mathscr{D}(G)}$ are respectively identified with i_G and $i_{\mathscr{D}(G)}$.*

(ii) *The k-torus $D_0 \subset G$ is contained in $\mathscr{D}(G)$, the root system $\Phi(\mathscr{D}(G), D_0) = \Phi(G, D_0)$ coincides with $\Phi \cup (1/2)\Phi_>$ of type* BC_n, *and* $(\mathbf{w}, 1)$ *represents the long Weyl element in $W(G, D_0)$. If moreover $1 \in V'$ then* $(\mathbf{w}, 1) \in \mathscr{D}(G)$ *and the k-subgroup* $\mathrm{Sp}_{2n} \subset \mathrm{R}_{K/k}(\mathrm{Sp}_{2n})$ *lifts to a Levi k-subgroup of $\mathscr{D}(G)$ containing D_0.*

(iii) *For multipliable $c \in \Phi(G, D_0)$,* $(\mathscr{U}_c\mathbf{w} \times \{1\})(\mathbf{w}, 1)$ *is the c-root group; this is identified with $\underline{V'} \times \underline{V}$ equipped with the evident action by $Z_G(D_0) = C = \mathrm{R}_{K/k}(D)$ over the C-action on $\mathrm{R}_{K/k}(U'_{2c})$. Moreover, $Z_{\mathscr{D}(G)}(D_0) = C_0$. Likewise,* $(\mathscr{U}\mathbf{w} \times \{1\})(\mathbf{w}, 1)$ *is the smooth connected unipotent k-subgroup of G generated by the D_0-root groups for roots in $\Phi^+ \cup (1/2)\Phi^+_>$.*

(iv) *The pseudo-reductive k-groups G and $\mathscr{D}(G)$ are of minimal type, and $G = \mathscr{D}(G)$ if $V_0 = K$.*

(v) *Consider a second triple $(\mathscr{V}', \mathscr{V}, \mathfrak{q})$ relative to K/k, and if $n = 1$ then assume $k\langle \mathscr{V}_0\rangle = K$ where $\mathscr{V}_0 := k \cdot \mathfrak{q}(\mathscr{V}) + \mathscr{V}'$. Let $\mathscr{G}$ be the associated k-group. The following are equivalent: $G \simeq \mathscr{G}$, $\mathscr{D}(G) \simeq \mathscr{D}(\mathscr{G})$, and there exists $\lambda \in K^\times$ such that $\mathscr{V}' = \lambda V'$ and $\mathscr{V}' + \mathfrak{q}(\mathscr{V}) = \lambda(V' + q(V))$.*

REMARK 8.3.2. Let us explain the necessity of the hypothesis $K_0 = K$ when $n = 1$. Two desired properties guided the construction: K/k should be the minimal field of definition for the geometric unipotent radical of $\mathscr{D}(G)$ (as is confirmed in (i)) and the linear algebra data (V', V, q) used in the construction of G should appear in a description of the root group for any multipliable root c in the spirit of Example 8.1.1 (as is confirmed by (iii) due to the construction of ϕ via the map f considered in Theorem 8.2.2). Since $\phi(\mathbf{w}, 1) = w$ by design, it follows from (i) that (as intended) the image of the c-root group under $i_{\mathscr{D}(G)} = i_G|_{\mathscr{D}(G)}$ is the k-subgroup $\underline{V_0}$ inside the root groups of $\mathrm{R}_{K/k}((\mathrm{Sp}_{2n})_{2c}) = \mathrm{R}_{K/k}(\mathrm{SL}_2)$ relative to its diagonal k-torus. In particular, if $n = 1$ then $i_{\mathscr{D}(G)}(\mathscr{D}(G)) = H_{V_0,K/k}$, so the minimal field of definition over k for the geometric unipotent radical of $i_{\mathscr{D}(G)}(\mathscr{D}(G))$ is $k\langle V_0\rangle =: K_0$.

But this field of definition over k must coincide with that for $\mathscr{D}(G)$, by Proposition 7.1.3(ii). Thus, we must assume $K_0 = K$ when $n = 1$.

REMARK 8.3.3. For arithmetic applications, the most basic case is $[k : k^2] = 2$ (such as when k is a local or global function field). In such cases the only purely inseparable finite extensions of k are $k^{1/2^m}$ for $m \geqslant 0$, so $kK^2 = K^2$ and V' is forced to be a K^2-line inside K. Hence, via $K^\times$-scaling we can assume $V' = K^2$. But $q(V)$ is a nonzero K^2-subspace of K meeting V' trivially, so it must be a complementary K^2-line; i.e., $V' + q(V) = K$. Thus, $G = \mathscr{D}(G)$ by (iv) above, so by (v) there is *only one* k-isomorphism class among the k-groups G produced by this construction for a given pair $(K/k, n)$ when $[k : k^2] = 2$.

Explicitly, by taking $V' = K^2$ in such cases, if we write $q(V) = K^2\alpha$ for some $\alpha \in K - K^2$ then we can say that the construction of G rests on the triple $(K/k, \alpha, n)$ (see Remark 8.2.1), but the isomorphism class does not depend on α.

In general, without any hypotheses on $[k : k^2]$, the identification of $Z_{\mathscr{D}(G)}(D_0)$ with C_0 in part (iii) shows that $G = \mathscr{D}(G)$ if and only if $(V_0)^*_{K_0/k} = \mathrm{R}_{K/k}(\mathrm{GL}_1)$, so a necessary condition for the perfectness of G is that $K_0 = K$ (as is required when $n = 1$, but generally fails otherwise when $[k : k^2] > 2$).

We have constructed pseudo-split absolutely pseudo-simple k-groups $\mathscr{D}(G)$ with root system BC_n in terms of linear-algebraic data $(K/k, V', V, q)$ (provided that $K_0 = K$ when $n = 1$), and in part (v) of Theorem 8.3.1 we characterized when such k-groups are isomorphic in terms of simple operations on this data. But is this construction exhaustive? There is a small complication: when $n = 2$ it is *not* exhaustive (if $[k : k^2] \geqslant 8$).

To understand what is special about the case $n = 2$, recall from Proposition 8.1.9 (with b as defined there) that if $n \geqslant 2$ then $K_b = K$, so $G_b = H_{V_b, K/k}$ for some nonzero kK^2-subspace $V_b \subset K$ satisfying $k\langle V_b \rangle = K$. However, we have provided no reason that necessarily $V_b = K$, or equivalently that G_b should be standard (whereas G_b is standard for every G as in Theorem 8.3.1)! Here is such a reason when $n \geqslant 3$: in such cases every short root in the C_n-diagram is adjacent to another short root, and together they generate a root system of type A_2, so we can use centralizers of codimension-2 tori and standardness for type-A_2 in all characteristics to conclude that G_b must be standard for all such b when $n \geqslant 3$.

But this reasoning does not work if $n = 2$, and in fact there *are* more k-groups to be built (when $[k : k^2] \geqslant 8$). In effect, we need to introduce additional linear-algebraic data, to play the role of the root space for non-multipliable non-divisible roots: a nonzero kK^2-subspace $V'' \subset K$ that satisfies $k\langle V'' \rangle = K$. This subspace must satisfy some conditions in relation to $q(V)$ and V' to ensure the necessary condition that the $c^\vee(\mathrm{GL}_1)$-centralizer in G_c normalizes the b-root group, where $\{2c, b\}$ is a basis of the root system of type C_2. Since $\langle b, (2c)^\vee \rangle = -1$, this normalizing property holds whenever V'' is a subspace of K over the subfield $K_0 := k\langle V_0 \rangle \subset K$ that contains kK^2 (where $V_0 := k \cdot q(V) + V'$); see [**CGP**, 9.8.3] for the calculations. The case $V'' = K$ recovers the construction of G in Theorem 8.3.1 for $n = 2$, so this is primarily of interest when $V'' \neq K$.

Consider the pseudo-split pseudo-reductive k-group G of minimal type with root system of type BC_2 provided by Theorem 8.3.1 using $n = 2$ and a 4-tuple $(K/k, V', V, q)$. Let V'' be a nonzero *proper* K_0-subspace of K. (In [**CGP**, 9.8.3] it is shown by elementary field-theoretic considerations that such V'' exists for some choice (V', V, q) if and only if $[K : kK^2] > 4$; this happens for some K/k if and only if $[k : k^2] \geqslant 8$, ruling out local and global function fields.) Fix a basis $\{c, b\}$ of $\Phi(G, D_0) = \mathrm{BC}_2$ with multipliable c, and let G'' be the k-subgroup of $\mathscr{D}(G)$ generated by G_c and $H_{V'',K/k} \subset \mathrm{R}_{K/k}(\mathrm{SL}_2) = G_b$. The following is established in [**CGP**, 9.8.3, Prop. 9.8.4, Prop. 9.8.9]:

PROPOSITION 8.3.4. *The k-group G'' is absolutely pseudo-simple of minimal type with D_0 as a split maximal k-torus, and K/k is the minimal field of definition for its geometric unipotent radical. Moreover,*

$$Z_{G''}(D_0) = (V_0)^*_{K_0/k} \times (V'')^*_{K/k}$$

inside $Z_G(D_0) = \mathrm{R}_{K/k}(D)$, $\phi|_{G''} = i_{G''}$ (so $(\ker i_{G''})(k_s) = 1$), and $(\mathbf{w}, 1) \in G''(k)$ if $1 \in V' \cap V''$.

If $(\mathscr{V}', \mathscr{V}, \mathfrak{q}, \mathscr{V}'')$ is another such 4-tuple with $\mathscr{V}'' \neq K$ then the associated k-group $\mathscr{G}''$ is isomorphic to G'' if and only if there exist $\lambda, \mu \in K^\times$ such that

$$\mathscr{V}'' = \mu V'', \;\; \mathscr{V}' = \lambda V', \;\; \mathscr{V}' + \mathfrak{q}(\mathscr{V}) = \lambda(V' + q(V)).$$

It is clear by consideration of the b-root group ($\simeq \underline{V}''$) that the k-groups produced by this result never arise among the pseudo-simple groups produced by Theorem 8.3.1. Fortunately, essentially by reversing the long path of theoretical reasoning that motivated the conditions imposed in our constructions (including the additional reasoning that explained why the case $n = 2$ might admit additional possibilities beyond Theorem 8.3.1 but cases with $n \neq 2$ cannot), one can show that we have constructed everything:

THEOREM 8.3.5. *Every pseudo-split absolutely pseudo-simple k-group of minimal type with minimal field of definition K/k for its geometric unipotent radical and root system BC_n is produced by the preceding constructions.*

If $[k : k^2] = 2$ then there is only one k-isomorphism class of such k-groups for a given pair $(K/k, n)$, and $i_G : G \to \mathrm{R}_{K/k}(G')$ is bijective on k-points for such G.

PROOF. See [**CGP**, Thm. 9.8.6] for a proof of exhaustiveness of the constructions. (The $K^\times$-scaling flexibility allows us to arrange further that $1 \in V'$, and also $1 \in V''$ for the additional rank-2 construction.) Uniqueness of the k-isomorphism class given $(K/k, n)$ when $[k : k^2] = 2$ is then immediate via Remark 8.3.3.

It remains to show that if $[k : k^2] = 2$ then i_G is bijective on k-points. Since $(\ker i_G)(k_s) = 1$ for all of our constructions (due to the minimal-type property), we just need to check that $i_G(G(k))$ generates $G'(K)$. But G' is a split connected semisimple K-group that is simply connected, so it is generated by the K-points of its root groups relative to a split maximal K-torus. Thus, it suffices to show that i_G is bijective between D_0-root groups. Such bijectivity is clear for roots that are neither multipliable nor divisible. For multipliable roots c the map $U_c \to \mathrm{R}_{K/k}(U'_{2c})$ is identified on k-points with the natural map $V' \oplus V \simeq V' \oplus q(V) \to K$. But as we noted in Remark 8.3.3, V' is a K^2-line in K and $q(V)$ is a complementary K^2-line since $[k : k^2] = 2$. □

In §7.4 the notion of root field and its associated properties (especially Proposition 7.4.5) have been discussed only for absolutely pseudo-simple k-groups G such that the irreducible root system Φ of G_{k_s} is *reduced*. We now extend this to cases for which Φ is non-reduced, associating a "root field" to the longest (i.e., divisible) roots over k_s, exactly as in cases with a reduced root system.

Let G be an absolutely pseudo-simple k-group *of minimal type* such that G_{k_s} has a non-reduced root system. For a maximal k-torus $T \subset G$, $\Phi := \Phi(G_{k_s}, T_{k_s})$ is of type BC_n for $n = \dim T$. If $a \in \Phi$ is divisible with root group denoted U_a then $(G_{k_s})_a := \langle U_a, U_{-a}\rangle$ is of minimal type and its maximal k_s-torus $S = a^\vee(\mathrm{GL}_1)$ satisfies $\Phi((G_{k_s})_a, S) = \{\pm a\}$ (Example 7.1.7). The reducedness of this root system implies that the notion of root field F'_a for $(G_{k_s})_a$ is already defined.

Since the Weyl group $W(G_{k_s}, T_{k_s})$ acts transitively on the set of roots in Φ with a given length, so it is transitive on the set of divisible roots, the purely inseparable finite extension F'_a/k_s is independent of a. Likewise, this extension is independent of the choice of T and is k_s-isomorphic to its $\mathrm{Gal}(k_s/k)$-twists. Such k_s-isomorphisms are unique and hence constitute a Galois descent datum, so there is a unique purely inseparable finite extension F/k such that $F \otimes_k k_s = F'_a$ over k_s for all such a (for all T).

DEFINITION 8.3.6. The *root field* of G is the extension F/k constructed above.

The possibilities for G_{k_s} are determined in Theorem 8.3.5, and the field $F \otimes_k k_s$ can be described in terms of the linear-algebraic and field-theoretic data entering into those constructions. This yields the following result (proved in [**CP**, 9.1.1–9.1.3]) that extends to such G what has been established (e.g., Proposition 7.4.5) for absolutely pseudo-simple k-groups with a reduced root system over k_s:

PROPOSITION 8.3.7. *Let G be an absolutely pseudo-simple k-group of minimal type such that G_{k_s} has a non-reduced root system. Let K/k be the minimal field of definition of $\mathscr{R}_u(G_{\overline{k}}) \subset G_{\overline{k}}$, and let F/k be the root field.*

Then $kK^2 \subset F \subset K$, G_F^{prmt} has a non-reduced root system over F_s and root field F, and the natural map $G \to \mathscr{D}(\mathrm{R}_{F/k}(G_F^{\mathrm{prmt}}))$ is an isomorphism

REMARK 8.3.8. By definition, G_F^{prmt} is the maximal quotient of G_F that is pseudo-reductive of minimal type (i.e., $G_F^{\mathrm{pred}}/\mathscr{C}_{G_F^{\mathrm{pred}}}$, with $G_F^{\mathrm{pred}} := G_F/\mathscr{R}_{u,F}(G_F)$). The non-reducedness of the root system over F_s for G_F^{prmt} in Proposition 8.3.7 is a special property of the subfield $F \subset K$ over k since in general when $[K : kK^2] > 2$ (as often occurs when $[k : k^2] > 2$) there are proper subfields $E \subset K$ over k such that G_E^{prmt} has a reduced root system over E_s; see [**CGP**, 9.8.17–9.8.18].

When $[k : k^2] = 2$ something remarkable happens: the "minimal type" and "pseudo-split" hypotheses in Theorem 8.3.5 are unnecessary. That is:

PROPOSITION 8.3.9. *Assume $[k : k^2] = 2$. Every absolutely pseudo-simple k-group G with a non-reduced root system over k_s is pseudo-split and of minimal type, and $\mathrm{H}^1(k, G) = 1$.*

PROOF. To verify that G is of minimal type, we may assume $k = k_s$. Now the pseudo-split property holds, so G is a central extension

$$1 \longrightarrow Z \longrightarrow G \longrightarrow \overline{G} \longrightarrow 1 \qquad (8.3.9.1)$$

where $Z = \mathscr{C}_G$ is a central unipotent k-subgroup scheme and $\overline{G}$ is of minimal type and hence is produced by the pseudo-simple construction in Theorem 8.3.1. Note that Z contains no nontrivial smooth connected k-subgroup (as G is pseudo-reductive). It suffices to show that the central extenson (8.3.9.1) is split.

The splitting criterion from [**CGP**, Prop. 5.1.3] used in the proofs of Proposition 6.2.17 and Corollary 7.5.11 reduces this splitting problem to a calculation with a

Cartan k-subgroup of $\overline{G}$. One such Cartan k-subgroup is described in Theorem 8.3.1(iii), and the linear algebra data there simplifies a lot since $[k:k^2]=2$ (e.g., we can assume $V' = K^2$ and $q(V)$ is a complementary K^2-line in K). Hence, although in general it appears to be very difficult to do calculations with $(V_0)^*_{K_0/k}$, in the present case it becomes very tractable; see [**CGP**, Prop. 9.9.1] for the details. This completes the proof that the "minimal type" property automatically holds!

Now we return to general k (not necessarily separably closed), and consider an absolutely pseudo-simple k-group H with root system BC_n over k_s and minimal field of definition K/k for its geometric unipotent radical. The k_s-group H_{k_s} is pseudo-split and of minimal type, so it is in the *unique* k_s-isomorphism class attached to the pair $(K_s/k_s, n)$ as in Theorem 8.3.5. But over k itself there is likewise a unique (up to k-isomorphism) pseudo-split absolutely pseudo-simple group G with root system BC_n and minimal field of definition K/k for its geometric unipotent radical. Thus, H is a k_s/k-form of G, so to prove H is pseudo-split it suffices to show that G has no nontrivial k_s/k-forms.

It remains to prove that $\mathrm{H}^1(k_s/k, \mathrm{Aut}(G_{k_s}))$ and $\mathrm{H}^1(k_s/k, G(k_s))$ vanish for pseudo-split G. By Theorem 8.3.5, the natural map $G(k_s) \to G'(K_s)$ is bijective. Since $Z_G = 1$ (Proposition 8.1.5), so $G(k_s) \subset \mathrm{Aut}(G_{k_s})$, if all automorphisms of G_{k_s} are inner then we would have $\mathrm{Aut}(G_{k_s}) = G(k_s) = G'(K_s)$, so the Galois cohomology sets of interest would coincide with $\mathrm{H}^1(K_s/K, G'(K_s))$. But $G' = \mathrm{Sp}_{2n}$ as K-groups, and Sp_{2n} has vanishing degree-1 Galois cohomology over every field (i.e., a symplectic space is determined up to isomorphism by its dimension).

Our task is reduced to showing when $k = k_s$ that every k-automorphism φ of G arises from a $G(k)$-conjugation. We can assume $\varphi(D_0) = D_0$, and since $N_G(D_0)(k)/Z_G(D_0)(k) = W(\Phi(G, D_0))$ we can assume φ preserves a positive system of roots Φ^+. But $\Phi(G_0, D_0)$ has no nontrivial automorphism preserving Φ^+, so φ acts trivially on D_0 and hence trivially on the commutative pseudo-reductive $Z_G(D_0)$. The effect of φ_K on G' coincides with the action of a K-point t of the adjoint torus $D/Z_{G'}$. But for the long simple root $2c$ in a basis for $\Phi(G', D)$, the action of t on $i_G(U_c(k)) = V' \oplus q(V)$ must preserve the K^2-line $i_G(U_{2c}(k)) = V'$ inside the K-line $U'_{2c}(K)$. Since t acts on $U'_{2c}(K)$ through scaling by $(2c)(t)$, such preservation means that $(2c)(t) \in (K^\times)^2$. This latter property characterizes the image of $D(K)$ inside $(D/Z_{G'})(K)$ for type-C_n, so t arises from a k-point of the Cartan k-subgroup $\mathrm{R}_{K/k}(D)$ of $\mathrm{R}_{K/k}(G')$. Since $i_G: G(k) \to G'(K)$ is bijective (Theorem 8.3.5), φ arises from $G(k)$. □

PROPOSITION 8.3.10. *Let k be a field of characteristic 2 such that $[k:k^2]=2$, and let G, K/k, and G' be as in Theorem 8.3.5.*

(i) *If k is complete for a fixed nontrivial non-archimedean absolute value and K is equipped with the unique extension of that absolute value then the bijection $G(k) \to G'(K)$ is a homeomorphism.*
(ii) *If k is a global function field then the natural map $G(\mathbf{A}_k) \to G'(\mathbf{A}_K)$ on adelic points is a homeomorphism.*

The proofs of these assertions reduce to a direct verification at the level of root groups and Cartan subgroups (using that the pseudo-reductive construction in Theorem 8.3.1 is perfect when $[k:k^2]=2$; see Remark 8.3.3). For example, on root groups for multipliable roots the map on points over a local field is $K^2 \oplus K \to K$ defined by $(x, y) \mapsto x + \alpha y^2$ (for $\alpha \in K - K^2$), and this is visibly a homeomorphism.

See [**CGP**, Prop. 9.9.4(2),(3)] for further details. It follows that for arithmetic computations with pseudo-reductive groups over such fields k we can often replace the intervention of such a k-group G with the associated symplectic K-group G'.

To conclude our discussion of absolutely pseudo-simple k-groups G of minimal type with a non-reduced root system over k_s, we consider their automorphisms. Recall from Proposition 6.1.2 that for any pseudo-reductive group H over a field k and any Cartan k-subgroup C, the functor $\underline{\mathrm{Aut}}_{H,C}$ classifying automorphisms of H restricting to the identity on C is represented by an affine k-group scheme $\mathrm{Aut}_{H,C}$ of finite type whose maximal smooth closed k-subgroup $Z_{H,C}$ is commutative and identity component $Z^0_{H,C}$ is pseudo-reductive.

EXAMPLE 8.3.11. If H is a connected reductive group and $T \subset H$ is a maximal k-torus then $Z_{H,T} = T/Z_G$.

In §9.1–§9.2 we will discuss the classification of Galois-twisted forms of pseudo-reductive groups H. This classification may appear to be a hopeless task, since for absolutely pseudo-simple H there is no concrete description of the structure of C in non-standard cases for types B, C, and BC in characteristic 2 when the universal smooth k-tame central extension of H is not of minimal type (as occurs in abundance for those root systems over every imperfect field k of characteristic 2 satisfying $[k : k^2] \geqslant 16$; see [**CP**, App. B] for the construction of such k-groups).

Since $Z_{H,C}$ is unaffected by passage to the central quotient H/Z_H that is always pseudo-reductive and of minimal type (see Lemma 9.1.9(ii)), and the universal smooth k-tame central extension of H/Z_H is of minimal type, analyzing minimal-type cases over k_s (where everything is pseudo-split) will yield that $Z_{H,C}$ is connected for *every* pseudo-reductive group H (see Proposition 9.1.13)! The structure of $Z_{H,C}$ will be crucial in §9.1–§9.2, especially the connectedness of $Z_{H,C}$ (relevant to the notion of "pseudo-inner form"; see Definition 9.1.3, Lemma 9.1.9(ii), and Proposition 9.1.15). In the pseudo-split minimal-type case with C containing a split maximal k-torus S of H, $Z_{H,C}$ will admit a concrete description as a direct product indexed by a basis of $\Phi(H, S)$ (generalizing the familiar direct product structure of the split maximal tori in the adjoint central quotient of a split connected reductive group). This requires case-analysis depending on the root system, after reducing to absolutely pseudo-simple H.

The case of root system BC_n requires separate treatement, and is settled as an application of our explicit description of all pseudo-split minimal type groups with root system BC_n in Theorem 8.3.1 and Proposition 8.3.4. To formulate this application, we need to set up some notation. Let G be an absolutely pseudo-simple k-group of minimal type with root system BC_n over k_s and minimal field of definition K/k for its geometric unipotent radical. Let $T \subset G$ be a maximal k-torus, $C = Z_G(T)$, and $G' = G_K/\mathscr{R}_{u,K}(G_K)$ (a K-form of Sp_{2n}). Define the "adjoint torus" T^{ad} over k to be the quotient of T corresponding to the quotient $T'/Z_{G'} \subset G'/Z_{G'}$ of the K-torus $T' := T_K$.

The action of the k-group $Z_{G,C}$ on G induces an action of $(Z_{G,C})_K$ on G_K and hence on G' (since $Z_{G,C}$ is smooth). This latter action is trivial on T', so it defines a k-homomorphism

$$f : Z_{G,C} \longrightarrow \mathrm{R}_{K/k}(Z_{G',T'}) = \mathrm{R}_{K/k}(T'/Z_{G'}) = \mathrm{R}_{K/k}(T^{\mathrm{ad}}_K).$$

It therefore makes sense to define the subfield $F \subset K$ to be the unique minimal subfield over k such that f factors through $\mathrm{R}_{F/k}(T^{\mathrm{ad}}_F)$.

PROPOSITION 8.3.12. *If $n \neq 2$ then $F = K$ and if $n = 2$ then $kK^2 \subset F \subset K$. There exists a K-finite subfield $F' \subset K^{1/2}$ for which $Z_{G,C}$ fits into a fiber square*

$$\begin{array}{ccc} Z_{G,C} & \xrightarrow{\theta} & \mathrm{R}_{F'/k}(T_{F'}) \\ {\scriptstyle f}\downarrow & & \downarrow \\ \mathrm{R}_{F/k}(T_F^{\mathrm{ad}}) & \longrightarrow & \mathrm{R}_{F'/k}(T_{F'}^{\mathrm{ad}}) \end{array}$$

using the natural maps along the bottom and right sides, and θ is uniquely determined. Moreover, among all F'/K there is a unique such minimal extension $F_\natural/K$. The formation of $F_\natural/K$ commutes with separable extension on k.

If T is split and Δ is a basis of $\Phi(G,T)$ with unique multipliable root c then via the identification of T with GL_1^Δ using coroots we have

(8.3.12.1)
$$Z_{G,C} = \mathrm{R}_{F_\natural/k}(\mathrm{GL}_1) \times \prod_{b \in \Delta - \{c\}} \mathrm{R}_{F/k}(\mathrm{GL}_1)$$

inside $\mathrm{R}_{F_\natural/k}(T_{F_\natural}) \supset \mathrm{R}_{K/k}(T_K)$.

The product expression (8.3.12.1) is always applicable over k_s (as T_{k_s} is split), so Proposition 8.3.12 implies that $Z_{G,C}$ is *connected.* The proof of Proposition 8.3.12 amounts to solving the following problem: given an automorphism $\varphi' \in Z_{G',T'}(K_s) = (T'/Z_{G'})(K_s)$, when does its action on $G'(K_s)$ preserve the subgroup $G(k_s)$? This is largely a matter of systematic (though sometimes delicate) calculations with root groups and Cartan subgroups. One first proves (8.3.12.1) via computation with k_s-points, and then recasts it in the fiber-square form that is better-suited to Galois descent. See [**CGP**, Prop. 9.8.15] for further details.

9. Classification of forms

9.1. Automorphisms and Galois-twisting. The Existence and Isomorphism Theorems for *split* connected semisimple groups G over a field k characterize isomorphism classes via root data. There are two approaches to classifying connected semisimple groups G beyond the split case, over a general field k: Galois cohomology associated to the split form, and the Tits classification that rests on relative root systems (and treats the k-anisotropic case as a black box). The latter is better-suited for generalization to the pseudo-semisimple case, but we first review the context for each of these approaches.

The Galois cohomological approach rests on viewing a connected semisimple k-group G as a k_s/k-form of the unique split connected semisimple k-group G_0 whose root datum coincides with that of G_{k_s}. The set of such G (up to k-isomorphism) for a fixed G_0 is in bijection with $\mathrm{H}^1(k, \mathrm{Aut}_{G_0/k})$, where $\mathrm{Aut}_{G_0/k}$ is the smooth automorphism scheme of G_0. The structure of $\mathrm{Aut}_{G_0/k}$ informs this classification: $\mathrm{Aut}_{G_0/k}$ is a smooth affine k-group with identity component G_0/Z_{G_0} (defining a notion of "inner form") and constant component group equal to the automorphism group of the based root datum (coinciding with the automorphism group of the Dynkin diagram when G_0 is simply connected or of adjoint type); e.g., $\mathrm{Aut}_{G_0/k} = G_0/Z_{G_0}$ when the Dynkin diagram does not admit a nontrivial automorphism. An outcome of this approach is that G admits a unique quasi-split inner form.

For example, if $n > 2$ then the A_{n-1}-diagram has automorphism group $\mathbf{Z}/2\mathbf{Z}$ and the non-split quasi-split k-forms of SL_n are the special unitary groups $SU(h_n)$ for the split hermitian form h_n on k'^n relative to a quadratic Galois extension k'/k, where $h_{2m}(x,y) = \sum_{j=1}^{m}(x_j\sigma(y_{-j}) + x_{-j}\sigma(y_j))$ for $m > 1$ and $h_{2m+1}(x,y) = h_{2m} + x_0\sigma(y_0)$ for $m \geqslant 1$, with σ denoting the nontrivial k-automorphism of k'.

The *Tits classification* (announced by Tits [**Ti1**, 2.7.1] and completed by Selbach [**Sel**]) rests on a choice of maximal split k-torus $S \subset G$ and the relationship between the relative root system $\Phi(G,S)$ of nontrivial S-weights on $\mathrm{Lie}(G)$ and the absolute root system $\Phi(G_{k_s}, T_{k_s})$ for a maximal k-torus $T \supset S$. The precise formulation involves the k-anisotropic group $\mathscr{D}(Z_G(S))$ (called the *semisimple anisotropic kernel*) and actions of $\mathrm{Gal}(k_s/k)$ on Dynkin diagrams associated to G and $\mathscr{D}(Z_G(S))$; we will review this when we generalize it to the pseudo-semisimple case. There does not exist a corresponding "existence theorem" for general k since usually there is no way to describe the semisimple anisotropic kernel; see [**Ti1**, Table II] for an existence theorem in the semisimple case over interesting fields k.

It is natural to ask if these approaches generalize to pseudo-semisimple k-groups H. There are some reasons for optimism:

(i) The functor $\underline{\mathrm{Aut}}_{H/k} : A \rightsquigarrow \mathrm{Aut}_A(H_A)$ on k-algebras is represented by an affine k-group scheme $\mathrm{Aut}_{H/k}$ of finite type [**CP**, Prop. 6.2.2]. Although this k-group is generally *not* smooth (see [**CP**, Ex. 6.2.3] for examples over any imperfect field), its maximal smooth closed k-subgroup $\mathrm{Aut}_{H/k}^{\mathrm{sm}}$ can be used to study k_s/k-forms of H. We call such a form *pseudo-inner* if it is obtained by twisting H against a class in $\mathrm{H}^1(k, (\mathrm{Aut}_{H/k}^{\mathrm{sm}})^0)$. (In general $(\mathrm{Aut}_{H/k}^{\mathrm{sm}})^0$ is larger than H/Z_H; see Remark 9.1.16.)

(ii) There is a robust theory of relative root systems and associated relative root groups for arbitrary pseudo-reductive groups [**CGP**, C.2.13–C.2.28].

REMARK 9.1.1. The existence of $\mathrm{Aut}_{H/k}$ for pseudo-semisimple H is not a formality since existence fails for many commutative pseudo-reductive H, such as $\mathrm{R}_{k'/k}(\mathrm{GL}_1)$ for any purely inseparable extension k'/k of degree $p = \mathrm{char}(k) > 0$ (ultimately because for $n \geqslant 1$ the automorphism functor of $\mathbf{G}_{\mathrm{a}}^n$ in characteristic p is not representable; see [**CP**, Ex. 6.2.1]). The key to the existence proof is that since (fiberwise) maximal tori in a smooth affine group scheme are conjugate fppf-locally on the base [**SGA3**, XI, Cor. 5.4], if $T \subset H$ is a maximal k-torus then the representability of $\underline{\mathrm{Aut}}_{H/k}$ is reduced to that of the T-stabilizer subfunctor of $\underline{\mathrm{Aut}}_{H/k}$.

Over a finite extension of k this stabilizer subfunctor is covered by translates of *finitely many* copies of the subfunctor $\underline{\mathrm{Aut}}_{H,T}$ classifying automorphisms of H restricting to the identity on T because the $\mathbf{Z}$-span of the finite set $\Phi(H_{k_s}, T_{k_s})$ has finite index inside $\mathrm{X}(T_{k_s})$ (as $H_{\overline{k}}^{\mathrm{red}}$ is semisimple). This subfunctor is represented by an affine k-group scheme of finite type since H is perfect [**CGP**, Cor. 2.4.4, Prop. A.2.11], so we are done. See [**CP**, Prop. 6.2.2] for further details.

Although in general there does not exist an automorphism scheme in the pseudo-reductive case, there is a reasonable notion of "pseudo-inner form" beyond the pseudo-semisimple case. This is inspired by the observation that for connected reductive G, the natural map $\mathrm{Aut}_{G/k}^0 \to \mathrm{Aut}_{\mathscr{D}(G)/k}^0$ is an isomorphism because it is inverse to the natural isomorphism $\mathscr{D}(G)/Z_{\mathscr{D}(G)} \simeq G/Z_G$. If G is a pseudo-reductive k-group then suitable use of the equality $C \cdot \mathscr{D}(G) = G$ for a Cartan k-subgroup $C \subset G$ yields [**CP**, Lemma C.2.3]:

LEMMA 9.1.2. *The natural* $(\mathrm{Aut}^{\mathrm{sm}}_{\mathscr{D}(G)/k})^0$*-action on* $\mathscr{D}(G)$ *uniquely extends to an action on* G.

This lemma motivates:

DEFINITION 9.1.3. Let G be a pseudo-reductive k-group. We say G is *quasi-split* if it admits a pseudo-parabolic k-subgroup B such that B_{k_s} is a minimal pseudo-parabolic k_s-subgroup of G_{k_s}. A k_s/k-form of G is *pseudo-inner* if it is classified by an element of the image of

$$\mathrm{H}^1(k_s/k, (\mathrm{Aut}^{\mathrm{sm}}_{\mathscr{D}(G)/k})^0(k_s)) \longrightarrow \mathrm{H}^1(k_s/k, \mathrm{Aut}(G_{k_s})).$$

There are difficulties with the Galois cohomological approach when studying k_s/k-forms of a pseudo-semisimple k-group H:

- (i′) The geometric component group of $\mathrm{Aut}^{\mathrm{sm}}_{H/k}$ is always naturally a subgroup of the automorphism group of the based root datum of H_{k_s} [**CP**, Rem. 6.3.6], but examples exist over every imperfect field for which it is a *proper* subgroup [**CP**, Ex. 6.3.8, Ex. C.1.6]. Hence, there can be subtleties when trying to characterize in Galois-cohomological terms those k_s/k-forms obtained via pseudo-inner twisting.
- (ii′) The existence of a pseudo-split k_s/k-form fails in every positive characteristic [**CP**, Ex. C.1.2, C.1.6].
- (ii″) If $\mathrm{char}(k) = 2$, $[k : k^2] \geqslant 8$, and k has sufficiently rich Galois theory then there exists a (non-standard) absolutely pseudo-simple k-group with root system over k_s of any type B_n, C_n, or BC_n ($n \geqslant 1$) which does not admit a quasi-split pseudo-inner k_s/k-form; see Example 9.1.5. (These counterexamples are optimal because a pseudo-reductive group H over a general field k admits a quasi-split pseudo-inner k_s/k-form except possibly when H is non-standard with $\mathrm{char}(k) = 2$ and $[k : k^2] \geqslant 8$ [**CP**, Thm. C.2.10]; the proof involves a degree-2 cohomological obstruction whose vanishing characterizes the existence of a quasi-split pseudo-inner k_s/k-form; this obstruction is never seen in the semisimple case.)

Both (i′) and (ii′) are caused by field-theoretic obstructions that do not arise in the absolutely pseudo-simple case away from characteristic 2 [**CP**, Prop. C.1.3(i), Prop. 6.3.10].

REMARK 9.1.4. In connection with (ii′) and (ii″), it is natural to ask if a pseudo-split k_s/k-form or quasi-split pseudo-inner k_s/k-form of a pseudo-reductive k-group is unique when it exists. There are well-known affirmative results in the connected reductive case, but the traditional proofs there rest on the Isomorphism Theorem and Galois cohomological considerations respectively and neither of those approaches works in the general pseudo-reductive case. (There is an Isomorphism Theorem in the pseudo-split pseudo-reductive case [**CP**, Thm. 6.1.1], but it goes beyond combinatorial invariants.) The Tits-style classification in §9.2 provides a way around these problems to give affirmative answers to the uniqueness questions in general; see Corollary 9.2.10.

In characteristic 2, the phenomena in (i′) and (ii′) persist in the *absolutely pseudo-simple case*. More precisely, (i′) occurs exactly for type D_{2n} with $n \geqslant 2$ (a priori, any absolutely pseudo-simple instance of (i′) must have an irreducible root system over k_s admitting a nontrivial diagram automorphism, so it is either A_m ($m \geqslant 2$) or D_m ($m \geqslant 4$) or E_6, and hence must be *standard* by Theorem 7.4.8); see Example 9.1.6 below. Likewise, the non-standard examples in (ii″) with type B_n, C_n, or BC_n for $n \geqslant 1$ certainly fulfill (ii′); these are discussed in the following example.

EXAMPLE 9.1.5. In [**CP**, §C.3–§C.4] there are examples in characteristic 2 of absolutely pseudo-split groups with any type B, C, or BC (with any rank $n \geqslant 1$) which do not admit a pseudo-split k_s/k-form; these also do not admit a quasi-split pseudo-inner form, due to the absence of nontrivial diagram automorphisms (see [**CP**, Lemma C.2.2]). All of these constructions ultimately rest on a single class of examples for type $A_1 = B_1 = C_1$, so we now sketch that core example; see [**CP**, Ex. C.3.1] for further details.

Let k be a field of characteristic 2 admitting a quadratic Galois extension k'/k with nontrivial automorphism σ, and let K/k be a nontrivial finite subextension of $k^{1/2}$. For $K' := k' \otimes_k K$ we seek a k'-subspace $V' \subset K'$ that generates K' as a k'-algebra and whose $K'^{\times}$-homothety class is stable under the action of σ but for which no $K'^{\times}$-multiple of V' is σ-stable. The idea is that $H_{V',K'/k'}$ is then k'-isomorphic to its σ-twist $H_{\sigma(V'),K'/k'}$, and if this k'-isomorphism can be chosen to satisfy the cocycle condition then $H_{V',K'/k'}$ admits a k-descent G. Such a k-group G *cannot* admit any pseudo-split k_s/k-form! Indeed, suppose $\mathscr{G}$ were such a form, so $\mathscr{G}_{k'}$ and $G_{k'} = H_{V',K'/k'}$ are pseudo-split k'_s/k'-forms of each other, and hence are k'-isomorphic (by the uniqueness of pseudo-split forms). The description of $\mathscr{G}$ in Theorem 7.2.5(ii) would then provide a σ-stable member of the $K'^{\times}$-homothety class of V' due to Proposition 7.2.3(ii), contradicting how V' was chosen.

To build $(K/k, V')$ satisfying the above properties, we assume $[k : k^2] \geqslant 8$ and that $\mathrm{Br}(k) \to \mathrm{Br}(k')$ has nontrivial kernel; e.g., $k = \kappa(x, y, z)$ and $k' = L(y, z)$ for a finite field κ of characteristic 2 and any quadratic Galois extension $L/\kappa(x)$. Via Tate cohomology, the nontrivial kernel of $\mathrm{Br}(k) \to \mathrm{Br}(k')$ is identified with $k^{\times}/\mathrm{N}_{k'/k}(k'^{\times})$. Choose $e \in k^{\times} - \mathrm{N}_{k'/k}(k'^{\times})$ and define $t_1 = \sqrt{e} \in k^{1/2} - k$; since $[k : k^2] \geqslant 8$, we can extend $\{t_1\}$ to a triple $\{t_1, t_2, t_3\}$ that is part of a 2-basis of k. Let $K = k(t_1, t_2, t_3)$. For a primitive element a of k'/k, the k'-subspace

$$V' = k' + k't_1 + k'(t_2 + at_3) + k't_1(t_2 + \sigma(a)t_3) \subset K'$$

is 4-dimensional and generates K' as a k'-algebra, and $\sigma(V') = t_1V'$. Moreover, the "root field" $\{c' \in K' \mid c'V' \subset V'\}$ is equal to k'. Since the σ-invariant element

$$\begin{pmatrix} 0 & t_1 \\ 1 & 0 \end{pmatrix} \in \mathrm{PGL}_2(K) = \mathrm{R}_{K/k}(\mathrm{PGL}_2)(k) \subset \mathrm{R}_{K'/k'}(\mathrm{PGL}_2)(k')$$

has order 2 and its action on $\mathrm{R}_{K'/k'}(\mathrm{SL}_2)$ carries $H_{\sigma(V'),K'/k'}$ to $H_{V',K'/k'}$, it defines the desired k'/k-descent datum. The property that no member of the $K'^{\times}$-homothety class of V' is σ-stable uses that V' has root field k' and that the square $t_1^2 = e \in k^{\times}$ is not a norm from k'.

EXAMPLE 9.1.6. The phenomenon of absolutely pseudo-simple groups over imperfect fields k of characteristic 2 without a pseudo-split k_s/k-form occurs in cases of type D_{2n} for any $n \geqslant 2$ whenever k admits a quadratic Galois extension k'/k.

These arise as pseudo-reductive central quotients $G = \mathrm{R}_{K/k}(\mathscr{G}_K)/Z$ for suitable purely inseparable finite extensions K/k and the quasi-split non-split k-group $\mathscr{G}$ of outer type D_{2n} that is split by k'. The idea is that by choosing K/k suitably, there are many k'-subgroups Z' between $\mathrm{R}_{K/k}(\mu_2 \times \mu_2)_{k'}$ and $\mu_2 \times \mu_2$ such that $\sigma^*(Z')$ is *distinct* from Z' yet is carried to Z' by an involution swapping the μ_2-factors (induced by a diagram involution). The k-subgroup $Z \subset \mathrm{R}_{K/k}((Z_{\mathscr{G}})_K)$ is a k'-descent of Z', and pseudo-reductivity holds if $\mathrm{R}_{K/k}(\mu_2 \times \mu_2)_{k'}/Z'$ contains no nontrivial smooth connected k'-subgroup.

See [**CP**, C.1.3–C.1.5] for details of this construction and why it accounts for essentially all standard absolutely pseudo-simple examples of (ii′). The hypotheses on Z' obstruct the standard diagram involution from arising in $\pi_0(\mathrm{Aut}^{\mathrm{sm}}_{G/k})(k_s)$. If $n = 2$ and k admits a cubic Galois extension then the preceding has a variant resting on triality. This yields essentially all absolutely pseudo-simple groups (standard or not) for which (i′) occurs; see [**CP**, Ex. 6.3.9, Prop. 6.3.10].

Overall, we see that the cohomological approach encounters problematic phenomena over specific classes of fields, especially in characteristic 2. Yet remarkably, a Tits-style classification theorem for pseudo-semisimple k-groups holds in complete generality, with a *characteristic-free proof.* An essential ingredient in the success of the Tits-style approach for general pseudo-semisimple k-groups is that if G is an arbitrary pseudo-reductive k-group then we can understand the structure of the k-group

$$Z_{G,C} \subset \mathrm{Aut}_{G,C}$$

introduced in Proposition 6.1.2 for Cartan k-subgroups $C \subset G$; e.g., using our work in the minimal-type case, we shall see that $Z_{G,C}$ is *always* connected. (The connectedness of $Z_{G,C}$ for absolutely pseudo-simple G of minimal type with a non-reduced root system over k_s was addressed in Proposition 8.3.12.)

Before we study $Z_{G,C}$, and then deduce consequences for the classification in the pseudo-semisimple case, we need to record a version of the Isomorphism Theorem for pseudo-split pseudo-reductive groups. Consider a pseudo-reductive k-group G admitting a split maximal k-torus T. The pseudo-split commutative pseudo-reductive k-group $Z_G(T)$ generally does not have a combinatorial description in terms of Galois lattices. Moreover, the rank-1 pseudo-split absolutely pseudo-simple k-subgroup G_a generated by root groups for opposite roots $\pm a$ generally *does not admit a notion of pinning* when $\mathrm{char}(k) = 2$. Indeed, $H_{V,K/k}$ in Proposition 7.2.3 only determines V up to $K^\times$-scaling, and when G_a is not of minimal type (as can happen for suitable pseudo-split absolutely pseudo-simple k-groups G with root system of any type B, C, or BC whenever $[k : k^2] \geqslant 16$ [**CP**, App. B]) then we do not even have a concrete description of G_a!

To circumvent the absence of an entirely combinatorial framework for describing pseudo-split groups, we shall work directly with k-isomorphisms between certain basic building blocks of the groups. Let G and G' be pseudo-split pseudo-reductive k-groups with respective split maximal k-tori T and T'. Fix a k-isomorphism $f : Z_G(T) \simeq Z_{G'}(T')$, as well as k-isomorphisms $f_a : G_a \simeq G'_{a'}$ for corresponding roots in chosen bases $\Delta \subset \Phi(G,T)$ and $\Delta' \subset \Phi(G',T')$ that are assumed to be carried to each other under the restriction $f_T : T \simeq T'$ of f.

THEOREM 9.1.7 (Isomorphism Theorem). *A k-isomorphism $\varphi : G \simeq G'$ recovering f and $\{f_a\}_{a\in\Delta}$ exists when this data satisfies the necessary compatibilities* *that* $f_T \circ a^\vee = a'^\vee$ *and* f_a *is equivariant with respect to* f_T *for all* $a \in \Delta$*;* φ *is then unique.*

REMARK 9.1.8. Since there is no uniform characteristic-free description of $\mathrm{Aut}(G_a)$ akin to $\mathrm{PGL}_2(k)$ in the reductive case, the characteristic-free proof of Theorem 9.1.7 in [**CP**, Thm. 6.1.1] uses a method entirely different from the classical proof via rational homomorphisms and structure constants. Adapting an idea of Steinberg in the reductive case, one builds the graph of the desired isomorphism as a pseudo-reductive k-subgroup of $G \times G'$ via Theorem 5.4.3. A variation on the same graph idea yields a pseudo-reductive version of the Isogeny Theorem in [**CP**, App. A].

Returning to our considerations with a general pseudo-reductive k-group G and Cartan k-subgroup $C \subset G$, the key to the structure of $Z_{G,C}$ is that its formation is extremely robust:

LEMMA 9.1.9. *Let* G *be a pseudo-reductive* k*-group.*

(i) *For a pseudo-reductive central quotient* $\overline{G}$ *of* G*, a Cartan* k*-subgroup* $C \subset G$*, and its image* $\overline{C} \subset \overline{G}$*, the natural map* $Z_{G,C} \to Z_{\overline{G},\overline{C}}$ *is an isomorphism.*
(ii) *For the Cartan* k*-subgroup* $C' := C \cap \mathscr{D}(G)$ *of* $\mathscr{D}(G)$*, the natural map* $Z_{G,C} \to Z_{\mathscr{D}(G),C'}$ *induced by* $\mathrm{Aut}_{G,C} \to \mathrm{Aut}_{\mathscr{D}(G),C'}$ *is an isomorphism.*

The proof of (i) is based on dynamic techniques with open cells and rational homomorphisms. For the proof of (ii), the idea is that if $k = k_s$ then at the level of an open cell, the root groups of G and $\mathscr{D}(G)$ relative to the unique maximal tori of C and C' coincide. An automorphism that restricts to the identity on a Cartan k-subgroup must carry each root group into itself and is determined by its effect on root groups. Hence, it is plausible that the map $Z_{G,C} \to Z_{\mathscr{D}(G),C'}$ between *smooth* k-groups is bijective on points valued in every separable extension of k, and so is a k-isomorphism. See [**CP**, Lemma 6.1.2] for a complete proof of Lemma 9.1.9.

To understand the structure of $Z_{G,C}$, by Lemma 9.1.9 we may assume G is pseudo-semisimple. In such cases C is generated by analogues for rank-1 groups when G is pseudo-split (such as when working over k_s):

LEMMA 9.1.10. *Let* G *be a pseudo-split pseudo-semisimple* k*-group with a split maximal* k*-torus* T*, and let* Δ *be a basis of* $\Phi(G,T)$*. Let* $C = Z_G(T)$*. For the Cartan* k*-subgroup* $C_a := C \cap G_a = Z_{G_a}(a^\vee(\mathrm{GL}_1))$ *inside* $G_a = \langle U_a, U_{-a}\rangle$*, the multiplication map*

$$m : \prod_{a \in \Delta} C_a \longrightarrow C$$

is surjective. If $G^{\mathrm{ss}}_{\overline{k}}$ *is simply connected and* G *is of minimal type then* m *is an isomorphism.*

PROOF. To show that $\{C_a\}_{a\in\Delta}$ generates C, we first note that $\{G_a\}_{a\in\Delta}$ generates G by perfectness since $W(\Phi(G,T))$ is generated by reflections r_a coming from $N_{G_a}(a^\vee(\mathrm{GL}_1))(k)$ for $a \in \Delta$. Thus, if C' is the k-subgroup of C generated by $\{C_a\}_{a\in\Delta}$ then the pseudo-reductive k-groups $C' \cdot G_a = (C' \ltimes G_a)/C_a$ $(a \in \Delta)$ generate G yet share the *same* Cartan k-subgroup $Z_{C'\cdot G_a}(T) = C'$. Hence, the k-group G that they generate satisfies $Z_G(T) = C'$ too, due to Theorem 5.4.3(i).

Now assume that G is of minimal type. This is inherited by each G_a (Example 7.1.7), so $i_G|_{C_a} = i_{G_a}|_{C_a}$ has trivial kernel for each a. This realizes C_a as a

k-subgroup of $\mathrm{R}_{K_a/k}(a^\vee(\mathrm{GL}_1)_{K_a}) \subset \mathrm{R}_{K/k}(a^\vee(\mathrm{GL}_1)_K)$ for each a. Assume furthermore that $G^{\mathrm{ss}}_{\overline{k}}$ is simply connected, so the multiplication map $\prod_{a\in\Delta} a^\vee(\mathrm{GL}_1) \to T$ is an isomorphism. Hence, applying $\mathrm{R}_{K/k}((\cdot)_K)$ implies that the composite map

$$\prod_{a\in\Delta} C_a \xrightarrow{m} C \xrightarrow{i_G} \mathrm{R}_{K/k}(T_K)$$

has trivial kernel, so $\ker m = 1$. □

9.1.11. For G, C, and Δ as in Lemma 9.1.10, the natural k-homomorphism

$$\varphi : \mathrm{Aut}_{G,C} \longrightarrow \prod_{a\in\Delta} \mathrm{Aut}_{G_a,C_a}$$

carries $Z_{G,C}$ into $\prod_{a\in\Delta} Z_{G_a,C_a}$. Since C is generated by the C_a's, it follows from Theorem 9.1.7 in the special case of the trivial automorphism of C that φ is *bijective* on k-points. But then by the same reasoning φ is bijective on k'-points for *every* separable extension field k'/k (as we may use everything above after first applying scalar extension up to such k'). Equivalently, the homomorphism

$$Z_{G,C} \longrightarrow \prod_{a\in\Delta} Z_{G_a,C_a} \tag{9.1.11.1}$$

between *smooth* k-groups is bijective on k'-points for all separable extensions k'/k. But it is a general fact (reviewed in the proof of [**CGP**, Prop. 8.2.6]) that a homomorphism between smooth groups of finite type over a field is an isomorphism whenever it is bijective on points valued in all separable extension fields. Thus, we have proved:

LEMMA 9.1.12. *The map* (9.1.11.1) *is an isomorphism.*

Note that the map in (9.1.11.1) makes sense without assuming G to be perfect, and in this generality it continues to be an isomorphism due to the identification of $Z_{G,C}$ and $Z_{\mathscr{D}(G),C'}$ in Lemma 9.1.9(ii). In the connected reductive case, Lemma 9.1.12 recovers a well-known fact: if G is a split connected reductive group and $T \subset G$ is a split maximal k-torus then $Z_{G,T} = T/Z_G \simeq \prod_{a\in\Delta} \mathrm{GL}_1$ via $t \bmod Z_G \mapsto (a(t))$. In general, Lemma 9.1.12 opens the door to using concrete calculations in rank-1 cases to prove:

PROPOSITION 9.1.13. *For a pseudo-reductive k-group G and Cartan k-subgroup $C \subset G$, the smooth commutative affine k-group $Z_{G,C}$ is connected and even pseudo-reductive.*

To prove Proposition 9.1.13 we may assume $k = k_s$ (so G is pseudo-split), and then Lemma 9.1.12 allows us to assume that G is absolutely pseudo-simple of rank 1. By Lemma 9.1.9(i) we may replace G with $G/\mathscr{C}_G$ so that G is of minimal type, and then pass to the universal smooth k-tame central extension (clearly also of minimal type) so that $G^{\mathrm{ss}}_{\overline{k}} \simeq \mathrm{SL}_2$.

The case of root system BC_1 is a special case of Proposition 8.3.12, so we may assume instead that the root system is A_1. Hence, $\ker i_G = 1$ (Example 7.1.4), so naturally $G \subset \mathrm{R}_{K/k}(G')$ for $G' := G_K/\mathscr{R}_{u,K}(G_K)$. But for a choice of maximal k-torus $T \subset G$ we may pick a Levi k-subgroup $L \subset G$ containing T (Theorem 5.4.4). Upon identifying L with SL_2 carrying T to the diagonal k-torus D, we get

$$\mathrm{SL}_2 \subset G \subset \mathrm{R}_{K/k}(\mathrm{SL}_2)$$

with $D = T$.

The possibilities for $G = i_G(G)$ are explicitly described in Theorem 7.2.5, so essentially by definition of the root field F of the pair (G, T) in Definition 7.4.3 we compute that the resulting natural map $\mathrm{Aut}_k(G) \to \mathrm{PGL}_2(K)$ carries $Z_{G,C}(k)$ into

$$T^{\mathrm{ad}}(F) = \mathrm{R}_{F/k}(T_F^{\mathrm{ad}})(k) \subset \mathrm{R}_{K/k}(T_K^{\mathrm{ad}})(k)$$

for $T^{\mathrm{ad}} := T/(T \cap Z_G)$. A more refined version of the calculation (working systematically over all separable extensions of k), given in [**CP**, Lemma 6.1.3], upgrades this inclusion to an isomorphism of k-groups

$$Z_{G,C} \simeq \mathrm{R}_{F/k}(T_F^{\mathrm{ad}}), \tag{9.1.13.1}$$

affirming the connectedness of $Z_{G,C}$. (The same technique even provides such a k-isomorphism for general k without assuming T is k-split.) This completes our sketch of the proof of Proposition 9.1.13.

REMARK 9.1.14. We can also prove a compatibility of $Z_{G,C}$ with Weil restriction. To make this precise, let k' be a nonzero finite reduced k-algebra, G' a smooth affine k'-group whose fiber over each factor field of k' is pseudo-reductive, and C' a Cartan k'-subgroup of G'. Define the pseudo-reductive k-group $G := \mathrm{R}_{k'/k}(G')$ and its Cartan k-subgroup $C := \mathrm{R}_{k'/k}(C')$. By passage to the rank-1 minimal-type absolutely pseudo-simple case with $G_{\overline{k}}^{\mathrm{ss}} \simeq \mathrm{SL}_2$, Theorem 9.1.7 and calculations using the explicit description of $Z_{G,C}$ (depending on whether the root system is BC_1 or A_1) yield a *canonical* isomorphism

$$\mathrm{R}_{k'/k}(Z_{G',C'}) \simeq Z_{G,C};$$

see [**CP**, Prop. 6.1.7] for further details.

We noted in Remark 9.1.1 that if G is pseudo-semisimple then its automorphism functor (on k-algebras) is represented by an affine k-group scheme $\mathrm{Aut}_{G/k}$ of finite type. Thus, it makes sense to consider its maximal smooth closed k-subgroup $\mathrm{Aut}_{G/k}^{\mathrm{sm}}$. More specifically, $\mathrm{H}^1(k, \mathrm{Aut}_{G/k}^{\mathrm{sm}})$ classifies isomorphism classes of k_s/k-forms of G. (The study of $\overline{k}/k$-forms of G is not of interest for imperfect k because such forms, though smooth and connected, are usually not pseudo-reductive.)

The structure of $(\mathrm{Aut}_{G/k}^{\mathrm{sm}})^0$ is of interest for Galois cohomological purposes (e.g., to define the notion of "pseudo-inner form") and will also play an essential role in the proof of a pseudo-semisimple Tits classification. In the semisimple case this identity component is G/Z_G, and for a maximal k-torus $T \subset G$ we can describe it as a quotient of a smooth group by a smooth central subgroup:

$$G/Z_G \simeq (G \rtimes (T/Z_G))/T,$$

where T is anti-diagonally embedded as a central k-subgroup of $G \rtimes (T/Z_G)$. The connectedness of $Z_{G,C}$ yields an analogous description in the pseudo-semisimple case:

PROPOSITION 9.1.15. *If C is a Cartan k-subgroup of a pseudo-semisimple k-group G then the natural k-subgroup inclusion*

$$(G \rtimes Z_{G,C})/C \hookrightarrow (\mathrm{Aut}_{G/k}^{\mathrm{sm}})^0$$

is an equality. In particular, $\mathscr{D}((\mathrm{Aut}_{G/k}^{\mathrm{sm}})^0) = G/Z_G$.

It suffices to prove equality on k_s-points for *general* k, and by connectedness it is enough to prove equality *up to finite index*. Since $G(k_s)$-conjugation is transitive on the set of maximal k_s-tori, and any automorphism of G_{k_s} preserving T_{k_s} must permute the finite set $\Phi(G_{k_s}, T_{k_s})$ whose $\mathbf{Z}$-span is of finite index in $X(T_{k_s})$ (as $G^{\rm red}_{\overline{k}}$ is semisimple), it suffices to analyze automorphisms of G_{k_s} that restrict to the identity on T_{k_s}. But such automorphisms act as the identity on $Z_{G_{k_s}}(T_{k_s}) = C_{k_s}$ [**CGP**, Prop. 1.2.2], and $\mathrm{Aut}_{G,C}(k_s) = Z_{G,C}(k_s)$. This completes the proof.

REMARK 9.1.16. In contrast with the connected reductive case, for which the smooth k-group $\mathrm{Aut}^0_{G/k} = G/Z_G$ is perfect, generally for pseudo-semisimple G the k-group $(\mathrm{Aut}^{\rm sm}_{G/k})^0$ is *not* perfect. For example, if $G = \mathrm{R}_{k'/k}(\mathrm{SL}_p)$ for a nontrivial purely inseparable finite extension k'/k in characteristic $p > 0$ then $(\mathrm{Aut}^{\rm sm}_{G/k})^0 = \mathrm{R}_{k'/k}(\mathrm{PGL}_p)$ due to the description of $Z_{G,C}$ provided by [**CGP**, Thm. 1.3.9]. The phenomenon of $(\mathrm{Aut}^{\rm sm}_{G/k})^0$ being strictly larger than G/Z_G is the reason that we speak of "pseudo-inner forms" rather than "inner forms" for pseudo-semisimple G.

REMARK 9.1.17. An immediate consequence of Proposition 9.1.15 and the invariance of $Z_{G,C}$ with respect to passage to a pseudo-reductive central quotient of G in Lemma 9.1.9(i) is that if G is pseudo-semisimple then $(\mathrm{Aut}^{\rm sm}_{G/k})^0$ is naturally invariant under replacing G with a pseudo-reductive central quotient. The same does not hold without restricting attention to the identity component, as we already see in the connected absolutely simple case, such as for type-D_{2n}.

9.2. Tits-style classification. The Tits classification of connected semisimple groups G over a field k reinterprets parts of the Galois cohomological formulation in terms of Dynkin diagrams with Galois action. We shall first review the relationship between the two viewpoints, and then see how the use of Galois actions on Dynkin diagrams sidesteps many difficulties seen in §9.1 when generalizing to pseudo-semisimple groups (e.g., the absence of a pseudo-split k_s/k-form, or of a quasi-split pseudo-inner k_s/k-form).

9.2.1. Let $R = (X, \Phi, X^\vee, \Phi^\vee)$ be a reduced semisimple root datum, and Δ a basis of Φ. Let $(G_0, B_0, T_0, \{X_a\}_{a\in\Delta})$ be a pinned split connected semisimple k-group with this based root datum (so for each $a \in \Delta$, X_a is a nonzero element in the a-root space of $\mathrm{Lie}(G_0)$); this pinned split connected semisimple k-group is unique up to isomorphism. The subgroup $\Gamma \subset \mathrm{Aut}_{G_0/k}(k)$ consisting of automorphisms of $(G_0, B_0, T_0, \{X_a\}_{a\in\Delta})$ maps isomorphically onto the geometric component group $\pi_0(\mathrm{Aut}_{G_0/k})(k_s)$, and the evident inclusion $\Gamma \hookrightarrow \mathrm{Aut}(R, \Delta)$ is an equality due to the Isomorphism Theorem for split connected semisimple groups. The subset

$$\mathrm{H}^1(k, \mathrm{Aut}(R, \Delta)) = \mathrm{H}^1(k, \Gamma) \subset \mathrm{H}^1(k, \mathrm{Aut}_{G_0/k}) \tag{9.2.1.1}$$

classifies the quasi-split k-forms of G_0.

Every connected semisimple k-group with root datum R over k_s has a unique quasi-split inner k-form, so we fix a quasi-split form G of G_0 and study its inner forms; this consists of the image of the map

$$f : \mathrm{H}^1(k, G/Z_G) = \mathrm{H}^1(k, \mathrm{Aut}^0_{G/k}) \longrightarrow \mathrm{H}^1(k, \mathrm{Aut}_{G/k}). \tag{9.2.1.2}$$

To describe $\mathrm{H}^1(k, G/Z_G)$ we may try to partition it according to the k_s-rational conjugacy class of minimal parabolic k-subgroups in an inner form of G, but this runs into a complication: as a map of sets, f is generally not injective even when it has trivial kernel as a map of pointed sets. For example, if $G = G_0 = \mathrm{SL}_n$ with

$n > 2$ then we get the map $\mathrm{H}^1(k, \mathrm{PGL}_n) \to \mathrm{H}^1(k, \mathrm{PGL}_n \rtimes (\mathbf{Z}/2\mathbf{Z}))$ whose fiber through the Brauer class $[A]$ of a central simple k-algebra A of dimension n^2 is a singleton if and only if $[A]$ is 2-torsion [**C3**, Ex. 7.1.12].

A key idea in Tits' approach is to use relative root systems and $\mathrm{Gal}(k_s/k)$-actions on Dynkin diagrams to keep track of minimal parabolic k-subgroups while interpreting $\mathrm{H}^1(k, \mathrm{Aut}^0_{G/k})$ in a useful manner, bypasing the failure of injectivity of (9.2.1.2). This requires Tits' "$*$-action" of $\mathrm{Gal}(k_s/k)$ on the Dynkin diagram $\mathrm{Dyn}(G_{k_s})$ for an arbitrary connected semisimple k-group G (possibly k-anisotropic).

We now extend these methods to the pseudo-reductive setting, initially working over k_s to define a notion of "canonical diagram" before we bring in k-structures and Galois actions. Let $\mathscr{G}$ be a pseudo-reductive k_s-group. For each maximal k_s-torus $\mathscr{T} \subset \mathscr{G}$ and minimal pseudo-parabolic k_s-subgroup $\mathscr{B} \supset \mathscr{T}$ in $\mathscr{G}$, we get a diagram $\mathrm{Dyn}(\mathscr{G}, \mathscr{T}, \mathscr{B})$ arising from the basis of the positive system of roots $\Phi(\mathscr{B}, \mathscr{T}) \subset \Phi(\mathscr{G}, \mathscr{T})$; we denote it as $\mathrm{Dyn}(\mathscr{T}, \mathscr{B})$ when $\mathscr{G}$ is understood from the context. If $(\mathscr{T}', \mathscr{B}')$ is another such pair in $\mathscr{G}$ then there exists an element $g \in \mathscr{G}(k_s)$ that carries $(\mathscr{T}, \mathscr{B})$ over to $(\mathscr{T}', \mathscr{B}')$ due to Proposition 4.1.3 and Theorem 4.2.9 (over k_s). Any such g induces an isomorphism of diagrams (i.e., respecting the pairings of roots and coroots)

$$\mathrm{Dyn}(g) : \mathrm{Dyn}(\mathscr{T}, \mathscr{B}) \simeq \mathrm{Dyn}(\mathscr{T}', \mathscr{B}').$$

Likewise, if $\mathscr{G}$ is pseudo-semisimple then we can choose $\varphi \in (\mathrm{Aut}^{\mathrm{sm}}_{\mathscr{G}/k_s})^0(k_s)$ carrying $(\mathscr{T}, \mathscr{B})$ to $(\mathscr{T}', \mathscr{B}')$, and consider the induced isomorphism $\mathrm{Dyn}(\varphi)$ between Dynkin diagrams.

LEMMA 9.2.2. *The isomorphism* $\mathrm{Dyn}(g)$ *is independent of the choice of* g*, and likewise for* $\mathrm{Dyn}(\varphi)$ *in the pseudo-semisimple case.*

PROOF. We treat the case of $\mathrm{Dyn}(\varphi)$ with pseudo-semisimple $\mathscr{G}$; the other case goes similarly (and is easier). Let $\mathscr{C} = Z_{\mathscr{G}}(\mathscr{T})$, so $\mathscr{C} \subset \mathscr{B}$. Any two choices of φ are related through composition against an element of the $(\mathscr{T}, \mathscr{B})$-stabilizer in

$$(\mathrm{Aut}^{\mathrm{sm}}_{\mathscr{G}/k})^0(k_s) = (\mathscr{G}(k_s) \rtimes Z_{\mathscr{G},\mathscr{C}}(k_s))/\mathscr{C}(k_s),$$

so we just need to prove that every automorphism in the $(\mathscr{T}, \mathscr{B})$-stabilizer acts trivially on $\mathrm{Dyn}(\mathscr{T}, \mathscr{B})$. The action of $Z_{\mathscr{G},\mathscr{C}}(k_s)$ on $\mathscr{G}$ restricts to the identity on $\mathscr{C}$ and hence likewise on both $\mathscr{T}$ and $\Phi := \Phi(\mathscr{G}, \mathscr{T})$, so the $Z_{\mathscr{G},\mathscr{C}}(k_s)$-action preserves $\mathscr{B}$ and acts as the identity on $\mathrm{Dyn}(\mathscr{T}, \mathscr{B})$. It therefore suffices to analyze the effect of the $(\mathscr{T}, \mathscr{B})$-stabilizer in $\mathscr{G}(k_s)$, which is to say elements $n \in N_{\mathscr{G}}(\mathscr{T})(k_s)$ that preserve $\mathscr{B}$. The class of such an n in $W(\mathscr{G}, \mathscr{T}) = W(\Phi)$ preserves the positive system of roots $\Phi(\mathscr{B}, \mathscr{T})$, so n has trivial image on $W(\Phi)$. In other words, $n \in Z_{\mathscr{G}}(\mathscr{T})(k_s) = \mathscr{C}(k_s)$. □

We have built canonical isomorphisms among all diagrams $\mathrm{Dyn}(\mathscr{T}, \mathscr{B})$, transitively with respect to choices of pairs $(\mathscr{T}, \mathscr{B})$. Define the *canonical diagram*

$$\mathrm{Dyn}(\mathscr{G})$$

as follows: a *vertex* consists of a compatible choice of vertex on $\mathrm{Dyn}(\mathscr{T}, \mathscr{B})$ for *every* pair $(\mathscr{T}, \mathscr{B})$ as above, where "compatibility" is meant in the sense of the preceding canonical isomorphisms, and its *edges*, etc. are defined in a similar manner.

Now consider a pseudo-reductive k-group G. By definition, the *$*$-action* of $\mathrm{Gal}(k_s/k)$ on $\mathrm{Dyn}(G_{k_s})$ makes each $\gamma \in \mathrm{Gal}(k_s/k)$ carry a vertex a to its image (denoted $\gamma * a$) under the composite isomorphism

$$\mathrm{Dyn}(G_{k_s}) = \mathrm{Dyn}(\mathscr{T}, \mathscr{B}) \simeq \mathrm{Dyn}({}^{\gamma}\mathscr{T}, {}^{\gamma}\mathscr{B}) = \mathrm{Dyn}(G_{k_s})$$

(using γ-twisting in the middle via the canonical isomorphism $G_{k_s} \simeq {}^{\gamma}(G_{k_s})$) for any $\mathscr{T}$ and $\mathscr{B}$ inside $\mathscr{G} := G_{k_s}$ as above. It is easily checked that $\gamma * a$ is independent of the choice of $(\mathscr{T}, \mathscr{B})$, so this definition is intrinsic to G (and k_s/k). Explicitly, if Δ is the basis of $\Phi(\mathscr{B}, \mathscr{T})$ then $\gamma * a = w_\gamma(\gamma(a))$ where $w_\gamma \in W(\Phi(G_{k_s}, \mathscr{T}))$ is the unique element carrying $\gamma(\Delta)$ to Δ.

The $*$-action of $\mathrm{Gal}(k_s/k)$ on $\mathrm{Dyn}(G) := \mathrm{Dyn}(G_{k_s})$ is continuous since the open subgroup $\mathrm{Gal}(k_s/k')$ acts trivially for k'/k such that $G_{k'}$ is pseudo-split. The explicit description using w_γ's shows that the evident equality of diagrams $\mathrm{Dyn}(\mathscr{D}(G)) = \mathrm{Dyn}(G)$ is $\mathrm{Gal}(k_s/k)$-equivariant.

EXAMPLE 9.2.3. If G is quasi-split, with $B \subset G$ a k_s-minimal pseudo-parabolic k-subgroup and $T \subset B$ a maximal k-torus, then there is a natural $\mathrm{Gal}(k_s/k)$-action on $\mathrm{X}(T_{k_s})$ that preserves the basis Δ of $\Phi(B_{k_s}, T_{k_s})$, so $w_\gamma = 1$ for all $\gamma \in \mathrm{Gal}(k_s/k)$. Hence, this natural action on $\Delta \subset \mathrm{X}(T_{k_s})$ coincides with the $*$-action.

Now assume G is *pseudo-semisimple*. There is an evident action of $\mathrm{Aut}(G_{k_s}) = \mathrm{Aut}^{\mathrm{sm}}_{G/k}(k_s)$ on the diagram $\mathrm{Dyn}(G)$, and by Lemma 9.2.2 points in $(\mathrm{Aut}^{\mathrm{sm}}_{G/k})^0(k_s)$ act trivially, so this defines an action of the finite geometric component group $\pi_0(\mathrm{Aut}^{\mathrm{sm}}_{G/k})(k_s)$ on $\mathrm{Dyn}(G)$. Let $T \subset G_{k_s}$ be a maximal k_s-torus. Since $\mathrm{X}(T)_{\mathbf{Q}}$ is spanned by $\Phi(G_{k_s}, T)$, and any automorphism of G_{k_s} that is the identity on T must be the identity on $C := Z_{G_{k_s}}(T)$ (i.e., it comes from $Z_{G,C} \subset (\mathrm{Aut}^{\mathrm{sm}}_{G/k})^0$), we see that the natural homomorphism

$$\pi_0(\mathrm{Aut}^{\mathrm{sm}}_{G/k}) \longrightarrow \mathrm{Aut}(\mathrm{Dyn}(G))$$

is injective. In particular, triviality on $\mathrm{Dyn}(G)$ *characterizes* the automorphisms of G_{k_s} that arise from the identity component of $\mathrm{Aut}^{\mathrm{sm}}_{G/k}$.

To give an application to the study of k_s/k-twists of G, for each $\gamma \in \mathrm{Gal}(k_s/k)$ let $c_\gamma : {}^{\gamma}(G_{k_s}) \simeq G_{k_s}$ be the canonical k_s-isomorphism arising from the k-descent G. A k_s/k-twist of G is built from k_s-isomorphisms ${}^{\gamma}(G_{k_s}) \simeq G_{k_s}$ of the form $f_\gamma \circ c_\gamma$ for k_s-automorphisms $f_\gamma \in \mathrm{Aut}_{k_s}(G_{k_s}) = \mathrm{Aut}_{G/k}(k_s)$ satisfying the cocycle condition $f_{\gamma'\gamma} = f_{\gamma'} \circ \gamma'(f_\gamma)$ and the "continuity" condition $f_\gamma = 1$ for all γ in some open subgroup of $\mathrm{Gal}(k_s/k)$. Our characterization of $(\mathrm{Aut}^{\mathrm{sm}}_{G/k})^0(k_s)$ inside $\mathrm{Aut}_{G/k}(k_s)$ via triviality of the action on $\mathrm{Dyn}(G)$ implies that we can analyze whether or not a given k_s/k-twist of G arises from a continuous 1-cocycle $\gamma \mapsto f_\gamma$ valued in $(\mathrm{Aut}^{\mathrm{sm}}_{G/k})^0(k_s)$ by keeping track of the $*$-action on the canonical diagram $\mathrm{Dyn}(G)$ throughout the twisting process. This has the following useful immediate consequence:

PROPOSITION 9.2.4. *The set* $\mathrm{H}^1(k, (\mathrm{Aut}^{\mathrm{sm}}_{G/k})^0)$ *classifies isomorphism classes of pairs* (H, φ) *consisting of a* k_s/k*-form* H *of* G *and a* $*$*-compatible isomorphism of diagrams* $\varphi : \mathrm{Dyn}(H) \simeq \mathrm{Dyn}(G)$ *induced by a* k_s*-isomorphism* $H_{k_s} \simeq G_{k_s}$. *(The trivial class corresponds to such pairs for which* φ *arises from a* k*-isomorphism.)*

Consider a maximal split k-torus $S \subset G$ and a minimal pseudo-parabolic k-subgroup $P \subset G$ containing S. The set of such pairs (S, P) is acted upon transitively by $G(k)$-conjugation (Theorems 4.2.9 and 5.1.3), and by minimality $P = M \ltimes U$ for $M := Z_G(S)$ and the k-split $U := \mathscr{R}_{u,k}(P)$ (Proposition 5.1.2). Fix one such (S, P, M). It will be convenient to keep track of the k-anisotropic $\mathscr{D}(M)$ only up to a central quotient. Since the central quotient M/Z_M is pseudo-reductive with trivial center (Proposition 6.1.1) and inherits k-anisotropicity from M (as for a central quotient of any k-anisotropic smooth connected affine k-group), the k-anisotropic pseudo-semisimple derived group $\mathscr{D}(M/Z_M)$ also has trivial center (by [**CGP**, Lemma 1.2.5(ii), Prop. 1.2.6]) and thus coincides with $\mathscr{D}(M)/Z_{\mathscr{D}(M)}$.

REMARK 9.2.5. The formation of M commutes with Weil restriction (so likewise for the formation of $\mathscr{D}(M)$ and $\mathscr{D}(M/Z_M)$). To state this precisely, suppose for simplicity that $G = \mathrm{R}_{k'/k}(G')$ for a finite extension field k'/k and pseudo-semisimple k'-group G'. There is a unique maximal split k'-torus $S' \subset G'$ such that S is the maximal split k-torus in $\mathrm{R}_{k'/k}(S')$, and $\mathrm{R}_{k'/k}(Z_{G'}(S')) = Z_G(\mathrm{R}_{k'/k}(S'))$ (see [**CGP**, Prop. A.5.15]), so we define $M' := Z_{G'}(S')$. The precise claim is that the obvious inclusion $\mathrm{R}_{k'/k}(M') \subset M$ is an equality.

The explicit description of the natural map $q : G_{k'} \to G'$ on points valued in k'-algebras (see [**CGP**, Prop. A.5.7]) shows that q carries $S_{k'}$ isomorphically onto S' Thus, $q(M_{k'}) \subset M'$. Since the composition of $\mathrm{R}_{k'/k}(q)$ and $G \hookrightarrow \mathrm{R}_{k'/k}(G_{k'})$ is the canonical equality $G = \mathrm{R}_{k'/k}(G')$, we obtain $M \subset \mathrm{R}_{k'/k}(M')$ as required.

For a maximal k_s-torus $\mathscr{T} \subset M_{k_s}$, the minimal pseudo-parabolic k_s-subgroups $\mathscr{B} \subset G_{k_s}$ containing $\mathscr{T}$ and contained in P_{k_s} are permuted transitively by the subgroup $N_{M_{k_s}}(\mathscr{T})(k_s) \subset M(k_s)$ since (i) pseudo-parabolic k_s-subgroups of P_{k_s} are pseudo-parabolic in G_{k_s} (Corollary 4.3.5), and (ii) the set of such k_s-subgroups of P_{k_s} is in bijective correspondence with the set of pseudo-parabolic k_s-subgroups of $(P/U)_{k_s} = M_{k_s}$ [**CGP**, Prop. 2.2.10]. Thus, the natural subdiagram inclusion $\mathrm{Dyn}(M_{k_s}, \mathscr{T}, \mathscr{B}/U_{k_s}) \hookrightarrow \mathrm{Dyn}(G_{k_s}, \mathscr{T}, \mathscr{B})$ defines a subdiagram inclusion

$$\iota : \mathrm{Dyn}(\mathscr{D}(M/Z_M)) = \mathrm{Dyn}(\mathscr{D}(M)) = \mathrm{Dyn}(M) \hookrightarrow \mathrm{Dyn}(G)$$

that is *independent of all choices* and so is $*$-compatible.

REMARK 9.2.6. Let $T \subset M$ be a maximal k-torus, and let Δ be the basis of $\Phi(G_{k_s}, T_{k_s})$ corresponding to a minimal pseudo-parabolic k_s-subgroup of P_{k_s} containing T_{k_s}. Denote by Δ_0 the set of roots in Δ with trivial restriction to S_{k_s}. The theory of relative root systems developed in [**CGP**, C.2.13ff.] ensures that P corresponds to a basis ${}_k\Delta$ of the relative root system $\Phi(G, S)$ and that restriction to S_{k_s} defines a surjection $\Delta - \Delta_0 \to {}_k\Delta$ whose fibers are the orbits for the $*$-action on $\Delta - \Delta_0$. Under the labeling of $G(k_s)$-conjugacy classes of pseudo-parabolic k_s-subgroups of G_{k_s} by subsets of $\Delta = \mathrm{Dyn}(G)$, P_{k_s} corresponds to Δ_0; i.e., Δ_0 is a basis $\Phi(M_{k_s}, T_{k_s})$. The inclusion ι thereby specifies the set Δ_0 of "non-distinguished" roots inside Δ as in Tits' notion of *index* defined in [**Ti1**, 2.3] for semisimple G.

Consider 4-tuples $(\mathscr{G}, \tau, \mathscr{M}, j)$ consisting of a pseudo-semisimple k_s-group $\mathscr{G}$, a continuous action τ of $\mathrm{Gal}(k_s/k)$ on the canonical diagram $\mathrm{Dyn}(\mathscr{G})$, a k-anisotropic pseudo-semisimple k-group $\mathscr{M}$ with trivial center, and a $\mathrm{Gal}(k_s/k)$-equivariant subdiagram inclusion $j : \mathrm{Dyn}(\mathscr{M}) \hookrightarrow \mathrm{Dyn}(\mathscr{G})$. For example, any pseudo-semisimple k-group G gives rise to such a 4-tuple $(G_{k_s}, *, \mathscr{D}(M/Z_M), \iota)$ as above.

DEFINITION 9.2.7. An *isomorphism* $(\mathscr{G}, \tau, \mathscr{M}, j) \simeq (\mathscr{G}', \tau', \mathscr{M}', j')$ consists of a k_s-isomorphism $f : \mathscr{G} \simeq \mathscr{G}'$ such that $\mathrm{Dyn}(f)$ intertwines τ and τ' and a k-isomorphism $f_0 : \mathscr{M} \simeq \mathscr{M}'$ such that $j' \circ \mathrm{Dyn}(f_0) = \mathrm{Dyn}(f) \circ j$.

The Tits-style classification is:

THEOREM 9.2.8. *The isomorphism class of a pseudo-semisimple k-group G is determined by the isomorphism class of the associated 4-tuple $(G_{k_s}, *, \mathscr{D}(M), \iota)$.*

The main task in the proof of Theorem 9.2.8 is to show that if G' is a second pseudo-semisimple k-group and we define $M' := Z_{G'}(S')$ for a maximal split k-torus $S' \subset G'$ then the existence of an isomorphism of 4-tuples

$$(f, f_0) : (G'_{k_s}, *', \mathscr{D}(M'/Z_{M'}), \iota') \simeq (G_{k_s}, *, \mathscr{D}(M/Z_M), \iota)$$

implies triviality of the class of G' in $\mathrm{H}^1(k, \mathrm{Aut}^{\mathrm{sm}}_{G/k})$. Inspired by Tits' simplification (and correction) of his proof in the semisimple case, we will repeatedly perform "reduction of the structure group" until we reach a structure group with trivial degree-1 Galois cohomology; the analysis of the final structure group involves some new problems that one does not encounter in the semisimple case. We sketch some key ideas, and refer the reader to [**CP**, 6.3.11–6.3.16] for complete details.

The requirement in Definition 9.2.7 that $\mathrm{Dyn}(f)$ is compatible with $*$-actions implies that the 1-cocycle $c : \gamma \mapsto f \circ ({}^{\gamma}f)^{-1}$ is valued in the group of k_s-automorphisms of G_{k_s} whose effect on $\mathrm{Dyn}(G_{k_s})$ is *trivial*, so c expresses G' as a pseudo-inner k_s/k-form of G. That is, we have achieved reduction of the structure group to

$$(\mathrm{Aut}^{\mathrm{sm}}_{C/k})^0 = (G \rtimes Z_{C,C})/C$$

with $C := Z_G(T)$ for a maximal k-torus $T \subset M$.

Let $P \subset G$ and $P' \subset G'$ be minimal pseudo-parabolic k-subgroups. By Remark 9.2.6 and the compatibility of (f, f_0) with ι and ι', the $G(k_s)$-conjugacy class of P_{k_s} corresponds to the $G'(k_s)$-conjugacy class of P'_{k_s}. In other words, by composing f with a $G(k_s)$-conjugation (i.e., changing c by a coboundary valued in the image of $G(k_s)$ in $(\mathrm{Aut}^{\mathrm{sm}}_{G/k})^0(k_s)$) it can be assumed that $f(P_{k_s}) = P'_{k_s}$. This achieves a further reduction of the structure group to the stabilizer of P in $(G \rtimes Z_{G,C})/C$. Since $N_G(P) = P$ by Proposition 4.3.6, this stabilizer coincides with $(P \rtimes Z_{G,C})/C$.

Since $P = M \ltimes U$ for a k-split smooth connected unipotent k-group U, the structure group can be reduced further still, to

$$(M \rtimes Z_{G,C})/C = (\mathscr{D}(M) \rtimes Z_{G,C})/C'$$

for the Cartan k-subgroup $C' := Z_{\mathscr{D}(M)}(T')$ in $\mathscr{D}(M)$ where $T' := T \cap \mathscr{D}(M)$ (and $Z_{G,C}$ preserves $M = Z_G(S)$ inside G since it acts trivially on $S \subset C$).

For the natural map

$$q : (\mathscr{D}(M) \rtimes Z_{G,C})/C' \longrightarrow (\mathrm{Aut}^{\mathrm{sm}}_{\mathscr{D}(M)/k})^0 = (\mathscr{D}(M) \rtimes Z_{\mathscr{D}(M),C'})/C',$$

via Proposition 9.2.4 applied to $\mathscr{D}(M)$ we see that $\mathrm{H}^1(q)$ carries the class of c to the class of the pair $(\mathscr{D}(M'), \mathrm{Dyn}((f_0)_{k_s}))$. This latter class is trivial since f_0 is a k-isomorphism, so we achieve a final reduction of the structure group to $\ker q$ *provided* that q is a smooth surjection.

In an evident manner, q arises from the natural restriction map

$$\rho : Z_{G,C} \longrightarrow Z_{\mathscr{D}(M),C'}.$$

This identifies $\ker q$ with $\ker \rho$, so we are reduced to proving:

PROPOSITION 9.2.9. *The map ρ is surjective and* $\ker \rho \simeq \mathrm{R}_{F/k}(\mathrm{GL}_1)$ *for a finite reduced k-algebra F.*

For a finite Galois splitting field E/k of T, Lemma 9.1.12 identifies ρ_E with the projection from a Δ-indexed product onto the Δ_0-indexed subproduct. Thus, ρ is surjective with smooth connected kernel. In particular, $\ker \rho$ is a commutative pseudo-reductive k-group.

The determination of $\ker \rho$ as a k-group (rather than just an E-group) is a delicate problem in Galois descent because the factor fields of the k-algebra F will not generally be separable over a purely inseparable extension of k (an issue that does not arise in the semisimple case). Proposition 6.1.1 and Lemma 9.1.9(i) allow us to assume $Z_G = 1$, so by Corollary 3.2.5 (and Remarks 9.1.14 and 9.2.5) *we may assume G is absolutely pseudo-simple.* The advantage of the absolutely pseudo-simple case is that in such cases F will turn out to be a product $\prod_{\alpha \in {}_k\Delta} F_\alpha$ where each F_α is a field that is a compositum of separable and purely inseparable extensions of k.

The crucial input over E is that for each $a \in \Delta$ we have

$$Z_{(G_E)_a,(C_E)_a} \simeq \mathrm{R}_{F'_a/E}(\mathrm{GL}_1)$$

for a purely inseparable finite extension F'_a/E (by Proposition 8.3.12 when $(G_E)_a$ is of type BC_1, and by (9.1.13.1) when $(G_E)_a$ is of type A_1). The extension F'_a/E depends only on the length of a since $N_G(T)(E)$ acts transitively on the set of roots with a given length (as G is now absolutely pseudo-simple). In particular, the *purely inseparable* extensions F'_a/E coincide for all a in an orbit for the $*$-action on Δ, so by Galois descent $F'_a = F_a \otimes_k E$ for a canonically determined purely inseparable finite extension F_a/k.

By using the link between the $*$-action and relative root systems, one finds (after some work) that if $\{a_i\}$ is a set of representatives for the $*$-action on $\Delta - \Delta_0$ then the E/k-descent $\ker \rho$ of $\ker \rho_E = \prod_{a \in \Delta - \Delta_0} \mathrm{R}_{F'_a/E}(\mathrm{GL}_1)$ is $\prod_i \mathrm{R}_{L_{a_i} F_{a_i}/k}(\mathrm{GL}_1)$ where $L_a \subset k_s$ corresponds to the stabilizer of a in $\mathrm{Gal}(k_s/k)$ under the $*$-action. Thus, we may take F to be $\prod_i F_i$ for $F_i := L_{a_i} F_{a_i}$ (so the index set is identified with ${}_k\Delta$). This completes our sketch of the proof of Theorem 9.2.8.

The following application of Theorem 9.2.8 and its proof circumvents problems that arose in Remark 9.1.4.

COROLLARY 9.2.10. *Let G be a pseudo-reductive group over a field k. Up to k-isomorphism, G admits at most one pseudo-split k_s/k-form and at most one quasi-split pseudo-inner form.*

In the pseudo-semisimple case, the pseudo-split assertion is an immediate consequence of how the pseudo-split property is expressed in the Tits-style classification in Theorem 9.2.8. Indeed, suppose G and G' are pseudo-split pseudo-semisimple k-groups that are k_s/k-forms of each other. The associated 4-tuples are $(G_{k_s}, \tau, 1, \iota)$ and $(G'_{k_s}, \tau', 1, \iota')$, where the $*$-actions τ and τ' on the respective canonical diagrams of G_{k_s} and G'_{k_s} are *trivial* and the diagram inclusions ι and ι' are vacuous since the canonical diagram of the trivial group is empty. Hence, these 4-tuples are isomorphic, so $G \simeq G'$! The general pseudo-reductive case can be deduced from the settled pseudo-semisimple case by close study of Cartan k-subgroups containing split maximal k-tori; see [**CP**, Prop. C.1.1].

The proof of the quasi-split assertion in Corollary 9.2.10 rests on the work with Galois descent in the proof of the Tits-style classification for the pseudo-semisimple $\mathscr{D}(G)$ (e.g., Proposition 9.2.9); for further details see [**CP**, Prop. C.2.8].

In view of Corollary 9.2.10, it is natural to ask for a characterization (e.g., in terms of field-theoretic or linear-algebraic data) for when a given absolutely pseudo-simple k_s-group admits a descent to a *pseudo-split* (absolutely pseudo-simple) k-group. This is addressed in [**CP**, Rem. C.2.13], and a satisfactory characterization is given there away from characteristic 2 (largely due to Theorem 7.4.8 and Proposition 7.5.10); this is not to be confused with the task of building a pseudo-split k_s/k-form (which can fail to exist in the absolutely pseudo-simple case in every positive characteristic) since no initial k-group has been given.

REMARK 9.2.11. The techniques in the proof of Theorem 9.2.8 can be used to analyze the relative rank of absolutely pseudo-simple groups of type F_4 over imperfect fields k of characteristic 2. In the connected semisimple case the only possible relative ranks are 0, 1, or 4 [**Spr**, 17.5.2(i)]. In the pseudo-semisimple case relative rank 3 remains impossible, but relative rank 2 can occur (and the possibilities with k-rank 2 are classified by conformal isometry classes of certain anisotropic quadratic forms over k); this is addressed in [**CP**, App. D].

10. Structural classification

10.1. Exceptional constructions. So far we have encountered three classes of non-standard absolutely pseudo-simple k-groups (of minimal type) with a reduced root system over k_s:

(i) the k-groups $\mathrm{SO}(q)$ in §7.3 for regular quadratic spaces (V, q) satisfying $1 < \dim V^{\perp} < \dim V$ over imperfect fields k of characteristic 2,

(ii) basic exotic k-groups in §7.5 for type G_2 in characteristic 3 and types B_n ($n \geqslant 1$), C_n ($n \geqslant 1$), and F_4 in characteristic 2,

(iii) the pseudo-split k-groups G in Theorem 7.5.14 over imperfect fields of characteristic 2; these have root system B_n or C_n with $n \geqslant 2$ and depend on some auxiliary field-theoretic and linear-algebraic data, and $G^{\mathrm{ss}}_{\overline{k}}$ is simply connected.

We shall now recall key features of the first two constructions so that we have some context for the additional constructions that remain to be given (recovering (iii) in the pseudo-split case).

The groups arising in (ii) are always non-standard and exist over every imperfect field of characteristic 2 or 3. Moreover, for types G_2 in characteristic 3 and F_4 in characteristic 2 we saw in Corollary 7.5.11 that (up to purely inseparable Weil restriction) they account for *all* deviations from standardness in the absolutely pseudo-simple case with those root systems over k_s. Hence, for the purpose of an exhaustive description of all non-standard groups, the main work is in characteristic 2 for types B, C, and BC. We studied type BC in the (pseudo-split) minimal-type case in §8, so in this section we largely focus on types B and C (with a minimal-type hypothesis).

In case (i) above, the k_s-group $\mathrm{SO}(q)_{k_s}$ has root system B_n where $\dim(V/V^{\perp}) = 2n$, and $\mathrm{SO}(q)$ is non-standard except precisely when $n = 1$ and $q(V^{\perp})^{1/2}$ is a line over a nontrivial extension field of k inside $k^{1/2}$ (so $\mathrm{SO}(q)$ is non-standard for some (V, q) over k with $n = 1$ if and only if $[k : k^2] > 2$); see Remark 7.3.2. Further work (see [**CP**, Prop. 7.2.2]) shows that isomorphisms between such $\mathrm{SO}(q)$ and $\mathrm{SO}(q')$ arise from conformal isometries between (V, q) and (V', q'), akin to the well-known case of connected semisimple groups of adjoint type B.

Since $\mathrm{GO}_{2n+1} = \mathrm{GL}_1 \times \mathrm{SO}_{2n+1}$ and $\mathrm{SO}_{2n+1} = \mathrm{Aut}_{\mathrm{SO}_{2n+1}/k}$, by Hilbert 90 the k_s/k-forms of SO_{2n+1} are the k-groups $\mathrm{SO}(q)$ for non-degenerate quadratic spaces (V, q) of dimension $2n+1$ such that q_{k_s} is conformal to the standard split quadratic space q_{2n+1}. Such conformality over k_s always holds since in the odd-dimensional non-degenerate case in characteristic 2 we can arrange for the quadratic form to be x^2 on the defect line by $k_s^\times$-scaling. In contrast, for regular degenerate q as above, $\mathrm{SO}(q)$ is generally a proper k-subgroup of $\mathrm{Aut}^{\mathrm{sm}}_{\mathrm{SO}(q)/k}$ (due to "extra" conformal isometries arising from an action of the short root field when it is larger than k; see Example 7.4.6). This suggests that for (V, q) over k_s, the k_s-group $\mathrm{SO}(q)$ may have a k-descent that is not an $\mathrm{SO}(q')$.

To motivate where to find such additional k-descents, recall that the automorphism scheme of a *smooth* quadric hypersurface in $\mathbf{P}^{2n}$ is a form of $\mathrm{GO}_{2n+1}/\mathrm{GL}_1 = \mathrm{SO}_{2n+1}$. This generalizes to certain non-smooth quadrics:

PROPOSITION 10.1.1. *Let D be a geometrically integral non-smooth quadric in a Severi–Brauer variety X over k. Assume D is regular at its k_s-points. Then $\mathrm{Aut}^{\mathrm{sm}}_{D/k}$ is connected and affine, and $G_D := \mathscr{D}(\mathrm{Aut}^{\mathrm{sm}}_{D/k})$ is absolutely pseudo-simple of type* B *with trivial center. The Cartan k-subgroups of G_D are tori.*

See [**CP**, Rem. 7.3.2, Prop. 7.3.3] for a proof of this result. If $X(k) \neq \emptyset$ then $D \simeq (q = 0) \subset \mathbf{P}(V^*) = X$ for some (V, q) as above (the injectivity of $q|_{V^\perp}$ is equivalent to the regularity hypothesis on D), so $G_D = \mathrm{SO}(q)$ by [**CP**, Prop. 7.3.3(iii)]. Using general X, the k-groups G_D generalize the $\mathrm{SO}(q)$-construction.

REMARK 10.1.2. For G_D with k_s-rank $n \geqslant 2$, the root field is equal to k and the minimal field of definition K/k for its geometric unipotent radical satisfies $K^2 \subset k$. To see this we may assume $k = k_s$, so $G_D \simeq \mathrm{SO}(q)$ for some (V, q) as above. The long root groups of $\mathrm{SO}(q)$ are 1-dimensional (see [**CP**, Prop. 7.1.3]), so the root field is k, and Theorem 7.4.7(iii) implies that K^2 is contained in the root field.

Remarkably, the k-groups G_D are *exactly* the non-reductive absolutely pseudo-simple groups G whose center is trivial and whose Cartan subgroups are tori. (Note that any such G is trivially of minimal type, since $Z_G = 1$.) The idea of the proof is that since D depends functorially (with respect to isomorphisms) on G_D by [**CP**, Prop. 7.3.3(ii), Rem. 7.3.2], it suffices to check the result after passing to a finite Galois extension on k to reduce to the case where G contains a split maximal k-torus T. In those cases the pseudo-split k-subgroups $G_a = \langle U_a, U_{-a}\rangle$ for $a \in \Phi(G, T)$ may have nontrivial center (unlike G) but inherit the minimal-type property from G (Example 7.1.7) and hence fall into the rank-1 classification scheme in Theorem 7.2.5. The Cartan subgroups of such G_a must be tori when the same holds for G, so this severely limits the possibilities for such G_a. By taking that into account, one finds that the constructions $\mathrm{SO}(q)$ for "pseudo-split" (V, q) with varying $q|_{V^\perp}$ are sufficient to exhaust all possibilities for G in accordance with the Isomorphism Theorem (i.e., Theorem 9.1.7) because of the assumption that the Cartan subgroups (whose structure is generally mysterious) are tori. See [**CP**, Prop. 7.3.7] for further details.

Within the class of k-groups of the form G_D, the ones arising as an $\mathrm{SO}(q)$ have an intrinsic characterization:

PROPOSITION 10.1.3. *For D as above, the following conditions are equivalent:*

(i) *$G_D \simeq \mathrm{SO}(q)$ for some (V, q),*

(ii) G_D *contains a Levi k-subgroup,*
(iii) *the Severi–Brauer variety X is trivial (i.e.,* $X = \mathbf{P}_k^N$ *for some N).*

The equivalence of (i) and (ii) is [**CP**, Prop. 7.3.5], and the equivalence of these with (iii) is (part of) [**CP**, Thm. 7.3.6].

REMARK 10.1.4. The construction of Levi k-subgroups of $\mathrm{SO}(q)$ implies that the non-empty non-smooth locus of D has codimension in D equal to $2n$, where $(G_D)_{k_s}$ has root system B_n. Indeed, we may assume $k = k_s$, so $D = (q = 0)$ with

$$q = x_1x_2 + \cdots + x_{2n-1}x_{2n} + c_0y_0^2 + \cdots + c_dy_d^2$$

in suitable projective coordinates on $\mathbf{P}_k^N$ for some $n \geqslant 1$ and $d = N - 2n \geqslant 0$; here, $c_0, \ldots, c_d$ are linearly independent over k^2. The root system for $\mathrm{SO}(q) = G_D$ is B_n, and the singularities in $D_{\overline{k}}$ are the points where $x_1, \ldots, x_{2n}$ vanish and $\sum \sqrt{c_j}y_j = 0$. Hence, the singular locus of $D_{\overline{k}}$ is a linear space in $\mathbf{P}_{\overline{k}}^N$ of codimension $2n+1$, so it has codimension $2n$ in the hypersurface $D_{\overline{k}}$. The formation of the non-smooth locus commutes with extension of the ground field, so this closed locus in D has codimension $2n$ as well.

The G_D-construction goes beyond the $\mathrm{SO}(q)$-construction even in arithmetically interesting cases: for every $n \geqslant 2$ and local function field k over a finite field of characteristic 2 there exist basic exotic k-groups $\mathscr{G}$ of type B_n that have k-rank $< n - 1$, and then $\mathscr{G}/Z_{\mathscr{G}}$ is such a G_D that is *not* of the form $\mathrm{SO}(q)$ (see [**CP**, Ex. 7.2.4]). The short root field of this G_D is $k^{1/2}$, and $\mathscr{G}$ admits a pseudo-split k_s/k-form (as does any basic exotic k-group!), so its maximal central quotient G_D does as well. This pseudo-split k_s/k-form of G_D admits a Levi k-subgroup by Theorem 5.4.4 and so must be an $\mathrm{SO}(q')$ by Proposition 10.1.3. Hence, this G_D is not obtained by the $\mathrm{SO}(q)$-construction but is related to it through k_s/k-twisting.

Over general imperfect fields k of characteristic 2, the class of k-groups G_D goes beyond even Galois-twists of the k-groups $\mathrm{SO}(q)$ (though that can only occur if $[k : k^2] \geqslant 8$, as we shall soon see). To understand this, first note that every $\mathrm{SO}(q)$ admits a pseudo-split k_s/k-form H [**CP**, Prop. 7.1.2]. Any pseudo-split pseudo-reductive k-group admits a Levi k-subgroup (Theorem 5.4.4), so $H = \mathrm{SO}(q')$ for some (V', q') by Proposition 10.1.3. Thus, it is equivalent to determine if G_D admits a pseudo-split k_s/k-form. A subtle degree-2 cohomological obstruction implies that such a pseudo-split form exists if $[k : k^2] \leqslant 4$ [**CP**, Cor. C.2.12]. This is optimal because if $[k : k^2] \geqslant 8$ and k has sufficiently rich Galois theory (more specifically, if k admits a quadratic Galois extension k'/k such that $\ker(\mathrm{Br}(k) \to \mathrm{Br}(k')) \neq 1$) then for every $n \geqslant 1$ there exist k-groups G_D with k_s-rank n which do not admit a pseudo-split k_s/k-form (e.g., the maximal central quotients of the type-B groups constructed in [**CP**, C.3.1, C.4.1] are such k-groups).

We shall now build upon the Severi–Brauer construction G_D via fiber products and universal smooth k-tame central extensions to make "exceptional" constructions that account for all non-standard absolutely pseudo-simple G of *minimal type* with root system B_n or C_n ($n \geqslant 1$) over k_s, with k any imperfect field of characteristic 2. By considering universal smooth k-tame central extensions, it suffices to make such k-groups G with $G_{\overline{k}}^{\mathrm{ss}}$ simply connected.

REMARK 10.1.5. In what follows, the "minimal type" hypothesis is an additional constraint precisely when $[k : k^2] \geqslant 16$. To explain this, let Φ be a root system of type B or C (with rank $n \geqslant 1$), and consider absolutely pseudo-simple k-groups G such that G_{k_s} has root system Φ and $G^{\rm ss}_{\overline{k}}$ is simply connected. If $[k : k^2] \leqslant 8$ then Gabber proved that G is *automatically* of minimal type [**CP**, Prop. B.3.1, Prop. 4.3.3(ii)]. (The core of his proof is a study of the possibilities for $V^*_{K/k}$ when $K^2 \subset k$ and $[K : k] \leqslant 8$ in order to use a splitting criterion for central extensions in [**CGP**, Prop. 5.1.3] to show that the minimal-type central quotient map $G \to G/\mathscr{C}_G$ is an isomorphism. This leads to the verification of a property introduced in Definition 10.2.9 that will be proved equivalent to "minimal type" in Proposition 10.2.10 when $G^{\rm ss}_{\overline{k}}$ is simply connected.) In contrast, if $[k : k^2] \geqslant 16$ then there exist such G over k that are not of minimal type; see [**CP**, B.1, B.2] for the construction.

Consider D as above, so G_D has root system B_n over k_s for some $n \geqslant 1$. If $n \geqslant 2$ then the root field of G_D is k by Remark 10.1.2. But if $n = 1$ then G_D of type A_1 can have a root field larger than k. This explains the additional condition on the root field imposed in the rank-1 case of the following definition (ensuring that in such cases there is no intervention of a nontrivial Weil restriction).

DEFINITION 10.1.6. For $n \geqslant 1$, a *type-*B_n *generalized basic exotic k-group* is the universal smooth k-tame central extension G of a k-group of the form G_D for a non-smooth geometrically integral quadric D in a Severi–Brauer variety over k, provided that G_D has root field k if the non-smooth locus in D has codimension 2.

In this definition, G/Z_G recovers G_D since the k-groups G_D and G/Z_G have trivial center (see Proposition 6.1.1), and the codimension-2 case at the end corresponds to G_D with k_s-rank 1 due to Remark 10.1.4. The k-groups G in Definition 10.1.6 are absolutely pseudo-simple and of minimal type (see [**CP**, Lemma 5.3.2] for the latter), and for any separable extension k'/k of fields a k-group G is generalized basic exotic of type B if and only if $G_{k'}$ is (as the same holds for groups arising from the G_D-construction, due to the characterization of that construction in terms of non-reductive absolutely pseudo-simple groups whose Cartan subgroups are tori).

For any absolutely pseudo-simple k-group H with a reduced root system over k_s, the root field and minimal field of definition over k for the geometric unipotent radical are unaffected by passage to a pseudo-reductive central quotient ([**CP**, Rem. 3.3.3] and Proposition 6.2.2). Hence, G has root field k and the minimal field of definition K/k for its geometric unipotent radical satisfies $K^2 \subset k$.

For $n \geqslant 2$, the basic exotic k-groups of type B_n are precisely the k-groups G in Definition 10.1.6 for which the short root field $F_<$ coincides with minimal field of definition K/k for the geometric unipotent radical (as may be checked over k_s via Proposition 7.2.3(ii) and Theorem 7.5.14 over k_s with $V = K_s$ and $V_>$ a k_s-line). The link to the basic exotic construction goes further: for any G as in Definition 10.1.6 with $n \geqslant 2$ and minimal field of definition K/k for its geometric unipotent radical, we shall soon give a *canonical* procedure to "fatten" its short root groups over k_s to become $\mathrm{R}_{K_s/k_s}(\mathbf{G}_{\mathrm{a}})$ and thereby obtain a basic exotic k-group containing the generalized basic exotic k-group G.

Put another way, for $n \geqslant 2$ a type-B_n generalized basic exotic k-group with minimal field of definition K/k for its geometric unipotent radical (so $K^2 \subset k$ by Remark 10.1.2) is obtained from a basic exotic one with the same K/k by "replacing" each short root group $\mathrm{R}_{K_s/k_s}(\mathbf{G}_{\mathrm{a}})$ with the k_s-subgroup corresponding to a nonzero k_s-subspace $V \subset K_s$ satisfying $k_s\langle V\rangle = K_s$. (No such $V \neq K_s$ exists when $[K : k] = 2$, as occurs whenever $[k : k^2] = 2$, so the need for Definition 10.1.6 occurs if and only if $[k : k^2] > 2$.) The type-B_n generalized basic exotic groups for $n \geqslant 2$ were initially found via this latter perspective in the pseudo-split case by inspecting the possibilities in Theorem 7.5.14. The desire to remove the pseudo-split condition eventually led to the discovery of the G_D-construction that underlies a more satisfactory definition (for all $n \geqslant 1$).

Here is a *canonical* link between the generalized basic exotic case and the basic exotic case, via a fiber-product construction in the spirit of Theorem 7.5.7.

PROPOSITION 10.1.7. *Let G be a type-B_n generalized basic exotic k-group with $n \geqslant 2$, and let K/k be the minimal field of definition for its geometric unipotent radical. Let G' be the connected absolutely simple K-group $G_K/\mathscr{R}_{u,K}(G)$ of type B_n.*

The K-group G' is simply connected, and if $\pi : G' \to \overline{G}'$ is the very special K-isogeny for G' then for $f := \mathrm{R}_{K/k}(\pi)$ the image $\overline{G} := f(G) \subset \mathrm{R}_{K/k}(\overline{G}')$ is a Levi k-subgroup of $\mathrm{R}_{K/k}(\overline{G}')$. In particular, the k-group $f^{-1}(\overline{G})$ containing G is a basic exotic k-group with associated invariants $(K/k, G')$.

The proof of this result is immediately reduced to the case $k = k_s$, so (by Theorem 5.4.4) G contains a Levi k-subgroup L that in turn contains a split maximal k-torus of G. Since $L_K \to G'$ is an isomorphism (by the definition of L being a Levi k-subgroup of G), everything can then be verified by computations with open cells; see [**CP**, Prop. 8.1.3] for the details.

There is a notion of "very special quotient" $G \to \overline{G}$ for any G as above, with $\overline{G}$ semisimple and simply connected of type C_n (see [**CP**, Def. 8.1.5]). This is an important ingredient in a "basic exceptional" construction for type B_2 given in [**CP**, §8.3] that goes beyond the generalized basic exotic construction when $n = 2$ and $[k : k^2] \geqslant 16$, but it will not be described here.

REMARK 10.1.8. The need for the additional (minimal type) *basic exceptional* construction when $n = 2$ is due to reasons similar to what we saw for BC_2 in Proposition 8.3.4 and the discussion immediately preceding it (namely, a field-theoretic invariant for $n > 2$ can be replaced with an appropriate vector subspace of the same field-theoretic invariant if $n = 2$ and $[k : k^2] \geqslant 8$). Such additional k-groups exist if and only if $[k : k^2] \geqslant 16$ (see [**CP**, Rem. 8.3.1]), and are studied in detail in [**CP**, §8.3].

For $n \geqslant 2$, a type-C_n analogue of Definition 10.1.6 uses a fiber-product construction similar to the type-C_n basic exotic case resting on Theorem 7.5.7. (If $n = 2$ then $\mathrm{C}_2 = \mathrm{B}_2$ but this variant of Definition 10.1.6 is new: relative to the pseudo-split basic exotic construction for a given $(K/k, n)$, it fattens the long root groups rather than shrinking the short root groups.)

This type-C analogue rests on data $(K/k, K'/k, G', \overline{G}, j)$ defined as follows. Fix a nontrivial purely inseparable finite extension K/k satisfying $K^2 \subset k$, and fix $n \geqslant 2$. Let G' be a connected semisimple K-group that is absolutely simple and simply connected of type C_n, with $\pi : G' \to \overline{G}'$ its very special isogeny.

Let $\overline{G}$ be a k-group that is either absolutely simple and simply connected of type B_n or is a type-B_n generalized basic exotic k-group whose geometric unipotent radical has minimal field of definition K'/k contained in K/k; define $K' = k$ when

$\overline{G}$ is semisimple. Assume there is given a k-subgroup inclusion

$$j : \overline{G} \hookrightarrow \mathrm{R}_{K/k}(\overline{G}')$$

such that the associated K-homomorphism $\overline{G}_K \to \overline{G}'$ identifies $\overline{G}'$ with the quotient $\overline{G}_K^{\mathrm{ss}}$ (recovering j via $i_{\overline{G}}$). Finally, we impose the most subtle hypothesis: assume $j(\overline{G})$ is contained in the image of $f := \mathrm{R}_{K/k}(\pi)$.

Note that if $\overline{G}$ is *absolutely simple* then such an inclusion j amounts to identifying $\overline{G}$ with a Levi k-subgroup of $\mathrm{R}_{K/k}(\overline{G}')$ via j (see [**CGP**, Lemma 7.2.1]), so in such cases we have specified exactly the setup for the type-C_n basic exotic construction.

PROPOSITION 10.1.9. *The fiber product*

$$G = \overline{G} \times_{\mathrm{R}_{K/k}(\overline{G}')} \mathrm{R}_{K/k}(G')$$

is absolutely pseudo-simple of minimal type, K/k is the minimal field of definition for its geometric unipotent radical, and the natural map $G_K \to G'$ identifies G' with G_K^{ss}. In particular, $G_{\overline{k}}^{\mathrm{ss}}$ is simply connected and G_{k_s} has root system C_n.

The main idea in the proof of Proposition 10.1.9 is to introduce an auxiliary basic exotic k-subgroup and use its properties to analyze G. More specifically, we may assume $k = k_s$, so $\overline{G}$ contains a Levi k-subgroup $\overline{L}$; this is also a Levi k-subgroup of $\mathrm{R}_{K/k}(\overline{G}')$. Since $\overline{L} \subset \mathrm{im}(f)$, it follows that $f^{-1}(\overline{L})$ is a basic exotic k-group, so $f^{-1}(\overline{L})$ is *smooth* and even absolutely pseudo-simple of type C_n. A choice of Levi k-subgroup L of $f^{-1}(\overline{L})$ then enables one to carry out calculations with root groups and open cells to establish the desired properties of G (e.g., smoothness); see [**CP**, Prop. 8.2.2] for the details.

REMARK 10.1.10. For any separable extension of fields k'/k, an absolutely pseudo-simple k-group G arises from the construction in Proposition 10.1.9 if and only if $G_{k'}$ does over k'. The implication "$\Rightarrow$" is obvious, and for the converse direction we shall use the procedures in Proposition 10.1.9 that reconstruct from the fiber product some of the data that enters into the construction.

Suppose $G_{k'}$ is a fiber product in the desired manner. Let K/k be the minimal field of definition over k for the geometric unipotent radical of G, and define $G' = G_K^{\mathrm{ss}}$. Note that the formation of K/k and G' are compatible with scalar extension along $k \to k'$, and so is the very special isogeny $\pi : G' \to \overline{G}'$. The following properties hold because they are all satisfied after scalar extension to k': $K^2 \subset k$, G' is absolutely simple and simply connected of type C_n with $n \geqslant 2$, the map $G \to \mathrm{R}_{K/k}(G')$ has trivial kernel, the image $\overline{G}$ of $f := \mathrm{R}_{K/k}(\pi)$ is either absolutely simple and simply connected of type B_n or is type-B_n generalized basic exotic, the minimal field of definiton K'/k for the geometric unipotent radical of $\overline{G}$ is a subextension of K/k (as purely inseparable extensions of k), and the K-homomorphism $\overline{G}_K \to \overline{G}'$ corresponding to the inclusion j of $\overline{G}$ into $\mathrm{R}_{K/k}(\overline{G}')$ is identified with $\overline{G}_K^{\mathrm{ss}}$ (using that $K' \subset K$ over k).

Consequently, the 5-tuple $(K/k, K'/k, G', \overline{G}, j)$ satisfies the conditions required in Proposition 10.1.9 (in particular, $\overline{G} \subset \mathrm{im}(f)$ by design). Hence, $f^{-1}(\overline{G})$ is as in Proposition 10.1.9, and this coincides with G inside $\mathrm{R}_{K/k}(G')$ because this can be checked over k'.

The description of G as in Proposition 10.1.9 yields (see [**CP**, 8.2.4]) that the short root groups of G_{k_s} have the form $\mathrm{R}_{K_s/k_s}(\mathbf{G}_\mathrm{a})$ and the long root groups are given by $\underline{V} \subset \mathrm{R}_{K_s/k_s}(\mathbf{G}_\mathrm{a})$ for some nonzero k_s-subspace $V \subset K_s$ such that $k_s\langle V\rangle = K'_s$ (so $\dim V > 1$ when $K' \neq k$, whereas long root groups for type-C_n generalized basic exotic groups are 1-dimensional). The root field F of G coincides with the long root field (as in Theorem 7.4.7(iii)), so F_s is the maximal subfield of K_s over which V is a subspace.

We have $G = \mathscr{D}(\mathrm{R}_{F/k}(G_F^{\mathrm{prmt}}))$ by Proposition 7.4.5. The F-group G_F^{prmt} can also be constructed as in Proposition 10.1.9. Indeed, by Remark 10.1.10 it suffices to check over $F_s = F \otimes_k k_s$, so we may (and do) assume $k = k_s$. We can use the F-vector space structure on V and Theorem 7.5.14 to make an F-group H' that is a fiber product as in Proposition 10.1.9 such that $G \simeq \mathscr{D}(\mathrm{R}_{F/k}(H'))$. But then the natural map $G_F \to H'$ identifies H' with G_F^{prmt} by [**CP**, (2.3.13)], so G_F^{prmt} arises as in Proposition 10.1.9 as claimed. Returning to general k (not necessarily separably closed), since G_F^{prmt} has root field F we see that the intervention of nontrivial Weil restrictions is avoided in the following definition via a condition on the root field:

DEFINITION 10.1.11. For $n \geqslant 2$, a *type-C_n generalized basic exotic k-group* is a k-group G arising as in Proposition 10.1.9 for which the root field is k.

REMARK 10.1.12. The centralizer of a split maximal k-torus in a pseudo-split generalized basic exotic k-group of type B_n or C_n ($n \geqslant 2$) can be described as a direct product similar to (7.5.1.1) by using Lemma 9.1.10; see [**CP**, Prop. 8.2.5].

By design, the condition of equality between the root field and the ground field holds for generalized basic exotic groups of types B and C, as well as for basic exotic groups of types F_4 or G_2 (and for the rank-2 basic exceptional groups addressed in Remark 10.1.8). Thus, we get a strictly larger class of groups by incorporating Weil restrictions:

DEFINITION 10.1.13. A *generalized exotic group* G over an imperfect field k of characteristic $p \in \{2, 3\}$ is a k-group isomorphic to $\mathscr{D}(\mathrm{R}_{k'/k}(G'))$ for a nonzero finite reduced k-algebra k' and a k'-group G' whose fiber G'_i over each factor field k'_i of k' is any of the following: basic exotic of type G_2, basic exotic of type F_4, type-B or type-C generalized basic exotic, or basic exceptional of type $\mathrm{B}_2 = \mathrm{C}_2$.

Any such k-group G is pseudo-semisimple, and if k' is a field purely inseparable over k then the root system of G_{k_s} coincides with the reduced and irreducible root system of $G'_{k'_s}$. In particular, for general k'/k as in Definition 10.1.13, the group G_{k_s} is a direct product of non-standard pseudo-semisimple k_s-groups with a reduced root system and G is absolutely pseudo-simple precisely when k' is a field purely inseparable over k. Moreover, G is of *minimal type* (see the discussion following Definition 7.1.2) and $G_{\overline{k}}^{\mathrm{ss}}$ is simply connected (by [**CGP**, Thm. 1.6.2(2), Prop. A.4.8]).

REMARK 10.1.14. By Proposition 9.1.13, the maximal smooth k-subgroup $Z_{G,C} = \mathrm{Aut}_{G,C}^{\mathrm{sm}}$ of the scheme of automorphisms of G restricting to the identity on C is always connected (and even pseudo-reductive). For certain *absolutely pseudo-simple* G we have given an explicit description of $Z_{G,C}$ exhibiting the connectedness: $\mathrm{R}_{k'/k}(T'/Z_{G'})$ in the standard case where $\mathrm{R}_{k'/k}(T')$ is the preimage of C in the central extension $\mathrm{R}_{k'/k}(G')$ of G (use Lemma 9.1.9(i), Remark 9.1.14, and Remark 6.1.3), (9.1.13.1) for G of minimal type with root system A_1 over k_s, and Proposition 8.3.12 for G of minimal type with root system BC_n over k_s. Such a description of $Z_{G,C}$ for absolutely pseudo-simple generalized exotic groups with k_s-rank $\geqslant 2$ is given in [**CP**, Prop. 8.5.4].

The property of being generalized exotic is insensitive to scalar extension to k_s. Indeed, this is an immediate consequence of Galois descent and the following result that proves the input data $(k'/k, G')$ is uniquely functorial with respect to isomorphisms among such k-groups $G = \mathscr{D}(\mathrm{R}_{k'/k}(G'))$.

PROPOSITION 10.1.15. *Let k' and ℓ' be nonzero finite reduced k-algebras and let $\mathscr{G}'$ and $\mathscr{H}'$ be groups over k' and ℓ' respectively such that the fiber $\mathscr{G}'_i$ and $\mathscr{H}'_j$ over each respective factor field k'_i and ℓ'_j of k' and ℓ' is absolutely pseudo-simple of minimal type. Assume the root field of $\mathscr{G}'_i$ is equal to k'_i for each i, and that the root field of $\mathscr{H}'_j$ is ℓ'_j for each j.*

Every k-isomorphism $\sigma : G := \mathscr{D}(\mathrm{R}_{k'/k}(\mathscr{G}')) \simeq \mathscr{D}(\mathrm{R}_{\ell'/k}(\mathscr{H}')) =: H$ arises uniquely from a pair (φ, α) consisting of a k-algebra isomorphism $\alpha : k' \simeq \ell'$ and a group isomorphism $\varphi : \mathscr{G}' \simeq \mathscr{H}'$ over α.

PROOF. By Galois descent we may assume $k = k_s$, so each factor field k'_i of k' and ℓ'_j of ℓ' is purely inseparable over k. Thus, the natural map $\mathrm{R}_{k'_i/k}(\mathscr{G}'_i)_{k'_i} \to \mathscr{G}'_i$ is a smooth surjection with connected unipotent kernel [**CGP**, Prop. A.5.11(1),(2)], so the maximal geometric reductive quotient of $\mathrm{R}_{k'_i/k}(\mathscr{G}'_i)$ is the same as that of the absolutely pseudo-simple k'_i-group $\mathscr{G}'_i$. This quotient is perfect, so it is also the maximal geometric reductive quotient of the derived group $G_i := \mathscr{D}(\mathrm{R}_{k'_i/k}(\mathscr{G}'_i))$ by [**CGP**, Prop. A.4.8]. Hence, each G_i is absolutely pseudo-simple over k by Lemma 3.2.1, and is of minimal type (by behavior under Weil restriction and passage to normal subgroups reviewed immediately after Definition 7.1.2); the same likewise holds for $H_j := \mathscr{D}(\mathrm{R}_{\ell'_j/k}(\mathscr{H}'_j))$.

Since $\prod G_i = G$ and $\prod H_j = H$, by Proposition 3.2.2 and dimension considerations $\{G_i\}$ is the set of minimal nontrivial smooth connected normal k-subgroups of G and $\{H_j\}$ is the analogous such set for H. Hence, each k-isomorphism $\sigma : G \simeq H$ arises uniquely from a bijection $\tau : I \simeq J$ and k-isomorphisms $\sigma_i : G_i \simeq H_{\tau(i)}$. This reduces our task to the case where k' and ℓ' are fields.

Using that k' is the root field for $\mathscr{G}'$ by hypothesis, we claim that k'/k is the root field for G. To compute the root field of G, consider the root system $\Phi = \Phi(G,T)$ for a maximal k-torus $T \subset G$; this is naturally identified with the irreducible root system $\Phi(\mathscr{G}', \mathscr{T}')$ for the unique maximal k'-torus $\mathscr{T}' \subset \mathscr{G}'$ such that $T \subset \mathrm{R}_{k'/k}(\mathscr{T}')$. Fix a root $a \in \Phi$ with maximal length (this is any root when Φ is simply laced, and is a divisible root when Φ is non-reduced). Let G_a be the rank-1 pseudo-simple k-subgroup generated by the $\pm a$-root groups of G, and define $\mathscr{G}'_a$ likewise; these have root system A_1 (since a is divisible when Φ is non-reduced). These rank-1 groups inherit the minimal-type property from G and $\mathscr{G}'$ respectively, so $\ker i_{G_a}$ and $\ker i_{\mathscr{G}'_a}$ are trivial (see Example 7.1.4). The root fields of G and $\mathscr{G}'$ coincide with those of G_a and $\mathscr{G}'_a$ respectively: if Φ is reduced then this is part of Theorem 7.4.7, and if Φ is non-reduced then this is Definition 8.3.6. Our task is reduced to showing that G_a has root field k'.

Open cell considerations imply that the roots for $\mathrm{R}_{k'/k}(\mathscr{G}'_a)$ relative to $a^\vee(\mathrm{GL}_1)$ are $\pm a$, and the root groups coincide with those of G_a. Hence, $G_a = \mathscr{D}(\mathrm{R}_{k'/k}(\mathscr{G}'_a))$ by Remark 3.1.11. By Theorem 7.2.5, the group $\mathscr{G}'_a$ with root system A_1 and root field k' has the form $H_{V',K'/k'}$ or $\mathrm{P}H_{V',K'/k'}$ with $\{c \in K' \mid cV' \subset V'\} = k'$ when k' is imperfect of characteristic 2, and $\mathscr{G}'_a$ is equal to SL_2 or PGL_2 otherwise. It is then immediate that G_a has root field k' (using the good behavior of $H_{V',K'/k'}$ and $\mathrm{P}H_{V',K'/k'}$ under $\mathscr{D} \circ \mathrm{R}_{k'/k}$ [**CP**, Ex. 3.1.6] when k is imperfect of characteristic 2).

We have proved that k'/k is the root field for G. Likewise, ℓ'/k is the root field for H. The existence of σ then implies that $k' = \ell'$ as purely inseparable extensions of k. Upon identifying k' and ℓ' uniquely in this manner, the natural maps $G_{k'} \to \mathscr{G}'$ and $H_{k'} \to \mathscr{H}'$ are the maximal pseudo-reductive quotients of minimal type (see [**CP**, Prop. 2.3.13]), so $\sigma_{k'}$ induces a k'-isomorphism φ that is the unique one which does the job. □

By Theorem 7.4.8, in the absolutely pseudo-simple case standardness can only fail over imperfect fields k of characteristic 2 or 3. For any such k, the generalized exotic k-groups account for all deviations from standardness over k in the minimal-type case when the root system over k_s is reduced:

THEOREM 10.1.16. *The non-standard absolutely pseudo-simple k-groups G of minimal type for which G_{k_s} has a reduced root system and $G^{\mathrm{ss}}_{\overline{k}}$ is simply connected are the generalized exotic k-groups that are absolutely pseudo-simple.*

Let us sketch the proof of Theorem 10.1.16. The main point is that Proposition 7.4.5 reduces the problem to the case of G whose root field is k, for which the aim is to show that such G are precisely the groups given by either the basic exotic construction for types F_4 or G_2, the generalized basic exotic construction of types B or C, or the rank-2 basic exceptional construction. We may assume $k = k_s$. Since G is of minimal type and has a reduced root system, $\ker i_G = 1$ (as noted in Example 7.1.4). The root field condition ensures that the minimal field of definition K/k for the geometric unipotent radical of G satisfies $K^p \subset k$ (use Theorem 7.4.7(ii),(iii) if G has rank $\geqslant 2$ and Theorem 7.2.5 in the rank-1 case).

Inspection of open cells shows that the known constructions exhaust all possibilities. More precisely, for types G_2 and F_4 we see via Theorem 7.4.7(ii),(iii) and consideration of a Levi k-subgroup $L \subset G$ that the image of $i_G : G \hookrightarrow \mathrm{R}_{K/k}(L_K)$ is a basic exotic k-group. (This recovers Proposition 7.5.10 with a new proof, but does not recover Corollary 7.5.11.) Theorem 7.2.5 and the explicit description at the end of Remark 7.3.2 settle the rank-1 case, and Theorem 7.5.14 settles types B_n and C_n for $n \geqslant 2$ (using some additional elementary calculations when $n = 2$ to show that the rank-2 basic exceptional construction – which we have not discussed – accounts for the cases that are not generalized basic exotic; see [**CP**, Thm. 8.4.5] for details). This completes our sketch of the proof of Theorem 10.1.16.

10.2. Generalized standard groups. Over every imperfect field k of characteristic $p \in \{2, 3\}$, we have built non-standard pseudo-split absolutely pseudo-simple k-groups realizing all of the exceptional possibilities for the root system. The most concrete non-standard absolutely pseudo-simple groups occur in characteristic 2: the centerless k-groups $\mathrm{SO}(q)$ of type B_n ($n \geqslant 1$) in §7.3, the k-groups $H_{V,K/k}$ of type A_1 in Definition 7.2.1, and the pseudo-split groups of minimal type given by Theorem 7.5.14.

Building on those groups and basic exotic k-groups (Definition 7.5.9), we defined generalized basic exotic k-groups in §10.1 (see Definitions 10.1.6 and 10.1.11). For $p = 2$, birational group laws were used in §8 to build all pseudo-split absolutely pseudo-simple k-groups of minimal type with root system BC_n for any $n \geqslant 1$. We shall combine all of the preceding constructions to generalize the standard construction, and characterize the k-groups obtained in this manner (yielding all pseudo-reductive k-groups except possibly when $\mathrm{char}(k) = 2$ and $[k : k^2] > 2$, as well as all pseudo-reductive k-groups of minimal type without restriction on k).

Consider pairs $(G', k'/k)$ consisting of a nonzero finite reduced k-algebra k' and a smooth affine k'-group G'. We defined the standard construction over k in terms of Weil restrictions $\mathrm{R}_{k'/k}(G')$ with $(G', k'/k)$ for which the fiber G'_i of G' over each factor field k'_i of k' is connected reductive. The pair $(G', k'/k)$ is not determined by the pseudo-reductive k-group G obtained from that construction (e.g., see (2.2.3.1), but recall from Proposition 2.2.7 that we can always arrange for such G'_i to be semisimple, absolutely simple, and simply connected without affecting the standard pseudo-reductive G). Limiting G' as follows will circumvent non-uniqueness.

DEFINITION 10.2.1. For a nonzero finite reduced k-algebra k' and smooth affine k'-group G', the pair $(G', k'/k)$ is *primitive* if the fiber G'_i over each factor field k'_i of k is in any of the following three classes of absolutely pseudo-simple k'_i-groups:

(i) connected semisimple, absolutely simple, and simply connected;
(ii) basic exotic, generalized basic exotic, or rank-2 basic exceptional (as defined in [**CP**, Def. 8.3.6]);
(iii) absolutely pseudo-simple of minimal type with a non-reduced root system over $k'_{i,s}$ and root field equal to k'_i.

(In case (iii), the notion of root field is as in Definition 8.3.6.)

If $(G', k'/k)$ is a primitive pair then the associated pseudo-semisimple k-group

$$\mathscr{G} := \mathscr{D}(\mathrm{R}_{k'/k}(G'))$$

satisfies some good properties: it is of minimal type since that property is preserved under Weil restriction [**CP**, Ex. 2.3.9] and is inherited by smooth connected normal subgroups [**CP**, Lemma 2.3.10], and $\mathscr{G}^{\mathrm{ss}}_{\overline{k}}$ is simply connected (as it is the direct product of the analogous such geometric quotients for the fibers G'_i over the factor fields k'_i of k', due to [**CP**, Prop. 2.3.13]). Moreover:

LEMMA 10.2.2. *The center $Z_{\mathscr{G}}$ is k-tame.*

The idea of the proof of Lemma 10.2.2 is to show that $Z_{\mathscr{G}} \subset \mathrm{R}_{k'/k}(Z_{G'}) = \prod_i \mathrm{R}_{k'_i/k}(Z_{G'_i})$ and $Z_{G'_i}$ is k'_i-tame for each i; see (the proof of) [**CP**, Prop. 9.1.6] for the details.

DEFINITION 10.2.3. A *generalized standard* pseudo-reductive k-group is a k-group that is either commutative pseudo-reductive or of the form

$$G = (\mathscr{G} \rtimes C)/\mathscr{C}$$

where: $\mathscr{G} := \mathscr{D}(\mathrm{R}_{k'/k}(G'))$ for a primitive pair $(G', k'/k)$, $\mathscr{C}$ is the Cartan k-subgroup $\mathscr{G} \cap \mathrm{R}_{k'/k}(C')$ of $\mathscr{G}$ for a Cartan k'-subgroup $C' \subset G'$, C is a commutative pseudo-reductive group fitting into a factorization diagram

$$\mathscr{C} \xrightarrow{\phi} C \xrightarrow{\psi} Z_{\mathscr{G},\mathscr{C}} \tag{10.2.3.1}$$

of the canonical map $\mathscr{C} \to Z_{\mathscr{G},\mathscr{C}}$ arising from the conjugation action of $\mathscr{C}$ on $\mathscr{G}$, and $\mathscr{C}$ is embedded anti-diagonally as a central k-subgroup of $\mathscr{G} \rtimes C$.

In the preceding definition, $Z_{\mathscr{G},\mathscr{C}}$ is the k-group defined as in Proposition 6.1.2; it is commutative and pseudo-reductive (see Proposition 9.1.13). Moreover, the k-group $(\mathscr{G} \rtimes C)/\mathscr{C}$ is automatically pseudo-reductive because it is an instance of the construction in Proposition 2.2.1, and C is a Cartan k-subgroup of G because it is easily seen to be its own centralizer in G (as $\mathscr{C}$ is its own centralizer in $\mathscr{G}$).

EXAMPLE 10.2.4. Every standard pseudo-reductive k-group is generalized standard, due to Proposition 2.2.7. In the context of the standard construction (so only case (i) in Definition 10.2.1), we have $\mathscr{G} = \mathrm{R}_{k'/k}(G')$, $\mathscr{C} = \mathrm{R}_{k'/k}(T')$ for a unique maximal k'-torus $T' \subset G'$, and $Z_{\mathscr{G},\mathscr{C}} = \mathrm{R}_{k'/k}(T'/Z_{G'})$.

PROPOSITION 10.2.5. *If G is a pseudo-reductive k-group then it is generalized standard if and only if $\mathscr{D}(G)$ is generalized standard. Likewise, if $\overline{G} = G/Z$ is a pseudo-reductive central quotient of G then G is generalized standard if and only if $\overline{G}$ is generalized standard. The same holds with "standard" in place of "generalized standard" throughout.*

The first equivalence is [**CP**, Cor. 9.1.14], but the proof below is much simpler.

PROOF. We treat "generalized standard", and the same arguments apply without change for "standard". To prove the first assertion, suppose G is generalized standard, arising from a 4-tuple $(G', k'/k, C', C)$ and factorization diagram as in Definition 10.2.3. Clearly $\mathscr{D}(G)$ arises from $(G', k'/k, C', \phi(\mathscr{C}))$ (with the evident factorization diagram).

Conversely, assume $\mathscr{D}(G)$ arises from a 4-tuple $(G', k'/k, C', C)$ and factorization diagram as in (10.2.3.1). Perfectness of $\mathscr{D}(G)$ forces $C = \phi(\mathscr{C})$, so $\mathscr{D}(G) = \mathscr{G}/(\ker\phi)$ with $\mathscr{G} := \mathscr{D}(\mathrm{R}_{k'/k}(G'))$. Note that $\mathscr{D}(G)$ is a *central* quotient of $\mathscr{G}$, since the conjugation action of $\ker\phi$ on $\mathscr{G}$ is classified by the homomorphism $\psi \circ \phi : \ker\phi \to Z_{\mathscr{G},\mathscr{C}}$ that is trivial.

The Cartan k-subgroup $\mathscr{C}/(\ker\phi) = \phi(\mathscr{C}) = C$ of $\mathscr{D}(G)$ uniquely extends to a Cartan k-subgroup $C^\natural$ of G (see [**CGP**, Lemma 1.2.5(ii),(iii)]), and $G = \mathscr{D}(G) \cdot C^\natural$. The conjugation action of $C^\natural$ on $\mathscr{D}(G)$ is classified by a homomorphism

$$C^\natural \longrightarrow Z_{\mathscr{D}(G),\mathscr{C}/(\ker\phi)} = Z_{\mathscr{G},\mathscr{C}}$$

(equality by Lemma 9.1.9(i)) extending the canonical homomorphism $\mathscr{C} \to Z_{\mathscr{G},\mathscr{C}}$. In this way we get a 4-tuple $(G', k'/k, C', C^\natural)$ and factorization diagram for the generalized standard k-group

$$(\mathscr{G} \rtimes C^\natural)/\mathscr{C} = (\mathscr{D}(G) \rtimes C^\natural)/\phi(\mathscr{C}) = G$$

(as $\mathscr{D}(G) \cap C^\natural = C = \phi(\mathscr{C})$).

Next, we show that the generalized-standard property for G is equivalent to the same for a pseudo-reductive central quotient $\overline{G}$. By the preceding, we may assume G (and hence $\overline{G}$) is perfect. Since a perfect generalized standard k-group is of the form $\mathscr{G}/(\ker\phi)$, in view of the bijection $\mathscr{C} \mapsto \overline{\mathscr{C}} := \mathscr{C}/\mathscr{Z}$ between the sets of Cartan k-subgroups of $\mathscr{G}$ and of any pseudo-reductive central quotient $\mathscr{G}/\mathscr{Z}$ we see immediately that G is generalized standard if and only if $\overline{G}$ is (using the same $(G', k'/k)$ for each). □

If a pseudo-reductive k-group G arises via the generalized standard construction for some 4-tuple $(G', k'/k, C', C)$ then C is identified with a Cartan k-subgroup of G and moreover $\mathscr{D}(G) = \mathscr{D}(\mathrm{R}_{k'/k}(G'))/Z$ for a central k-subgroup $Z := \ker\phi$ that is k-tame by Lemma 10.2.2. In particular, $\mathscr{D}(\mathrm{R}_{k'/k}(G'))$ is *uniquely* determined by G: it is the universal smooth k-tame central extension of $\mathscr{D}(G)$! Taking into consideration Proposition 10.1.15, we obtain:

COROLLARY 10.2.6. *The triple $(G', k'/k, j)$ incorporating the k-homomorphism $j : \mathscr{D}(\mathrm{R}_{k'/k}(G')) \to G$ (a central quotient map onto $\mathscr{D}(G)$) is uniquely determined by G up to unique isomorphism.*

If G is generalized standard then for any 4-tuple $(G', k'/k, C', C)$ giving rise to G via the generalized standard construction, not only is the resulting triple $(G', k'/k, j)$ uniquely determined by G up to unique isomorphism, but it can be arranged that C is *any* Cartan k-subgroup of G that we wish. To prove this, first note that the proof of the initial assertion in Proposition 10.2.5 reduces this to the pseudo-semsimple case by passing to $\mathscr{D}(G)$ because $C \mapsto C \cap \mathscr{D}(G)$ is a bijection between the sets of Cartan k-subgroups of G and of $\mathscr{D}(G)$ (due to [**CGP**, Lemma 1.2.5(ii),(iii)]). Now we may assume G is perfect, so by Lemma 10.2.2 it is generalized standard *if and only if* the universal smooth k-tame central extension $\widetilde{G}$ of G has the form $\mathscr{D}(\mathrm{R}_{k'/k}(G'))$ for a primitive pair $(G', k'/k)$. The set of Cartan k-subgroups C of such a G is in bijective correspondence with the set of Cartan k'-subgroups C' of G' via the condition $C = \mathscr{C}/Z$ for $\mathscr{C} := \widetilde{G} \cap \mathrm{R}_{k'/k}(C')$ and $Z := \ker(\widetilde{G} \twoheadrightarrow G)$, in which case

$$G = \widetilde{G}/Z = (\widetilde{G} \rtimes C)/\mathscr{C}$$

gives a generalized standard description (where $C \to Z_{\mathscr{G},\mathscr{C}}$ arises from the conjugation action of $C = \mathscr{C}/Z$ on $\mathscr{G}$).

We summarize these conclusions as follows, recording the extent of the non-uniqueness of the data giving rise to a specified generalized standard pseudo-reductive k-group.

PROPOSITION 10.2.7. *If G is a generalized standard pseudo-reductive k-group and $(G', k'/k)$ is the associated pair as in Corollary* 10.2.6 *then for every Cartan k-subgroup C of G there exists a unique Cartan k'-subgroup C' of G' such that the central quotient map*

$$\mathscr{G} := \mathscr{D}(\mathrm{R}_{k'/k}(G')) \twoheadrightarrow \mathscr{D}(G)$$

carries $\mathscr{C} := \mathrm{R}_{k'/k}(C') \cap \mathscr{G}$ onto $C \cap \mathscr{D}(G)$. Moreover, for any such C' and C the 4-tuple $(G', k'/k, C', C)$ and the factorization diagram

$$\mathscr{C} \longrightarrow C \longrightarrow Z_{\mathscr{D}(G), C\cap\mathscr{D}(G)} = Z_{\mathscr{G},\mathscr{C}} \tag{10.2.7.1}$$

arising from C-conjugation on $\mathscr{D}(G)$ give rise to G via the generalized standard construction.

The equality in (10.2.7.1) is a special case of Lemma 9.1.9(i). Since we have the flexibility to require that a "generalized standard" description of a given pseudo-reductive k-group rests on a chosen Cartan k-subgroup $C \subset G$ (even prior to knowing that G is generalized standard!), we immediately deduce from Proposition 10.2.7 the following result via Corollary 10.2.6 and Galois descent:

COROLLARY 10.2.8. *A pseudo-reductive k-group G is generalized standard if and only if G_{k_s} is generalized standard, and likewise for "standard" in place of "generalized standard".*

The relevance of Corollary 10.2.6 in the proof of Corollary 10.2.8 is that for a Cartan k-subgroup $C \subset G$, the triple $(H', K'/k_s, i)$ with a primitive pair $(H', K'/k_s)$ corresponding to the generalized standard presentation of G_{k_s} relative to C_{k_s} via Proposition 10.2.7 has a canonically associated k-descent $(G', k'/k, j)$ that together with C underlies a generalized standard presentation for G.

To characterize when a pseudo-reductive k-group is generalized standard, we require a notion that refines "minimal type":

DEFINITION 10.2.9. A pseudo-reductive k-group G is *locally of minimal type* if the subgroup of G_{k_s} generated by any pair of opposite root groups (relative to a maximal k_s-torus) is a central quotient of an absolutely pseudo-simple k_s-group of minimal type.

This condition depends on G only through its derived group (as $\mathscr{D}(G)_{k_s}$ contains all root groups of G_{k_s} relative to a maximal k_s-torus). As the terminology suggests, if G is of minimal type then it is locally of minimal type. Indeed, we may assume $k = k_s$, and then for a maximal k-torus $T \subset G$ the minimal-type property for G is inherited by G_a for each $a \in \Phi(G,T)$ due to Proposition 7.1.5 and the explicit description of G_a in terms of centralizers and derived groups in Remark 3.2.8. Here is a partial converse, refining Lemma 9.1.10.

PROPOSITION 10.2.10. *Let G be a pseudo-semisimple k-group locally of minimal type such that $G^{\rm ss}_{\overline{k}}$ is simply connected. Then G is of minimal type.*

PROOF. We may and do assume $k = k_s$. For a maximal k-torus $T \subset G$, Cartan k-subgroup $C := Z_G(T)$, minimal field of definition K/k for $\mathscr{R}_u(G_{\overline{k}}) \subset G_{\overline{k}}$, and $G' := G_K/\mathscr{R}_{u,K}(G_K)$, we want to prove the triviality of $\mathscr{C}_G := C \cap \ker i_G$ where $i_G : G \to \mathrm{R}_{K/k}(G')$ is the natural map. For any $a \in \Phi(G,T)$, let G_a denote the rank-1 pseudo-simple k-subgroup $\langle U_a, U_{-a}\rangle \subset G$.

The Cartan k-subgroup $C_a := Z_{G_a}(a^\vee(\mathrm{GL}_1))$ of G_a is equal to $C \cap G_a$ (since the isogeny complement $(\ker a)^0_{\rm red} \subset T$ to $a^\vee(\mathrm{GL}_1)$ centralizes G_a), and $G_a \cap \ker i_G = \ker i_{G_a}$ (see Example 7.1.7), so $C_a \cap \ker i_G = \mathscr{C}_{G_a}$. For a basis Δ of $\Phi(G,T)$ we have $T = \prod_{a\in\Delta} a^\vee(\mathrm{GL}_1)$ since $G^{\rm red}_{\overline{k}} = G^{\rm ss}_{\overline{k}}$ is *simply connected.* Hence, the composition

$$\textstyle\prod_{a\in\Delta} C_a \xrightarrow{\ \pi\ } C \xrightarrow{\ i_G|_C\ } \mathrm{R}_{K/k}(T_K) \hookrightarrow \mathrm{R}_{K/k}(G')$$

has kernel $\prod_{a\in\Delta} \mathscr{C}_{G_a}$. The map π is surjective by Lemma 9.1.10, so $\prod_{a\in\Delta} \mathscr{C}_{G_a} \to \mathscr{C}_G$ is surjective. It therefore suffices to prove that each $\mathscr{C}_{G_a}$ is trivial. Since $(G_a)^{\rm ss}_{\overline{k}}$ is generated by a pair of opposite root groups in $G^{\rm ss}_{\overline{k}}$ (Example 7.1.7), and the latter group is simply connected, $(G_a)^{\rm ss}_{\overline{k}} = \mathrm{SL}_2$. Thus, we may replace G with G_a to reduce to the case that G is absolutely pseudo-simple of rank 1.

Now by hypothesis $G = H/Z$ for an absolutely pseudo-simple H of minimal type and $Z \subset Z_H$. Since $G^{\rm ss}_{\overline{k}} = \mathrm{SL}_2$ is simply connected, so the central quotient map $H^{\rm ss}_{\overline{k}} \twoheadrightarrow G^{\rm ss}_{\overline{k}}$ is an isomorphism, the equality of the minimal fields of definition over k for the geometric unipotent radicals of G and H (see Proposition 6.2.2) identifies i_H with $i_G \circ q$ for the *central* quotient map $q : H \twoheadrightarrow G$. It follows that $\mathscr{C}_G = q(\mathscr{C}_H)$ is trivial (as H is of minimal type). □

The "minimal type" property can fail in the standard absolutely pseudo-simple case over every imperfect field (see Example 6.2.6), but "locally minimal type" is truly ubiquitous: it is an immediate consequence of Theorem 7.2.5, Proposition

6.2.15, and Proposition 8.3.9 that every pseudo-reductive k-group G is locally of minimal type *except possibly* when $\mathrm{char}(k) = 2$ and $[k : k^2] > 2$. Here is a large supply of pseudo-reductive groups locally of minimal type over arbitrary fields:

EXAMPLE 10.2.11. If G is a generalized standard pseudo-reductive k-group then it is locally of minimal type. To prove this we may assume $k = k_s$, and by Proposition 10.2.5 we may replace G with $\mathscr{D}(G)$ so that G is perfect. Now $G = \mathscr{G}/Z$ for $\mathscr{G} := \mathscr{D}(\mathrm{R}_{k'/k}(G'))$ with a primitive pair $(G', k'/k)$ and central closed k-subgroup $Z \subset \mathscr{D}(\mathrm{R}_{k'/k}(G'))$. For each factor field k'_i of k', let G'_i be the k'_i-fiber of G'. Consider the pseudo-simple normal k-subgroups $G_i := \mathscr{G}_i/Z_i$ of G, where $\mathscr{G}_i = \mathscr{D}(\mathrm{R}_{k'_i/k}(G'_i))$ and $Z_i = Z \cap \mathscr{G}_i$. Each G_i is generalized standard by normality in G, and these pairwise commute and generate G, so we may treat each G_i separately to reduce to the case that k' is a field.

For a maximal k-torus $T \subset G$, the Cartan k-subgroup $Z_G(T)$ has the form $(\mathscr{G} \cap \mathrm{R}_{k'/k}(C'))/Z$ for a unique Cartan k-subgroup $C' \subset G'$ (argue as in the handling of Cartan subgroups in the proof of Proposition 10.2.5). Thus, upon writing $C' = Z_{G'}(T')$ for a unique maximal k'-torus $T' \subset G'$, we have canonically $\Phi(G, T) = \Phi(G', T')$ since pseudo-reductive central quotients of pseudo-reductive k-groups have the same root system (as we see via consideration of an open cell, for instance).

For each $a \in \Phi(G, T)$ and the corresponding $a' \in \Phi(G', T')$, clearly G_a is a central quotient of $H_a := \mathscr{D}(\mathrm{R}_{k'/k}(G'_{a'}))$. It is therefore enough to show that H_a is of minimal type, and by Proposition 7.1.5 that reduces to $\mathrm{R}_{k'/k}(G'_{a'})$ being of minimal type. Since G' is of minimal type (by inspection of the possibilities for G' in the definition of the generalized standard construction!), so the same holds for $G'_{a'}$ (Example 7.1.7), it suffices to check that $\mathrm{R}_{k'/k}$ preserves the property of being of minimal type, and that in turn is an elementary verification with the definitions (see [**CP**, Ex. 2.3.9] for the details).

The preceding discussion of ubiquity of the "locally minimal type" property is optimal because if $\mathrm{char}(k) = 2$ and $[k : k^2] > 2$ then for any $n \geqslant 1$ there exist pseudo-split absolutely pseudo-simple k-groups G with non-reduced root system BC_n such that G is *not* locally of minimal type. Examples of such G are given in [**CP**, B.4] (built as quotients of k-subgroups of Weil restrictions of symplectic groups, without any appeal to birational group laws).

Continuing to assume $\mathrm{char}(k) = 2$, if we consider pseudo-reductive k-groups G with a *reduced* root system then one can do better with the degree bounds. To be precise, such a G is locally of minimal type whenever $[k : k^2] \leqslant 8$ (see [**CP**, Prop. B.3.1]), but whenever $[k : k^2] \geqslant 16$ there exist pseudo-split absolutely pseudo-simple k-groups G *not* locally of minimal type (see [**CP**, 4.2.2] for examples with root system A_1, and those are used to make others with root system B_n or C_n for any $n \geqslant 2$ in [**CP**, B.1, B.2]).

REMARK 10.2.12. According to Proposition 6.2.9, for a non-reductive absolutely pseudo-simple k-group G, i_G is an isomorphism if and only if G is standard and the order of the fundamental group of $G^{\mathrm{ss}}_{\overline{k}}$ is not divisible by the characteristic p of k. If i_G is an isomorphism then G is clearly of minimal type. If moreover G is standard (which is the case if i_G is an isomorphism), then the root system of G_{k_s} is reduced. Now we will assume that the root system Φ of G_{k_s} is reduced and the order of the fundamental group of $G^{\mathrm{ss}}_{\overline{k}}$ is not divisible by p, and explore when G

fails to be standard (equivalently, i_G fails to be an isomorphism). So let us assume that G is not standard.

Since the root system Φ of G_{k_s} has been assumed to be reduced, Theorem 7.4.8 gives that k is imperfect with $p \in \{2, 3\}$ and that if $p = 3$ then Φ is of type G_2 whereas if $p = 2$ then Φ is of type F_4 or B_n or C_n with some $n \geqslant 1$ (where B_1 and C_1 mean A_1). The group $G_{\overline{k}}^{\mathrm{ss}}$ must be simply connected: this is obvious for types F_4 or G_2, and it holds for types B and C (with $p = 2$) since we assumed that the fundamental group of $G_{\overline{k}}^{\mathrm{ss}}$ has order not divisible by p.

If G is of minimal type then $\ker i_G = 1$ (as noted in Example 7.1.4), so in such cases G must be generalized exotic due to Theorem 10.1.16 and thus i_G is not surjective. The minimal-type property is automatic for types G_2 and F_4 by Corollary 7.5.11. On the other hand, if G_{k_s} has a root system of type B or C (of some rank $n \geqslant 1$), so $p = 2$, then G is locally of minimal type by [**CP**, Prop. B.3.1] (and hence is of minimal type by Proposition 10.2.10 unless $[k : k^2] \geqslant 16$).

As a special case, we recover (with an entirely different proof) Tits' result in [**Ti3**, Cours 1992-93, II] that i_G is an isomorphism when Φ has trivial fundamental group (i.e., types E_8, F_4, and G_2) assuming $p \neq 2$ for F_4 and $p \neq 3$ for G_2.

The appearance of the root systems B_n, C_n, and BC_n ($n \geqslant 1$) in examples of pseudo-split absolutely pseudo-simple groups not locally of minimal type is natural, in view of Theorem 7.4.8 and Corollary 7.5.11. The main reason for our interest in the "locally of minimal type" property is due to:

THEOREM 10.2.13. *A pseudo-reductive group G is generalized standard if and only if it is locally of minimal type.*

PROOF. In Example 10.2.11 we established the implication "$\Rightarrow$". For the converse result we may assume $k = k_s$ (Corollary 10.2.8) and G is perfect (Proposition 10.2.5). Letting $q : \widetilde{G} \twoheadrightarrow G$ be the (pseudo-semisimple) universal smooth k-tame central extension, it suffices to show that $\widetilde{G} = \mathscr{D}(\mathrm{R}_{k'/k}(G'))$ for some primitive pair $(G', k'/k)$. We first check that $\widetilde{G}$ is locally of minimal type.

For a maximal k-torus $\widetilde{T} \subset \widetilde{G}$ and its isogenous image $T \subset G$ we have naturally $\Phi(G, T) = \Phi(\widetilde{G}, \widetilde{T})$ since G is a central pseudo-reductive quotient of $\widetilde{G}$. For each $a \in \Phi(G, T)$ the centrality of $\ker q$ implies (via consideration of open cells) that the a-root group of $\widetilde{G}$ maps onto that of G, so q carries $\widetilde{G}_a$ onto G_a with kernel that is k-tame (since $\ker q$ is k-tame by design). But $(\widetilde{G}_a)_{\overline{k}}^{\mathrm{ss}} = \mathrm{SL}_2$ since this group is generated by a pair of opposite root groups in the connected semisimple group $\widetilde{G}_{\overline{k}}^{\mathrm{ss}}$ that is simply connected (due to the characterization of $\widetilde{G}$). Hence, $\widetilde{G}_a$ is the universal smooth k-tame central extension of G_a.

The k-group G_a is absolutely pseudo-simple of rank 1, and it admits a pseudo-simple central extension of minimal type (as G is assumed to be locally of minimal type). Thus, by a systematic study of the structure of rank-1 pseudo-simple k-groups (via Theorem 7.2.5(i) and Proposition 6.2.2 for root system A_1), the universal smooth k-tame central extension $\widetilde{G}_a$ of G_a is of minimal type; see [**CP**, Lemma 5.3.2] for the details.

We have shown that $\widetilde{G}$ is locally of minimal type, so it is of minimal type by Proposition 10.2.10. By replacing G with $\widetilde{G}$ we may arrange that $G_{\overline{k}}^{\mathrm{ss}}$ is simply connected, and aim to find a primitive pair $(G', k'/k)$ such that $G = \mathscr{D}(\mathrm{R}_{k'/k}(G'))$. The pseudo-simple normal k-subgroups G_i of G are of minimal type, and by Proposition

3.2.4(ii) multiplication $\pi : \prod G_i \to G$ is a surjective homomorphism with central kernel. But $G^{\rm ss}_{\overline{k}}$ is simply connected, so π is an isomorphism by Lemma 9.1.10. We may therefore treat each G_i separately, so now G is (absolutely) pseudo-simple.

Consider the irreducible root system Φ of G. Since G is absolutely pseudo-simple and *of minimal type*, if Φ is non-reduced then G has the desired form (using Weil restriction from its root field) due to Proposition 8.3.7.

Assume instead that Φ is reduced. We shall separately treat the cases that G is standard or not standard. Suppose G is standard. By inspection, G is a central (pseudo-reductive) quotient of $\mathrm{R}_{k'/k}(G')$ for a finite extension k'/k and connected semisimple k'-group G' that is *simply connected.* But then the minimal fields of definition over k for the geometric unipotent radicals of G and $\mathrm{R}_{k'/k}(G')$ coincide by Proposition 6.2.2, so this common field is equal to k'/k (see [**CGP**, Thm. 1.6.2(2)]). Hence, $G^{\rm ss}_{k'}$ is a central quotient of G' yet is simply connected (as it is a k'-descent of $G^{\rm ss}_{\overline{k}}$), so naturally $G' \simeq G^{\rm ss}_{k'}$ and we get a factorization

$$\mathrm{R}_{k'/k}(G') \twoheadrightarrow G \stackrel{i_G}{\hookrightarrow} \mathrm{R}_{k'/k}(G')$$

of the identity map. This forces i_G to be an isomorphism.

Finally, assume G is not standard. In this case k must be imperfect of characteristic 2 or 3 (by Theorem 7.4.8) and the non-standard absolutely pseudo-simple k-groups G of minimal type with $G^{\rm ss}_{\overline{k}}$ simply connected are given as in Theorem 10.1.16. Hence, G has the desired form due to Definition 10.2.1. $\square$

An application of Theorem 10.2.13 and our preceding discussion of all cases of failure of the "locally minimal type" property, we see that if G_{k_s} has a reduced root system then G is generalized standard except possibly when $\mathrm{char}(k) = 2$ and $[k : k^2] \geqslant 16$, and that whenever $\mathrm{char}(k) = 2$ and $[k : k^2] \geqslant 16$ there exist pseudo-split absolutely pseudo-simple k-groups that are *not* generalized standard (with any desired root system of type B or C with any rank $n \geqslant 1$). In particular:

Corollary 10.2.14. *Every pseudo-reductive k-group is standard except possibly if k is imperfect with* $\mathrm{char}(k) = p \in \{2, 3\}$ *and one of the following holds: the root system Φ of G_{k_s} is non-reduced (only possible when $p = 2$), some irreducible component of Φ has an edge of multiplicity p, or $p = 2$ with $[k : k^2] > 2$ and Φ has an irreducible component of type* A_1.

The proof of Corollary 10.2.14 when $\mathrm{char}(k) \neq 2, 3$ or in characteristic $p \in \{2, 3\}$ with $[k : k^p] = p$ has nothing to do with non-reduced root systems or the explicit non-standard constructions that occupied much of §7–§8.

Acknowledgements

The authors are grateful to the Tulane University Mathematics Department for its warm hospitality in hosting the Clifford Lectures, to Michel Brion and Mahir Can for providing us with the opportunity to speak on our work, and to the referee for careful reading and valuable comments and corrections on an earlier version. B.C. was partially supported by NSF grant DMS-1100784 and G.P. was partially supported by NSF grant DMS-1401380.

References

[Bo1] Armand Borel, *Some finiteness properties of adele groups over number fields*, Inst. Hautes Études Sci. Publ. Math. **16** (1963), 5–30. MR0202718

[Bo2] Armand Borel, *Linear algebraic groups*, 2nd ed., Graduate Texts in Mathematics, vol. 126, Springer-Verlag, New York, 1991. MR1102012

[BP] Armand Borel and Gopal Prasad, *Finiteness theorems for discrete subgroups of bounded covolume in semi-simple groups*, Inst. Hautes Études Sci. Publ. Math. **69** (1989), 119–171. MR1019963

[BS] A. Borel and J.-P. Serre, *Théorèmes de finitude en cohomologie galoisienne* (French), Comment. Math. Helv. **39** (1964), 111–164. MR0181643

[BoTi1] Armand Borel and Jacques Tits, *Groupes réductifs* (French), Inst. Hautes Études Sci. Publ. Math. **27** (1965), 55–150. MR0207712

[BoTi2] Armand Borel and Jacques Tits, *Homomorphismes "abstraits" de groupes algébriques simples* (French), Ann. of Math. (2) **97** (1973), 499–571. MR0316587

[BoTi3] Armand Borel and Jacques Tits, *Théorèmes de structure et de conjugaison pour les groupes algébriques linéaires* (French, with English summary), C. R. Acad. Sci. Paris Sér. A-B **287** (1978), no. 2, A55–A57. MR0491989

[BLR] Siegfried Bosch, Werner Lütkebohmert, and Michel Raynaud, *Néron models*, Ergebnisse der Mathematik und ihrer Grenzgebiete (3) [Results in Mathematics and Related Areas (3)], vol. 21, Springer-Verlag, Berlin, 1990. MR1045822

[Bou] Nicolas Bourbaki, *Lie groups and Lie algebras. Chapters 4–6*, Elements of Mathematics (Berlin), Springer-Verlag, Berlin, 2002. Translated from the 1968 French original by Andrew Pressley. MR1890629

[BrTi] F. Bruhat and J. Tits, *Groupes réductifs sur un corps local* (French), Inst. Hautes Études Sci. Publ. Math. **41** (1972), 5–251. MR0327923

[Chev] C. Chevalley, *Une démonstration d'un théorème sur les groupes algébriques* (French), J. Math. Pures Appl. (9) **39** (1960), 307–317. MR0126447

[C1] Brian Conrad, *A modern proof of Chevalley's theorem on algebraic groups*, J. Ramanujan Math. Soc. **17** (2002), no. 1, 1–18. MR1906417

[C2] Brian Conrad, *Finiteness theorems for algebraic groups over function fields*, Compos. Math. **148** (2012), no. 2, 555–639, DOI 10.1112/S0010437X11005665. MR2904198

[C3] Brian Conrad, *Reductive group schemes* (English, with English and French summaries), Autour des schémas en groupes. Vol. I, Panor. Synthèses, vol. 42/43, Soc. Math. France, Paris, 2014, pp. 93–444. MR3362641

[CGP] Brian Conrad, Ofer Gabber, and Gopal Prasad, *Pseudo-reductive groups*, 2nd ed., New Mathematical Monographs, vol. 26, Cambridge University Press, Cambridge, 2015. MR3362817

[CP] Brian Conrad and Gopal Prasad, *Classification of pseudo-reductive groups*, Annals of Mathematics Studies, vol. 191, Princeton University Press, Princeton, NJ, 2016. MR3379926

[SGA3] M. Demazure, A. Grothendieck, *Schémas en groupes* I, II, III, Lecture Notes in Math **151, 152, 153**, Springer-Verlag, New York (1970).

[GQ] Philippe Gille and Anne Quéguiner-Mathieu, *Formules pour l'invariant de Rost* (French, with English and French summaries), Algebra Number Theory **5** (2011), no. 1, 1–35, DOI 10.2140/ant.2011.5.1. MR2833783

[EGA] A. Grothendieck, *Eléments de Géométrie Algébrique*, Publ. Math. IHES **4, 8, 11, 17, 20, 24, 28, 32**, 1960–7.

[Ha1] G. Harder, *Minkowskische Reduktionstheorie über Funktionenkörpern* (German), Invent. Math. **7** (1969), 33–54. MR0284441

[Ha2] G. Harder, *Über die Galoiskohomologie halbeinfacher algebraischer Gruppen. III* (German), J. Reine Angew. Math. **274/275** (1975), 125–138. Collection of articles dedicated to Helmut Hasse on his seventy-fifth birthday, III. MR0382469

[Hum1] James E. Humphreys, *Introduction to Lie algebras and representation theory*, Springer-Verlag, New York-Berlin, 1972. Graduate Texts in Mathematics, Vol. 9. MR0323842

[Hum2] James E. Humphreys, *Linear algebraic groups*, Springer-Verlag, New York-Heidelberg, 1975. Graduate Texts in Mathematics, No. 21. MR0396773

[Kem] George R. Kempf, *Instability in invariant theory*, Ann. of Math. (2) **108** (1978), no. 2, 299–316, DOI 10.2307/1971168. MR506989

[Oes] Joseph Oesterlé, *Nombres de Tamagawa et groupes unipotents en caractéristique p* (French), Invent. Math. **78** (1984), no. 1, 13–88, DOI 10.1007/BF01388714. MR762353

[P] Gopal Prasad, *Weakly-split spherical Tits systems in quasi-reductive groups*, Amer. J. Math. **136** (2014), no. 3, 807–832, DOI 10.1353/ajm.2014.0017. MR3214277

[PR] Gopal Prasad and M. S. Raghunathan, *Tame subgroup of a semisimple group over a local field*, Amer. J. Math. **105** (1983), no. 4, 1023–1048, DOI 10.2307/2374303. MR708372

[R] Bertrand Rémy, *Groupes algébriques pseudo-réductifs et applications (d'après J. Tits et B. Conrad–O. Gabber–G. Prasad)* (French, with French summary), Astérisque **339** (2011), Exp. No. 1021, viii–ix, 259–304. Séminaire Bourbaki. Vol. 2009/2010. Exposés 1012–1026. MR2906357

[Sel] Martin Selbach, *Klassifikationstheorie halbeinfacher algebraischer Gruppen*, Mathematisches Institut der Universität Bonn, Bonn, 1976. Diplomarbeit, Univ. Bonn, Bonn, 1973; Bonner Mathematische Schriften, Nr. 83. MR0432776

[Ser] Jean-Pierre Serre, *Local fields*, Graduate Texts in Mathematics, vol. 67, Springer-Verlag, New York-Berlin, 1979. Translated from the French by Marvin Jay Greenberg. MR554237

[Spr] T. A. Springer, *Linear algebraic groups*, 2nd ed., Progress in Mathematics, vol. 9, Birkhäuser Boston, Inc., Boston, MA, 1998. MR1642713

[St] Robert Steinberg, *The isomorphism and isogeny theorems for reductive algebraic groups*, J. Algebra **216** (1999), no. 1, 366–383, DOI 10.1006/jabr.1998.7776. MR1694546

[Ti1] J. Tits, *Classification of algebraic semisimple groups*, Algebraic Groups and Discontinuous Subgroups (Proc. Sympos. Pure Math., Boulder, Colo., 1965), Amer. Math. Soc., Providence, R.I., 1966, 1966, pp. 33–62. MR0224710

[Ti2] J. Tits, *Lectures on algebraic groups*, Yale Univ., New Haven, 1967.

[Ti3] Jacques Tits, *Résumés des cours au Collège de France 1973–2000* (French), Documents Mathématiques (Paris) [Mathematical Documents (Paris)], 12, Société Mathématique de France, Paris, 2013. MR3235648

Index

Stanford University, Stanford, California 94305
E-mail address: conrad@math.stanford.edu

University of Michigan, Ann Arbor, Michigan 48109
E-mail address: gprasad@umich.edu

Proceedings of Symposia in Pure Mathematics
Volume **94**, 2017
http://dx.doi.org/10.1090/pspum/094/01624

Invariants of algebraic groups and retract rationality of classifying spaces

Alexander S. Merkurjev

1. Introduction

Let G be an algebraic group over a field F, V a generically free representation of G (i.e., the stabilizer of the generic point in V is trivial) and $U \subset V$ a G-equivariant open subset such that there is a G-torsor $f : U \longrightarrow U/G$. This is a *versal* G-torsor, in particular, every G-torsor over a field extension K/F with K infinite is isomorphic to the fiber of f over a K-point of U/G. Thus, the K-points of U/G parameterize all G-torsors over $\mathrm{Spec}(K)$.

We think of U/G as an approximation of the classifying space $\mathrm{B}G$. The stable birational and retract rational equivalence classes of U/G are independent of the choice of V and U. We simply say that $\mathrm{B}G$ is *stably rational* (respectively, *retract rational*) if so is U/G. In these cases all the G-torsors over field extensions of F can be parameterized by algebraically independent variables.

The stable (retract) non-rationality of $\mathrm{B}G$ can be detected by *cohomological invariants* which were introduced by J.-P. Serre in [**27**]. A cohomological invariant of an algebraic group G over a field F with coefficients in a Galois module M over F assigns naturally to every G-torsor over a field extension K/F a Galois cohomology class of K with coefficients in M. A cohomological invariant I is called *unramified* if all values of I over a field extension K/F are unramified with respect to all discrete valuations of K over F. A relation between retract rationality of $\mathrm{B}G$ and unramified invariants is given by the following statement:

If G admits a non-constant unramified cohomological invariant, then the classifying space $\mathrm{B}G$ is not retract rational.

For example, this was used by D. Saltman who disproved Noether's Conjecture on the rationality of classifying spaces of finite groups by showing that certain finite groups admit a non-constant degree 2 unramified cohomological invariant.

It is still an open problem whether there exists a connected algebraic group G over an algebraically closed field such that $\mathrm{B}G$ is not retract rational.

In the present paper we review and slightly improve some classical results on the properties of classifying spaces and cohomological invariants. We don't impose any restrictions on the ground field F. The coefficient module $\mathbb{Q}/\mathbb{Z}(j)$ includes

---

2010 *Mathematics Subject Classification.* Primary 14E08; Secondary 11R34, 20G15.
The work has been supported by the NSF grant DMS #1160206.

nontrivial p-primary component if $p = \mathrm{char}(F) > 0$. In particular, Galois cohomology groups with values in $\mathbb{Q}/\mathbb{Z}(j)$ do not form a cycle module of M. Rost if $\mathrm{char}(F) > 0$ since the residue homomorphisms are not always defined. Moreover, the étale cohomology groups with coefficients in $\mathbb{Q}_p/\mathbb{Z}_p(j)$ are not homotopy invariant. All this makes some proofs more involved. We also present a new simpler proof of Rost's theorem on the determination of an invariant by its value at the generic torsor. In particular, non-smooth algebraic groups are allowed. We also consider retract rational varieties over arbitrary fields and give a classification of degree 1 invariants.

A *variety* over a field F in the paper is an integral separated scheme of finite type over F. If R is a commutative F-algebra and X a variety over F, we write $X(R)$ for the set $\mathrm{Mor}_F(\mathrm{Spec}\, R, X)$ of R-points of X.

An *algebraic group* is an affine group scheme of finite type over a field, not necessarily smooth or connected.

2. Galois cohomology

2.1. The complexes $\mathbb{Q}/\mathbb{Z}(j)$. For every $j \in \mathbb{Z}$, the complex $\mathbb{Q}/\mathbb{Z}(j)$ in the derived category of sheaves of abelian groups on the big étale site of $\mathrm{Spec}\, F$ is defined as the direct sum of two complexes. The first complex is the sheaf

$$\operatorname*{colim}_n(\mu_n^{\otimes j})$$

placed in degree 0, where $\mu_n^{\otimes j}$ is the jth tensor power of the Galois module μ_n of nth roots of unity. The second complex is nontrivial only in the case $p = \mathrm{char}(F) > 0$ and it is defined via logarithmic de Rham-Witt differentials (see [**16**, I.5.7] or [**17**]). In particular, $\mathbb{Q}/\mathbb{Z}(0) = \mathbb{Q}/\mathbb{Z}$ and the p-part of $\mathbb{Q}/\mathbb{Z}(j)$ is trivial if $j < 0$.

For a scheme X over F, we write $H^n\big(X, \mathbb{Q}/\mathbb{Z}(j)\big)$ for the degree n étale cohomology group of X with values in $\mathbb{Q}/\mathbb{Z}(j)$. For example,

$$H^2\big(X, \mathbb{Q}/\mathbb{Z}(1)\big) = H^2_{\text{ét}}(X, \mathbf{G}_{\mathrm{m}}) = \mathrm{Br}(X)$$

is the (cohomological) Brauer group of X.

2.2. Residue homomorphisms. For a variety X over F and a closed subscheme $Z \subset X$ we write $H^n_Z\big(X, \mathbb{Q}/\mathbb{Z}(j)\big)$ for the étale cohomology group of X with support in Z and values in $\mathbb{Q}/\mathbb{Z}(j)$ (see [**23**, Ch. III, §1]). Let $X^{(i)}$ be the set of points in X of codimension i. For a point $x \in X^{(1)}$ set

$$H^n_x\big(X, \mathbb{Q}/\mathbb{Z}(j)\big) = \operatorname*{colim}_{x \in W} H^n_{\overline{\{x\}} \cap W}\big(W, \mathbb{Q}/\mathbb{Z}(j)\big),$$

where the colimit is taken over all open subsets $W \subset X$ containing x. Write

$$\partial_x : H^n\big(F(X), \mathbb{Q}/\mathbb{Z}(j)\big) \longrightarrow H^{n+1}_x\big(X, \mathbb{Q}/\mathbb{Z}(j)\big)$$

for the *residue* homomorphisms arising from the *coniveau spectral sequence* [**8**, §1.2].

REMARK 2.1. If l is a prime integer different from $\mathrm{char}(F)$, then by purity [**23**, Chapter VI, §5], the primary l-component of $H^{n+1}_x\big(X, \mathbb{Q}/\mathbb{Z}(j)\big)$ is isomorphic to $H^{n-1}\big(F(x), \mathbb{Q}_l/\mathbb{Z}_l(j-1)\big)$ and the l-component of ∂_x is the standard residue homomorphism of a cycle module (see [**24**]).

If X is a smooth variety over F and $x \in X^{(1)}$, the sequence

$$0 \longrightarrow H^n\big(O_{X,x}, \mathbb{Q}/\mathbb{Z}(j)\big) \longrightarrow H^n\big(F(X), \mathbb{Q}/\mathbb{Z}(j)\big) \xrightarrow{\partial_x} H^{n+1}_x\big(X, \mathbb{Q}/\mathbb{Z}(j)\big)$$

is exact (see [**8**, Proposition 2.1.2]).

2.3. The sheaves $\mathcal{H}^n(\mathbb{Q}/\mathbb{Z}(j))$. Let X be a smooth variety over F. Write $\mathcal{H}^n(\mathbb{Q}/\mathbb{Z}(j))$ for the Zariski sheaf on X associated with the presheaf

$$U \mapsto H^n(U, \mathbb{Q}/\mathbb{Z}(j))$$

of the étale cohomology groups. Pulling back to the generic point of X yields an exact sequence

$$0 \longrightarrow H^0_{\mathrm{Zar}}(X, \mathcal{H}^n(\mathbb{Q}/\mathbb{Z}(j))) \longrightarrow H^n(F(X), \mathbb{Q}/\mathbb{Z}(j)) \xrightarrow{\partial} \coprod_{x \in X^{(1)}} H^{n+1}_x(X, \mathbb{Q}/\mathbb{Z}(j)),$$

where $\partial = \coprod \partial_x$ (see [**8**, §2.1]).

2.4. Unramified cohomology. Let K/F be a field extension and v a discrete valuation of K over F with valuation ring R. Following [**5**] and [**7**], we say that an element $a \in H^n(K, \mathbb{Q}/\mathbb{Z}(j))$ is *unramified with respect to* v if a belongs to the image of the map $H^n(R, \mathbb{Q}/\mathbb{Z}(j)) \longrightarrow H^n(K, \mathbb{Q}/\mathbb{Z}(j))$. We write $H^n_{\mathrm{nr}}(K, \mathbb{Q}/\mathbb{Z}(j))$ for the subgroup of all elements in $H^n(K, \mathbb{Q}/\mathbb{Z}(j))$ that are unramified with respect to all discrete valuations of K over F. We have a natural homomorphism

$$H^n(F, \mathbb{Q}/\mathbb{Z}(j)) \longrightarrow H^n_{\mathrm{nr}}(K, \mathbb{Q}/\mathbb{Z}(j)). \tag{1}$$

The following statement is well-known (see [**5**, Theorem 4.1.5]) if one deletes the p primary component of $\mathbb{Q}/\mathbb{Z}(j)$ in the case $\operatorname{char}(F) = p > 0$.

PROPOSITION 2.2 ([**2**, Proposition 5.1]). *Let K/F be a purely transcendental field extension. Then the homomorphism (1) is an isomorphism.*

LEMMA 2.3. *Let K be an extension of an infinite field F and $a \in H^n(K, \mathbb{Q}/\mathbb{Z}(j))$ such that $a_{K(t)} \in H^n_{\mathrm{nr}}(K(t), \mathbb{Q}/\mathbb{Z}(j))$. Then $a \in H^n_{\mathrm{nr}}(K, \mathbb{Q}/\mathbb{Z}(j))$.*

PROOF. We simply write $H(-)$ for $H^n(-, \mathbb{Q}/\mathbb{Z}(j))$. Let v be a discrete valuation of K over F, $R \subset K$ the discrete valuation ring of v, M maximal ideal of R and $\widetilde{R}$ the localization of the polynomial ring $R[t]$ at the prime ideal $M[t]$. Then $\widetilde{R}$ is a discrete valuation ring with quotient field $F(t)$. By assumption, $a_{K(t)}$ belongs to the image of $H(\widetilde{R}) \longrightarrow H(K(t))$. As the étale cohomology commutes with colimits by [**23**, Ch. III, Lemma 1.16], there is a polynomial $f \in R[t] \setminus M[t]$ such that the image of a in $H(K[t]_f)$ belongs to the image of $\gamma^* : H(R[t]_f) \longrightarrow H(K[t]_f)$, where γ is the embedding of $R[t]_f$ into $K[t]_f$.

Since the residue field R/M contains F, it is infinite by the assumption. Therefore, there exists an element $r \in R$ such that $f(r) \in R^\times$. Consider the diagram

$$\begin{array}{ccc} R[t]_f & \xrightarrow{\ \gamma\ } & K[t]_f \\ {\scriptstyle \alpha_R}\uparrow\downarrow{\scriptstyle \beta_R} & & {\scriptstyle \alpha_K}\uparrow\downarrow{\scriptstyle \beta_K} \\ R & \xrightarrow{\ \delta\ } & K \end{array}$$

with the evaluation homomorphisms $t \mapsto r$ from the top row to the bottom row and the other homomorphisms the natural inclusions. We have $\alpha^*_K(a) = \gamma^*(b)$ for some $b \in H(R[t]_f)$ and therefore,

$$a = \beta^*_K(\alpha^*_K(a)) = \beta^*_K(\gamma^*(b)) = \delta^*(\beta^*_R(b)) \in \operatorname{Im}(\delta^*).$$

Hence a is unramified with respect to v. □

Let X be a smooth variety over F and $x \in X^{(1)}$. Then the local ring $O_{X,x}$ is a discrete valuation ring and let v_x be the corresponding discrete valuation on $F(X)$. It follows from the exact sequence in §2.2 that an element $a \in H^n\big(F(X), \mathbb{Q}/\mathbb{Z}(j)\big)$ is unramified with respect to the discrete valuation v_x if and only if $\partial_x(a) = 0$.

As shown in §2.3, $H^0_{\mathrm{Zar}}\big(X, \mathcal{H}^n(\mathbb{Q}/\mathbb{Z}(j))\big)$ is identified with the subgroup of all elements in $H^n\big(F(X), \mathbb{Q}/\mathbb{Z}(j)\big)$ that are unramified with respect to v_x for all $x \in X^{(1)}$. Moreover, we have

$$H^n_{\mathrm{nr}}\big(F(X), \mathbb{Q}/\mathbb{Z}(j)\big) \subset H^0_{\mathrm{Zar}}\big(X, \mathcal{H}^n(\mathbb{Q}/\mathbb{Z}(j))\big) \subset H^n\big(F(X), \mathbb{Q}/\mathbb{Z}(j)\big).$$

3. Retract rational varieties

The notion of a retract rational field was introduced in [**26**]. Retract rational varieties were defined in [**9**, §1]. We extend the definition and the properties of retract rational varieties to the case of an arbitrary base field (not necessarily infinite).

Let X be a variety over a field F. For a (commutative) local F-algebra R with residue field $\overline{R}$, we say that X *has the R-lifting property* if there is a nonempty open subset $W \subset X$ such that the map $W(R) \longrightarrow W(\overline{R})$ is surjective.

Note that if $W' \subset W$ are two open subsets of X and w in $W'(\overline{R})$ belongs to the image of $W(R) \longrightarrow W(\overline{R})$, then $w \in \operatorname{Im}\big(W'(R) \longrightarrow W'(\overline{R})\big)$ since R is a local ring.

It follows that if X has the R-lifting property and a variety Y is birationally isomorphic to X, then Y also has the R-lifting property.

The following statement is a slight generalization of [**9**, Proposition 1.2].

PROPOSITION 3.1. *Let X be a variety over F. Then the following conditions are equivalent:*

1. *X has the R-lifting property for all local F-algebras R.*
2. *X has the R-lifting property for all R with infinite residue field $\overline{R}$.*
3. *For every local F-algebra R with residue field $F(X)$, the generic point of X belongs to the image of the map $X(R) \longrightarrow X(F(X))$.*
4. *There is a morphism $f : Y \longrightarrow X$, where Y is an open subset of the affine space $\mathbb{A}^n_F$ for some n, and a rational map $g : X \dashrightarrow Y$ such that $f \circ g = 1_X$.*
5. *There is a morphism $f : Y' \longrightarrow W$, where Y' is an open subset of the affine space $\mathbb{A}^n_F$ for some n and $W \subset X$ a nonempty open subset such that the map $Y'(K) \longrightarrow W(K)$ is surjective for every field extension K/F.*

PROOF. (1) $\Rightarrow$ (2) is trivial.

(2) $\Rightarrow$ (3): We may assume that $\overline{R} = F(X)$ is a finite field. In this case $X = \operatorname{Spec}(\overline{R})$. If $\overline{R} = F$, i.e., $X = \operatorname{Spec}(F)$, then X has the R-lifting property. We show that the case $\overline{R} \neq F$ does not occur. Let S be a localization of the polynomial ring $F[x_1, \dots, x_n]$ with respect to a prime ideal such that the residue field is isomorphic to $\overline{R}(t)$. If $\overline{R} \neq F$, i.e., $\overline{R}/F$ is a nontrivial finite field extension, we have $X(\overline{R}(t)) \neq \emptyset$ and $X(S) = \emptyset$ as F is algebraically closed in S. Hence X does not have the S-lifting property for the local ring S with infinite residue field, a contradiction.

(3) $\Rightarrow$ (4): Choose an F-algebra homomorphism $\alpha : F[x_1, \dots, x_n] \longrightarrow F(X)$ such that the quotient field of the image of α is equal to $F(X)$. Let $P = \operatorname{Ker}(\alpha)$. The F-algebra $R = F[x_1, \dots, x_n]_P$ is a local F-algebra with residue field $F(X)$.

Let $W \subset X$ be the open subset in the definition of the R-lifting property. By assumption, the generic point of X is in the image of the map $W(R) \longrightarrow W(F(X)) \subset X(F(X))$. Therefore, there exists a morphism $\beta : \operatorname{Spec}(R) \longrightarrow W$ such that the composition

$$\operatorname{Spec} F(X) \xrightarrow{\gamma} \operatorname{Spec}(R) \xrightarrow{\beta} W \hookrightarrow X$$

is the generic point of X. The map β yields a morphism

$$f : Y \longrightarrow W \hookrightarrow X,$$

where $Y = \operatorname{Spec} F[x_1, \ldots, x_n]_h$ for some $h \in F[x_1, \ldots, x_n] \setminus P$. The map γ gives a rational map $g : X \dashrightarrow Y$ with the required property.

(4) $\Rightarrow$ (1): Let $W \subset X$ be the domain of definition of g. The composition $W \xrightarrow{g} Y \xrightarrow{f} X$ is the inclusion. Let R be a local F-algebra. In the commutative diagram

$$\begin{array}{ccccc} W(R) & \xrightarrow{g} & Y(R) & \xrightarrow{f} & X(R) \\ {\scriptstyle a}\downarrow & & {\scriptstyle b}\downarrow & & {\scriptstyle c}\downarrow \\ W(\overline{R}) & \xrightarrow{g} & Y(\overline{R}) & \xrightarrow{f} & X(\overline{R}) \end{array}$$

the map b is surjective since $\mathbb{A}^n_F(R) \longrightarrow \mathbb{A}^n_F(\overline{R})$ is surjective, Y is open in $\mathbb{A}^n_F$ and R is local. It follows that every point in $W(\overline{R})$ is in the image of c and hence is in the image of a since R is local.

(4) $\Rightarrow$ (5): Let W be the domain of definition of g and let $Y' = f^{-1}(W)$. Then the composition $W \longrightarrow Y' \longrightarrow W$ is the identity. It follows that the map $Y'(K) \longrightarrow W(K)$ is surjective for every field extension K/F.

(5) $\Rightarrow$ (4): A lift in $Y'(F(W))$ of the generic point from $W(F(W))$ yields a rational map $X \dashrightarrow Y'$ such that the composition $X \dashrightarrow Y' \longrightarrow W \hookrightarrow X$ is the identity. $\square$

A variety X is called *retract rational* if X satisfies the equivalent conditions of Proposition 3.1. If X is retract rational and a variety Y is birationally isomorphic to X, then Y is also retract rational.

LEMMA 3.2. *A variety X over F is retract rational if and only if $X \times \mathbb{A}^1_F$ is retract rational.*

PROOF. Suppose $X \times \mathbb{A}^1_F$ is retract rational and let $U \subset X \times \mathbb{A}^1_F$ be a nonempty subset such that the map $U(R) \longrightarrow U(\overline{R})$ is surjective for all local F algebras R. Let W be the image of U under the projection $X \times \mathbb{A}^1_F \longrightarrow X$. As the projection is a flat morphism, W is a nonempty open subset of X.

Let R be a local F-algebra with infinite residue field $\overline{R}$. The fiber of the projection $U \longrightarrow W$ over a point $x \in W(\overline{R})$ is a nonempty open subset of $\mathbb{A}^1_{\overline{R}}$. As the field $\overline{R}$ is infinite, the fiber has an $\overline{R}$-point. It follows that there is a point in $U(\overline{R})$ over x.

The top map in the diagram

$$\begin{array}{ccc} U(R) & \longrightarrow & U(\overline{R}) \\ \downarrow & & \downarrow \\ W(R) & \longrightarrow & W(\overline{R}) \end{array}$$

is surjective. It follows that x is in the image of the bottom map. By Proposition 3.1(2), X is retract rational. □

COROLLARY 3.3. *Let X be a retract rational variety and Y a variety stably birationally isomorphic to X. Then Y is also retract rational. A stably rational variety is retract rational.*

PROPOSITION 3.4. *Let X be a smooth variety over F. Consider the following properties of X:*

(1) *X is a rational variety.*
(2) *X is a stably rational variety.*
(3) *X is a retract rational variety.*
(4) *The natural homomorphism*

$$H^n(F, \mathbb{Q}/\mathbb{Z}(j)) \longrightarrow H^n_{\mathrm{nr}}(F(X), \mathbb{Q}/\mathbb{Z}(j))$$

is an isomorphism.

Then (1) $\Rightarrow$ (2) $\Rightarrow$ (3) $\Rightarrow$ (4).

PROOF. (1) $\Rightarrow$ (2) is obvious.
(2) $\Rightarrow$ (3) is proved in Corollary 3.3.
(3) $\Rightarrow$ (4): Since X is retract rational, there is a (dominant) morphism $f : Y \longrightarrow X$, where Y is an open subset of the affine space $\mathbb{A}^n_F$ for some n, and a rational map $g : X \dashrightarrow Y$ such that $f \circ g = 1_X$. Let $W \subset X$ be the domain of definition of g. We have the following commutative diagram with all vertical maps injective:

$$\begin{array}{ccccc}
H^n_{\mathrm{nr}}(F(X), \mathbb{Q}/\mathbb{Z}(j)) & \xrightarrow{\alpha} & H^n_{\mathrm{nr}}(F(Y), \mathbb{Q}/\mathbb{Z}(j)) & & \\
\downarrow & & \downarrow & & \\
H^0_{\mathrm{Zar}}(X, \mathcal{H}^n(\mathbb{Q}/\mathbb{Z}(j))) & \xrightarrow{\beta} & H^0_{\mathrm{Zar}}(Y, \mathcal{H}^n(\mathbb{Q}/\mathbb{Z}(j))) & \longrightarrow & H^0_{\mathrm{Zar}}(W, \mathcal{H}^n(\mathbb{Q}/\mathbb{Z}(j))) \\
\downarrow & & & & \downarrow \\
H^n(F(X), \mathbb{Q}/\mathbb{Z}(j)) & & = & & H^n(F(W), \mathbb{Q}/\mathbb{Z}(j)).
\end{array}$$

By diagram chase, β is injective. It follows that α is injective. By Proposition 2.2, $H^n_{\mathrm{nr}}(F(Y), \mathbb{Q}/\mathbb{Z}(j)) = H^n(F, \mathbb{Q}/\mathbb{Z}(j))$. Therefore, $H^n_{\mathrm{nr}}(F(X), \mathbb{Q}/\mathbb{Z}(j)) = H^n(F, \mathbb{Q}/\mathbb{Z}(j))$. □

4. Retract rational classifying spaces

4.1. Versal torsors. Let G be an algebraic group over a field F. A G-torsor E over a variety X is called *weakly versal* if every G-torsor $T \longrightarrow \mathrm{Spec}(K)$ for a field extension K/F with K infinite is isomorphic to the pull-back of E with respect to a point $\mathrm{Spec}(K) \longrightarrow X$, or equivalently, there is a fiber product square

$$\begin{array}{ccc}
T & \longrightarrow & E \\
\downarrow & & \downarrow \\
\mathrm{Spec}(K) & \longrightarrow & X
\end{array}$$

with the top G-equivariant morphism.

A G-torsor $E \longrightarrow X$ is called *versal* if for every nonempty open subset $W \subset X$, the restriction $E_W \longrightarrow W$ of the torsor $E \longrightarrow X$ is weakly versal.

4.2. Standard torsors. Let G be an algebraic group, V a generically free G-representation and $U \subset V$ a G-equivariant open subset together with a G-torsor $U \longrightarrow U/G$. We call this torsor a *standard* G-torsor.

By [**13**, Part 1, §5.4], a standard G-torsor $U \longrightarrow U/G$ is versal.

EXAMPLE 4.1. Embed G as a subgroup into $\mathbf{GL}(W)$ for a finite dimensional vector space W over F. Then G acts generically freely on the affine space of $V := \operatorname{End}_F(W)$ and taking $U = \mathbf{GL}(W) \subset V$, we have $U/G = \mathbf{GL}(W)/G$.

We think of U/G as an approximation of the classifying space BG (which we don't define). The stable birational equivalence class (and hence retract rational equivalence class by Corollary 3.3) of U/G is well defined by the no-name Lemma [**4**]. We simply say that BG is *stably rational* (respectively, *retract rational*) if so is U/G.

The implication (2) $\Rightarrow$ (1) in the following statement was proved in [**9**, Proposition 3.15].

PROPOSITION 4.2. *Let G be an algebraic group over F. The following conditions are equivalent.*

(1) *The classifying space* BG *is retract rational.*
(2) *For every local F-algebra R with infinite residue field $\overline{R}$, every G-torsor over $\overline{R}$ can be lifted to a G-torsor over R.*
(3) *There is a weakly versal G-torsor over an open subset $Y \subset \mathbb{A}^n_F$ for some n.*

PROOF. (1) $\Rightarrow$ (3): Let $U \longrightarrow U/G$ be a standard G-torsor. As U/G is retract rational, there are nonempty open subsets $W \subset U/G$ and $Y \subset \mathbb{A}^n_F$, and morphisms

$$W \longrightarrow Y \longrightarrow U/G$$

with the composition the inclusion. Let $E \longrightarrow Y$ be the pull-back of the versal torsor $U \longrightarrow U/G$. The pull-back $J \longrightarrow W$ of $E \longrightarrow Y$ to W is the restriction $U_W \longrightarrow W$ of $U \longrightarrow U/G$ and hence is weakly versal. Therefore, $E \longrightarrow Y$ is also weakly versal.

(3) $\Rightarrow$ (2): Let R be a local F-algebra with infinite residue field $\overline{R}$. The top map in the commutative diagram

$$\begin{array}{ccc} Y(R) & \longrightarrow & Y(\overline{R}) \\ \downarrow & & \downarrow \\ \operatorname{Tors}_G(R) & \longrightarrow & \operatorname{Tors}_G(\overline{R}) \end{array}$$

is surjective as Y is an open subset of an affine space. The right vertical map is surjective since $\overline{R}$ is infinite. Therefore, the bottom map is surjective.

(2) $\Rightarrow$ (1): Embed G as a subgroup into $\mathbf{GL}_N$ for some N and consider the variety $X = \mathbf{GL}_N/G$. Let R be a local F-algebra with infinite residue field $\overline{R}$. We will show that X has the R-lifting property. Let $x \in X(\overline{R})$. By assumption, the

corresponding G-torsor over $\overline{R}$ can be lifted to a G-torsor J over R. As R is local, by Hilbert Theorem 90, the map

$$X(R) \longrightarrow \operatorname{Tors}_G(R)$$

is surjective, therefore, there is $\tilde{x} \in X(R)$ mapping to J. The image $x' \in X(\overline{R})$ of $\tilde{x}$ and x give the same G-torsor over $\overline{R}$. Since

$$\operatorname{Tors}_G(\overline{R}) \simeq X(\overline{R})/\operatorname{GL}_N(\overline{R}),$$

we have $x = gx'$ for some $g \in \operatorname{GL}_N(\overline{R})$. Since the map $\operatorname{GL}_N(R) \longrightarrow \operatorname{GL}_N(\overline{R})$ is surjective, there is $\tilde{g} \in \operatorname{GL}_N(R)$ over g. The image of $\tilde{g}\tilde{x}$ under the map $X(R) \longrightarrow X(\overline{R})$ is $gx' = x$. □

4.3. Classifying spaces of spinor groups. We consider an example here. Let F be a field of characteristic different from 2. For every n, consider the quadratic form

$$q_n = \begin{cases} m\mathbb{H}, & \text{if } n = 2m; \\ \langle 1 \rangle \perp m\mathbb{H}, & \text{if } n = 2m+1, \end{cases}$$

where $\mathbb{H}$ is the hyperbolic plane. Write $\mathbf{O}_n^+$, $\mathbf{Spin}_n$ and $\mathbf{\Gamma}_n^+$ for the (split) special orthogonal, spinor and even Clifford groups of q_n, respectively.

Let R be a local F-algebra. The set $\operatorname{Tors}_{\mathbf{O}_n^+}(R) = H^1_{\text{ét}}(R, \mathbf{O}_n^+)$ is identified with the set of isomorphism classes of non-degenerate quadratic forms of rank n over R and determinant $(-1)^m$, if $n = 2m$ or $n = 2m+1$. The connecting map $H^1_{\text{ét}}(R, \mathbf{O}_n^+) \longrightarrow H^2_{\text{ét}}(R, \mathbf{G}_{\text{m}}) = \operatorname{Br}(R)$ for the exact sequence

$$1 \longrightarrow \mathbf{G}_{\text{m}} \longrightarrow \mathbf{\Gamma}_n^+ \longrightarrow \mathbf{O}_n^+ \longrightarrow 1$$

takes a form q to the *Clifford invariant* of q which is the class of the Clifford algebra $C(q)$ in $\operatorname{Br}(R)$ if n is even, and to the class of the even Clifford algebra $C_0(q)$ if n is odd. It follows that the set $\operatorname{Tors}_{\mathbf{\Gamma}_n^+}(R) = H^1_{\text{ét}}(R, \mathbf{\Gamma}_n^+)$ is identified with the set of isomorphism classes of non-degenerate quadratic forms of rank n over R of determinant $(-1)^m$ and trivial Clifford invariant.

LEMMA 4.3. *The space* $\mathrm{B}\mathbf{\Gamma}_{2m}^+$ *is retract rational if and only if* $\mathrm{B}\mathbf{\Gamma}_{2m-1}^+$ *is retract rational.*

PROOF. Let R be a local F-algebra with infinite residue field. By Proposition 4.2, it suffices to show that $\mathrm{B}\mathbf{\Gamma}_{2m-1}^+$ has the R-lifting property if and only if so does $\mathrm{B}\mathbf{\Gamma}_{2m}^+$. Let $q \in \operatorname{Tors}_{\mathbf{\Gamma}_{2m-1}^+}(\overline{R})$ be a non-degenerate quadratic form of rank n over $\overline{R}$ of determinant $(-1)^{m-1}$ and trivial even Clifford invariant. Consider the form $q' = (-q) \perp \langle 1 \rangle$. We have $\det(q') = (-1)^m$ and by [**11**, Proposition 11.4],

$$[C(q')] = [C_0(q' \perp \langle -1 \rangle)] = [C_0(-q \perp \mathbb{H})] = [C_0(q)] = 0 \in \operatorname{Br}(\overline{R}).$$

It follows that $q' \in \operatorname{Tors}_{\mathbf{\Gamma}_{2m}^+}(\overline{R})$. Let Q' be a lift of q' in $\operatorname{Tors}_{\mathbf{\Gamma}_{2m}^+}(R)$. Write $Q' = P \perp \langle r \rangle$ for some $r \in R^\times$ with $\bar{r} = 1$ and a $(2m-1)$-dimensional form P over R. Let $Q := -rP$. Then $Q \in \operatorname{Tors}_{\mathbf{\Gamma}_{2m-1}^+}(R)$ and we have

$$(-Q) \perp \langle 1 \rangle = rP \perp \langle 1 \rangle = r(P \perp \langle r \rangle) = rQ'.$$

It follows that

$$(-\overline{Q}) \perp \langle 1 \rangle = \bar{r}\overline{Q}' = \overline{Q}' = q' = (-q) \perp \langle 1 \rangle.$$

Hence Q is a lift of q.

Now let $q' \in \operatorname{Tors}_{\mathbf{\Gamma}^+_{2m}}(\overline{R})$. Write $q' = q \perp \langle a\rangle$ for some $a \in \overline{R}$ and q a $(2m-1)$-dimensional form over $\overline{R}$. Then $aq' = aq + \langle 1\rangle$, hence $-aq \in \operatorname{Tors}_{\mathbf{\Gamma}^+_{2m-1}}(\overline{R})$. Choose $A \in R^\times$ with $\overline{A} = a$. Let P be a lift in $\operatorname{Tors}_{\mathbf{\Gamma}^+_{2m-1}}(R)$ of $-aq$ and set $Q' := A(-P \perp \langle 1\rangle)$. Then $Q' \in \operatorname{Tors}_{\mathbf{\Gamma}^+_{2m}}(R)$ and

$$\overline{Q}' = \overline{A}(-\overline{P} \perp \langle 1\rangle) = a(aq \perp \langle 1\rangle) = q \perp \langle a\rangle = q'.$$

It follows that Q' is a lift of q'. □

THEOREM 4.4 (cf., [**9**, Theorem 4.15]). *The classifying spaces* $\mathrm{B}\mathbf{\Gamma}^+_n$ *and* $\mathrm{B}\mathbf{Spin}_n$ *are retract rational if* $n \leq 14$.

PROOF. We first show that $\mathrm{B}\mathbf{\Gamma}^+_n$ is retract rational. By Lemma 4.3 it suffices to assume that n is even. We will prove the R-lifting property for a local F-algebra R. We use the classification of quadratic forms of dimension $n = 2m \leq 14$ of determinant $(-1)^m$ and trivial even Clifford invariant given in [**14**].

- If $n \leq 6$, then $\operatorname{Tors}_{\mathbf{\Gamma}^+_n}(\overline{R})$ is trivial.
- $\operatorname{Tors}_{\mathbf{\Gamma}^+_8}(\overline{R})$ consists of all multiples of 3-fold Pfister forms.
- The map $\operatorname{Tors}_{\mathbf{\Gamma}^+_8}(\overline{R}) \longrightarrow \operatorname{Tors}_{\mathbf{\Gamma}^+_{10}}(\overline{R})$ taking q to $q \perp \mathbb{H}$ is a bijection.
- $\operatorname{Tors}_{\mathbf{\Gamma}^+_{12}}(\overline{R})$ consists of tensor products of a binary form and a 6-dimensional form of determinant -1.
- $\operatorname{Tors}_{\mathbf{\Gamma}^+_{14}}(\overline{R})$ consists of the corestriction (with respect to the trace map) of the form $\sqrt{d}\ \varphi'$ for a quadratic extension $\overline{R}(\sqrt{d})/\overline{R}$, where φ' is the 7-dimensional pure subform of a 3-fold Pfister form φ over $\overline{R}(\sqrt{d})$.

The proofs of the lifting property in all the cases are similar. We consider only the case $n = 14$. Consider a form q in $\operatorname{Tors}_{\mathbf{\Gamma}^+_{14}}(\overline{R})$. Let $D \in R^\times$ be a lift of d and set $S = R(\sqrt{D}) = R[t]/(t^2 - D)$. Let Φ be a 3-fold Pfister form over S lifting φ. Then the corestriction of the form $\sqrt{D}\ \Phi'$ in the extension S/R is a lift of q in $\operatorname{Tors}_{\mathbf{\Gamma}^+_{14}}(R)$. By Proposition 4.2, $\mathrm{B}\mathbf{\Gamma}^+_n$ is retract rational.

The exact sequence

$$1 \longrightarrow \mathbf{Spin}_n \longrightarrow \mathbf{\Gamma}^+_n \xrightarrow{\mathrm{Sn}} \mathbf{G}_{\mathrm{m}} \longrightarrow 1,$$

where Sn is the spinor norm homomorphism, yields a $\mathbf{G}_{\mathrm{m}}$-torsor $\mathrm{B}\mathbf{Spin}_n \longrightarrow \mathrm{B}\mathbf{\Gamma}^+_n$. It follows that $\mathrm{B}\mathbf{Spin}_n$ is stably birational to $\mathrm{B}\mathbf{\Gamma}^+_n$ and hence is retract rational for $n \leq 14$ by Corollary 3.3. □

CONJECTURE 4.5. *If* $n \geq 15$, *the space* $\mathrm{B}\mathbf{Spin}_n$ *is not retract rational.*

An (indirect) evidence in the support of the conjecture is that the essential dimension of $\mathbf{Spin}_n$ (the smallest number of parameters needed to give a $\mathbf{Spin}_n$-torsor) goes up from 7 parameters for $\mathbf{Spin}_{14}$ to 24 parameters for $\mathbf{Spin}_{16}$ (see [**20**, §7]).

5. Cohomology of classifying spaces

Let E be a G-torsor over a variety X. Write E^n for the product of n copies of E. We have a G-torsor $E^n \to E^n/G$ for all n and therefore a simplicial scheme $E^\bullet/G$.

Let H be a contravariant functor from the category of smooth schemes over F to the category of abelian groups. Set

$$H(E^{\bullet}/G) := \operatorname{Ker}(p_1^* - p_2^*) \subset H(X),$$

where p_1^* and p_2^* are induced by the two projections (the face maps of $E^{\bullet}/G$) $p_i : E^2/G \longrightarrow E/G = X$.

REMARK 5.1. If H is a sheaf on the big Zariski site over F, the group $H(E^{\bullet}/G)$ coincides with the group of sections $H^0_{\mathrm{Zar}}(E^{\bullet}/G, H)$ of H over $E^{\bullet}/G$ (see [**10**, 5.1.3]).

Let $E' \to X'$ be another G-torsor and $f_1, f_2 : E' \longrightarrow E$ two G-equivariant morphisms. Then f_1 and f_2 induce morphisms of simplicial schemes $f_1^{\bullet}, f_2^{\bullet} : E'^{\bullet}/G \longrightarrow E^{\bullet}/G$.

LEMMA 5.2. *The homomorphisms* $f_1^*, f_2^* : H(E^{\bullet}/G) \longrightarrow H(E'^{\bullet}/G)$ *are equal.*

PROOF. It is standard that the morphisms $f_1^{\bullet}$ and $f_2^{\bullet}$ are (canonically) homotopic, and therefore, yield the same homomorphisms $f_1^* = f_2^*$. Precisely, f_1 and f_2 yield a morphism (homotopy) $h = (f_1, f_2) : X = E'/G \longrightarrow E^2/G$ such that $f_i = p_i \circ h$. Therefore, $f_1^* = h^* \circ p_1^*$ coincides with $f_2^* = h^* \circ p_2^*$ on $H(E^{\bullet}/G) = \operatorname{Ker}(p_1^* - p_2^*)$. □

Now let $E \longrightarrow X$ be a versal G-torsor and $E' \longrightarrow X'$ another G-torsor with the generic fiber $T \longrightarrow \operatorname{Spec} F(X')$. If $F(X')$ is infinite, by definition of the weak versality, there is a G-equivariant morphism $T \longrightarrow E$. Hence there exist a nonempty open subset $W \subset X'$ and a G-equivariant morphism $E'_W \longrightarrow E$. It follows that there is a canonical homomorphism $H(E^{\bullet}/G) \longrightarrow H({E'_W}^{\bullet}/G)$.

Consider the following "purity condition" on a G-torsor $E \longrightarrow X$ (and the functor H):

(PC): *The restriction homomorphism* $H(E^{\bullet}/G) \longrightarrow H(E^{\bullet}_W/G)$ *is an isomorphism for every nonempty open subset* $W \subset X$.

Therefore, if we assume that $E \longrightarrow X$ is a versal G-torsor and the torsor $E' \longrightarrow X'$ satisfies the condition (PC), we get a canonical homomorphism

$$H(E^{\bullet}/G) \longrightarrow H(E'^{\bullet}/G). \tag{2}$$

Symmetrically, if we assume in addition that the torsor $E' \longrightarrow X$ is also versal and the torsor $E \longrightarrow X$ with $F(X)$ infinite satisfies the condition (PC), then the homomorphism (2) is an isomorphism, i.e., the group $H(E^{\bullet}/G)$ does not depend on the versal torsor $E \longrightarrow X$ up to canonical isomorphism.

LEMMA 5.3. *The condition* (PC) *holds for every standard* G*-torsor* $U \longrightarrow U/G$ *and the functor* $H(Y) = H^0_{\mathrm{Zar}}\big(Y, \mathcal{H}^n(\mathbb{Q}/\mathbb{Z}(j))\big)$ *for every* n *and* j.

PROOF. Let $X = U/G$ and $W \subset X$ be a nonempty open subset. By [**2**, Proposition A 9], we have $H(U^{\bullet}_W/G) \subset \operatorname{Ker}(\partial)$, where ∂ is the residue homomorphism in the commutative diagram

$$\begin{array}{ccccccc}
0 & \longrightarrow & H(X) & \longrightarrow & H(W) & \xrightarrow{\ \partial\ } & \coprod_{x \in X^{(1)} \setminus W^{(1)}} H^{n+1}_x\big(X, \mathbb{Q}/\mathbb{Z}(j)\big) \\
 & & \Big\downarrow{\scriptstyle p_1^* - p_2^*} & & \Big\downarrow{\scriptstyle p_1^* - p_2^*} & & \\
0 & \longrightarrow & H(U^2/G) & \longrightarrow & H(U^2_W/G). & &
\end{array}$$

The rows are exact by §2.3. The statement follows by diagram chase. □

It follows from Lemma 5.3 that the group $H^0_{\mathrm{Zar}}\big(U^\bullet/G, \mathcal{H}^n(\mathbb{Q}/\mathbb{Z}(j))\big)$ does not depend on the choice of the standard G-torsor up to canonical isomorphism. We denote this group by $H^0_{\mathrm{Zar}}\big(\mathrm{B}G, \mathcal{H}^n(\mathbb{Q}/\mathbb{Z}(j))\big)$.

6. Invariants of algebraic groups

Let G be an algebraic group over a field F. Consider the functor

$$\mathrm{Tors}_G : \mathsf{Fields}_F \longrightarrow \mathsf{Sets},$$

where Fields_F is the category of field extensions of F, taking a field K to the set $\mathrm{Tors}_G(K)$ of isomorphism classes of G-torsors over $\operatorname{Spec} K$. Let

$$H : \mathsf{Fields}_F \longrightarrow \mathsf{Abelian\ Groups}$$

be another functor. As defined in [**13**], an *H-invariant* of G is a morphism of functors

$$I : G\text{- torsors} \longrightarrow H,$$

viewed as functors to Sets. In other words, an invariant is a natural in K collection of maps of sets $\mathrm{Tors}_G(K) \longrightarrow H(K)$ for every field extension K/F.

We write $\mathrm{Inv}(G, H)$ for the group of H-invariants of G.

An invariant $I \in \mathrm{Inv}(G, H)$ is called *normalized* if $I(E) = 0$ for every trivial G-torsor E. The normalized invariants form a subgroup $\mathrm{Inv}(G, H)_{\mathrm{norm}}$ of $\mathrm{Inv}(G, H)$ and

$$\mathrm{Inv}(G, H) \simeq H(F) \oplus \mathrm{Inv}(G, H)_{\mathrm{norm}}.$$

Let $h : E \longrightarrow X$ be a weakly versal torsor defined in §5. The generic fiber E^{gen} of h is the *generic torsor* over $\operatorname{Spec} F(X)$. The evaluation at the generic torsor yields a homomorphism

$$\theta_G : \mathrm{Inv}(G, H) \longrightarrow H^n(F(X), H), \quad I \mapsto I(E^{\mathrm{gen}}).$$

Consider the following "injectivity condition" on the functor H:

(IC): *The homomorphism $H(K) \longrightarrow H(K(Y))$ is injective for every smooth variety Y over a field extension K/F with a K-rational point.*

The following statement was previously known in the case G is smooth (see [**13**]).

PROPOSITION 6.1. *Suppose that a functor H satisfies the condition (IC). Then the map θ_G is injective.*

PROOF. Let $I \in \mathrm{Inv}(G, H)$ be an invariant such that $I(E^{\mathrm{gen}}) = 0$. We will show that for every G-torsor $T \longrightarrow \mathrm{Spec}(K)$ for a field extension K/F, the value $I(T)$ is trivial. Replacing K by $K(t)$ if necessary, we may assume that K is infinite.

Consider the following commutative diagram

$$\begin{array}{ccccc} T & \xleftarrow{\ g\ } & T \times E & \longrightarrow & E \\ \downarrow & & \downarrow{\scriptstyle f} & & \downarrow \\ \mathrm{Spec}(K) & \longleftarrow & Y & \longrightarrow & X \end{array}$$

with $Y = (T \times E)/G$, the two fiber product squares and the vertical morphisms the G-torsors. Let $K' = K(Y)$ be the function field of Y and T' the generic fiber of f. Then the G-torsors

$$T_{K'} := T \times_K \operatorname{Spec}(K') \quad \text{and} \quad (E^{\text{gen}})_{K'} := (E^{\text{gen}}) \times_{F(X)} \operatorname{Spec}(K')$$

are both isomorphic to T'. It follows that

$$I(T)_{K'} = I(T_{K'}) = I(T') = I\big((E^{\text{gen}})_{K'}\big) = I(E^{\text{gen}})_{K'} = 0.$$

The torsor E is weakly versal, hence, there is a G-equivariant morphism $T \longrightarrow E$. Therefore, the morphism g in the diagram has a G-equivariant section. It follows that Y has a K-rational point. By assumption, the map $H(K) \longrightarrow H(K')$ is injective and hence $I(T) = 0$. □

We will be considering the cohomology functors H taking a field K/F to the Galois cohomology $H^n\big(K, \mathbb{Q}/\mathbb{Z}(j)\big)$ and write $\operatorname{Inv}^n\big(G, \mathbb{Q}/\mathbb{Z}(j)\big)$ for the group of *cohomological invariants of G of degree n with coefficients in $\mathbb{Q}/\mathbb{Z}(j)$*.

LEMMA 6.2. *The functors $H(K) = H^n\big(K, \mathbb{Q}/\mathbb{Z}(j)\big)$ satisfy the condition (IC) for all n and j.*

PROOF. Let Y be a variety over a field K with a K-rational point y. The completion of the local ring $O_{Y,y}$ at y is the ring of formal power series $K[[t_1, t_2, \dots, t_d]]$, where $d = \dim(Y)$. Therefore, $K(Y)$ is a subfield of the iterated power series field $K((t_1))\dots((t_d))$. Thus, it suffices to show that for every field L, the map $H(L) \longrightarrow H\big(L((t))\big)$ is injective. This was proved in [**13**, Part 2, Proposition A.9]. □

Consider a standard G-torsor $U \longrightarrow U/G =: X$ and let $E_{\text{gen}} \longrightarrow \operatorname{Spec} F(X)$ be its generic fiber. Since the pull-back of $U \longrightarrow U/G$ with respect to any of the two projections $U^2/G \longrightarrow X$ coincides with the G-torsor $U^2 \longrightarrow U^2/G$, the pull-backs of the generic G-torsor $E_{\text{gen}} \longrightarrow \operatorname{Spec} F(X)$ with respect to the two morphisms $\operatorname{Spec} F(E^2/G) \longrightarrow \operatorname{Spec} F(X)$ induced by the projections are isomorphic. Hence for every invariant $I \in \operatorname{Inv}\big(G, H^n(\mathbb{Q}/\mathbb{Z}(j))\big)$ we have

$$p_1^*\big(I(E_{\text{gen}})\big) = I\big(p_1^*(E_{\text{gen}})\big) = I\big(p_2^*(E_{\text{gen}})\big) = p_2^*\big(I(E_{\text{gen}})\big)$$

in $H^n\big(F(E^2/G), \mathbb{Q}/\mathbb{Z}(j)\big)$. It follows that the image of θ_G is contained in the subgroup

$$H^0_{\text{Zar}}\big(BG, \mathcal{H}^n(\mathbb{Q}/\mathbb{Z}(j))\big) \subset H^0_{\text{Zar}}\big(X, \mathcal{H}^n(\mathbb{Q}/\mathbb{Z}(j))\big) \subset H^n\big(F(X), \mathbb{Q}/\mathbb{Z}(j)\big).$$

THEOREM 6.3. *Let G be an algebraic group over a field F. Then θ_G yields an isomorphism*

$$\operatorname{Inv}^n\big(G, \mathbb{Q}/\mathbb{Z}(j)\big) \xrightarrow{\sim} H^0_{\text{Zar}}\big(BG, \mathcal{H}^n(\mathbb{Q}/\mathbb{Z}(j))\big).$$

PROOF. The inverse isomorphism was constructed in [**2**, Theorem 3.4] in the case F is an infinite field as follows. Let $u \in H^0_{\text{Zar}}\big(BG, \mathcal{H}^n(\mathbb{Q}/\mathbb{Z}(j))\big)$. We define an invariant in $\operatorname{Inv}^n\big(G, \mathbb{Q}/\mathbb{Z}(j)\big)$ by taking a G-torsor $J \longrightarrow \operatorname{Spec}(K)$ to the image of u under the pull-back homomorphism

$$H^0_{\text{Zar}}\big(BG, \mathcal{H}^n(\mathbb{Q}/\mathbb{Z}(j))\big) \longrightarrow H^0_{\text{Zar}}\big(\operatorname{Spec}(K), \mathcal{H}^n(\mathbb{Q}/\mathbb{Z}(j))\big) = H^n(K, \mathbb{Q}/\mathbb{Z}(j)),$$

induced by a G-equivariant morphism $f : J \longrightarrow U$. By Lemma 5.2, the result is independent of the choice of f.

Now let F be a finite field. It suffices to prove that for a given prime integer p, the p-primary components $A(F)$ and $B(F)$ of the groups in the statement are isomorphic. Let l be a prime integer different from p and F'/F an infinite Galois field extension with the Galois group Δ a pro-l-group. Since the field F' is infinite, the map $A(F') \longrightarrow B(F')$ is an isomorphism. By a restriction-corestriction argument, the map $A(F) \longrightarrow B(F)$ is identified with the homomorphism $A(F')^{\Delta} \longrightarrow B(F')^{\Delta}$ of the groups of Δ-invariant elements and hence is an isomorphism. □

6.1. Unramified invariants. Let G be an algebraic group over F. An invariant $I \in \operatorname{Inv}^n(G, \mathbb{Q}/\mathbb{Z}(j))$ is called *unramified* if for every field extension K/F and every $T \in \operatorname{Tors}_G(K)$, we have $I(T) \in H^n_{\mathrm{nr}}(K/F, \mathbb{Q}/\mathbb{Z}(j))$. We will write $\operatorname{Inv}^n_{\mathrm{nr}}(G, \mathbb{Q}/\mathbb{Z}(j))$ for the subgroup of all unramified invariants in $\operatorname{Inv}^n(G, \mathbb{Q}/\mathbb{Z}(j))$.

PROPOSITION 6.4. *Let G be an algebraic group over a field F. Then an invariant $I \in \operatorname{Inv}^n\big(G, \mathbb{Q}/\mathbb{Z}(j)\big)$ is unramified if and only if*

$$I(E_{\mathrm{gen}}) \in H^n_{\mathrm{nr}}\big(F(U/G), \mathbb{Q}/\mathbb{Z}(j)\big).$$

PROOF. By an argument similar to the one in the proof of Theorem 6.3 we may assume that F is infinite. Suppose $I(E_{\mathrm{gen}}) \in H^n_{\mathrm{nr}}\big(F(U/G), \mathbb{Q}/\mathbb{Z}(j)\big)$ and let $T \in \operatorname{Tors}_G(K)$. Consider the variety $Y = (T \times U)/G$ and the natural morphisms $Y \longrightarrow \operatorname{Spec}(K)$ and $Y \longrightarrow U/G$. As in the proof of Proposition 6.1, the torsors E_{gen} and T are isomorphic over the field $K(Y)$. It follows that

$$I(T)_{F(Y)} = I(E_{\mathrm{gen}})_{F(Y)} \in H^n_{\mathrm{nr}}\big(K(Y), \mathbb{Q}/\mathbb{Z}(j)\big).$$

Since $K(Y)/K$ is a purely transcendental field extension, we have $I(T) \in H^n_{\mathrm{nr}}\big(K, \mathbb{Q}/\mathbb{Z}(j)\big)$ by Lemma 2.3. □

We write $H^n_{\mathrm{nr}}\big(F(\mathrm{B}G), \mathbb{Q}/\mathbb{Z}(j)\big)$ for $H^n_{\mathrm{nr}}\big(F(U/G), \mathbb{Q}/\mathbb{Z}(j)\big)$.

COROLLARY 6.5. *Let G be an algebraic group over a field F. Then θ_G yields an isomorphism $\operatorname{Inv}^n_{\mathrm{nr}}\big(G, \mathbb{Q}/\mathbb{Z}(j)\big) \xrightarrow{\sim} H^n_{\mathrm{nr}}\big(F(\mathrm{B}G), \mathbb{Q}/\mathbb{Z}(j)\big)$.*

Proposition 3.4 then implies the following corollary.

COROLLARY 6.6. *If an algebraic group over F admits a non-constant unramified cohomological invariant with values in $\mathbb{Q}/\mathbb{Z}(j)$ for some j, then the classifying space $\mathrm{B}G$ is not retract rational over F.*

7. Degree 1 invariants with coefficients in Galois module

Let G be an algebraic group over a field F. Write $\pi_0(G)$ for the factor group of G modulo the connected component of the identity G° of G. It is an étale algebraic group over F.

Let M be a discrete Γ_F-module, where $\Gamma_F = \operatorname{Gal}(F_{\mathrm{sep}}/F)$ and let

$$\alpha : \pi_0(G_{\mathrm{sep}}) \longrightarrow M$$

be a Γ_F-homomorphism. For every field extension K of F we then have the composition

$$I^{\alpha}_K : H^1(K, G) \longrightarrow H^1\big(K, \pi_0(G)\big) \xrightarrow{\alpha^*} H^1(K, M),$$

where the first map is induced by the canonical surjection $G \longrightarrow \pi_0(G)$ and the second one by α. We can view I^{α} as a normalized H-invariant of the group G with the functor H defined by $H(K) = H^1(K, M)$. We write $\operatorname{Inv}^1(G, M)$ for the group of H-invariants.

Note that the functor H satisfies the condition (IC). Indeed, if Y is a smooth variety over a field extension K/F with a K-rational point, then K is algebraically closed in $K(Y)$ and hence the restriction homomorphism $\Gamma_{K(Y)} \longrightarrow \Gamma_K$ is surjective. Therefore, the inflation homomorphism $H^1(K, M) \longrightarrow H^1(K(Y), M)$ is injective.

The following proposition was mentioned in [**18**, §31.15] without proof.

PROPOSITION 7.1. *The map*

$$\varphi : \operatorname{Hom}_\Gamma\big(\pi_0(G_{\mathrm{sep}}), M\big) \longrightarrow \operatorname{Inv}^1(G, M)_{\mathrm{norm}} \quad \textit{given by } \alpha \mapsto I^\alpha$$

is an isomorphism.

PROOF. Consider the case when G is connected. We need to show that every invariant $I \in \operatorname{Inv}^1(G, M)_{\mathrm{norm}}$ is trivial. Let E be a G-torsor over a field extension K/F. Since G is connected, the variety of G and hence the one of E are geometrically irreducible. Therefore, the separable closure of K in the function field $K(E')$ of the associated reduced scheme $E' = E_{\mathrm{red}}$ coincides with K. It follows that the restriction homomorphism $\Gamma_{K(E')} \longrightarrow \Gamma_K$ is surjective and hence the inflation homomorphism

$$H^1(K, M) \longrightarrow H^1(K(E'), M)$$

is injective. As E is trivial over $K(E')$, we have $I(E)_{K(E')} = 0$ and hence $I(E) = 0$ by the injectivity.

In the general case, we construct a map ψ inverse to φ. Let $U \longrightarrow U/G$ be a standard versal G-torsor. Its generic fiber E^{gen} is the *generic* torsor over the field $F(U/G)$.

Let $I(E^{\mathrm{gen}}) \in H^1(F(U/G), M)$ be the value of a normalized invariant I at the generic torsor E^{gen}. Since $U/G^\circ \longrightarrow U/G$ is a surjective morphism, we can view $F(U/G)$ as a subfield of $F(U/G^\circ)$. The image of $I(E^{\mathrm{gen}})$ in $H^1(F(U/G^\circ), M)$ is the value of the restriction of I on G°. By the first part of the proof, the latter value is trivial.

For a field extension K/F, write

$$N(K) := \operatorname{Ker}\big(H^1(K(U/G), M) \longrightarrow H^1(K(U/G^\circ), M)\big).$$

We have proved that $I(E^{\mathrm{gen}}) \in N(F)$.

The field extension $F_{\mathrm{sep}}(U/G^\circ)$ of $F_{\mathrm{sep}}(U/G)$ is Galois with Galois group $\pi_0(G_{\mathrm{sep}})$. By exactness of the inflation-restriction sequence, we have

$$N(F_{\mathrm{sep}}) = H^1(\pi_0(G_{\mathrm{sep}}), M) = \operatorname{Hom}(\pi_0(G_{\mathrm{sep}}), M)$$

since $\pi_0(G_{\mathrm{sep}})$ acts trivially on M. We define the map

$$\psi : \operatorname{Inv}^1(G, M) \longrightarrow \operatorname{Hom}_\Gamma\big(\pi_0(G_{\mathrm{sep}}), M\big)$$

by taking I to the image of $I(E^{\mathrm{gen}})$ under the homomorphism

$$N(F) \longrightarrow N(F_{\mathrm{sep}})^\Gamma = \operatorname{Hom}_\Gamma(\pi_0(G_{\mathrm{sep}}), M).$$

To prove that the composition $\psi \circ \varphi$ is the identity, we may assume that F is separably closed. Let $I = \varphi(\alpha)$ for some $\alpha \in \operatorname{Hom}(\pi_0(G_{\mathrm{sep}}), M)$. By construction, the element $I(E^{\mathrm{gen}})$ in $H^1(K, M) = \operatorname{Hom}(\Gamma_K, M)$, where $K = F(U/G)$, is equal to the composition $\Gamma_K \longrightarrow \pi_0(G) \xrightarrow{\alpha} M$. It follows that $\psi(I) = \alpha$.

Now we prove that the composition $\varphi \circ \psi$ is the identity. For a field extension F'/F consider a homomorphism

$$\beta_{F'} : \mathrm{Hom}_\Gamma(\pi_0(G), M) \longrightarrow H^1(F'(U/G), M)$$

as follows. The element $\beta_{F'}(\alpha)$ is defined as the image of the class of the generic G-torsor under the composition

$$H^1(F'(U/G), G) \longrightarrow H^1(F'(U/G), \pi_0(G)) \xrightarrow{\alpha^*} H^1(F'(U/G), M).$$

Taking $F' = F_{\mathrm{sep}}$ we get a commutative diagram

$$\begin{array}{ccc} \mathrm{Hom}_\Gamma(\pi_0(G), M) & \xrightarrow{\beta_F} & H^1(F(U/G), M) \\ \downarrow & & \downarrow{\scriptstyle j} \\ \mathrm{Hom}(\pi_0(G), M) & \xrightarrow{\beta_{F'}} & H^1(F'(U/G), M). \end{array}$$

Note that the bottom map $\beta_{F'}$ is the inflation.

Let $I \in \mathrm{Inv}^1(G, M)$, $\alpha = \psi(I) \in \mathrm{Hom}_\Gamma(\pi_0(G), M)$ and $I' = \varphi(\alpha)$. We need to show that $I = I'$. By construction, $\beta_F(\alpha) = I'(E)$, where E is the generic G-torsor over $F(U/G)$. By the definition of ψ, $j(I(E))$ is the image of α under the diagonal map in the diagram. It follows that $I'(E) - I(E) \in \mathrm{Ker}(j)$. By the inflation-restriction sequence, $\mathrm{Ker}(j) = H^1(F, M)$, hence $I'(E) - I(E) = b_{F(U/G)}$ for an element $b \in H^1(F, M)$. The torsor E is trivial over $F(U)$, hence b vanishes over $F(U)$. It follows that $b = 0$ as the map $H^1(F, M) \longrightarrow H^1(F(U), M)$ is injective by (IC).

We proved that I and I' are equal at the generic torsor and hence $I = I'$ by Proposition 6.1. □

8. Brauer invariants

The invariants with values in the Brauer group are the degree 2 cohomological invariants: $\mathrm{Inv}\big(G, \mathrm{Br}\big) = \mathrm{Inv}^2\big(G, \mathbb{Q}/\mathbb{Z}(1)\big)$. Let G be a (connected) semisimple group over F and

$$1 \longrightarrow C \longrightarrow \widetilde{G} \longrightarrow G \longrightarrow 1$$

an exact sequence with $\widetilde{G}$ a simply connected semisimple group and C finite central subgroup of $\widetilde{G}$ of multiplicative type. For every character $\chi \in C^* := \mathrm{Hom}(C, \mathbf{G}_{\mathrm{m}})$, consider the push-out diagram

$$\begin{array}{ccccccccc} 1 & \longrightarrow & C & \longrightarrow & \widetilde{G} & \longrightarrow & G & \longrightarrow & 1 \\ & & \downarrow & & \downarrow & & \| & & \\ 1 & \longrightarrow & \mathbf{G}_{\mathrm{m}} & \longrightarrow & G' & \longrightarrow & G & \longrightarrow & 1. \end{array}$$

Let K/F be a field extension. By Hilbert Theorem 90, the sequence

$$1 \longrightarrow \mathbf{G}_{\mathrm{m}}(K_{\mathrm{sep}}) \longrightarrow G'(K_{\mathrm{sep}}) \longrightarrow G(K_{\mathrm{sep}}) \longrightarrow 1$$

is exact. Therefore, we have the connecting map

$$\delta_K(\chi) : \mathrm{Tors}_G(K) = H^1(K, G) \longrightarrow H^2(K, \mathbf{G}_{\mathrm{m}}) = \mathrm{Br}(K).$$

This collection of maps $\delta_K(\chi)$ over all K/F is an invariant of G (depending on χ) with values in the Brauer group. Thus, we have a homomorphism

$$\delta : C^* \longrightarrow \operatorname{Inv}(G, \operatorname{Br})_{\mathrm{norm}}.$$

THEOREM 8.1 ([**2**, Theorem 2.4]). *The map $\delta : C^* \longrightarrow \operatorname{Inv}(G, \operatorname{Br})_{\mathrm{norm}}$ is an isomorphism.*

The following theorem was proved by F. Bogomolov [**3**, Lemma 5.7] in characteristic zero and in [**2**, Theorem 5.10] in general.

THEOREM 8.2. *Let G be a semisimple group over a field F. Then $\operatorname{Inv}_{\mathrm{nr}}(G, \operatorname{Br})_{\mathrm{norm}} = 0$ and $\operatorname{Br}_{\mathrm{nr}}\big(F(BG)\big) = \operatorname{Br}(F)$.*

EXAMPLE 8.3 (see [**30**], [**12**], [**19**] and [**13**]). Let G be a cyclic group of order 8 and $F = \mathbb{Q}$. A G-torsor over a field extension K/F is a G-Galois cyclic algebra L over K. The class of the central cyclic K-algebra $(L/K, 16)$ of degree 8 in $\operatorname{Br}(K)$ is a non-constant unramified degree 2 invariant ([**13**, Proposition 33.15]). It follows that $\mathrm{B}G$ is not retract rational. In other words, if V is a faithful representation of G over $\mathbb{Q}$, then the variety V/G is not retract rational.

EXAMPLE 8.4. Let G be a cyclic group of prime order p. By [**26**], $\mathrm{B}G$ is retract rational. But it was shown in [**28**] and [**29**] that $\mathrm{B}G$ is not stably rational if $p = 47$ and $F = \mathbb{Q}$. The relationship of the papers cited in this example and Example 8.3 with the work [**26**] is explained in [**6**, §7].

EXAMPLE 8.5. Let F be an algebraically closed field. D. Saltman in [**25**] constructed a finite group G such that $\mathrm{B}G$ is not retract rational over F (a counter-example to Noether's problem) by exhibiting a nontrivial unramified invariant of G with values in the Brauer group. It was proved in [**3**] that $\operatorname{Br}_{\mathrm{nr}}\big(F(BG)\big)$ for a finite group G is isomorphic to the subgroup of $H^3(G, \mathbb{Z})$ consisting of all classes having trivial restrictions to all bicyclic subgroups of G. In [**15**] examples of finite groups of order p^5, p odd and 2^6 with nontrivial group $\operatorname{Br}_{\mathrm{nr}}\big(F(BG)\big)$ were given.

9. Invariants of degree 3 with coefficients in $\mathbb{Q}/\mathbb{Z}(2)$

Let G be a split semisimple group over F and C the kernel of the universal cover of G. The cohomological cup-product

$$H^2(K, \mathbb{Q}/\mathbb{Z}(1)) \otimes K^\times \longrightarrow H^3(K, \mathbb{Q}/\mathbb{Z}(2))$$

for any field extension K/F yields a pairing

$$\operatorname{Inv}^2\big(G, \mathbb{Q}/\mathbb{Z}(1)\big)_{\mathrm{norm}} \otimes F^\times \longrightarrow \operatorname{Inv}^3\big(G, \mathbb{Q}/\mathbb{Z}(2)\big)_{\mathrm{norm}}.$$

By Theorem 8.1, the group $\operatorname{Inv}^2\big(G, \mathbb{Q}/\mathbb{Z}(1)\big)_{\mathrm{norm}}$ is isomorphic to C^*, so we get a pairing

$$\tau : C^* \otimes F^\times \longrightarrow \operatorname{Inv}^3\big(G, \mathbb{Q}/\mathbb{Z}(2)\big)_{\mathrm{norm}}.$$

The image of τ is the group of *decomposable* invariants. We write $\operatorname{Inv}^3\big(G, \mathbb{Q}/\mathbb{Z}(2)\big)_{\mathrm{ind}}$ for the cokernel of τ.

Let $T \subset G$ be a split maximal torus. The Weyl group W of G acts on the character group T^* of T and therefore, on the symmetric powers $S^n(T^*)$ of T^*. Let $x \in T^*$ and $\{x_1, \dots, x_m\}$ the W-orbit of x in T^*. Then the element $c_2(x) := \sum_{i<j} x_i x_j$ belongs to $S^2(T^*)^W$. Write $\operatorname{Dec}(G)$ for the subgroup of $S^2(T^*)^W$ generated by the elements $c_2(x)$ over all $x \in T^*$, so $\operatorname{Dec}(G)$ is the subgroup of the "obvious" elements in $S^2(T^*)^W$.

THEOREM 9.1 ([**21**, Theorem 3.9]). *Let G be a split semisimple group over F, $T \subset G$ a split maximal torus and C the kernel of the universal cover of G. Then there is an exact sequence*

$$0 \longrightarrow C^* \otimes F^\times \xrightarrow{\tau} \mathrm{Inv}^3\big(G, \mathbb{Q}/\mathbb{Z}(j)\big)_{\mathrm{norm}} \longrightarrow S^2(T^*)^W / \mathrm{Dec}(G) \longrightarrow 0.$$

It follows from this theorem that

$$\mathrm{Inv}^3\big(G, \mathbb{Q}/\mathbb{Z}(2)\big)_{\mathrm{ind}} \simeq S^2(T^*)^W / \mathrm{Dec}(G).$$

This group was computed when G is simply connected (M. Rost, see [**13**]), when G is adjoint (see [**21**]) and when G is (almost) simple in [**1**].

The unramified invariants of semisimple groups over an algebraically closed fields were considered in [**22**].

THEOREM 9.2 ([**22**, Proposition 8.1. and Theorems 8.4 and 11.3]). *Let G be a semisimple group over an algebraically closed field F and p a prime integer different from* $\mathrm{char}(F)$. *Then the p-primary component of* $\mathrm{Inv}^3_{\mathrm{nr}}(G, p)$ *is trivial in the following cases:*

- *G is a simply connected or adjoint group.*
- *G is a simple group.*
- *p is odd.*

References

[1] Hernando Bermudez and Anthony Ruozzi, *Degree 3 cohomological invariants of split simple groups that are neither simply connected nor adjoint*, J. Ramanujan Math. Soc. **29** (2014), no. 4, 465–481. MR3284049

[2] Sam Blinstein and Alexander Merkurjev, *Cohomological invariants of algebraic tori*, Algebra Number Theory **7** (2013), no. 7, 1643–1684, DOI 10.2140/ant.2013.7.1643. MR3117503

[3] F. A. Bogomolov, *The Brauer group of quotient spaces of linear representations* (Russian), Izv. Akad. Nauk SSSR Ser. Mat. **51** (1987), no. 3, 485–516, 688; English transl., Math. USSR-Izv. **30** (1988), no. 3, 455–485. MR903621

[4] F. A. Bogomolov and P. I. Katsylo, *Rationality of some quotient varieties* (Russian), Mat. Sb. (N.S.) **126(168)** (1985), no. 4, 584–589. MR788089

[5] J.-L. Colliot-Thélène, *Birational invariants, purity and the Gersten conjecture*, K-theory and algebraic geometry: connections with quadratic forms and division algebras (Santa Barbara, CA, 1992), Proc. Sympos. Pure Math., vol. 58, Amer. Math. Soc., Providence, RI, 1995, pp. 1–64. MR1327280

[6] Jean-Louis Colliot-Thélène and Jean-Jacques Sansuc, *Principal homogeneous spaces under flasque tori: applications*, J. Algebra **106** (1987), no. 1, 148–205, DOI 10.1016/0021-8693(87)90026-3. MR878473

[7] Jean-Louis Colliot-Thélène and Manuel Ojanguren, *Variétés unirationnelles non rationnelles: au-delà de l'exemple d'Artin et Mumford* (French), Invent. Math. **97** (1989), no. 1, 141–158, DOI 10.1007/BF01850658. MR999316

[8] J.-L. Colliot-Thélène, R. T. Hoobler, and B. Kahn, *The Bloch-Ogus-Gabber theorem*, Algebraic K-theory (Toronto, ON, 1996), Amer. Math. Soc., Providence, RI, 1997, pp. 31–94.

[9] Jean-Louis Colliot-Thélène and Jean-Jacques Sansuc, *The rationality problem for fields of invariants under linear algebraic groups (with special regards to the Brauer group)*, Algebraic groups and homogeneous spaces, Tata Inst. Fund. Res. Stud. Math., Tata Inst. Fund. Res., Mumbai, 2007, pp. 113–186. MR2348904

[10] Pierre Deligne, *Théorie de Hodge. III* (French), Inst. Hautes Études Sci. Publ. Math. **44** (1974), 5–77. MR0498552

[11] Richard Elman, Nikita Karpenko, and Alexander Merkurjev, *The algebraic and geometric theory of quadratic forms*, American Mathematical Society Colloquium Publications, vol. 56, American Mathematical Society, Providence, RI, 2008. MR2427530

[12] Shizuo Endô and Takehiko Miyata, *Invariants of finite abelian groups*, J. Math. Soc. Japan **25** (1973), 7–26. MR0311754

[13] Skip Garibaldi, Alexander Merkurjev, and Jean-Pierre Serre, *Cohomological invariants in Galois cohomology*, University Lecture Series, vol. 28, American Mathematical Society, Providence, RI, 2003. MR1999383

[14] Skip Garibaldi, *Cohomological invariants: exceptional groups and spin groups*, Mem. Amer. Math. Soc. **200** (2009), no. 937, xii+81, DOI 10.1090/memo/0937. With an appendix by Detlev W. Hoffmann. MR2528487

[15] Akinari Hoshi, Ming-Chang Kang, and Boris E. Kunyavskii, *Noether's problem and unramified Brauer groups*, Asian J. Math. **17** (2013), no. 4, 689–713, DOI 10.4310/AJM.2013.v17.n4.a8. MR3152260

[16] Luc Illusie, *Complexe de de Rham-Witt et cohomologie cristalline* (French), Ann. Sci. École Norm. Sup. (4) **12** (1979), no. 4, 501–661. MR565469

[17] Bruno Kahn, *Applications of weight-two motivic cohomology*, Doc. Math. **1** (1996), No. 17, 395–416. MR1423901

[18] Max-Albert Knus, Alexander Merkurjev, Markus Rost, and Jean-Pierre Tignol, *The book of involutions*, American Mathematical Society Colloquium Publications, vol. 44, American Mathematical Society, Providence, RI, 1998. With a preface in French by J. Tits. MR1632779

[19] H. W. Lenstra Jr., *Rational functions invariant under a finite abelian group*, Invent. Math. **25** (1974), 299–325. MR0347788

[20] Alexander S. Merkurjev, *Essential dimension: a survey*, Transform. Groups **18** (2013), no. 2, 415–481, DOI 10.1007/s00031-013-9216-y. MR3055773

[21] Alexander Merkurjev, *Degree three cohomological invariants of semisimple groups*, J. Eur. Math. Soc. (JEMS) **18** (2016), no. 3, 657–680, DOI 10.4171/JEMS/599. MR3463420

[22] A. Merkurjev, *Unramified degree three invariants of reductive groups*, Adv. Math. **293** (2016), 697–719, DOI 10.1016/j.aim.2016.02.027. MR3474332

[23] James S. Milne, *Étale cohomology*, Princeton Mathematical Series, vol. 33, Princeton University Press, Princeton, N.J., 1980. MR559531

[24] Markus Rost, *Chow groups with coefficients*, Doc. Math. **1** (1996), No. 16, 319–393 (electronic). MR1418952

[25] David J. Saltman, *Noether's problem over an algebraically closed field*, Invent. Math. **77** (1984), no. 1, 71–84, DOI 10.1007/BF01389135. MR751131

[26] David J. Saltman, *Retract rational fields and cyclic Galois extensions*, Israel J. Math. **47** (1984), no. 2-3, 165–215, DOI 10.1007/BF02760515. MR738167

[27] Jean-Pierre Serre, *Cohomologie galoisienne: progrès et problèmes* (French, with French summary), Astérisque **227** (1995), Exp. No. 783, 4, 229–257. Séminaire Bourbaki, Vol. 1993/94. MR1321649

[28] Richard G. Swan, *Invariant rational functions and a problem of Steenrod*, Invent. Math. **7** (1969), 148–158. MR0244215

[29] V. E. Voskresenskiĭ, *The birational equivalence of linear algebraic groups* (Russian), Dokl. Akad. Nauk SSSR 188 (1969), 978–981; erratum, ibid. **191** (1969), no. nos., 1, 2, 3, vii. MR0252403

[30] V. E. Voskresenskiĭ, *The geometry of linear algebraic groups* (Russian), Trudy Mat. Inst. Steklov. **132** (1973), 151–161, 266. Proc. Internat. Conf. Number Theory (Moscow, 1971). MR0342521

Department of Mathematics, University of California, Los Angeles, California 90095-1555

E-mail address: merkurev@math.ucla.edu

Selected Published Titles in This Series

94 **Mahir Bilen Can, Editor,** Algebraic Groups: Structure and Actions, 2017

93 **Vincent Bouchard, Charles Doran, Stefan Méndez-Diez, and Callum Quigley, Editors,** String-Math 2014, 2016

92 **Kailash C. Misra, Daniel K. Nakano, and Brian J. Parshall, Editors,** Lie Algebras, Lie Superalgebras, Vertex Algebras and Related Topics, 2016

91 **V. Sidoravicius and S. Smirnov, Editors,** Probability and Statistical Physics in St. Petersburg, 2016

90 **Ron Donagi, Sheldon Katz, Albrecht Klemm, and David R. Morrison, Editors,** String-Math 2012, 2015

89 **D. Dolgopyat, Y. Pesin, M. Pollicott, and L. Stoyanov, Editors,** Hyperbolic Dynamics, Fluctuations and Large Deviations, 2015

88 **Ron Donagi, Michael R. Douglas, Ljudmila Kamenova, and Martin Rocek, Editors,** String-Math 2013, 2014

87 **Helge Holden, Barry Simon, and Gerald Teschl, Editors,** Spectral Analysis, Differential Equations and Mathematical Physics: A Festschrift in Honor of Fritz Gesztesy's 60th Birthday, 2013

86 **Kailash C. Misra, Daniel K. Nakano, and Brian J. Parshall, Editors,** Recent Developments in Lie Algebras, Groups and Representation Theory, 2012

85 **Jonathan Block, Jacques Distler, Ron Donagi, and Eric Sharpe, Editors,** String-Math 2011, 2012

84 **Alex H. Barnett, Carolyn S. Gordon, Peter A. Perry, and Alejandro Uribe, Editors,** Spectral Geometry, 2012

83 **Hisham Sati and Urs Schreiber, Editors,** Mathematical Foundations of Quantum Field Theory and Perturbative String Theory, 2011

82 **Michael Usher, Editor,** Low-dimensional and Symplectic Topology, 2011

81 **Robert S. Doran, Greg Friedman, and Jonathan Rosenberg, Editors,** Superstrings, Geometry, Topology, and C^*-algebras, 2010

80 **D. Abramovich, A. Bertram, L. Katzarkov, R. Pandharipande, and M. Thaddeus, Editors,** Algebraic Geometry, 2009

79 **Dorina Mitrea and Marius Mitrea, Editors,** Perspectives in Partial Differential Equations, Harmonic Analysis and Applications, 2008

78 **Ron Y. Donagi and Katrin Wendland, Editors,** From Hodge Theory to Integrability and TQFT, 2008

77 **Pavel Exner, Jonathan P. Keating, Peter Kuchment, Toshikazu Sunada, and Alexander Teplyaev, Editors,** Analysis on Graphs and Its Applications, 2008

76 **Fritz Gesztesy, Percy Deift, Cherie Galvez, Peter Perry, and Wilhelm Schlag, Editors,** Spectral Theory and Mathematical Physics: A Festschrift in Honor of Barry Simon's 60th Birthday, 2007

75 **Solomon Friedberg, Daniel Bump, Dorian Goldfeld, and Jeffrey Hoffstein, Editors,** Multiple Dirichlet Series, Automorphic Forms, and Analytic Number Theory, 2006

74 **Benson Farb, Editor,** Problems on Mapping Class Groups and Related Topics, 2006

73 **Mikhail Lyubich and Leon Takhtajan, Editors,** Graphs and Patterns in Mathematics and Theoretical Physics, 2005

72 **Michel L. Lapidus and Machiel van Frankenhuijsen, Editors,** Fractal Geometry and Applications: A Jubilee of Benoît Mandelbrot, 2004

71 **Gordana Matić and Clint McCrory, Editors,** Topology and Geometry of Manifolds, 2003

For a complete list of titles in this series, visit the
AMS Bookstore at **www.ams.org/bookstore/pspumseries/**.

ISBN 978-1-4704-2601-9
9 781470 426019
PSPUM/94